Agricultural Biotechnology

This book presents strategies and techniques highlighting the sustainability and application of microbial and agricultural biotechnologies to ensure food production and security. This book includes different aspects of applications of Artificial Intelligence in agricultural systems, genetic engineering, human health and climate change, recombinant DNA technology, metabolic engineering and so forth. Post-harvest extension of food commodities, environmental detoxification, proteomics, metabolomics, genomics, bioinformatics and metagenomic analysis are discussed as well.

Features:

- Reviews technological advances in microbial biotechnology for sustainable agriculture using Artificial Intelligence and molecular biology approach.
- Provides information on the fusion between microbial biotechnology and agriculture.
- Specifies the influence of climate changes on livestock, agriculture and environment.
- Discusses sustainable agriculture for food security and poverty alleviation.
- Explores current biotechnology advances in food and agriculture sectors for sustainable crop production.

This book is aimed at researchers and graduate students in agriculture, food engineering, metabolic engineering and bioengineering.

Current Developments in Agricultural Biotechnology and Food Security

Series Editor:
Charles Oluwaseun Adetunji

This series intends to provide comprehensive coverage of sustainable modern technologies, aimed at the improvement of food production via agriculture and food biotechnological techniques. The proposed suite of books focusses on topics in agricultural microbiology, biotechnology, food science, crop production, post-harvest management, aimed at both basic and advance food and agricultural biotechnology. The series seeks to discuss and provide foundational content from bench to bedside in food microbiology, agricultural and food biotechnology. The Series' goal is to enhance knowledge, and present update on hot topics in the field of the Agricultural Biotechnology specifically, the series aims to translate results and recent findings of studies into enhanced food production. This is primarily intended for researchers, students in Food and Agricultural Biotechnology at the graduate level and above, including those working in academic, corporate, or non-profit settings.

Agricultural Biotechnology
Food Security Hot Spots
Edited by Charles Oluwaseun Adetunji, Deepak Gopalrao Panpatte and
Yogeshvari Kishorsinh Jhala

For more information about this series, please visit: www.routledge.com/Current-Developments-in-Agricultural-Biotechnology-and-Food-Security/book-series/CRCABFS

Agricultural Biotechnology
Food Security Hot Spots

Edited by
Charles Oluwaseun Adetunji
Deepak Gopalrao Panpatte
Yogeshvari Kishorsinh Jhala

CRC Press
Taylor & Francis Group
Boca Raton London New York

CRC Press is an imprint of the
Taylor & Francis Group, an **informa** business

MATLAB® is a trademark of The MathWorks, Inc. and is used with permission. The MathWorks does not warrant the accuracy of the text or exercises in this book. This book's use or discussion of MATLAB® software or related products does not constitute endorsement or sponsorship by The MathWorks of a particular pedagogical approach or particular use of the MATLAB® software.

Designed cover image: Shutterstock

First edition published 2023
by CRC Press
6000 Broken Sound Parkway NW, Suite 300, Boca Raton, FL 33487-2742

and by CRC Press
4 Park Square, Milton Park, Abingdon, Oxon, OX14 4RN

CRC Press is an imprint of Taylor & Francis Group, LLC

© 2023 selection and editorial matter, Charles Oluwaseun Adetunji, Deepak Gopalrao Panpatte and Yogeshvari Kishorsinh Jhala; individual chapters, the contributors

ISBN: 978-1-032-21446-7 (hbk)
ISBN: 978-1-032-21448-1 (pbk)
ISBN: 978-1-003-26846-8 (ebk)

DOI: 10.1201/9781003268468

Typeset in Times
by codeMantra

Contents

Preface

The global population has been forecasted to rise drastically to 9 billion in the year 2050. Therefore, there is a clarion call to us as a scientist to come up with an innovative solution that will help in resolving diverse global challenges such as food insecurity, malnutrition, polluted environment and numerous health challenges. The application of Agricultural Biotechnology has been identified as sustainable Food Security hot spot that could be applied in the management of diverse problems of food insecurity.

Moreover, it has been observed that agricultural biotechnology has been operated by private industry for farmers in developed countries and the product developed has not considered the farmers in the developing countries. The application of biotechnology will a long way in developing a more quality food and this will also improve export and trade of agricultural products to be more profitable. Furthermore, the application of agricultural biotechnology will also boost the gross domestic product of numerous countries by minimizing hunger and increasing food security in developing countries and boost the development of several sectors such as fisheries, animal, crop, forestry and towards more robust and towards more robust food security in developing countries. Therefore, this book provides several innovative techniques that could help in the achievement of a more secure food and provision of a more nutritious food. Typical examples of such techniques include Genome engineering in agriculturally beneficial microorganisms using CRISPR-Cas9 technology, A computational approach for prediction and modeling of agricultural crop, Crop Monitoring and aquaculture using Artificial Intelligence, and application of Nanosensor Technology for Smart Intelligent Agriculture. Relevance of microbial biotechnology in the processing of several food products was also highlighted. The application of Bioinformatics, Genomics and proteomics in the management of post-harvest diseases and pests in Crops was also highlighted. The relevance of Natural bioresources and their application in pharmaceutical, Environmental and Agricultural sector were also elaborated. The role of beneficial microorganisms in the actualization of food security and their modes of action was also highlighted.

In conclusion, this book summarizes numerous potentials of Agricultural Biotechnology as a sustainable technological innovation that could assist farmers in resolving numerous agricultural challenges. Moreover, there are many controversial issues surrounding the acceptability of Agricultural Biotechnology, but the economic evaluation of their influence has established that both the consumers and the producers most especially from developing countries could benefit substantially. Also, Agricultural biotechnology could help in linking and meeting the actual needs of the consumers and farmers in adequate time.

Charles Oluwaseun Adetunji

Deepak Gopalrao Panpatte

Yogeshvari Kishorsinh Jhala

MATLAB® is a registered trademark of The MathWorks, Inc. For product information, please contact:
The MathWorks, Inc.
3 Apple Hill Drive
Natick, MA 01760-2098 USA
Tel: 508-647-7000
Fax: 508-647-7001
E-mail: info@mathworks.com
Web: www.mathworks.com

Editors

Charles Oluwaseun Adetunji is presently a faculty member at the Microbiology Department, Faculty of Sciences, Edo State University Uzairue (EDSU), Edo State, Nigeria, where he utilized the application of biological techniques and microbial bioprocesses for the actualization of sustainable development goals and agrarian revolution, through quality teaching, research, and community development. He was formally the Acting Director of Intellectual Property and Technology Transfer, the Head of department of Microbiology, Sub Dean for Faculty of Science and currently the Chairman Grant Committee and the Ag Dean for Faculty of Science, at EDSU. He is a Visiting Professor and the Executive Director for the Center of Biotechnology, Precious Cornerstone University, Ibadan. He is presently an external examiner to many academic institutions around the globe most especially for PhD and MSc Students. He has won several scientific awards and grants from renowned academic bodies like Council of Scientific and Industrial Research (CSIR) India, Department of Biotechnology (DBT) India, The World Academy of Science (TWAS) Italy, Netherlands Fellowship Programme (NPF) Netherlands, The Agency for International Development Cooperation; Israel, Royal Academy of Engineering, UK among many others. He has published many scientific journal articles and conference proceedings in refereed national and international journals with over 405 manuscripts. He was ranked recently as number 20 among the top 500 prolific authors in Nigeria between 2019 till date by SciVal/SCOPUS. His research interests include Microbiology, Biotechnology, Post-harvest management, Food Science, Bioinformatics and Nanotechnology. He was recently appointed as the President and Chairman Governing Council of the Nigerian Bioinformatics and Genomics Network Society. He was recently appointed as the Director for International Affiliation and Training Centre for Environmental and Public Health, Research and Development, Zaria. He is presently a series editor with Taylor and Francis, USA editing several textbooks on Agricultural Biotechnology, Nanotechnology, Pharmafoods, and Environmental Sciences. He is an editorial board member of many international journals and serves as a reviewer to many double-blind peer review journals like Elsevier, Springer, Francis and Taylor, Wiley, PLOS One, Nature, American Chemistry Society, Bentham Science Publishers etc. He is a member of many scientific and professional including bodies like American Society for Microbiology, Biotechnology Society of Nigeria, and Nigerian Society for Microbiology, and he is presently the General/Executive Secretary of Nigerian Young Academy. He has won a lot of international recognition and also acted as a keynote speaker delivering invited talk/ position paper at various Universities, research institutes and several centers of excellence which span across several continent of the globe. He has over the last fifteen years built strong working collaborations with reputable research groups in numerous and leading Universities across the globe. He is the convener for Recent Advances in Biotechnology, which is an annual international conference where renown Microbiologist and Biotechnologist come together to share their latest discoveries. He is the president and founder of the Nigerian Post-Harvest and Food Biotechnology Society.

Deepak Gopalrao Panpatte (Ph. D. in Agricultural Microbiology) is an Assistant Professor having 9 years of research experience in the field of Agricultural Microbiology. His research interests include agriculturally beneficial microorganisms viz. biofertilizers, biopesticides and biodegraders. He has done pioneering work for the development of fortified biocontrol bacterial consortium with phyto-extracts for management of phytopathogenic nematodes and fungi. He has received 5 awards for the presentation of research outcomes in International conferences and Rashtriya Gaurav Award and the Bharat Gaurav award for outstanding contribution in agriculture. He has got an appreciation certificate from Government of Gujarat for his work on biological control of nematodes. He has also got the Best Ph.D. thesis award by GAAS. His publication profile includes 15

research papers, 8 books with Springer-Nature, 1 with scientific publishers India, 15 book chapters in various books published with springer publishing house, 1 practical manual, 35 popular articles and 2 editorial pages. He is also an active member of Association of Microbiologist of India and Gujarat Association for Agricultural Sciences. He is also serving on the editorial board of various peer-reviewed journals and agricultural magazines.

Yogeshvari Kishorsinh Jhala (Assistant professor, Agricultural Microbiology) has 14 years of teaching and research experience. Her research interests include biofertilizers and biodegraders. She had world over first time reported five unique strains of methanotrophic bacteria. For her outstanding research work of methanotrophic bacteria, she was honored with All India Best Research Award and Young Faculty Award. She has also registered herself as an Indian Record Holder for discovery of unique methane degrading bacteria in 2018. She is handling research projects funded by Government of India on bioremediation of methane flux from rice field. Her publications include 25 research papers, 10 book chapters, 9 books, 2 teaching manuals, 17 popular articles, 2 editorial pages. She is also working as an editorial board member of various scientific journals and magazines.

Contributors

Temitayo Adedeji Adedoyin
Department of International Relations
Obafemi Awolowo University
Ile-Ife, Nigeria

Juliana Bunmi Adetunji
Nutritional and Toxicological Research
  Laboratory
Department of Biochemistry
Osun State University
Osogbo, Nigeria

A. Adewumi
Department of Agricultural and Biosystems
  Engineering
College of Engineering
Landmark University
Omu-Aran, Nigeria

Ayodeji Anthony Aduloju
Department of International Relations
Obafemi Awolowo University
Ile-Ife, Nigeria
and
Department of History and International
  Studies
Edo State University Uzairue
Iyamho, Nigeria

Muhammad Akram
Department of Eastern Medicine
Government College University Faisalabad
Faisalabad, Pakistan

Rumaisa Ansari
Department of Eastern Medicine
Government College University Faisalabad
Faisalabad, Pakistan

Hina Anwar
Department of Eastern Medicine
Government College University Faisalabad
Faisalabad, Pakistan

Oluwaseun Peter Bamidele
Department of Consumer and Food Sciences
University of Pretoria
Pretoria, South Africa

Ehis-Eriakha Chioma Bertha
Environmental and Molecular biology
  laboratory
Edo University Iyamho
Uzairue, Nigeria

Ashok Bhattacharyya
Department of Plant Pathology
Assam Agricultural University
Jorhat, India

Ruth Ebunoluwa Bodunrinde
Department of Microbiology
Federal University of Technology Akure
Akure, Nigeria

Darshan T. Dharajiya
Bio Science Research Centre
Sardarkrushinagar Dantiwada Agricultural
  University (SDAU)
Sardarkrushinagar, India

K. Dinesh
Department of Plant Pathology
University of Agricultural Sciences
Dharwad, India

G.O. Egharevba
Department of Physical Sciences
Industrial Chemistry Programme
College of Pure and Applied Sciences
Landmark University
Omu Aran, Nigeria

Nabil Ibrahim Elsheery
Faculty of Agriculture, Agricultural Botany
  Department
Tanta University
Tanta, Egypt

Nabil Ibrahim Elsheery
Agricultural Botany Department, Faculty of
    Agriculture
Tanta University
Tanta, Egypt

Oluwakemi Christianah Erinle
Department of Food Science and Microbiology
College of Pure and Applied Sciences
Landmark University
Omu-Aran, Nigeria

Temitope Omolayo Fasuan
Department of Food Science and Technology
Obafemi Awolowo University
Ile-Ife, Nigeria

Krishi Godhani
Department of Information and
    Communication Technology
School of Technology
Pandit Deendayal Petroleum University
Gandhinagar, India

Daniel Ingo Hefft
University Centre Reaseheath
Reaseheath College
Nantwich, United Kingdom

O.P. Ikhimalo
Department of Biological Sciences
Plant Biology and Biotechnology Unit
Edo University Iyamho
Uzairue, Nigeria

Areeba Imtiaz
Department of Eastern Medicine
Government College University Faisalabad
Faisalabad, Pakistan

A.A. Inyinbor
Department of Physical Sciences
Industrial Chemistry Programme
College of Pure and Applied Sciences
Landmark University
Omu Aran, Nigeria

Mehwsih Iqbal
Institute of Management Sciences
Dow University of Health Sciences
Karachi, Pakistan

Yetunde Mary Iranloye
Department of Food Science and Microbiology
College of Pure and Applied Sciences
Landmark University
Omu-Aran, Nigeria

Oseni Kadiri
Faculty of Basic Medical Sciences, Department
    of Biochemistry
Edo State University Uzairue
Uzairue, Nigeria

Suresh Kaushik
Division of Soil Science and Agricultural
    Chemistry
Indian Agricultural Research Institute
New Delhi, India

Ayodeji Samuel Makinde
Informatics and Cyber Physical Systems
    Laboratory
Department of Computer Science
Edo University Iyamho
Uzairue, Nigeria

Sherine F. Mansour
Agricultural Economics
Desert Research Center
Cairo, Egypt

Ashir Mehta
Department of Computer Engineering
Indus University
Ahmedabad, India

Olugbenga Samuel Michael
Cardiometabolic Research Unit, Department of
    Physiology
College of Health Sciences, Bowen University
Iwo, Nigeria

Binny Naik
Department of Computer Engineering
Indus University
Ahmedabad, India

Wilson Nwankwo
Informatics and CyberPhysical Systems
    Laboratory
Department of Computer Science
Edo University Iyamho
Uzairue, Nigeria

Clinton Emeka Okonkwo
Department of Food Science and Microbiology
College of Pure and Applied Sciences
Landmark University
Omu-Aran, Nigeria

Abiola Folakemi Olaniran
Department of Food Science and Microbiology
College of Pure and Applied Sciences
Landmark University
Omu-Aran, Nigeria

Oluwaseyi Paul Olaniyan
Nutritional and Toxicological Research
    Laboratory
Department of Biochemistry
Osun State University
Osogbo, Nigeria

Olugbemi T. Olaniyan
Laboratory for Reproductive Biology and
    Developmental Programming
Department of Physiology
Rhema University, Aba, Nigeria

Victoria Akinyemi Omolara
Department of International Relations
Obafemi Awolowo University
Ile-Ife, Nigeria

Omorefosa Osarenkhoe Osemwegie
Department of Food Science and Microbiology
College of Pure and Applied Sciences
Landmark University
Omu-Aran, Nigeria

Wadzani Dauda Palnam
Crop Science Unit, Department of Agronomy
Federal University Gashua
Gashua, Nigeria

Deepak Panpatte
Department of Plant Pathology
College of Agriculture, Dongarshelki Tanda
Udgir, India
and
Vasantrao Naik Marathwada Agricultural
    University
Parbhani, India

Pranav Parekh
Department of Computer Engineering
Nirma University
Ahmedabad, India

L. D. Parmar
Bio Science Research Centre
Sardarkrushinagar Dantiwada Agricultural
    University (SDAU)
Sardarkrushinagar, India

Adit Patel
Department of Information and
    Communication Technology
School of Technology
Pandit Deendayal Petroleum University
Gandhinagar, India

Het K. Patel
Department of Computer Engineering
Vellore Institute of Technology
Vellore, India

Nivedita Patel
Department of Computer Engineering
Nirma University
Ahmedabad, India

Priya Patel
Department of Computer Engineering
School of Technology
Pandit Deendayal Petroleum University
Gandhinagar, India

Shireen Patel
Department of Computer Engineering
Nirma University
Ahmedabad, India

Yogesh R. Patel
Department of Microbiology
College of Basic Science and Humanities
Sardarkrushinagar Dantiwada Agricultural
  University (SDAU)
Sardarkrushinagar, India

E. Rajeswari
Department of Plant Pathology
PJTSAU
Hyderabad, India

Osahon Itohan Roli
Department of Anatomy
College of Basic Medical Science
Edo State University Uzairue
Uzairue, Nigeria

Bandana Saikia
Department of Plant Pathology
Assam Agricultural University
Jorhat, India

Neera Bhalla Sarin
Laboratory of Genetic Manipulation for Stress
  Alleviation and Value Addition in Plants
School of Life Sciences, Jawaharlal Nehru
  University
New Delhi, India

Ajit kumar Savani
Department of Plant Pathology
Assam Agricultural University
Jorhat, India

Manan Shah
Department of Chemical Engineering, School
  of Technology
Pandit Deendayal Petroleum University
Gandhinagar, India

Khurram Shahzad
Department of Eastern Medicine
Government College University Faisalabad
Faisalabad, Pakistan

Arish Sohail
Department of Eastern Medicine
Government College University Faisalabad
Faisalabad, Pakistan

Dineesha Soni
Department of Information and
  Communication Technology
School of Technology
Pandit Deendayal Petroleum University
Gandhinagar, India

Devdutt Thakkar
Department of Information and
  Communication Technology
School of Technology
Pandit Deendayal Petroleum University
Gandhinagar, India

Kapil K. Tiwari
Bio Science Research Centre
Sardarkrushinagar Dantiwada Agricultural
  University (SDAU)
Sardarkrushinagar, India

Benjamin Ewa Ubi
Department of Biotechnology
Ebonyi State University
Abakaliki, Nigeria

A.M. Ugbenyen
Department of Biochemistry
Edo University Iyamho
Uzairue, Nigeria

Kingsley Eghonghon Ukhurebor
Climatic/Environmental/Telecommunication
  Physics Unit
Department of Physics
Edo University Iyamho
Uzairue, Nigeria

Rabia Zahid
Department of Eastern Medicine
Government College University Faisalabad
Faisalabad, Pakistan

1 Agricultural System Modeling and Analysis

Sherine F. Mansour
Desert Research Center

Nabil Ibrahim Elsheery
Tanta University

CONTENTS

1.1 INTRODUCTION

In recent years, the environment has become more complicated due to many factors including our rising population and their demands for more food, water, and energy, the limited arable land for expanding food production, and increasing natural resource pressures. Such factors are further exacerbated by climate change, which as we have learned would result in many changes in the environment (e.g., Wheeler and von Braun, 2013). Why can science help overcome these complexities? On the one hand, the quantity of published knowledge and data contributions from every field of science is continuing to be explosive. On the other hand, the issue of handling all this expertise and supporting data becomes more complicated, and risks overloading information. The knowledge explosion leads to a greater understanding of the interconnectedness of what previously may have been viewed as separate elements and processes. We now know that interactions between components can have a major impact on system responses, so the analysis of components in isolation automatically suffices to draw conclusions regarding an overall structure (Hieronymi, 2013). Those experiences cross conventional limits of discipline. While a strong focus remains on disciplinary

science that leads to a greater understanding of components and individual processes, there is also an increasing focus on system science.

Systems science is the study of real-world "systems" which consists of specialist-defined components. To assess overall device behavior, these components communicate with each other and their environment (Wallach et al., 2018). Such interacting components are exposed to an external environment that may influence system component behavior, but the environment itself will not be influenced by changes occurring within the system boundary. While systems are real-world abstractions established for specific purposes, they are of great use in science and engineering in all fields, including agriculture. The science of agricultural systems is an interdisciplinary field that studies the behavior of complex farming systems. Although it is useful to research agricultural systems in nature using data gathered that describe how a particular system behaves under different conditions, in certain cases, it is difficult or impractical to do so. Scientific analysis of an agroecosystem includes a component system model and their interactions, taking into account agricultural production, natural resources, and human factors. Therefore, for specific purposes, models are required to understand and predict the overall performance of the agro-ecosystems.

Agricultural system models are playing an increasingly important role in developing sustainable land management across complex agroecological and socioeconomic environments, as field and farm experiments require large amounts of capital and do not yet have adequate space and time knowledge to define acceptable and successful management practices (Vries et al., 1993). Models can help define management options for optimizing sustainability objectives through space and time for land managers and policymakers as long as the appropriate soil, management, environment, and socioeconomic information is available.

1.2   BRIEF HISTORY

The history of agricultural system modeling is characterized by a number of key events and drivers that led scientists from different disciplines to develop and use models for different purposes (Figure 1.1). Some of the earliest agricultural systems modeling were done by Earl Heady and his students to optimize decisions at a farm scale and evaluate the effects of policies on the economic benefits of rural development (Heady, 1957; Heady and Dillon, 1964). This early work during the 1950s through the 1970s inspired additional economic modeling.

Dent and Blackie (1979) included models of farming systems with economic and biological components; their book provided an important source for different disciplines to learn about agricultural systems modeling. Soon after agricultural economists started modeling farm systems, the International Biological Program (IBP) was created. This led to the development of various ecological models, including models of grasslands during the late 1960s and early 1970s, which were also used for studying grazing by livestock. The IBP was inspired by forward-looking ecological scientists to create research tools that would allow them to study the complex behavior of ecosystems as affected by a range of environmental drivers (Worthington, 2009; Van Dyne and Anway, 1976).

The IBP initiative brought together scientists from different countries, different types of government, and different attitudes toward science. Before this program, systems modeling and analysis were not practiced in scientific efforts to understand complex natural systems. IBP left a legacy of thinking and conceptual and mathematical modeling that contributed strongly to the evolution of systems approaches for studying natural systems and their interactions with other components of more comprehensive, managed systems (Coleman et al., 2004).

Models of agricultural production systems were first conceived in the 1960s. One of the pioneers of agricultural system modeling was a physicist, C. T. de Wit of Wageningen University, who, in the mid-1960s, believed that agricultural systems could be modeled by combining physical and biological principles. Another pioneer was a chemical engineer, W. G. Duncan, who had made a fortune in the fertilizer industry and returned to graduate school to obtain his PhD degree in Agronomy at age 58. His paper on modeling canopy photosynthesis (Duncan et al., 1967) is an enduring development

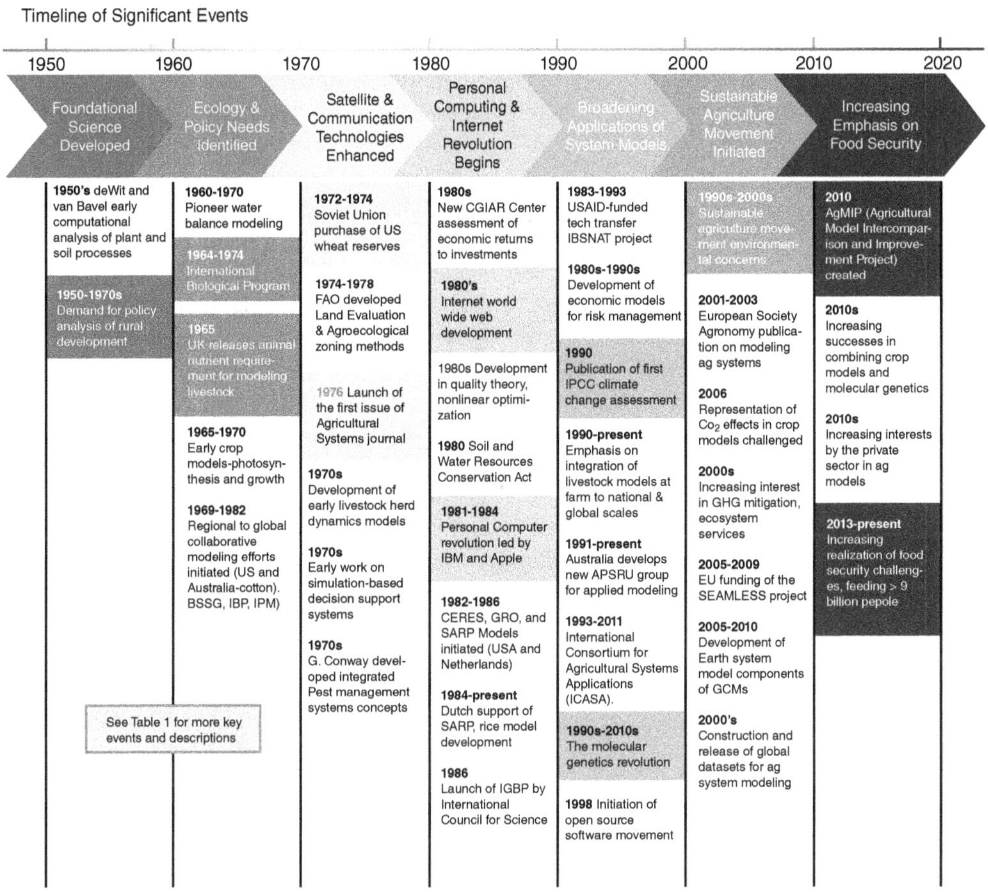
Timeline of Significant Events
1950
1960
1970
1980
1990
2000
2010
2020
Foundational Science Developed
Ecology & Policy Needs Identified
Satellite & Communication Technologies Enhanced
Personal Computing & Internet Revolution Begins
Broadening Applications of System Models
Sustainable Agriculture Movement Initiated
Increasing Emphasis on Food Security
1950's deWit and van Bavel early computational analysis of plant and soil processes
1950-1970s Demand for policy analysis of rural development
1960-1970 Pioneer water balance modeling
1964-1974 International Biological Program
1965 UK releases animal nutrient requirement for modeling livestock
1965-1970 Early crop models-photosynthesis and growth
1969-1982 Regional to global collaborative modeling efforts initiated (US and Australia-cotton). BSSG, IBP, IPM)
See Table 1 for more key events and descriptions
1972-1974 Soviet Union purchase of US wheat reserves
1974-1978 FAO developed Land Evaluation & Agroecological zoning methods
1976 Launch of the first issue of Agricultural Systems journal
1970s Development of early livestock herd dynamics models
1970s Early work on simulation-based decision support systems
1970s G. Conway developed integrated Pest management systems concepts
1980s New CGIAR Center assessment of economic returns to investments
1980's Internet world wide web development
1980s Development in quality theory, nonlinear optimization
1980 Soil and Water Resources Conservation Act
1981-1984 Personal Computer revolution led by IBM and Apple
1982-1986 CERES, GRO, and SARP Models initiated (USA and Netherlands)
1984-present Dutch support of SARP, rice model development
1986 Launch of IGBP by International Council for Science
1983-1993 USAID-funded tech transfer IBSNAT project
1980s-1990s Development of economic models for risk management
1990 Publication of first IPCC climate change assessment
1990-present Emphasis on integration of livestock models at farm to national & global scales
1991-present Australia develops new APSRU group for applied modeling
1993-2011 International Consortium for Agricultural Systems Applications (ICASA).
1990s-2010s The molecular genetics revolution
1998 Initiation of open source software movement
1990s-2000s Sustainable agriculture movement environmental concerns
2001-2003 European Society Agronomy publication on modeling ag systems
2006 Representation of Co2 effects in crop models challenged
2000s Increasing interest in GHG mitigation, ecosystem services
2005-2009 EU funding of the SEAMLESS project
2005-2010 Development of Earth system model components of GCMs
2000's Construction and release of global datasets for ag system modeling
2010 AgMIP (Agricultural Model Intercomparison and Improvement Project) created
2010s Increasing successes in combining crop models and molecular genetics
2010s Increasing interests by the private sector in ag models
2013-present Increasing realization of food security challenges, feeding > 9 billion pepole

FIGURE 1.1 Summary timeline of selected key events and drivers that influenced the development of agricultural system models.

that has been cited and used by many crop modeling groups since its publication. After his PhD degree, he began creating some of the first crop-specific simulation models (for corn, cotton, and peanut, Duncan, 1972). Bouman et al. (1996) intrigued many scientists and engineers who started developing and using crop models. In 1969, a regional research project was initiated in the USA to develop and use production system models for improving cotton production, building on the ideas of de Wit, Duncan, and Herb Stapleton (Stapleton et al., 1973), an agricultural engineer in Arizona. Thus, some of the first crop models were curiosity-driven with scientists and engineers from different disciplines developing new ways of studying agricultural systems that differed from traditional reductionist approaches, and inspiring others to get involved in a new, risky research approach. During this early time period, most agricultural scientists were highly skeptical of the value of quantitative, systems approaches and models. In 1972, the development of crop models received a major boost after the US government was surprised by large purchases of wheat by the Soviet Union, causing major price increases and global wheat shortages (Pinter et al., 2003). New research programs were funded to create crop models that would allow the USA to use them with newly available remote sensing information to predict the production of major crops that were grown anywhere in the world and traded internationally. This led to the development of the CERES-Wheat and CERES-Maize crop models by Joe Ritchie and his colleagues in Texas (Ritchie and Otter, 1984; Jones and Kiniry, 1986). These two models have continually evolved and are now contained in the DSSAT suite of crop models (Jones et al., 2003; Hoogenboom et al., 2012).

During much of the time since the 1960s, only small fractions of agricultural research funding were used to support agricultural system models, although the Dutch modeling group of C. T. de Wit was a notable exception (Bouman et al., 1996). Thus, most of those who were modeling cropping systems, for example, struggled to obtain financial support for the experimental and modeling research needed to develop new models or to evaluate and improve existing ones. Instead, there were other "crisis" events or realizations of key needs fueling model development, each typically leading to an infusion of additional financial support over short durations of time for model development or use.

The concept of Integrated Pest Management emerged in the 1970s, in particular from the work of Gordon Conway on the pests and diseases of plantation crops in Malaysia (Conway, 1987). In 1972, the so-called Huffaker Integrated Pest Management (IPM) project was funded in the USA to address the major problems associated with increasing pesticide use and development of resistance to pesticides by many of the target insects and diseases (Pimentel and Peshin, 2014). Mathematical models of insect pests and crop and livestock diseases had been developed starting during the first half of the 20th century, though the success of synthetic agrichemicals led to a shift in attention to other control measures in the years after the Second World War. The Huffaker project infused funds for developing insect and disease models of several crops, combined with experimental efforts aimed at reducing pesticide use and more effective use of all measures to prevent economic damage to major crops in the USA. This project continued until 1985 (as the Consortium for IPM after 1978). Coincident with this project was a major increase in the sophistication of population dynamic models in ecology and a growing appreciation of the importance of nonlinearities and the problems for forecasting they imply (May, 1976).

Lively debate about the appropriate way to model ecological interactions in agricultural settings characterized these decades (Dempster, 1983; Hassell, 1986; Gutierrez et al., 1994; Murdoch, 1994).

1.3   DEFINITION OF AGRICULTURAL SYSTEM MODEL

System models provide a simplified description of important system components and their interactions. Schoemaker (1982) identifies four purposes for systems models: (i) description, (ii) prediction, (iii) postdiction, and (iv) prescription. Descriptive models are used to characterize the system; their performance, in turn, allows modelers to evaluate whether they have adequately described the important aspects. Predictive models forecast future system behavior. Descriptive models may serve a predictive purpose, but many predictive models are much simpler than descriptive ones, especially when certain system patterns repeat themselves systematically, obviating the need to describe the underlying mechanisms. For example, seasonal temperature patterns can be predicted fairly reliably from historical data, without describing the revolution of the Earth around the sun and the attendant changes in insolation, ocean currents, and jet stream activity. Postdictive models tend to be human logical constructions that allow us to explain after-the-fact what system constraints or special phenomena caused a given outcome. Prescriptive models are normative ones that offer guidance on how a system should be managed to meet some goal. Many agricultural models serve more than one of these purposes. A secondary, but often very important, reason for modeling agricultural systems is to improve knowledge of the system. Knowledge of any given agricultural system is often uneven. Areas where knowledge of the system is sparse or missing tend to become apparent either (i) in the process of designing the model structure, or (ii) in the process of finding parameters that can make empirical models operational. For example, one recent exercise in developing a weed management model revealed that in the past 30 years, North American weed scientists have focused their research so heavily on herbicide performance, that little is known about weed biology and ecology; the modeling process helped to instigate a new research effort in this area (Forcella et al., 1992). Model design experiences often lead to revised priorities for future data collection research, based on data gaps defined (Dalton, 1982). Hence, systems modeling may provide value not just through the end-product model developed, but also through the development process itself.

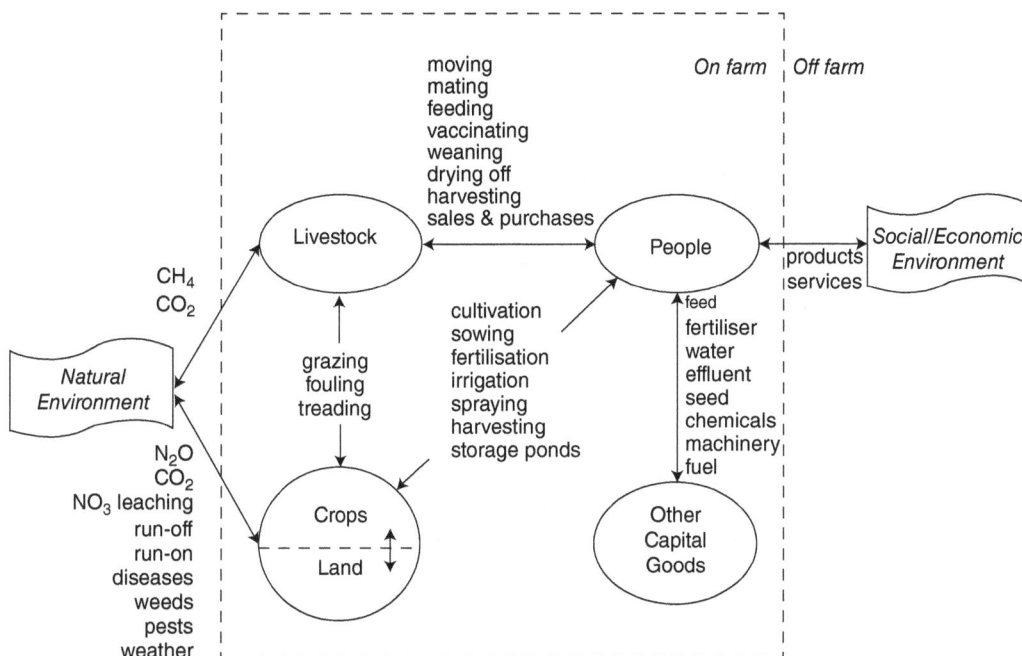

FIGURE 1.2 The main resources (land, crops, livestock, other capital goods, and people) that go into making up a farming system. Arrows and small types indicate possible interactions between these resources and their environments (natural, social, and economic). Note that besides various kinds of vegetation ("crops"), land may support other uses such as silos, waterways, buildings, and tracks. (Jones et al. 1997.)

1.4 FARMING SYSTEMS INNOVATION

Farming systems can be defined as arrangements of land, crops, livestock, other capital goods and labor put together for the primary purpose of producing plant and animal products for consumption (Figure 1.2; Jones et al., 1997). While farming systems are primarily businesses that operate within an economic environment, they are also communities that operate within a sociopolitical environment, and ecosystems that operate within a natural environment. The "system" refers to the particular pattern of arrangement of these interacting resources for the purpose of producing particular products or outcomes. Farming systems innovation is concerned with improving outcomes across one or more individual farms of a given "class" (Spedding, 1976). Examples range from assessing beef intensification options for a particular site (Ogle and Tither, 2000), through a group of local farmers seeking improved lamb and ewe performance (Webby, 2002), to designing resource-efficient dairy technologies for application on a national scale (Clark, 2002). A "farming system" therefore potentially touches many individual farms, farm families, communities, businesses, and regulatory stakeholders, all of whom may have an interest in improving the multiple physical, biological, economic, and social outcomes of farming. Farming systems innovation, then, is the pursuit of technical, managerial, and social means to improve the outcomes of farming systems for their stakeholders (Spedding 1990; Mueller, 1993; McRae, 1993; Barlow et al., 2002).

A key feature of farming systems is that many of the important outcomes are influenced by factors beyond immediate managerial control. These external factors include farm location, farm resource conditions in the past, and farm future environment (Menz and Knioscheer, 1981), as well as most aspects of the physical, biological, economic, and social processes operating within the farm and its environment. This means that farming systems are complex dynamic systems whose products and impacts are difficult to measure, let alone predict or control. Figure 1.3 illustrates

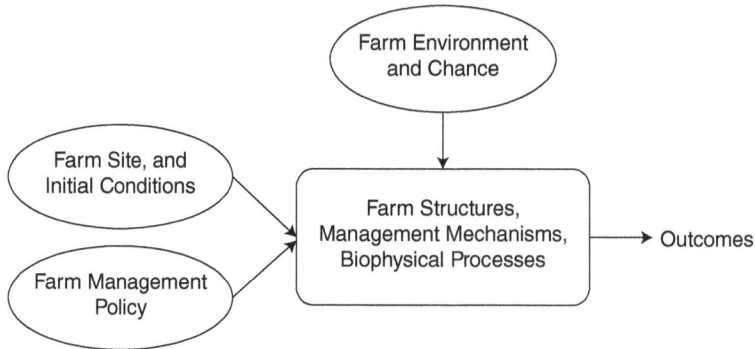

FIGURE 1.3 The factors that determine the outcomes of a farming system.

schematically how the various controllable and uncontrollable factors interact to produce farming outcomes. With this understanding, farming systems innovators have developed a range of methods to assist clients to (i) identify the "problem" (i.e., the relevant class of systems, its stakeholders and their issues), (ii) understand the key interactions between system components, (iii) predict system-level consequences of proposed management changes, (iv) design management systems to deliver desired improvements, and (v) promote the implementation of these improved systems. The process and criteria for deciding which systems research, development, and education methods are most suitable in a given project have been discussed in detail by Barlow et al. (2002). The key point is that systems problems require the use of systems methods.

Agriculture can be regarded as a system with inputs that have physical, cultural, economic and behavioral elements. In areas where farming is less developed, physical factors are usually more important, but as human inputs increase, these physical controls become less significant. This system model can be applied to all types of farming, regardless of scale or location. It is the variations in the inputs which are responsible for the different types and patterns of agriculture around the world (Figure 1.4). This leads to classifications of agriculture in which contrasts between the different types of farming are clear.

1.5 THE USERS OF AGRICULTURAL SYSTEMS MODELS

The users of agricultural systems models can be grouped by the purposes of the models themselves. Researchers are the main users of descriptive and postdictive models, for these are the two classes of models whose role is to enhance understanding of the system. A much wider group of individuals seeking decision support uses the other two types of models. Predictive models are useful to those whose decisions depend upon good forecasts of future outcomes. Many farm management practices rely on good predictions of what outcomes are likely to ensue. All farmers have in their heads some heuristic predictive model of what results to expect from, say, changing a livestock feed ration or taking a position in the futures market. More sophisticated, numerical predictive models are designed with the intent of formalizing and improving upon managers' subjective predictions. At a broader level, system models may be used by policymakers to predict the social welfare outcomes of proposed policies. Prescriptive models (most of which include a predictive component) have a similar audience—one which seeks to make decisions based on model recommendations.

1.6 TYPES OF AGRICULTURAL SYSTEMS

Before examining in greater detail how agricultural systems are modeled, consider first how they can be classified. One approach is to classify them in space or time. Other ways are by hierarchical

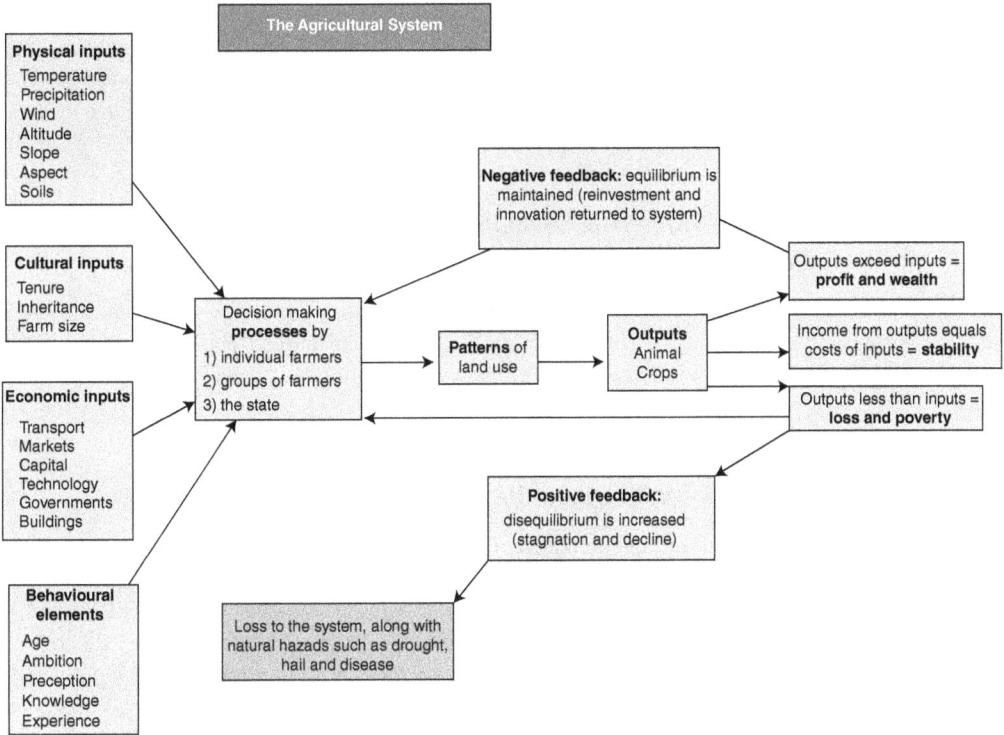

FIGURE 1.4 The agricultural system.

system level or by subject matter. In space, the hierarchy of agricultural biophysical systems ranges in scale from micro-level plant and animal components to the individual organism to the field, to the whole farm, to multiple farm enterprises and multiple farm businesses (sometimes going beyond production to processing and some form of marketing), to larger scales such as watersheds and environmental zones. In parallel with the physical system boundaries are social systems, including rural communities and links to the larger society and macro-economy (of which the farming sector is just one part). Some of the more complicated system environments lie at the intersection of different systems. For example, the character and density of human communities affect the level of concern with the quality of their biophysical environment and public policies developed to ensure that minimum environmental quality levels are maintained. These policies, in turn, affect the management practices of farmers and others who manipulate the biophysical environment for their livelihood. In time, agricultural systems can be viewed statically or dynamically. In some systems, we care about relationships in an atemporal fashion. The comparative statics models of microeconomic theory are illustrative.

A supply curve, for example, models the aggregate willingness of producers to change the quantity produced in response to price changes. This model captures a relationship of predictive interest even when the inherent time lags are not explicit. Of course, time is central to evolutionary processes. Examples would include plant and animal growth, pest demographics, and disease epidemiology, as well as how humans respond to previous events as well as current events. Questions of system stability and sustainability are often of special interest in dynamic models (Conway, 1987). Ridder (1997) proposes agro-ecosystem hierarchies that integrate space, time, and organization elements. He observes that in the same space and time, different (even overlapping) organizations may coexist. For example, livestock or crop individuals may be organized into crop or animal husbandry systems, just as individual people are organized into households. Likewise, larger-scale landscape

areas bound both physiographic natural resource units and human administrative units. Subject matter may be the most common way for people to think about systems, at least judging from our language. Contemporary discourse is rife with systems: ecological systems, economic systems, political systems, social systems, and information systems, to name just a few. Within each of these, the issues of space, time, hierarchy, and complexity can be explored

1.7 CHALLENGES TO SYSTEMS MODELING FOR FARMING SYSTEMS INNOVATION

Following an extensive review of the systems modeling literature, especially in agriculture, ecology, and business management, four particular challenges to conventional hard systems modeling approaches were identified, as were several techniques that might be effective in addressing them.

1.7.1 INVOLVING THE RIGHT PEOPLE IN THE RIGHT WAY TO ENSURE COMPATIBILITY WITH USER NEEDS AND PROCESSES

The first challenge relates to the implementation of modeling outcomes into farming policies and structures. In the late 1970s, Roberts (1977) pointed out the endemic failure of management model recommendations to be implemented in business practice. Similarly in agriculture, while modeling competence and computer ownership have increased immensely over the past 30 years, "this has not generated conspicuous or sustained enthusiasm among farmers or their advisors" for model-based interventions into farming practice (McCown, 2001). He argued that the lack of adoption has been a result of the prevailing paradigm of scientific intervention which sees science as a form of vicarious problem solving, where researchers solve disembodied problems on behalf of investors, farm managers, and other stakeholders while remaining disconnected from the world of practice in which these intended users and beneficiaries operate (Mueller, 1993; McCown, 2001, 2002a). In response, Lynch et al. (2000) and McCown (2002b) argued that "development methods such as participatory or adopter-based approaches will lead to systems that are perceived as more useful," compatible with user needs and processes, being "easy-to-use and thus, adopted more readily" (Lynch et al. 2000; Rogers, 2010). Reports from recent case studies where participatory approaches have been trialed show promise (Hochman et al., 2001; Meinke et al., 2001; Webby, 2002; Hare, et al., 2003). Such approaches require that social context, communication, and extension planning be addressed from the very beginning of the project, rather than as an afterthought at the end (Roberts, 1977).

1.7.2 DETERMINING WHAT SYSTEM TO MODEL TO REMAIN RELEVANT TO STAKEHOLDER CONCERNS

The second challenge relates to problem articulation. Researchers commonly insist that any research project must begin by defining the research problem (Bunge, 1998). Problem definition in farming systems research, however, is somewhat subjective. Multiple persons, interests, and issues surround practice decisions even on an individual farm, and farmers do not have absolute control over their land—the wider community finds ways to impose its values through mechanisms such as regulations, taxes, and social pressure (Valentine et al., 1993). In addition, people and communities are evolving organizations, whose perceptions, values, and interests are a moving target for any problem-solving project.

Given these challenges to problem definition, Checkland (1985) argued that the methodology of systems engineering, based on defining goals or objectives, simply does not work when applied to messy, ill-structured, real world problems. When applied in such situations, models have tended to be too problem-specific (i.e., they address artificially narrow problems that soon cease to be relevant), or conversely, not problem-oriented enough (addressing scientific questions which are

of academic interest only). In other words, the inability to define objectives is usually part of the problem. The solution to this impasse is to recognize that problems cannot be separated from their stakeholders (Smith, 1989). Clients need to be intimately and continually involved in the research process, and models need to encompass as wide a range of their issues as possible.

1.7.3 REPRESENTING IN MODELS WHAT FARM MANAGERS MIGHT DO

The third challenge relates to system boundary selection. In particular, this is the question of whether (and if so, how) to represent farmers within models, in order to recreate and evaluate what farmers might actually do in different situations (Sorrenson and Kristensen, 1992; Jones et al., 1997; Edwards-Jones et al., 1998a). In many simulation models this is achieved through specifying a pre-determined sequence of management actions ("calendar-based management", Romera 2004), which is too rigid (Cros et al., 1997), or alternatively by attempting to model the thought processes of farmers (Edwards-Jones et al., 1998b), which is too speculative if improved farm management is the goal. This suggests a need to simulate farm management policies and practices in a flexible, but idealized, way that is aligned closely with the concepts and options commonly available to farmers. The use of flexible decision rules to specify farm management may be a useful approach toward achieving this (Romera, 2004). In addition, some important issues that influence decision-making by farmers, such as practical skill levels, family goals, cultural constraints, habits, and changing personal worldviews, values, and interests, are difficult to represent in a computer model. It is unlikely that these factors could be modeled satisfactorily, indicating again the necessity of working closely with practitioners when exploring farming systems problems.

1.7.4 MAKING SOUND COMPARISONS BETWEEN ALTERNATIVE FARM MANAGEMENT POLICIES

The fourth challenge relates to model use in policy evaluation. The basic use of a simulation model involves the comparison of two or more scenarios. Romera (2004) argued that since simulation models are simplified representations of reality, outputs should be interpreted by comparison with other model outputs (based on different inputs), rather than in absolute terms. Furthermore, it is well known that the performance of farm management policies is heavily dependent on the initial conditions of the farm under study (which are partially known), and on the future weather (which is almost completely unknown) (Romera 2004). Management decisions must be robust under these uncertainties. However, the majority of farm systems simulation models simulate only a single farm, without replication or evaluation of uncertainty, and policy comparisons must consequently be done outside the model software itself. That is, users must devise their own means of comparing scenarios and estimating the risks associated with alternative options. It would seem preferable that models be intentionally designed to provide the ability to compare alternative farm management policies, in terms of both average performance and risk, and possibly across a number of farms. These four challenges having been identified, the next phase of the study was to explore in more detail the approaches that have proved effective (or show promise) in addressing these four concerns. These approaches are grouped under two headings:

- Establishing project aims and stakeholder relationships.
- Model design and development.

1.8 EVOLUTION OF MODELING APPROACHES IN FARMING SYSTEMS

Several authors have provided helpful conceptual models of the farm (Dillon, 1992; Sorrensen and Kristensen, 1992; Dent, 1994). The relatively simple model of Sorrensen and Kristensen, which distinguishes a Production System from a Management System, is sufficiently comprehensive to assist in our review of historical changes in farming systems analysis and intervention (Figure 1.5a). This

(a)

(b)

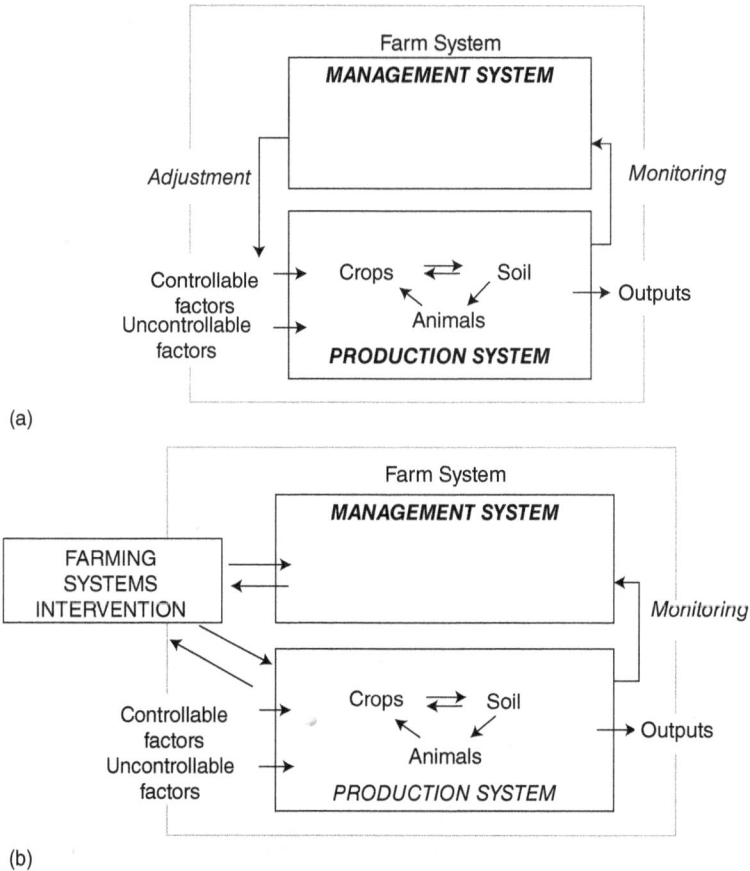

FIGURE 1.5 A 'cybernetic' framework for thinking about a farm as a purposeful, managed system. (a) Highlighting the concepts of monitoring and adjustment linking production and management systems (b) Highlighting the place for a 'Systems Analysis and Intervention' element. (Sorrensen and Kristensen, 1992.)

is a classic data flow diagram from the field of structured systems analysis (Jayaratna, 1986). The key aspect is the cybernetic relationship by which the production system is monitored and controlled to achieve management purposes. Intervention is the rationale for analysis, and the focal point for intervention in Figure 1.5a is 'adjustment'.

Analysis is deemed warranted when a decision about 'adjustment' is problematic and intervention might 'help farmers make more rational decisions'. In Figure 1.5b, a scientifically rational Systems Analysis and Intervention element is introduced —a 'notional system' in the data flow diagram conventions of Jayaratna (1986). To date, we recognize six types of systems of analysis and intervention having been used in farming systems (Table 1.1). Type 1, economic decision analysis, was underway prior to the advent of farm-competent production simulation models. The strength of this type lay in the fact of the unit of analysis is the whole farm or enterprise. As shown in Table 1.1, these models treated production as simple, static, mathematical functions of inputs and outputs. This assumed away any stochasticity and any sensitivity to timing in operations. In the late 1960s, some agricultural economists saw the advent of dynamic production models (Type 2), e.g. crop models, as an opportunity to overcome this deficiency in the way production processes were represented in farm economic models (Anderson and Dent, 1971; Dent and Anderson, 1971; Anderson, 1974).

Prominent efforts were sustained in bio-economic modeling (Type 3) for most of 20 years (and optimism for much of this period), by Dent and his students (Blackie and Dent, 1974; Dent, 1975;

TABLE 1.1

Approaches to Systems Analysis and Intervention that Have Been Applied to Farming Systems

Type of Systems Analysis/ Intervention	Characteristics of Systems Analysis and Intervention	Operational Model of Production System	Operational Model of Management System
1 Economic decision analysis using production functions	Recommendations based on whole farm, or enterprise, optimization of production inputs	Static input–output transformations	Suite of notional decision problems Categorized initial conditions for economic model
2 Dynamic simulation of production processes	Recommendations based on pseudo-optimization of simulations	Dynamic model of production processes Categorized initial conditions for simulation	Suite of notional technical problems Socio-economic 'filter' of technical recommendations
3 Economic decision analysis using dynamic simulation of production processes	Enhanced recommendations based on optimization of production inputs	Dynamic model of production processes	Suite of notional decision problem Categorized initial conditions for economic model
4 Decision support system	Decision support system on farmer's computer	Dynamic model of production processes	Source of notional problem User of decision aids
5 Expert system	Recommendations management actions based on conditional rules	Table of action–outcomes	'If. . ., then. . .' model of expert manager User of decision aids
6 Simulation-aided discussions about management	Localized simulation by intermediary in response to farmer's felt problems as input to farmer learning and decision making	Dynamic model of production processes	Farmer as rational manager with cognitive limits and continuing learning 'needs'

Source: Keating and McCown (2001).

Thornton and Dent, 1984a, 1984b). Of particular interest today, in the light of recent developments in model use, is the recognition in 1971 by Anderson and Dent, that there existed two major impediments to applying simulation models to actual farming. These were the costs of customization and the costs of validation. Achieving reduction in these costs became a major focus of the research of Dent and his students for the next decade (Blackie and Dent, 1974), considered two approaches to making simulation in farm management more adorable:

The cost of developing a simulation model for a particular enterprise can be reduced on a 'per farm' basis by constructing the model in such a manner that it can be used by a number of farms. There are two alternative applications of the approach. The first involves the development of a 'representative' farm or enterprise model which can be used to examine the effects of differing management policies. This type of model is largely confined to examining the implications of major management changes. The results from such models cannot be applied directly to an individual farm and therefore are unable to provide specific management guidance. The second approach relates to the construction of a 'skeleton' model which represents the logical structure and includes only the basic parameters of the real system. Such a model becomes functional only when 'coupled' with data from an individual farm and, in its 'coupled' state, is unique to that farm. The model must be capable of reflecting both the sequence and timing of feasible decisions in order to reflect individual management policies. Systems may appear similar except with regard to their detail; the model must have the capability to adequately distinguish and mimic all such systems.

These authors opted for the skeleton model. They saw private consultants as important players in generating farm data that formed individual farm information systems and farmers acting to update these systems through low-cost enhancements of their normal monitoring of the production system and the external environment (Figure 1.5a). Thus, at the interface between the Production system and the Management system in Figure 1.5.

Continuous comparison of projected targets with the present state of the systems provides a systematic procedure for management control. The action taken will depend on the estimated outcomes from the various alternatives (which may be explored using the model).

As regards costs of validation, Blackie and Dent (1974) point out, that there are real advantages to skeleton models coupled with information systems vis-a`-vis validation of models that simulate the general or hypothetical, and which can be tested only based on plausibility.

By contrast, validation of skeleton models is a more straightforward procedure. The system to be modeled is comparatively small and the interactive relationships between the various parts of the systems can be acceptably defined. The model is intended to mimic existing real systems under defined management policies. In this circumstance, real system data are available and can be directly compared with the model predictions.

In 1975, in a review of systems applications that featured skeleton models and coupled information systems, Dent stated: 'The application of skeleton models for management purposes on individual farms can be confidently expected in the not too distant future.' But, judging from lack of mention by Dent in subsequent reviews, it never happened.

Charlton and Street (1975) took a somewhat different tack. Their complex models of both financial and production aspects of pig and dairy enterprises were burden-some, but 'very much simpler ones would have been incapable of being applied to specific farm problems. The complexity of the models arose not from the introduction of sophisticated relationships but from the need to provide detail and adequate flexibility'. But the high overheads of their approach led them to conclude:

Models should be constructed to meet limited, well-defined objectives and there has to be a greater recognition of this need for relative simplicity. There is, in fact, a strong argument for producing less general programs than the ones which have been described here. By restricting each package to a single specific enterprise or problem, such as, for example, the expansion of a pig fattening herd, many of the problems of providing generality with a single program would be overcome.

This simpler approach was characteristic of the 'decision support system' (DSS) that had appeared in non-agricultural fields by this time (Little, 1970; Keen, 1975) and was to become prominent in agriculture. But attempts to overcome the conflict between the desirability and the feasibility of using the relatively comprehensive simulation models to assist farm management were far from over. Doyle (1990) bemoans 'the failure of systems concepts and simulation models to have any practical impact on farming', and found disturbingly, the reasons remain the same as those outlined by Dent (1975) some 15 years ago. In the first place, the failure of systems researchers to liaise with farm decision makers has meant that farmers are rightly suspicious of computer-generated predictions of optimal resource use. In the second place, the preoccupation of systems researchers with model-building rather than application has greatly limited the practical use of most models. This echoes the critique of Musgrave (1976) as well.

Another type of response to the failure using comprehensive models to deliver the previously envisioned intervention in important aspects of design and planning in farm management (Figure 1.5) was the scaling down of aims and expectations to what seemed more achievable.

The very complexity of biological systems and their susceptibility to unplanned variations make it di cult to design adequate representations of the real world. Nevertheless, the systems approach to analyzing processes and resource decisions on farms potentially opens up the prospects of using models as aids to control individual farm processes (Doyle, 1990).

Vaguely echoing the earlier quoted call of Charlton and Street (1975), for a less general, problem-focused approach, this flagged a class of alternative approaches for using models to

intervene in farm management, the DSS. Before considering DSS, we present what could be viewed as an epitaph on Farm Management modeling provided by Malcolm (1990):

But over time emerged an increasingly commonly-held unease, and occasionally conviction, that these were trails which, if followed, soon led from the complex and di cult whole-farm pastures of plenty to simpler and easier analyses characterized by incomplete and inappropriate disciplinary balances and resulting in work which was not really about farm management.

It may be that both farm management and systems research which manages to generate information about general principles and theory relating to the management of farms is more about research in one of the disciplines involved in the management of farms such as agronomy, agricultural economics, animal science, (rural) sociology psychology, engineering, than it is about farm management. This view has the merit of making explicit the gap which inevitably exists between the findings of research and the management of farms, and reminding researchers that agricultural science and agricultural economics are not directly about farming.

The concept for decision support systems (Type 4 in Table 1.1) in the field in which it originated, Management Science, was articulated by Keen (1987): [The DSS] meshes human judgment and the power of computer technology in ways that can improve the effectiveness of decision makers, without intruding on their autonomy. Traditional DSS provides a computerized [proxy for a] assistant. The manager's judgment selects alternatives and assesses results (Keen, 1987).

In agriculture, an indication of Keen's 'autonomy' was residence of the DSS on the farmer's personal computer. The model of the Production System was engineered around a crop model. To the interventionists, the Management System was notion-ally the source of the developer-construed farming 'problem'. It was assumed that the Management System of 'modern' farms would naturally be increasingly equip-ped with such aids to decision making as computer ownership increased.

It has taken some considerable time for it to become clear, but there is now little doubt that decision support systems, as originally conceived, have not generally found a significant place in farm management of even 'progressive' Management Systems. (Seligman, 1990; Ascough and Deer-Ascough, 1994; Hoag et al., 1999; Parker, 1999). These authors highlight the fact that farmers have not used DSSs that have been available. The reasons for this are not well researched or documented, but Webster (1990) ordered an economist's view:

The DSS adoption problem was the result of a gross oversupply by enthusiastic, commercially-unaccountable, publicly-funded research organizations of a technology which had a potential to benefit only a very small proportion of farms (Webster, 1990).

This economist's view may be a little harsh and may have been developed with the benefit of considerable hindsight. One notable exception to the lack of reflection on the usefulness of DSS is the report of Zadoks (1989) on EPIPRE, a computer based DSS on pest and disease control in wheat in Europe. This ground-breaking DSS e ort began in 1976, reached peak impact around 1982–1983, and appears to have fallen away up until 1986 when this report was made. Zadoks (1989) reviews a number of sources of evidence for the impact of EPIPRE in farming practice. The evidence of financial benefits was limited, evidence of environmental benefits in terms of reduced chemical usage stronger, and there was almost universal appreciation of the 'learning effect'. Interestingly, the standard recommendations coming from extension appeared to converge with the recommendations from EPIPRE over a 5-year period. The significant point to note here is while the science behind a crop– pest–weather DSS like EPIPRE may be complex, the management decision is simple — basically to spray or not spray. Even in such a well-defined management situation and decision problem, the benefit of the DSS tool appears to be the learning, not the decision support information itself. Once the lessons have been captured, the tool itself appears to be less important. So while the use of EPIPRE may have fallen away, it appears to have still delivered benefit.

In our own emerging analysis, central to the explanation of low adoption is the prevalent view of scientist-developers that the DSS is a way of 'packaging' information or a model that 'should' be useful to managers and that, for development to be justified, this aid must be generally applicable. But it has become clear that the key to a DSS being used is its localized, or situated, in practice

(McCown, 2001; Berg, 1997). The latter author found in a study of medical DSSs in the United States that only a 'handful' of the hundreds of products available were actually in use. These few had common histories of intimate, intensive co-development by 'tool-makers' and practitioners in a workplace. Painful compromises on both sides resulted in a 'transformed tool in a transformed practice', and use did not spread from the practice situation in which it was produced. Our own experience in using a cropping systems simulator with farmers who own and use computers, even when the usefulness of the tool is discovered through intensive interaction, farmer preference is almost always for accessing benefits via a consultant skilled in using the tool rather than farmer use of the software.

Expert Systems (Type 5 in Table 1.1) have been envisaged as a way of providing the model of the farm Management System generally missing in DSS (Dent, 1994). We will not discuss these further because, (i) in the main, they do not use a process model of the Production System and (ii) they user from problems of 'lack of fit' to specific real-world management situations leading to non-use except for very narrow 'context-free' technical problems (Jones, 1989). In spite of a history of minimal achievement of impact on farming, optimism about the potential for models in farm management has remained perennial.

The degree of success of agricultural enterprises depends to a large extent on the quality of tactical decision-making in response to a variable and uncertain environment. Tactical decisions are aimed at optimizing management practices in such a way that the objectives of the farmer are achieved as completely as possible. Decision support systems that allow the analysis of alternative management could be value-able aids in tactical decision making. Such systems, based on crop growth models, that quantitatively describe the relations between environmental factors and crop performance are useful tools in this respect. However, the dynamic nature of the environment (weather, soil conditions), which often appears di cult to predict, limits the applicability of these models, or at least the margins of uncertainty remain relatively large. Therefore it is, in almost all cases necessary to combine these models with field observations that allow adjustment of the models in the course of the growing season. Combination of these models with optimization techniques should provide the basics for such decision support systems (Van Keulen and Penning de Vries, 1993).

An enlightening history (spanning several centuries) of this 'typical' view of the way models are supposed to aid decision-making has been provided by Ulrich (1983). The limited relevance of such 'decisionism' lies in its insistence on treating the social Management System 'objectively'. The final category of systems analysis aimed at intervention (Type 6 in Table 1.1) in the next section departs from this tradition.

1.8.1   SUBSISTENCE AGRICULTURE

- Subsistence agriculture occurs when a plot of land produces only enough food to feed the family working it or the local community (group, tribe, etc.), pay taxes and sometimes leaves a little surplus for barter or to sell in better years.
  - The main priority is self-sufficiency, which is achieved by growing a wide range of crops wherever possible.
  - Improvements to the system are held back by a lack of capital to provide fertilizers, pesticides, and other farming technology.
  - Animals are kept, although where land is limited it is generally too valuable to allow grazing or growth of fodder crops.
  - Where the climate is too extreme to support permanent settled agriculture, farmers become pastoral nomads, moving in search of food for their animals.
  - Depending on their location, animals provide milk, meat and blood for consumption; wool and skins for shelter and clothing; dung for fuel; bones for utensils and weapons; and mounts for transport.

- Other examples of subsistence farming are shifting cultivation, which is practiced in parts of the Amazon basin and in southeast Asia, and wet rice agriculture, also in southeast Asia and the Indian sub-continent.

1.8.2 Commercial Agriculture

Commercial agriculture usually takes place on a large, profit-making scale. It may be carried out by individual farmers or by companies, with both groups trying to maximize the return on inputs and seeking maximum yields per unit of land.

- This is often achieved by growing a single crop or by raising one type of animal.
- Commercial agriculture develops in places where there are good communications and markets are large, often both domestically and on a global scale.
- Europeans have developed large-scale plantations in the tropics to supply the markets of Europe and North America with crops that include rubber, sugar cane, coffee, tea, palm oil, bananas, pineapples, and tobacco.
- Other types of commercial agriculture include cattle ranching, commercial grain farming, and the intensive cultivation of fruits, flowers, and vegetables (sometimes referred to as market gardening).
- A growing number of farmers throughout the world are now abandoning the growth of staple food crops in order to produce for the emerging biofuels market.

1.8.3 Extensive and Intensive Agriculture

Extensive and intensive refer to the relationship of inputs to each other, particularly labor, capital, and land. Extensive agriculture is carried out on a large scale, whereas intensive agriculture is usually relatively small scale.

- Extensive agriculture occurs when: (i) The amounts of capital and labor are small in relation to the amounts of land being farmed. Shifting cultivation is an example of farming in which labor and capital are both low but large areas are covered. (ii) Labour is limited and capital higher. For example, cattle ranching and extensive grain cultivation in the USA, Canada, and Australia.
- Intensive agriculture occurs when: (i) The amount of labor is high, even if the amount of capital is low in relation to the area being farmed. An example is intensive wet rice cultivation. (ii) Labour input can be low but capital input high, allowing high levels of mechanization and technology input. This occurs in intensive fruit, flower, and vegetable production in the Netherlands (Figure 1.6).

It should be remembered that:

- There is no widely accepted consensus on how the major types of farming should be recognized and classified.
- Boundaries between farming types, as drawn on a map, are usually very arbitrary.
- One type of farming merges gradually with a neighboring type: there are few rigid boundaries
- Several types of farming may occur within each broad area - as in West Africa, where sedentary cultivators live alongside nomadic herdsmen.
- A specialized crop may be grown locally - e.g. a plantation crop in an area otherwise used by subsistence farmers.

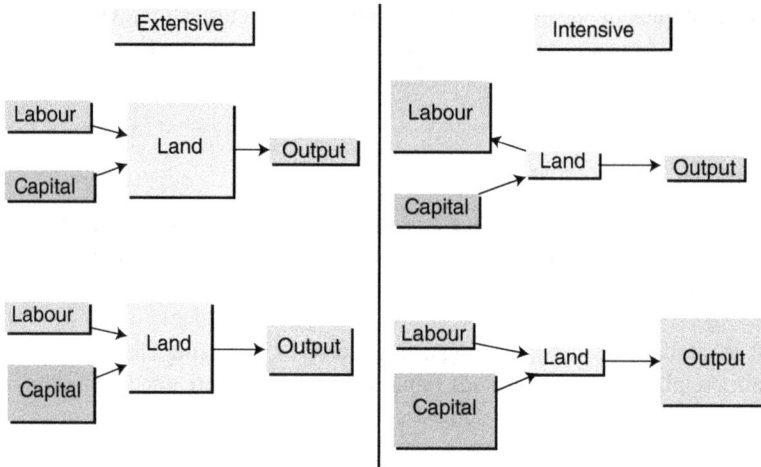

FIGURE 1.6 Global Distribution of farming types.

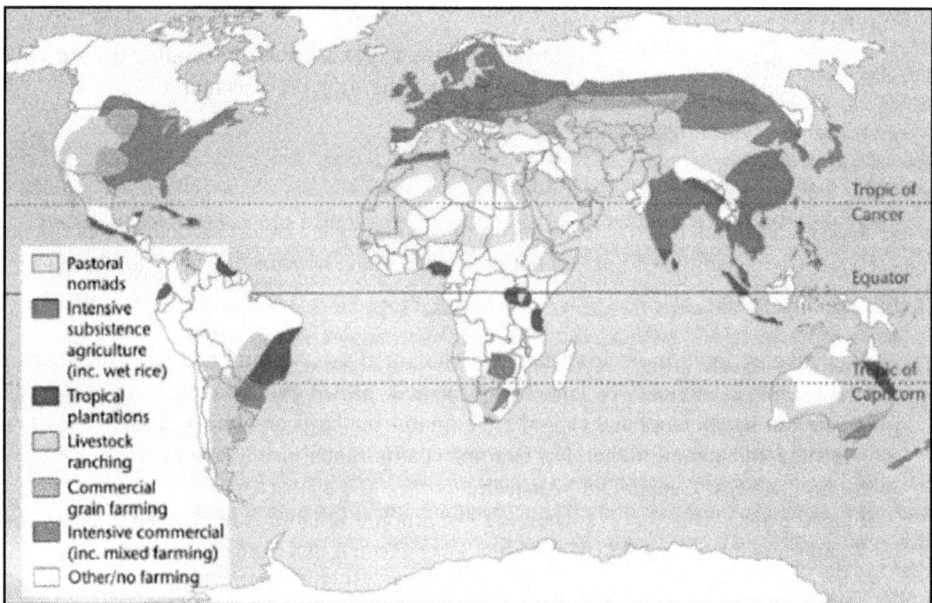

FIGURE 1.7 Global distribution of agricultural types.

- Types of farming alter over a period of time with changes in economics, rainfall, soil characteristics, behavioral patterns, and politics (Figures 1.7 and 1.8).

1.9 CONCLUSION

The history of modeling of agricultural systems reveals that major contributions were made by various disciplines, discussing specific production processes from field to farm, landscape, and beyond. Furthermore, there are excellent examples of integrating component models from various disciplines in different ways to create more robust system models that address biophysical, socioeconomic, and environmental responses. There are several examples where crop, livestock,

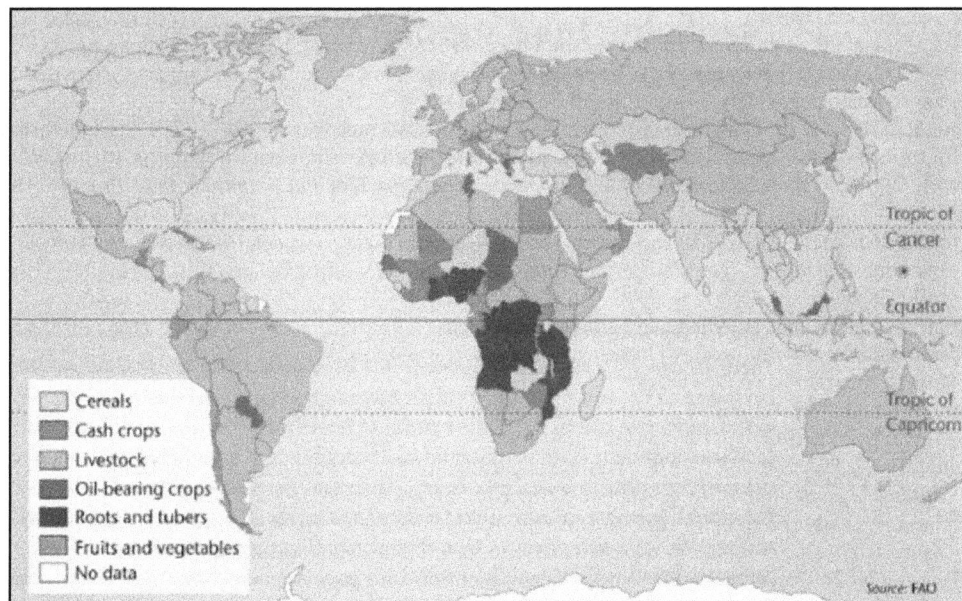

FIGURE 1.8 The highest value agricultural production by commodity group for each country, as recognized by the Food and Agriculture Organisation of the United Nations.

and economic models have been combined to research farming systems and evaluate the national and global impacts of climate change, policies or alternative technologies, as seen in the accompanying paper on the state of agricultural system science. This background also demonstrates that the production of models of agricultural systems continues to evolve through efforts by a growing number of research organizations worldwide and through numerous global initiatives, showing that researchers in these institutions are increasingly involved in contributing to scientific communities. Modularity, and interoperativeness. The research group of agricultural systems needs to provide standards and guidelines so that they can access and use the same "in the cloud" data sources from different sources, and run various models, information items, and decision support systems. Different models and approaches are important, but we need to develop standards and protocols in order to get the maximum benefits from these innovations. We now know that striving for just one "good" model doesn't pay off. Instead, we can strive for component models that are organized as modules that can be used on their own to resolve specific issues (such as when to apply a chemical or irrigation) and, more importantly, where those modules can be incorporated into systemic biophysical and economic models to solve more comprehensive problems. Modular models are needed to ensure successful scientific advancement, as well as viability and sustainability of the model. The user-driven creation of data and models. The history of data and model creation indicates that for research purposes many existing models were developed and then modified to meet user needs. Many models remain "user unfriendly," and although certain models are connected to DSS software, accessing model outputs or even using models is still difficult for many users. With the rapid developments in information and communications technology, it is now evident that there is substantial unrealized potential for more efficient use of data and models across different forms of "intelligence items," like computer simulation software and mobile technology.

REFERENCES

Anderson, J.R. (1974). Simulation: methodology and application in agricultural economics. *Review of Marketing and Agricultural Economics*, 42, 3–55.

Anderson, J.R. and Dent, J.B. (1971). Agricultural systems analysis: Retrospect and prospect. In: Dent, J.B. and Anderson, J.R. (Eds.), *Systems Analysis in Agricultural Management*. John Wiley, Sydney, pp. 383–388.

Ascough, J.C. and Deer-Ascough, L.A. (1994). Integrating evaluation into decision support system development. American Society of Agricultural Engineers (No. 943029). St Joseph, USA.

Barlow, R., Clark, D.A., Crawford, A., Paine, M., Sheath, G.W., and Weatherley, J. (2002). Guidelines for farming systems research and learning. Dairy Research and Development Corporation, Melbourne.

Berg, M. (1997). *Rationalizing Medical Work: Decision-Support Techniques and Medical Practices*. MIT Press, Cambridge.

Blackie, M. J. and Dent, J. B. (1974). The concept and application of skeleton models in farm business analysis and planning. *Journal of Agricultural Economics*, 25(2), 165–175. https://doi.org/10.1111/j.1477-9552.1974.tb00538.x.

Bouman, B., Van Keulen, H., Van Laar, H., and Rabbinge, R. (1996). The 'School of de wit' crop growth simulation models: A pedigree and historical overview. *Agricultural Systems*, 52(2–3), 171–198. https://doi.org/10.1016/0308-521x(96)00011-x.

Bunge, M. (1998). *Philosophy of Science: From Problem to Theory*. Transaction Publications, Routledge, Taylor & Francis Group, London and New York.

Charlton, P. and Street, P. (1975). The practical application of bio-economic models. In: Dalton, G.E. (Ed.), *Study of Agricultural Systems*. Applied Science Publishers, London, pp. 235–265.

Checkland, P. (1985). From optimizing to learning: A development of systems thinking for the 1990s. *The Journal of the Operational Research Society*, 36(9), 757. https://doi.org/10.2307/2582164.

Clark, D.A. (2002). Dexcel sees RED to help the environment. *Dexcelink Autumn*, 4–5.

Coleman, D.C., Swift, D.M., and Mitchell, J.E. (2004). From the frontier to the biosphere: A brief history of the USIBP grasslands biome program and its impacts on scientific research in North America. *Rangelands*, 26(4). https://doi.org/10.2458/azu_rangelands_v26i4_coleman.

Conway, G.R. (1987). The properties of agroecosystems. *Agricultural Systems*, 24(2), 95–117. https://doi.org/10.1016/0308-521x(87)90056-4.

Cros, M.J., Duru, M., Garcia, F., and Martin-Clouaire, R. (1997). Characterizing and simulating a rotational grazing strategy. In: Kure, H., Thysen, I., and Kristensen, A.R. (Eds.) *Proceedings of the First European Conference for Information Technology in Agriculture*, 15–18 June 1997, The Royal Veterinary and Agricultural University, Copenhagen, Denmark. pp. 379–382. Available online from http://www.dina.kvl.dk/-efita-confprogram.htm (date accessed 30 June 2008).

Dalton, G.E. (1982). *Managing Agricultural Systems*. Springer Science & Business Media.

Dempster, J.P. (1983). The natural control of populations of butterflies and moths. *Biological Reviews*, 58(3), 461–481. https://doi.org/10.1111/j.1469-185x.1983.tb00396.x.

Dent, J.B. and Blackie, M.J. (1979). *Systems Simulation in Agriculture*. Applied Science Publishers, London, p. 180.

Dent, J.B. (1975). The application of systems theory in agriculture. In: Dalton, G.E. (Ed.), *Study of Agricultural Systems*. Applied Science Publishers, London, p. 123.

Dent, J.B. (1994). The human response. Enabling technologies for land use and resource management. In: *Proceedings, Fifth International Congress for Computer Technology in Agriculture*, 29 June–5 July 1994 at Churchill College, University of Cambridge and the Royal International Agricultural Exhibition, National Agricultural Centre, StoneleighPark, Warwickshire, UK. Royal Agricultural Society of England, National Agricultural Centre, Kenilworth, pp. 40–45.

Dent, J.B. and Anderson, J.R. (1971). Systems, management, and agriculture. In: Dent, J.B. and Anderson, J.R. (Eds.), *Systems Analysis in Agricultural Management*. John Wiley, Sydney, pp. 3–14.

Dillon, J.L. (1992). The farm as a purposeful system. Miscellaneous publication: Department of Agricultural Economics and Business Management, University of New England (No. 10, iii).

Doyle, C. (1990). Application of systems theory to farm planning and control: Modelling resource allocation. In: Jones, J.G.W. and Street, P.R. (Eds.), *Systems Theory Applied to Agriculture and the Food Chain*. Elsevier Applied Science, London, pp. 89–112.

Duncan, W.G. (1972). SIMCOT: A simulation of cotton growth and yield. In: Murphy, C.M. (Ed.), *Proceedings of a Workshop for Modeling Tree Growth*. Duke University, Durham, North Carolina, pp. 115–118

Duncan, W.G., Loomis, R.S., Williams, W.A., and Hanau, R. (1967). A model for simulating photosynthesis in plant communities. *Hilgardia*, 38(4), 181–205. https://doi.org/10.3733/hilg.v38n04p181.

Edwards-Jones, G., Deary, I., and Willock, J. (1998b). Incorporating psychological variables in models of farmer behaviour: Does it make for better predictions? *Etudes et Recherches Sur Les Systemes Agraires et Le Developpement*, 31, 153–173. Versailles, France, Institute National de la Recherche Agronomique.

Edwards-Jones, G., Dent, J.B., Morgan, O., and McGregor, M.J. (1998a). Incorporating farm household decisionmaking within whole farm models. In: Tsuji, G.Y. et al. (Eds.), *Understanding Options for Agricultural Production*. Kluwer Academic, Dordrecht, pp. 347–365.

Forcella, F., Wilson, R.G., Renner, K.A., Dekker, J., Harvey, R.G., Alm, D.A., Buhler, D.D., and Cardina, J. (1992). Weed Seedbanks of the U.S. corn belt: Magnitude, variation, emergence, and application. *Weed Science*, 40(4), 636–644. https://doi.org/10.1017/s0043174500058240.

Gutierrez, A.P., Mills, N.J., Schreiber, S.J., and Ellis, C.K. (1994). A physiologically based Tritrophic perspective on bottom-up-Top-Down regulation of populations. *Ecology*, 75(8), 2227. https://doi.org/10.2307/1940879.

Hare, M., Letcher, R., and Jakeman, A. (2003). Participatory modelling in natural resource management: A comparison of four case studies. *Integrated Assessment*, 4(2), 62–72. https://doi.org/10.1076/iaij.4.2.62.16706.

Hassell, M.P. (1986). Parasitoids and population regulation. In: Waage, J.K. and Greathead, D. (Eds.), *Insect Parasitoids — 13th Symposium of the Royal Entomolgical Society of London*. Blackwell Scientific Publications, Oxford, pp. 201–224.

Heady, E.O. (1957). An econometric investigation of the technology of agricultural production functions. *Econometrica*, 25(2), 249. https://doi.org/10.2307/1910253.

Heady, E.O. and Dillon, J.L. (1964). *Agricultural Production Functions*. Ames Iowa State University Press, USA.

Hieronymi, A. (2013). Understanding systems science: A visual and integrative approach. *Systems Research and Behavioral Science*, 30(5), 580–595. https://doi.org/10.1002/sres.2215.

Hoag, D.L., Ascough, J.C., and Frasier, W.M. (1999). Farm computer adoption in the Great Plains. *Journal of Agricultural and Applied Economics*, 31(1), 57–67. https://doi.org/10.1017/s0081305200028776.

Hochman, Z.H., Carberry, P., McCown, R., Dalgliesh, N., Foale, M., and Brennan, L. (2001). Apsim in the marketplace: A tale of kitchen tables, boardrooms and courtrooms. *Acta Horticulturae*, 566, 21–33. https://doi.org/10.17660/actahortic.2001.566.1.

Hoogenboom, G., Jones, J.W., Wilkens, P.W., Porter, C.H., Boote, K.J., Hunt, L.A., Singh, U., Lizaso, J.L., White, J.W., Uryasev, O., Royce, F.S., Ogoshi, R., Gijsman, A.J., Tsuji, G.Y., and Koo, J. (2012). *Decision Support System for Agrotechnology Transfer (DSSAT) Version 4.5 [CD-ROM]*. University of Hawaii, Honolulu, HI. https://foodsupplyissues.weebly.com/agricultural-systems.html.

Jayaratna, N. (1986). Normative information model-based systems analysis and design (NIMSAD): A framework for understanding and evaluating methodologies. *Journal of Applied Systems Analysis*, 13, 73–87.

Jones, C.A. and Kiniry, J.R. (1986). *CERES-Maize: A Simulation Model of Maize Growth and Development*. Texas A & M University Press, College Station, TX.

Jones, F.T. (1989). Feed quality control in poultry production. *Korean Journal of Animal Nutrition & Feedstuffs*. 13(1), 25–37.

Jones, J.W, Hoogenboom, G., Porter, C., Boote, K., Batchelor, W., Hunt, L., Wilkens, P., Singh, U., Gijsman, A., and Ritchie, J. (2003). The DSSAT cropping system model. *European Journal of Agronomy*, 18(-3–4), 235–265. https://doi.org/10.1016/s1161-0301(02)00107-7.

Jones, J.W., Thornton, P.K., and Hansen, J.W. (1997). Opportunities for systems approaches at the farm scale. In: Teng, P.S. et al. (Eds.), *Application of Systems Approaches at the Farm and Regional Levels*. Kluwer Academic, Dordrecht. pp. 1–18.

Keen, P.G. (1975). Computer-based decision aids: The evaluation problem. *Sloan Management Review*, 17–29.

Keen, P.G. (1987). Decision support systems: The next decade. *Decision Support Systems*, 3(3), 253–265. https://doi.org/10.1016/0167-9236(87)90180-1.

Little, J.D. (1970). Models and managers: The concept of a decision calculus. *Management Science*, 16(8), B-466–B-485. https://doi.org/10.1287/mnsc.16.8.b466.

Lynch, T., Gregor, S., and Midmore, D. (2000). Intelligent support systems in agriculture: How can we do better? *Australian Journal of Experimental Agriculture*, 40(4), 609. https://doi.org/10.1071/ea99082.

Malcolm, L.R. (1990). Fifty years of farm management in Australia: Survey and review. *Review of Marketing and Agricultural Economics*, 58, 24–55.

McCown, R.L. (2001). Learning to bridge the gap between science-based decision support and the practice of farming: Evolution in paradigms of model-based research and intervention from design to dialogue. *Australian Journal of Agricultural Research*, 52(5), 549. https://doi.org/10.1071/ar00119.

McCown, R.L. (2002a). Changing systems for supporting farmers' decisions: Problems, paradigms, and prospects. *Agricultural Systems*, 74(1), 179–220. https://doi.org/10.1016/s0308-521x(02)00026-4.

McCown, R.L. (2002b). Locating agricultural decision support systems in the troubled past and socio-technical complexity of 'models for management'. *Agricultural Systems*, 74(1), 11–25. https://doi.org/10.1016/s0308-521x(02)00020-3.

McRae, A.F. (1993). As applied scientists we serve...? *Proceedings of the New Zealand Society of Animal Production*, 53, 107–110.

Meinke, H., Baethgen, W., Carberry, P., Donatelli, M., Hammer, G., Selvaraju, R., and Stöckle, C. (2001). Increasing profits and reducing risks in crop production using participatory systems simulation approaches. *Agricultural Systems*, 70(2–3), 493–513. https://doi.org/10.1016/s0308-521x(01)00057-9.

Menz, K., and Knioscheer, H. (1981). The location specificity problem in farming systems research. *Agricultural Systems*, 7(2), 95–103. https://doi.org/10.1016/0308-521x(81)90033-0.

Mueller, R.A.E. (1993). The product of science. In: *Proceedings of the XVII International Grassland Congress*, 8–21 February, 1993, Palmerston North, New Zealand. New Zealand Grassland Association. pp. 607–614.

Murdoch, W.W. (1994). Population regulation in theory and practice. *Ecology*, 75(2), 271–287. https://doi.org/10.2307/1939533.

Musgrave, W.F. (1976). Problems of change in Australian agricultural economics. *Australian Journal of Agricultural Economics*, 20(3), 133–143. https://doi.org/10.1111/j.1467-8489.1976.tb00184.x.

Ogle, G. and Tither, P. (2000). An analysis of the risks and benefits of beef intensification. *Proceedings of the New Zealand Grassland Association*, 25–29. https://doi.org/10.33584/jnzg.2000.62.2392.

Parker, C. (1999). Decision support systems: Lessons from past failures. *Farm Management*, 10, 273–289.

Pimentel, D. and Peshin, R. (2014). *Integrated Pest Management: Pesticide Problems*. Springer Science & Business Media, USA.

Pinter, Jr. P.J., Ritchie, J.C., Hatfield, J.L., and Hart, G.F. (2003). The agricultural research service's remote sensing program. *Photogrammetric Engineering & Remote Sensing*, 69(6), 615–618. https://doi.org/10.14358/pcrs.69.6.615.

Ridder, N. (1997). Hierarchical levels in agro-ecosystems: Selective case studies on water and nitrogen [Unpublished doctoral dissertation]. Wageningen Agricultural University, Wageningen, Netherlands.

Ritchie, J.T. and Otter, S. (1984). Description and performance of CERES-wheat: A user-oriented wheat yield model. In: Wheat Yield Project, ARS (Ed.), ARS-38. National Technical Information Service, Springfield, MI, pp. 159–175.

Roberts, E.B. (1977). Strategies for effective implementation of complex corporate models. *Interfaces*, 8(1), 26–33. https://doi.org/10.1287/inte.8.1.26.

Rogers, E.M. (2010). *Diffusion of Innovations* (4th ed.). Simon & Schuster, United Kingdom.

Romera, A.J. (2004). Simulation of cow-calf systems in the Salado region of Argentina [Unpublished doctoral dissertation]. Massey University, Palmerston North, New Zealand.

Romera, A.J., Morris, S.T., Hodgson, J., Stirling, W.D., and Woodward, S.J. (2006). The influence of replacement policies on stability of production in a simulated cow-calf farm system. *New Zealand Journal of Agricultural Research*, 49(1), 35–44. https://doi.org/10.1080/00288233.2006.9513691.

Romero, C. and Rehman, T. (2003). *Multiple Criteria Analysis for Agricultural Decisions*. Elsevier.

Schoemaker, P. (1982). The expected utility model: its variants, purposes, evidence and limitations. *Journal of Economic Literature*, 20, 529–563.

Seligman, N.G. (1990). The crop model record: Promise or poor show? In: Rabbinge, R., Goudriaan, J., Van Keulen, H., Penning de Vries, F.W.T., and Van Laar, H.H. (Eds.), *Theoretical Production Ecology: Reflections and Prospects*. Simulation Monographs 34, Pudoc, Wageningen, pp. 249–263.

Smith, G.F. (1989). Defining managerial problems: A framework for prescriptive theorizing. *Management Science*, 35(8), 963–981. https://doi.org/10.1287/mnsc.35.8.963.

Sorensen, J. T., and Kristensen, E. S. (1992). Systemic modelling: A research methodology in livestock farming. In A. Gibon & B. Matheron (Eds.), *Global Appraisal of Livestock Farming Systems and Study on their Organisational Levels: Concept, Methodology and Results: Proceedings of a Symposium*. Commission of European Communities, Toulose, France, pp. 45–57.

Spedding, C.R.W. (1976). Editorial. *Agricultural Systems*, 1, 1–3.

Spedding, C.R.W. (1990). Agricultural production systems. In: Rabbinge, R, Goudriaan, J, van Keulen, H., Penning de Vries, F.W.T., van Laar, H.H. (Eds.), *Theoretical Production Ecology: Reflections and Prospects*. Pudoc, Wageningen, pp. 239–248.

Stapleton, H.N., Buxton, D.R., Watson, F.L., Notling, D.J., and Baker, D.N. (1973). COTTON: A computer simulation of cotton growth. University of Arizona, agricultural experiment station. Technical Bulletin. 206.

Thornton, P. and Dent, J. (1984a). An information system for the control of puccinia hordei: II—Implementation. *Agricultural Systems*, 15(4), 225–243. https://doi.org/10.1016/0308-521x(84)90010-6.

Thornton, P. and Dent, J. (1984b). An information system for the control of puccinia hordei: I—Design and operation. *Agricultural Systems*, 15(4), 209–224. https://doi.org/10.1016/0308-521x(84)90009-x.

Ulrich, W. (1983). *Critical Heuristics of Social Planning: A New Approach to Practical Philosophy*. Bern, Switzerland, and Stuttgart, Germany: Haupt. Paperback reprint version, Wiley, Chichester, UK, 1994.

Valentine, I., Hurley, E.W., and Glass, W. (1993). Goals and management strategies of dairy farmers. *Proceedings of the New Zealand Society of Animal Production*, 53, 111–113.

Van Dyne, G.M. and Anway, J.C. (1976). A research program for and the process of building and testing grassland ecosystem models. *Journal of Range Management*, 29(2), 114. https://doi.org/10.2307/3897406.

Van Keulen, H. and Penning de Vries, F.W. (1993). Farming under uncertainty: Terminology and techniques. *International Crop Science I*, 139–144. https://doi.org/10.2135/1993.internationalcropscience.c24.

Vries, F.P., Teng, P., & Metselaar, K. (1993). Systems approaches for agricultural development. In: *Proceedings of the International Symposium on Systems Approaches for Agricultural Development*, 2–6 December 1991, Bangkok, Thailand. Springer Science & Business Media.

Wallach, D., Makowski, D., Jones, J.W., and Brun, F. (2018). *Working with Dynamic Crop Models: Methods, Tools and Examples for Agriculture and Environment*. Academic Press.

Webby, R. (2002). The value of decision support models for farmer learning. *Proceedings of the New Zealand Grassland Association*, 45–47. https://doi.org/10.33584/jnzg.2002.64.2472.

Webster, J. (1990). Reflections on the economics of Decision Support Systems. In: Kuhlmann, F. (Ed.), *Integrated Decision Support Systems in Agriculture: Successful Practical Applications. Third International Congress for Computer Technology*, 27–30 May 1990, Frankfurt, Bad Soden. Deutsche Landwirtschafts-Gesellschaft, Frankfurt, pp. 307.

Wheeler, T. and Von Braun, J. (2013). Climate change impacts on global food security. *Science*, 341(6145), 508–513. https://doi.org/10.1126/science.1239402.

Worthington, E. (2009). *The evolution of IBP*. Cambridge University Press.

Zadoks, J.C. (1989). EPIPRE: A computer based decision support system for pest and disease control in wheat: Its development and implementation in Europe. In: Leonard, K.J. and Fry, W.E. (Eds.), *Plant Disease Epidemiology*, (Vol. 2). Genetics, Resistance and Management. McGraw-Hill, New York, pp. 3–29.

2 A Discourse beyond Food Security and Environmental Security

Dear Epistemic Community, Should We Consider Agro-Security?

Ayodeji Anthony Aduloju
Obafemi Awolowo University
Edo State University Uzairue

Temitayo Adedeji Adedoyin
Obafemi Awolowo University

CONTENTS

2.1   INTRODUCTION

Security is an allusive term that has remained long-contested within the broader scheme of International Relations. Without dwelling much on the ontological turf of security as a dynamic concept, this study narrows its searchlight on food security, environmental security, and the nucleus, -agro-security. It is plausible that countless countries have taken domestic production of food, very seriously, but, a lot of them are, yet, short of the needed institutional capacity to securitize their environment. Thus, this undoing of effective environmental safeguards is indicative of health risks, climatic problems, and other ecosystemic challenges that people are battling (Aduloju and Pratt, 2014; Aduloju and Okwechime, 2016). For example, the indiscriminate disposal of hazardous chemicals, storage of obsolete pesticides, and concurrent improvement in food production levels in Ghana- a prototypical situation in many other nations- are pure counterproductive activities (FAO, 2017). Despite the multiple enactments and policies pull by the epistemic communities to secure the environment from food sufficiency-induced disasters, no remarkable improvement has been reported on human security (Ratner, 2018). While many have blamed the complexity of food security efforts, others have tagged the environmental security policies as expensive. Thus, this paper emerges at this center of indecision, and the northing recommendation appears to be "agro-security". That is, if

sustaining humanity is the essence of food security and protecting humanity is the quintessence of environmental security, both can only be simply and cheaply achieved if the epistemic communities explore agricultural security.

2.2 THEORETICAL CONTEXTUALIZATION

The theoretical analysis of this study blends both human security theory and security dilemma. Surprisingly, we are discussing conflicts or wars, but, the extant issues in this paper are silent killers, with similar ends, that conflicts and crises would attract (Bustic et al., 2015). First, human security places more value on individual safety and not on collective states. The multiple variations in human security dimensions all speak of protection, civil liberties, and freedom. Security dilemma underscores the suspicion and uncertainties that force states to step up their defense or response capacities. Similarly, the environmental uncertainties persist in natural disasters and unforeseen contingencies, for example, governments worldwide have forced their farmers indoors, unproductive and the farms lay fallow, breeding weeds, as the Covid-19 pandemic thwarted almost every human agenda (Martin and Romei, 2020).

Before this, however, the well-to-do nations have always been suspicious of migration trends from struggling countries, which are both food and environmentally insecure. This dilemma, in part, is responsible for some centrifugal benefits that flow towards the third world nations from the developed countries. The worst dilemma is that abandoning these poor nations to luck would create more quandary for the wealthy ones, whereas, 'human lives' is all at risk (Díaz et al., 2006). Therefore, this security dilemma suggests the combination of human security elements. If environmental promotes environmental safety and food security exalts food accessibility, political security permits individual liberties as well. Thus, individual experts through epistemic meet-ups are well placed to balloon agro-security; a simple solution against the security dilemma, that virtually all nations can equally afford.

2.3 HUMAN SECURITY CONCERNS: THE SIMULTANEOUS EQUATION BETWEEN FOOD SECURITY AND ENVIRONMENTAL SECURITY

Chronologically, food security predates environmental security, as the latter only became a global issue, amidst the Cold War, and more obvious, after the war. In corroboration with the apex worldwide food summit, Aduloju and Adedoyin et al. (2020) comprehends food security as the availability and accessibility of food at first, and then, the utility and stability of food production. Besides, Heidhues et al. (2004) argued that food security would be better off not defined globally, but regionally, nationally, and at individual levels. This distraction, perhaps, led to the modification of food security and hence, defined as a constant situation, whereby, all individuals can sufficiently afford their food preferences and nutritional value for healthy and active living.

However, to Ratner (2018), environmental security is an element that affects human security. Westing (1989) cited America's environmental warfare as tactics that adversely affected our normal natural environment. For example, the intentional salinization of freshwater reservoirs, arable lands, use of concussion bombs, chemical bombs, herbicides, forest fires, and dams breaking created extensive toxic contamination that affected the surroundings. But, then, to ignite more attention and seriousness about this security imbroglio, argued for environmental security to be seen as a national agenda, that each country must closely monitor. Consequently, developed countries, such as the United States (US), for example, have installed several environmental agendas for national security. The US paraphrased this environmental security exercise as 'greening of its defense' and spends over five billion dollars for pollution prevention, at home and America's cleanup missions abroad (Käkönen, 1994).

At their best, most third-world nations could only afford to setting up 'Environmental Protection Agencies' created to identify and control developmental triggers that can alter environmental

security, leaving out core issues such as environmental scarcity, social circulation of resources and environmental degradation that have more penchant for environmental conflicts and human security(Kahl, 2006). For example, self-acclaimed developing China is facing huge environmental problems, while multiple challenges abound in several other third-world countries.

Previously, despite acknowledging the huge impact of technology on food sufficiency, disadvantaged countries have clamored for a slowdown in technology-induced environmental threats. But, categorically, technological advancements have licensed the possibility of pursuing both sides of development while mitigating the challenges created by technology and innovation. In other words, the challenges of technology, captured in climate change conversations can be defeated without compromising human security, a joint end of both environmental security and food security (Schwartz and Randall, 2003).

Therefore, this section holds the premise that environmental security and food security are twin issues, with the complementary influence of one on the other. From a global perspective, food security enables environmental security (Falvey, 2005). But, beyond food security as a mere precursor, this section equates food security with environmental security. Although, developed few places more value on both, but, the developing more, prioritize the race towards food security to the detriment of environmental security. Realizing this tact that, shortage or lack of food in the poor zones encourages migration to the developed spaces, and this asserts more pressure and threats on both environments; the developed ones have resorted to the provision of palliatives for developing countries to boost their food production efforts. Also, the conception of food accessibility as one of the key universal rights has lubricated the entry of food exports (though, with rising prices) into developing countries (FAO, 2008)

However, with the global population expected to clinch 11 billion by 2050, technology is expected to elevate food security efforts and relegate environmental manipulation outcomes arising from increasing food production and human developmental activities. The major challenge, nevertheless, is that such a shift in agenda is politically imperative, and with visionary leadership eluding many third world governments, the outcomes of this paper will remain untapped. While the developed countries could boast of envisioned leaders, the corona virus outbreak in early 2020 has dipped all countries of the world into a wrecking ship of an economic quagmire (Evans et al., 2020). As infected people towers in millions, nations have embarked on lockdown policies that have prevented virtually all productive activities, including food production and farming. While the poor nations have almost emptied their little food reserves, it does not make any economic sense to export food from the global north to the south, as the international focus currently tilts towards ending corona virus.

Having synchronized the prevalent circumstances of countries in both global developmental halves (North and South), this section in essence, inaugurates the links between food security and environmental security. The worldwide measurement of accessible lands equals almost 14 billion hectares. Woodlands and pastures occupy 8.3 billion hectares or 61% while agricultural usage occupies 1.7 billion hectares. Also, irrigated cultivation is considered suitable within 2.1 billion hectares; covered by vegetation, woodland and swampland that absorbs carbon dioxide and preserves ecosystem biodiversity. The implication is that population projections in the coming years will attract a 90% increase in food production. To match the requirements of the world food agencies, arable lands of about 121 million hectares must be available in the developing world, more so, in Latin America and Sub-Saharan Africa. As 95% of the arable zones in Asia are presently in use, their cultivable landmass cannot add more. Thus, food production will be competing with rapid industrialization and so on, including biofuels markets, and even more, as oil prices have fallen significantly. For example, Brazil is the second biggest global producer and consumer of biofuels. Despite its teeming population's food needs, over 2.6 million hectares or 4.5% of Brazil's arable land have been dedicated to food crops such as sugar cane, palm oil, and corn (Carneiro and Hector, 2014). But, then, all of those food crops investments are tailored toward non-food purposes such as biofuels and gasoline production.

Although, the positives include employment spin-offs from Brazil's biofuel sector, where about a million unemployed people have been engaged and over 300,000 job outlets provided in Brazil's manufacturing industry. But, conversely, the attendant challenges also include land conversion for non-food production usage and inflated prices on limited available food products (FAO, 2008). Beyond food insecurity problems, global warming continues to blossom, resultant of nitrogen-protoxide emissions and other biofuel cycles. Although, the entry of lingo-cellulosic feedstock, a recent invention is expected to thrash off the biofuels emission, yet, lingo-cellulosic biomass is not commercially available.

Additionally, further threats to environmental security include land degradation, salinization, erosion, pollution, desertification, and other man-made factors within our ecosystem. The degenerative outcome of all of these is the impoverishment of land fertility, with direct repercussions for human security (Nkonya et al., 2016). For example, about 20,000–30,000 km^2 of available landmass becomes unusable for food production annually. Compared with Europe and North America, Latin America, Africa, and Asia, suffer that soil infertility six times more. Painfully, one-fifth of world population or 1.7 billion people (rural population) depend on agricultural dividends to sustain their lives (Zavatta, 2014). Just a reminder; food security exists when there is an unceasing availability and accessibility of food and water resources, at first, and also, the efficient utility and stability of nutritious food production and water. If the above points on food and environmental security are extrapolated and analyzed vis-à-vis the implications for human security, we can realistically conclude that food security equates to environmental security. In this equation, however, the coefficient of both variables remains human security while the constant variable is the environment- the platform for all human undertakings.

2.4 DEAR "EPISTEMIC COMMUNITY", SHOULD WE CONSIDER AGRO-SECURITY?

The previous section has brought to fore, human security, as the mid-point between food security and environmental security. The revelations from that section also show most countries (the poor and developing) may not afford food security agenda, given the default nation-building challenges facing them (such as endemic poverty, population explosion, visionless leadership, and the likes). As noted, the advanced or fairly-developed few countries that can relatively afford food security agenda may not be able to effectively control environmental insecurities (such as global warming), emanating from technology improvements, innovation advancements, industrialization, modernization and other human-made and unannounced disasters. More recently, the corona virus outbreak has not only deteriorated the situations of many developing nations but has also placed overwhelming pressure on the food security of the developed ones. For once, let us moderate the intimidations and assume the epistemic communities are immune against the looming dangers facing food security, environmental security and humanity, now, we hail and write thee, 'dear epistemic communities', should we consider agricultural security?

Notwithstanding the sovereign standing of countries worldwide, epistemic communities are esteemed as supranational gang of experts with substantive influence on policy makers regarding specific global issues of transnational concerns (Toke, 1999; Dunlop, 2017). Technically speaking, the reliant outcomes from epistemic platforms are topnotch compared with resolutions of national influencers that decide from a restricted information base, or regional agencies that may not be entirely objective in their perceptions. Aduloju and Adedoyin (2020) sanctioned epistemic platforms as international experts with evidence-based findings, standpoints, and innovations that appropriately tackle transnational problems and professionally submit their feedback for global utility, subsequently. The indispensable principles of such epistemic communities subsume selection of specific issues, innovation of workable solutions, and circulation of recommendations for nation-states. The final principle, perhaps, the most pivotal is the inclusion of academia and experts for validating their hypotheses.

2.5   THE WAY FORWARD: THE DYNAMICS OF LEVERAGING AGRO-SECURITY AS AN ALTERNATIVE

Truly, the coordinated efforts on global food security, by the epistemic communities, the FAO and other bodies have yielded immense improvements, not only in the third world, but, across board. For example, the universal 'freedom from hunger' campaigns transformed into model initiatives in several countries; for instance, Mexico adopted it as a Crusade-against-Hunger, while Grenada modified it as Zero-Hunger challenge. Also, Chile formed the Choose-Healthy-Living program, Brazil adopted it as Zero-Hunger agenda, while the broader 'Community of Latin American and Caribbean States' (CELAC) adopted the epistemic spin-off recommendations as 'Nutrition and Hunger Eradication' in the whole sub-region. These instances are repetitive of situations in Africa and elsewhere. In fact, some global existing global initiatives such as the MDGs/SDGs, or the South-South cooperation modified their policy framework to integrate the epistemic recommendations. But, then, the implementation in poor countries appears as a hard job. Yet, there was no significant scale of impact in many, but, in few of them. For instance, data of hungry people declined in the Caribbean and Latin America shows a fall from the 1990s 14.7% to about 5.7% by 2014.

On environmental security, the reflexive contributions from epistemic communities and governments have sustained the greenspace considerably and subdued several proposed dangerous environmental policies, that are void of fact-based investigations or research. Japan's Fukuoka City of 1.7 million people enjoyed enormous contributions from epistemic communities' techno-scientific findings. In collaboration with Japanese local actors, the evidence-based recommendations shaped Japan's urban environmental management, and hitherto- unforeseen threats, like pollution, flooding, and its multiple implications, were tactically avoided (Hanakata, 2017). The key argument here is the varying capacities of the countries worldwide to implement these environmental security agendas and that of food security, in terms of simplicity, technical strength, affordability, and sustainability (Barnett and Parnell, 2016). But, then, agricultural security (agro-security) resonates with simplicity, local ownership, and affordability.

Bearing the ends of human security at heart, agro-security is an all-inclusive chain of many players with an entitlement mentality that enables them to function effectively to sustain food production and preserve their green environment from turning grey. Shortly after the famous September 2001 attacks, America's Health Secretary stated as follows, "I, for the life of me, cannot understand why the terrorists have not attacked our food supply, because it is so easy to do" (Swerdloff, 2016). Thus, if any action that truncates fiber and food production is equivalent of war, their call-to-war (security) implies preventing any disruptions that will affect America's chain of food supply.

Simply put, agro-security is the combination of resources and programs to safeguard agriculture, natural resources, food sustainability and centrally, humanity. Such resources and programs include an assemblage of experts for intensive preparations, prevention, threats detection, diagnosis of dangers, response, and lastly, recovery.

Accordingly, agro-security is compatible with the political structures, socio-economic templates and multilateral interventionist frameworks in nearly all countries. Specifically, in financial terms, conventional market mechanisms have not allowed food security agenda thrive, while protectionists guidelines on environmental ownership allow governments to divert land use, unchecked (Díaz et al., 2006).

Interestingly, securitizing agriculture captures 75% of the poor or dwellers in rural areas. Quite revealing, the lack of epistemic-community influence on general agricultural policies for some decades have dwindled interests from private investors and public stakeholders. When investors are not clear about productivity losses or gains, they often resort to green technology, whereas the green revolution should be an appendage of agro-security. Meanwhile, the green revolution implies the widespread usage of technological innovation and scientific techniques in agriculture. But, then, the gap here is that human security is the opportunity cost of such overhauling scientific adaptations

in agriculture; not only in terms of increasing food prices and dangerous emissions from machines, but the replacement of human capital with pure machinery is also self-inflicting.

In fact, warlike situations like the corona virus palaver have further exposed the insufficiency of reliance on technology alone. For example, as international attention swings toward inventing vaccines that cure corona virus, most farms are emptied for safety reasons. Yet, the approaching shortage of available groceries in food silos has the potency of inviting another hunger-induced virus of all kinds. So, dear epistemic communities, agricultural security presents simple, locally owned, affordable and sustainable solutions for human security, the ends, that food and environmental security are designed to provide. The nucleus of agro-security interventions is encrypted as follows;

Firstly, agro-security strengthens global governance strategies in the agricultural sector. Unlike food security's unparalleled implementation in all nations or environmental security's expensive outlook, agro-security is reduced to the community level. The implementation model is self-regulatory and unilateral protectionist measures are maximally reduced. Agro-security creates a shared space for discussions and dialogue on the restoration of food's nutritional value, availability and sustainability, as the common goals. The epistemic community as a global player in agro-security permits effective response to socio-economic commotions and political irritations that have a transversal impact on food production and environmental sustainability.

Additionally, agro-security lubricates agricultural productivity and motivates economic development. Agro-security tends to facilitate the spreading of credible solutions, applicable tools and outfits to all countries, more importantly, the developing ones. Meeting the expected global food consumption in the coming decades requires a compound system of solutions. As population figures rise, pressures on increasing food baskets, as, climatic change impacts also persist. As producing more food requires continuous innovations, agro-security's local entitlement-mentality guarantees reduced environmental impacts, without compromising the food standards. Consequently, this attention on biotechnology and bio-sustainability diffuses best global agricultural practices to the grassroots and gaps would be bridged accordingly. Agro-security is designed to change the, hitherto, reliance on world agro-food exports that comes with a tax surge on the poor and developing countries. However, its success relies on an epistemic community or independent authority acting as surveillance and placing "position limits" on importation undertakings and exportation of resources (human and materials).

Lastly, agro-security appears suitable for managing dietary habits. Mechanisms have been installed for forecasting climatic change dynamics, while objectivity uncertainties have evolved with data on food security prospects and challenges. But, agro-security agenda enables differentiation in dietary choices and consumption patterns. For example, Asian diet differs from Western diet, African diet or that of the Mediterranean (Robotti, 2020). Thus, the varying environmental ingredients reflect on their dietary value and consumption habits- quality and quantity, respectively. The ability to forecast food consumption patterns in emerging populations assist policy-makers in making enormous economic decisions. In developing nations, with weak health systems, agro-security agenda also helps reduce health emergencies such as cardiovascular diseases, metabolic disorders and tumors derived from inadequate consumption or poor dietary habits.

This study does not seek to appear as a perfectionist research. Thus, there are some fundamental criticisms. Foremost, agro-security looks prepared for emergency situations, whereby, responses are readymade when any nation may require sudden intervention. However, agro-security seems too weak for disaster situations, whereby the damage and hardship are already affected and probably, and lives are also lost (Chapman, 2009).

Also, the promotion of agro-security agenda by the epistemic communities is open to three basic threats; natural threats, accidental threats, and intentional threats. Agro-security is affected, generally, by natural threats like hurricanes, ice storms, wild fire, floods, earthquake, dust storms, and droughts. When natural threats occur, plural productive activities suffer, ranging from wild life to livestock and agronomics crops. Likewise, agro-security is affected by accidental threats, consequent of non-natural occurrences like pesticide or chemical spills, irrigation system contamination or facility failure. When

accidental threats occur, epistemic communities are largely incapacitated from effective response to save neither the animals nor crops, as the damage would have been inflicted, beyond repair.

The last category is an intentional threat, also branded as agro-terrorism. Intentional threat entails deliberate disruption of economic welfare, livestock, animals, crops and food production within a country or regional setting. Terrorists may exploit the importance of agricultural sectors or programs to strengthen their bargaining posture. Such eco-terrorists can also explore the media to spread misinformation on national husbandry practices, processing or marketing facilities, all, to harm the agricultural sector. For example, some foreign-based terrorists disclosed their plans to attack America's farms and agricultural sector in 2007. When the terrorists, disguising as Mexicans were accosted by America's Homeland Security agents, vials of highly-infectious and contaminated blood were seized from terrorists. When interrogated, the terrorists confessed their aim to disrupt meat supply in southwestern America through the exposure of a poisonous disease-causing agent (FBI, 2008).

2.6 CONCLUSION

In conclusion, the best time to prepare for war is in peace time. Agro-security is the combination of resources and programs to safeguard agriculture, natural resources, food sustainability and centrally, humanity. Such resources and programs include assemblage of experts for intensive preparations, prevention, threats detection, diagnosis of dangers, response, and lastly, recovery. The revelations from that section also show most countries (the poor and developing) may not afford food security agenda, given the default nation-building challenges facing them (such as endemic poverty, population explosion, visionless leadership, and the likes). As noted, the advanced or fairly-developed few countries that can relatively afford food security agenda may not be able to effectively control environmental insecurities (such as global warming), emanating from technology improvements, innovation advancements, industrialization, modernization and other human-made and unannounced disasters.

This paper has established that food insecurity and ecosystem degradation aggravate conflict risks, human security, and vulnerability. For example, some previous studies revealed that, from 1946 to 2006, conflicts over administration of environmental issues like land control, natural resources, usage rights, and revenue distribution rose steadily; 44% of conflicts in African Sub-Saharan, 39% in North America and the Middle East, 56% in South Asia, 60% in the Pacific and East Asia (Rustad and Binningsbo, 2012). Furthermore, a 2017 evaluation of 1,800 resource-induced conflicts globally showed a larger fraction were results of improperly managed extractive activities that has damaged or polluted land, water, air, forests, and livelihoods in several communities (Akokpari, 2012).

As these avoidable threats to human security continue to culminate, it is imperative that new forms of human security administration in epistemic communities wave in. But then, the implementation vacuums, identified in food security networks and environmental security templates is suggestive of agro-security. As noted, reducing the plenty of risks to human security depends on the improvement of resource governance worldwide and building of a collegiate response squad to combat the stresses or shocks that face agricultural security. That is, sustaining the earth's ecosystem requires, not just the avoidance of conflicts alone, but, the governance of simple and common resources like air, water, food, and environment that connects humanity together. However, agro-security is the conceptual envelope that strategically embraces the food and environmental security pointers deemed central to human security, as critically conversed above. Finally, having premised the agro-security debate on its simplicity, affordability, and proximity (local ownership), dear epistemic communities, we ask again, should we consider agro-security?

Long live, the epistemic communities!!!

REFERENCES

Aduloju, A. A., & Okwechime, I. (2016). Oil and human security challenges in the Nigeria's Niger delta. *Critique*, 44(4), 505–525. https://doi.org/10.1080/03017605.2016.1236495.

Aduloju, A. A., & Pratt, O. O. (2014). Human security and developmental crisis in the contemporary West Africa. *Journal of Human Security*, 10(1), 46–58. https://doi.org/10.12924/johs2014.10010046.

Aduloju, A. A., & Adedoyin, T. A. (2020). The tech-novation pathway from pandemic to prosperity: a post Covid-19 roadmap for African economy. *Economic Consultant*, 31(3), 4–23. https://doi.org/10.46224/ecoc.2020.3.1.

Akokpari, J. (2012). Environmental degradation and human insecurity in sub-Saharan Africa. *Journal of Human Security*, 8(1), 24–46.

Barnett, C., & Parnell, S. (2016). Ideas Implementation and Indicators: Epistemologies of the Post-2015 Urban Agenda. *Sage Journals*. https://doi.org/10.1177/0956247815621473. Accessed 17 December 2019.

Carneiro, C. and Hector, N. (2014). An Economic Analysis of Land Use Changes and Biofuel Feedstock Production in Brazil: The Role of Irrigation Water, World Congress of Environmental and Resource, Turkey. https://doi.org/10.13140/2.1.1318.3362. Accessed 20 March 2020.

Chapman, K. (2009). Agrosecurity – Protecting America's Food Supply; An Introduction to Agrosecurity Challenges. https://digitalcommons.usu.edu/cgi/viewcontent.cgi?referer=https://www.google.com.ng/&httpsredir=1&article=1048&context=extension_curall. Accessed 20 March 2020.

Díaz, S., Fargione, J., Chapin, F. S., & Tilman, D. (2006). Biodiversity loss threatens human well-being. *PLoS Biology*, 4(8), e277. https://doi.org/10.1371/journal.pbio.0040277.

Dunlop, A. C. (2017). The irony of epistemic learning: Epistemic communities, policy learning and the case of Europe's hormones saga. *Journal of Policy and Society*, 36(2), 215–232. https://doi.org/10.1080/14494035.2017.1322260.

Evans, J., Terazono, E., & Abboud, L. (2020). Farmers Warn over Food Supply with Harvest Workers Shut Out. *Financial Times*, March 27, 2020. Available on https://www.ft.com/content/e27a9395-db47-4e7b-b054-3ec6ba4cbba3. Accessed 30 March 2020.

Falvey, L. (2005). Reconceiving food security and environmental protection University of Melbourne, Australia. *Asian Journal of Agriculture and Development*, 1, 2.

FAO (2008). The State of Food and Agriculture. http://www.fao.org/3/a-i0100e.pdf. Accessed 28 March 2020.

FAO (2017). Improving Food and Environmental Security in Ghana: Support for the safe disposal of obsolete pesticides, FAO Representation in Ghana.

FBI (2008). Federal Bureau of Investigation and Joint Terrorism Task Force. *Presented at International Symposium on Agro-Terrorism*. http://www.fao.org/partnerships/resource-partners/investing-for-results/news-article/en/c/1181597/. Accessed 28 December 2019.

Hanakata, N. (2017). The Production of Differences in the Tokyo Metropolitan Complex, Ph. D Thesis, ETH Zurich. https://fcl.ethz.ch/people/Researchers/NaomiHanakata/publications.html?batch_name=publications&page=0. Accessed 02 February 2019.

Heidhues, F., Atsain, A., & Vallee, L. (2004). Development Strategies and Food and Security in Africa: An Assessment. www.semanticsholar.com/paper. https://doi.org/10.22004/ag.econ.42270. Accessed 02 February 2019.

Hove, M., Ngwerume, E. T., & Muchemwa, C. (2013). The urban crisis in sub-Saharan Africa: A threat to human security and sustainable development. *Stability: International Journal of Security and Development*, 2(1), 7. https://doi.org/10.5334/sta.ap.

Kahl, C. (2006). *States, Scarcity, and Civil Strife in the Developing World*. Princeton, NJ: Princeton University Press.

Käkönen, J. (Ed.) (1994). Green security or militarized environment: An introduction. In: *Green Security or Militarised Environment*. Hanover, NH: Dartmouth Publishing Company.

Martin, A., & Romei, V. (2020). Business Activity Crashes to Record Low in Eurozone. *Financial Times*, March 24, 2020. https://www.ft.com/cotent/f5ebabd4-6dad-11ea-89df-41bea055720bAccessed 02 February 2019.

Nkonya, E., Mirzabaev, A., & von Braun, J. (Eds.) (2016). *Economics of Land Degradation and Improvement – A Global Assessment for Sustainable Development*. Heidelberg: Springer Cham. https://doi.org/10.1007/978-3-319-19168-3.

Ratner, B. (2018). *Environmental Security: Dimensions and Priorities*. Washington, DC: Global Environment Facility.

Robotti, S. (2020). What's the Healthiest Diet? www.medshadow.org/mediterranen-diet-versus-indian-african-mexican-foods. Accessed 02 February 2019.

Rustad, S. A., & Binningsbo, H. M. (2012). From Fragility to Resilience–Managing Natural Resources in Fragile Situations in Africa. https://www.afdb.org/fileadmin/uploads/afdb/Documents/Project-and-Operations/From_Fragility_to_Resilience_Managing_Natural_Resources_in_Fragile_States_in_Africa_-_Summary_Report.pdf. Accessed 02 February 2019.

Schreckenberg, K., Mace, G., & Poudyal, M. (2018). *Ecosystem Services and Poverty Alleviation: Trade-offs and Governance*. London: Routledge, Taylor & Francis Group. https://www.econstor.eu/bitstream/10419/181977/1/648753.pdf.

Schwartz, P. and Randall, D. (2003). An Abrupt Climate Change Scenario and Its Implications for United States National Security. http://www.edf.org/documents/3566_AbruptClimateChange.pdf. Accessed 02 February 2019.

Swerdloff, A. (2016). The FDA is Finally Concerned about Terrorist Attacks on Our Food Supply. www.vice.com/en_us/article/xym75a/the-fda-is-finally-concerned-about-terrorist-attacks-on-our-food-supply. Accessed 02 February 2020

Toke, D. (1999). Epistemic communities and environmental groups. *Politics*, 19(2), 97–102. https://doi.org/10.1111/1467-9256.00091.

Westing, A. H. (1989). The environmental component of comprehensive security. *Bulletin of Peace Proposals*, 20(2), 129–134. https://doi.org/10.1177/096701068902000203.

Zavatta, G. (2014). Agriculture Remains Central to the World Economy. www.expo2015.org/magazine/en/economy/agriculture-remains-central-to-the-world-economy.html. Accessed 02 February 2020.

3 The Epistemic Communities, Food and Agriculture Organisation (FAO) and Food Security in the Third World

Ayodeji Anthony Aduloju
ObafemiAwolowo University
Edo State University Uzairue

Victoria Akinyemi Omolara and Temitayo Adedeji Adedoyin
Obafemi Awolowo University

CONTENTS

3.1 INTRODUCTION

Over the years, the epistemic communities have been at the centre of providing technical know-how to policymakers on issues that affect the world in general. Their advisory roles in international organisations have been crucial for global problem identification and solving. One of such organisations where these roles are evident is the FAO. The FAO on its part provides information and knowledge about how to defeat hunger in the world and also doubles as an agenda-setting organisation on how the information and knowledge to defeat hunger can be governance inclined for global action. Apart from serving the problem-solving needs of its member states both developed, developing and underdeveloped it has prioritised the challenges of the food crisis, hunger and famine in the Third World as its core objectives. This may however not be one of its stated objectives, yet the body language of the organisation and its various programmes are navigated towards addressing food insecurity problems in the Third World.

FAO prioritising the Third World food insecurity is not farfetched, owing to how food security gets to impact upon the problems of poverty, conflict, famine, underdevelopment, health issues, political instability, natural disasters and food production in the Third World (UK Parliamentary Office of Science and Technology, 2006). Meanwhile, at different fora, issues surrounding food

insecurity in the Third World and their attendant problems have been the subject of discussions and actions at the level of the FAO framework. The epistemic communities in this regard have also simplified in a way, the process by which the FAO achieve its problem-solving role on the global food crisis. As touching the Third World Food Security crises, the epistemic communities and the FAO have worked to identify the problems of hunger, malnutrition, food scarcity, food supplies and famine through formulating policies to help people in Third World, most especially farmers, so as to meet up with the standard the organisation has set for achieving food security.

Consequently, regardless of the tireless efforts of the epistemic communities and the FAO, food security in the Third World seems to look like a mirage. In the area of agriculture, the FAO has assisted to transfer technology and eco-friendly materials to boost agricultural productivity and ensure food security. Despite all these, most of the Third World countries still languish and experience food shortage and scarcity. The epistemic communities have also unravelled some of the possible factors to these severe and protracted food crises in the Third World, which range from the problem of climate change, political instability and poverty. Although the epistemic communities have advised the FAO to help the Third World change to best practices in food and agricultural production, which the FAO has done appreciably, however, it seems that there is more to the food security problem in the Third World than what has been previously mentioned.

Also, it is important to make the point about how global warming and its attendant implications for food production and conflict in the Third World are restraining the FAO to record appreciable success in the Third World. Moreover, the epistemic communities have over the years emphasis in their research, how climate change has the tendency to impact negatively on a global natural resources such as the nutrient in soil and water, which agriculture depends on, couple with momentous consequences for global food security (Hoffmann, 2011). Climate change could also significantly constrain economic development in those developing countries that largely rely on agriculture and can contribute to global hunger which the FAO has to an extent contained. Meanwhile the problem of insecurity and conflict in some states in Africa, for instance, has expanded the scope of the FAO in ensuring food security by dragging the organisation to focus on advocacy for peace and conflict mediation in some of these countries first and then work towards setting agenda for food security.

As easy as it is to talk about, that is, the conflict factor of food insecurity or crises in the Third World, the FAO has been faced with problems in its interest to ameliorate food insecurity in the Third World. Also, the epistemic communities seem to be bugled with confusion on how it pinpoints on its technical advisory role to the FAO in making an edgeway to ensuring an end to hunger and food insecurity. This seems to be an Aquilian task for the epistemic communities and the FAO. Now, this study seeks to do a thorough review of how the roles of the epistemic communities and the FAO in the Third World has been met with little or no success. It looks critically at how the collaborations of both actors are working in other places of the world and relatively failing in the Third World. Finally, the study tests the hypothesis that hunger and food insecurity is synonymous and peculiar to the Third World, which could be an area for the epistemic community to look at for proper solution by the FAO and its members state.

3.2   FAO AND THE EPISTEMIC COMMUNITIES: A SUCCINCT INTRODUCTION

As organised as the FAO is, the organisation still relies on the epistemic communities for knowledge transfer, sharing and for policy formulation. The community has grown to be the life wire or put differently, a very important part of global policy formulation in international organisations and in the FAO to be precise. Since the notion of epistemic communities has been developed in academia, there has been the problem of ambiguity in the way it is applied to either discourse in research and how it is used in public debate. In order to clear this noticeable grey area, this study takes a succinct background of what the epistemic communities are how it will be used in this study.

3.2.1   FOOD AND AGRICULTURE ORGANISATION: A BACKGROUND

Seven decades ago and precisely on October 16, 1945, the FAO was established as one of the specialised agencies of the United Nations (UN). The organisation was the outcome of the Hot Springs Conference in Virginia, United States and attended by forty nations in May and June in 1943(OECD/FAO, 2016). The period in which the FAO was instituted and when the conference was held coincided with the end of the Second World War. At inception, the FAO was charged with the responsibility of ending the food crisis caused by the devastating impact of the Second World War on states in Europe (FAO, 2003). Since this period, the organisation has been active when it comes to tackling global food problems. The organisation has the mandate to promote and strengthen cooperation amongst member states, in the areas of food and agriculture. These areas cut across all agricultural products such as crops, poultry, fisheries, seafood, forestry, etc. The FAO also has an expanded role to monitor how these areas go into the production process, consumption and commercial distribution.

As an intergovernmental organisation within the UN framework with a large number of members states up to 191, it renders assistance to its members in the areas of formulating development-friendly policies and increase the level of interdependence between its members. In recent times, the organisation has offered a platform for states to deliberate on issues that border on food availability and affordability. More importantly, its noticeable agenda is the fight against hunger. In this, it has shown commitment by drawing its plans on how to fight hunger. The FAO's five steps to fight hunger are:

a. Contribute to the eradication of hunger, food insecurity and malnutrition;
b. Increase and improve the provision of goods and services from agriculture, forestry and fisheries in a sustainable manner;
c. Reduce rural poverty;
d. Enable more inclusive and efficient agricultural and food systems at local, national and international levels; and
e. Increase the resilience of livelihoods to threats and crises (FAO, 2016:15).

With a close look at the above goals of the FAO, it is glaring that there are challenges facing most Third World countries. One could then say that for the developed and countries in the first world, the FAO has completed and achieved its purpose. What the organisation mainly does today is to nature states in the Third World against or out of the food security associated challenges affecting them. Food security has become one of the cardinal objectives of FAO and its activities in recent times have been geared towards ensuring the actualisation of food secured world. Either by supervising and ensuring the quality of food produced as the case maybe in the developed world or ensuring affordability and availability of food in the Third World. With the noticeable plans of the FAO on food security, it is clear that most of the problems to be solved as regard food security are rampant in the Third World. It is as a result of this that this study will examine how the FAO has been able with the collaboration of the epistemic communities being able to ameliorate the problem in the Third World.

3.2.2   EPISTEMIC COMMUNITIES: A THEORY AND CONCEPT

As a concept, the epistemic communities have been used to understand and explain the roles of different actors involved in global governance and how these actors deploy their roles to addressing global extremely difficult problems. Whatever may be the nature of these problems, they are complicated in such a way that they could spread across bounds like wide fire and lead to more severe issues with uncertain outcomes. The onus of solving these problems is within the confines of the duties and roles of policymakers, who sometimes are incapacitated to find lasting solutions to them, thereby relying on the epistemic communities. This narrative of the helper (epistemic communities) and the helped (the states or policymakers) has been embossed and well nuanced through the robust works are done by Peter Haas (Haas, 1989, 1990 and 1992).

Ruggie (1975), invented the conceptual framework of 'epistemic communities.' To him, the way global policy is reached to solve problems is not hinged on the decisions of policymakers only, but on how knowledge is transformed and harnessed to build a channel of the relationship between science and politics. He then defines the concept as "a dominant way of looking at social reality, a set of shared symbols and references, mutual expectations and mutual predictability of intention" (p: 570). Haas (1992) on his own part, also push forward an argument about the link between science–politics of knowledge and decision making, which encapsulates the term epistemic communities. Morisse-Schilbach (2015), while also relying on Haas's foundational argument, sees the concept as a global network of convergence for technical experts, scientists and international bureaucrats on the one hand, and political and societal actors, on the other hand. This points to the fact that the epistemic communities build in conjunction with other relevant professionals, a relationship that thrives on collectively identifying and solving problems.

To be precise, the problem of global food security, in particular, has placed the world at a dependency level of relying on epistemic communities for research-driven policy advice. Even as world leaders are working towards achieving the goals of ending hunger and eradicating malnourishment in the Third World, for instance, some of the challenges (climate change, conflict, poverty, underdevelopment, etc.) toward achieving these goals bring about the complexity in successfully containing the challenges. Faleg (2012) sees this complexity as uncertainty that drift decision-makers to seek technical advice, which in turn influence the way global decision is made. In addition, Haas (1992) further demonstrates that the epistemic communities' framework rests on four cardinal principles. They are policy innovation, diffusion, selection and evolution as learning. All these four principles validated the relevance of the epistemic communities in providing evidence, know-how, new ideas and understanding about real-life scenarios for policymakers to make workable choices.

Furthermore, the epistemic communities could be seen from the lens of being international actors with revered collective values and a common policy project. The epistemic communities theory has been used and operationalised at various levels and spheres, either within the academia and public space to understand the roles of professional experts in developing and executing ideas for, most times, global gains. The question now to ask is in this regard is not whether the epistemic communities have made great in ensuring global food security, but to know if, in reality, they have been living up to what is expected of them in the way global problems are contained and addressed. Within the context of this study specifically, have the epistemic communities been performing their roles when it comes to achieving the FAO goals for the Third World countries? The reality in these countries shows that the FAO has faced challenges in attending to food security-related problems, it is pertinent therefore to know that this study is not in any way interrogating the roles of the epistemic communities buy bumping into the conclusion that they may have failed on the issues bordering on food security in the Third World, but to underscore some of the challenges they face when it comes the ever complex food problems facing the Third World.

As a result of the above position, the framework of epistemic communities also gives broad and crucial knowledge on global policy decision-making in the international arena is not solely based on the ideas of the decision-makers but on what the epistemic communities have recommended to be the solution. Form this, the foundation and basis for understanding the framework of epistemic communities, what we should be asking are why has the Third World entities not being freed from the strings of food insecurity? To attempt this, this study draws ideas from the high-level politics played in international organisations to have contributed to it. As it stands at the moment, the arguments around it are still sketchy thereby needing a well-constructed narrative.

3.3   FOOD SECURITY IN THE THIRD WORLD COUNTRIES

Undoubtedly, the plurality of studies and perspectives on food security does not provide food for the lots of hungry individuals we have in our world today. While having access to an abundance of food in terms of quantity and quality is referred to as food security, food insecurity is the lack or shortage

of food for the populace (FAO, 1996). Furthermore, saw food security as the capability to provide for the people of a country or region- access to food, that is nutritious, adequate and sufficient at any period of time- on and offseason. On the flip side, however, USDA (2003) perceives food insecurity as the limited access to adequate food and resources, thus, placing the people on a difficult lifeline to live, in physical, psychological, economic and social terms.

Postnote (2006) explains that to achieve an improved food security level, a larger percentage of our population must have access to nutritious food, enjoy an economically active life in good health and must be physically free and secured within their domestic jurisdiction. Furthermore, Postnote stated that some of the factors that affect food security include the level of poverty, health, political stability, food production, basic infrastructures, natural hazards and climate change issues. At the same time, an improved food security assessment in the Third World is a precursor for the global decline in starvation, hunger and poverty.

Among other problems, one major challenge facing our world today is putting an end to malnutrition and food insecurity. For instance, the number of people suffering from chronic hunger amplified from about 805 million in 1997 to above one billion in the last few years. Although the study focuses on the Third World, new indicators reveal that, even in middle-income and developed countries, a sizeable slice of the people lacks steady access to sufficient and nutritious food; for example, eight out of a hundred of people in Europe and Northern America are calculated to be affected by food insecurity at moderate levels. But, then, a lot of Third World countries, mostly in Africa and Asia are still grossly food insecure, while many countries in Latin America and the Caribbean are rising sluggishly, but largely malnourished (FAO, 2019a).

As indicated by the Food Insecurity Experience Scale (FIES), food insecurity in the Third World is more than just hunger in the land. The 2009 World Summit on Food Security acknowledged the availability of food, access to food, utilization of food and stability of food production as four key dimensions of examining food security in the Third World and that would be intelligently discussed in the next section.

3.4   FOUR DIMENSIONS OF FOOD SECURITY IN THE THIRD WORLD

Generally speaking, there are four key dimensions of food security within the context of Third World countries. The first is the availability of food. The availability of food to a very large extent depends on the production of food. As a complement, the overwhelming production of food can be deposited, warehoused and transported when necessary to address local shortages and unavailability of food. As the global population figures are expected to surpass 9 billion people in 2050, the level of food production must rise above 55% to cater for the Third World. (FAO, 2017). Thus, the paramount concern, therefore, is not only to provide nutritious and energetic food for the teeming population in the Third World but, to also enhance food production without degrading the natural ecosystem for future production of food, in tandem with efforts to sustain our environment.

For example, a study conducted in Zimbabwe reveals that naturally dry environments have higher tendencies of experiencing food deficit. To put it differently, the availability of rainfall or otherwise directly affects the availability of food for the people. In fact, farmers consider the profitability of production in dryland zones- which defines an extended land area and what transpires in the Third World countries at large. Some of the countries in the semi-developed and the developed ones have alternative irrigation systems that enable food production with or without rainfall. Therefore, this study identifies the irrigation system as one of the panaceas to ameliorate the problems of food insecurity in the Third World.

For other categories of agricultural undertakings, sustaining the forest biodiversity and having a balanced ecosystem could be consolidated through the different internal mechanisms developed by states. For example, the pollination services in Bangladesh yield a considerable increase in the production of mustard seed and some other pollinated crops. Likewise, the ecosystem services in India support food production, pest regulation and nutrient cycling. Also, the coastal water mangroves in

Tanzania yield an extensive increase in fish production . Therefore, the availability of food in the Third World depends, not only on the improved level of technology to produce on and off-season, but also enhanced strategies to sustain the ecosystem for the generations, yet unborn.

The second dimension in food security is access to food. While sufficient food must be provided at the local and national levels, the nutrients, content and quality of food provided must not be compromised. This element of food security, therefore, depends on the biophysical facets of producing the food, processing it, storage and distribution. Additionally, access to food also depends on the efficient security of lives at individual and household levels, together with the established political, economic, social and legal considerations at the municipal, national and transnational levels (FAO, 2015).

In some very harsh environments, where individuals and households were unable to access inputs such as pesticides, livestock feeds, fertilizer, veterinary medicines, and so on, the production and availability of food were drastically affected. In some inaccessible and remote locations, draught animals play important roles in food transportation (FAO, 2015a). So, access to food in many urban areas constitutes a big issue in the Third World. Although access to food in some areas is mostly determined by the income of individuals living in a particular area, many urban populations, due to other social factors, depend on restaurants, street-food sellers and community food outlets to access food (Lang, Barling and Caraher, 2009). Recent food-consumption data shows that people are consuming more and more processed foods at the expense of diverse fruits and vegetables (FAO, 2017e).

A study by Adato and Besset (2009) in Malawi, South Africa, Zambia and, Mozambique, to examine the effectiveness of cash transfers in alleviating barriers to access food, reveals that cash transfers could increase access to food and lessen food insecurity, with Malawi benefitting greatly from the cash transfer initiatives. But, Shiferaw (2003), using the Korodegaga peasants in Ethiopia as an example, argued that, other determinants of access to food include, access to markets, labour, land area, livestock, weather conditions, technology and education; aspects, in which the majority of the countries in the Third World are functionally crude and deficient.

Apart from that, the third dimension of food security refers to the way people utilize food to create nutritional value for themselves (FAO, 2006). Consuming a healthy diet requires a broad range of diverse, foods, plants and animals. Appropriate utilization of food requires knowledge on how food is processed, prepared and stored. In many instances, the poor people in the Third World depend on supplies from the ecosystem biodiversity for their nutritional wellbeing. For example, Zambia mentions vitamin A-rich varieties of maize and sweet potato, and iron- and zinc-rich varieties of beans as their main source of nutritional wealth. Nepal recently remarked that the various minor fish and other marine species that were once considered as "pests" are progressively being acknowledged for the various nutrients, contents and hence the potential nutritive importance in them. India noted that their livestock is significant as a spring of food options that can aid the deficiencies in vitamins, protein and other minerals.

Thus. we can rationally conclude, that a large number of Third World countries maximize the nutritional values in the food they get from their immediate environments. In a survey conducted, while many developing countries identified the availability of food and access to food as the main challenges to food security in the Third World, a similar percentage mentioned that, crop varieties with great concentrations of specific nutrients have been vital for their survival (FHI, 2016; FAO, 2017).

In the same vein, the final dimension of food security is the stability of food availability. This largely depends on the availability of adequate food for the people all the time, regardless of the season; that is, no seasonal famines, shortages and poor harvest intervals (FAO, 2006). At the household level or at the national level, stability is also regarded as the capability to diversify, such that, a wide array of different food-producing channels and species that have diverse life sequences and different adaptation characteristics that helps to sustain the supply of food throughout the different seasons of the year. For non-food products that are raised or reaped for sales, diversity can help to preserve the stability of revenue, in spite of the risks associated with market forces.

Also, stability, supported by attendant biodiversity factors helps to diminish the effects of unruly events (such as droughts, floods, pest outbreaks and diseases) that have the potency to affect the production, distribution and storage of food. Also, micro-organisms can be used for food preservation and help to tackle the barriers to food supplies, caused by seasonal variations (Thondhlana and Muchapondwa, 2014). For example, India again noted the importance of livestock, as it acts as buffer to mitigate crop failures. At the same time, both Zimbabwe and Zambia reported that small-holders have responded to regular drought issues by adopting some more sturdy crops like sorghum, millet, cassava, and sweet potato and by diversifying their systems of production. Both countries also mentioned that small ruminants can be used to respond to the growing effects of disease on cattle and drought.

Having examined the four dimensions of food security, and how the Third World features on each, there are some emerging constraints to food security that are deemed peculiar to the Third World. The next section fully starts the conversation.

3.5   EMERGING LIMITATIONS TO FOOD SECURITY FROM THE THIRD WORLD

It is not surprising that countries in the Third World face a lot of constraints to the attainment of food security. The population figures as an important variable in shaping efforts towards food security. This is because the alarming growth of the total population rises far above the supply of food, therefore, leaving a bourgeoning supply gap, most especially, in the Third World. Unlike the developed world, with fairly stagnant population growth, the supply and availability of food in the Third World must sharply increase to surpass the potential population growth.

Since the Third World mostly depends on agriculture to optimize their food security networks, the climate change issue is a critical consideration for individuals, households and states towards that goal. Due to the changing weather patterns, some regions that were once tagged appropriately for crop production are becoming increasingly unsuitable for planting and other agricultural activities. For example, Ringler (2010) the low adaptation capabilities of the Third World countries to climate change impact will induce a high level of poverty and higher vulnerability levels for several countries.

In 2008, for instance, drought-affected wheat production in Russia and consequently led to a sharp increase in the cost of wheat in most developing countries. By extension, the supply or export of wheat was also restricted to selected developed countries, thereby, creating shortages in the Third World, particularly in Africa. Invariably, World Bank (2012) opined that the dominant debt predicament across Europe aggravated the problems of food insecurity among the Third World countries. This is because, whenever an economic downturn occurs on a global scale, developed countries tend to look inwards and reduce the level of support they provide for the developing world to combat food insecurity.

Therefore, in view of the existing improved international economic conditions -unlike the recent economic meltdown era- that coincides with the 70th anniversary of the FAO, there has been a sharp decline in the chances of food insecurity in the Third World. For example, the protection and nutritional concerns of the food we consume have become frequently topical in recent times, and consequently, a lot of emphasis was placed on the availability of satisfactory food supply without further critical analysis of its nutritional status. Before then, WHO in 2010 had noted that above 60% of the children in the Third World were malnourished. Just as the Ethiopian experience suggests, many developing nations do not establish stern conditionalities that will force the developed countries to conform to policies that enhance food and commodity security in the Third World, due to the widening supply gaps that exist. Therefore, developing countries are expected to increase their agricultural food production and put measures in place to ensure that they restrict unsafe food production or importation from the developed world.

Apart from that, accessibility of water in the developing world has also become a germane issue to achieving food security; not just the quality, but the availability of hygiene water has become a

central consideration towards achieving food security. Hence, Ringler (2010) stated that the high rate of pollution in the developing and underdeveloped affects the quality and safety of water in circulation, thus having a catastrophic impact on people's health.

Ringler (2010) further argued that water forms an essential aspect of food security and safety is of overriding importance to guarantee a more noticeable improvement in food security. For example, WHO (2011) added that the deteriorating quality of water for Zimbabwe's urban populace has led to several cases of typhoid and cholera. Therefore, as a key component of food security, developing countries must strive to deal with the quality and availability of water as well, to achieve food security.

As critically compiled above, the limitations to food security remain dynamic and multi-faceted; they are seen from different angles including political, economic, social and environmental factors. To this end, the Third World is presented with massive emerging issues that call for strategic and explicit solutions to nip them in the bud. It is therefore imperative, to examine the several actions or inactions of the epistemic community, the FAO and other concerned stakeholders on the state of food security in the Third World.

3.6 THE EPISTEMIC COMMUNITY, FAO AND THE STATE OF FOOD SECURITY IN THE THIRD WORLD

Over the past three decades, scholars of International Relations, Public Policy Analysis and Comparative Politics have acknowledged the term 'epistemic community' as an arena where key actors converge to engage in critical deliberations and make key decisions for regional and global utility. Haas (1992) framed the epistemic community as a knowledge-based platform of individuals and professionals who are specialized in transnational policy-making. Epistemic communities, therefore, are assemblies of specialists, from different disciplines, that develop policy recommendations and solutions about issues that are technically complex (Haas, 1992).

However, there are four elements that are attributed to the body of knowledge produced from epistemic communities: one, there is a collective set of principles and normative beliefs, that provides a value-based justification for the actions and recommendations of the community members. Secondly, the community derives a shared set of assumptions about the cause of identified problems, which then function as a background for explaining the links between the policy recommended and the expected outcomes. Thirdly, the community seeks validity and objectivity of their notions and expertise through inter-subjective analysis. Lastly, epistemic communities direct their professional proficiency towards problems that are within the scope of their mission, with a vision of an associated or joint policy enterprise (Haas, 1992).

The four principles were designed to ensure that epistemic communities have control over their information, the proposed policies and knowledge produced after systematic debates, -that articulate the cause and impact relationships- on specific or general problems. With the structure and outlines appearing in form of International Organization, Haas (1992) empirically articulated the relevance of epistemic communities to a broad range of transnational problems including, security, trade, political, economic, social, agriculture and even, our commonwealth- the environment.

In the same vein with the transnational processes of epistemic communities, the FAO has been continuously involved in specific areas of global governance, such as promotion of agriculture (to reduce hunger worldwide), regulating safety standards for food and all consumables, monitoring the food security agenda and acting as a global watchdog for agricultural practices that can undermine sustainable development (Djelic and Quack, 2010).

As stated in the reports of FAO in 2009, the increasing production of food in the world in the past few decades has not convincingly reduced the level of hunger and malnutrition, most especially, in the Third World. That same year, the Committee on World Food Security (CFS) reviewed their policies to attract the status of a multinational-stakeholder, multi-sectoral and intergovernmental

platform (FAO, 2009a,b). Unlike the previous incarnation of the CFS, epistemic communities are more open, not only to the governments and stakeholders alone but, also engages the private sectors and the civil society; all, working towards ensuring adequate nutrition and food security for humanity. It should be added that this reformation was propelled by the 2008 sharp rise in the prices of food which strongly affected an overwhelming lot of Third World countries.

Therefore, in several attempts to understand the devastating impacts of the price hike and how to curtail similar occurrences, in essence, the CFS was reformed. Currently, the Committee consists of many nations within the UN; UN specialized agencies with mandates that relate to nutrition and food security; civil society and NGOs working on similar agenda; transnational agricultural research agencies; the private sector, philanthropic foundations; and lastly, the international financial institutions (FAO, 2015). Occasionally, the CFS invites intellectual groups and institutes to act as observers or join particular deliberations at its sittings. This assemblage of participants has been providing evidence-based studies and reports to strengthen the policy recommendations negotiated within the CFS.

In 2009, the High-Level Panel of Experts for Food Security and Nutrition (HLPE) was instituted to serve as the scientific policy interface of the CFS. The HLPE provides in-depth study and commendable findings on food security policy strategies for the CFS. The HLPE also supports the CFS in advancing its perception based on evidence, including an outline of the controversies, and detecting the emerging issues across the world (HLPE, 2010). A combination of experts from different backgrounds, knowledge systems and disciplines produce the reports implemented by the CFS. In the eventual analysis, the shared knowledge creates a single fact-based document that bridges the gaps among diverse perspectives. Such a model of shared understanding strengthens inclusiveness and commitment among the Committee members.

Since 2011, nine reports have been released by the HLPE which has continuously incited several debates at the floor of CFS; on matters that range from, climate change, -price volatility, social protection, food production, biofuels, waste management, investments in small-holder agriculture, aquaculture- to the provision of adequate water for the people (Gitz and Meybeck, 2011). To put it in another way, the HLPE reports have led to the acceptance of strategic recommendations or policy frameworks by the CFS, which confirmed the commendable impact of the epistemic communities on food security in the Third World.

In the foregoing, the overriding mandate of FAO is to assist and work with member states, -not just in the Third World alone- but, across the world, to ensure food security; that is, where all the people, at all times, have physical and economic access to sufficient safe and nutritious food that meets their dietary needs and food preferences for active and healthy living (FAO, 1996). Using the global scale, the production of food across the world is expected to be sufficient for human consumption. But, the percentage of undernourished people in the Third World peaked in 2009 and has recently grown to reach over one billion people, implying that one-out-of-every-seven-persons are food insecure.

As an appendage to the workings of the epistemic community, the target of FAO is to ensure that hunger and poverty are uniquely reduced by FAO's epistemic resource base; its multidisciplinary expertise, impartial analysis and global statistical chart, its depository of treaties, countless international committees and commissions for setting standards and lawful policy suggestions on food security (Duncan and Claeys, 2018). Therefore, the FAO, through intelligible sharing and transfer of knowledge has demonstrated its support for initiatives that promote sustainable development. FAO also maintains a firm international awareness about the unwavering impact of agriculture for global development and improvement on food security levels in the underdeveloped world.

However, despite the plethora of initiatives FAO has constantly introduced over the last two decades, the turn of the 21st century has brought with it, some challenges, -which at the same time- are interrelated and complex. For instance, the global population is expanding rapidly and is predicted that by 2050, the figures will rise to 9 billion people (DESA, 2019). With the possibility of a larger growth rate in the Third World, the implications for FAO member states and its partners

seem gloomy. In addition, migration from rural to urban areas is rising considerably, and by 2050, it may account for over 75% of countries in the developing world, as against the 48%, we currently have (DESA, 2019). Also, urbanization, expansion of industries and globalization are changing the consumption patterns of some kinds of food.

While intensive industrialization efforts in the Third World are placing unprecedented pressures on our natural resources, environmental and climatic changes are forcing out some technical emergencies on food availability as a result of frequent environmental disasters. Apart from the dangers posed by globalization, the dwindling development assistance funds, committed to the agricultural sphere have discouraged private firms willing to invest in the food sector. To consolidate that point, the uncertainties that surround the agricultural sectors and small-scale farmers in most Third World countries could be regarded as reflections of, not just some weak policy recommendations suggested by the epistemic communities, but, also failed to implement some trusted or tested policy frameworks.

As critically stated in the previous paragraphs, the epistemic community and the FAO are making coordinated efforts to ensure that a measurable improvement is recorded on food security in the Third World. Nevertheless, the collaboration of the FAO and the epistemic communities on food security have indeed, laid a solid foundation for improvement in Latin American and the Caribbean nations food sufficiency ratio- a prototype of what is expected in other Third World regions in Africa and elsewhere. Selectively, this section, in essence, reflects on instances within the Caribbean, Latin America and Africa, amongst others.

Notably, the fight against hunger in the Caribbean and Latin America is an offshoot of the political mobilization to eliminate undernutrition. Also, "the global 'freedom from hunger' movements have translated into model national initiatives like the Zero Hunger project in Brazil, or the recently launched Plan for Food Security, Nutrition and Hunger Eradication of CELAC (the Community of Latin American and the Caribbean States". Josué Castro believes that"hunger and war do not respect any natural rule, but, they are creations of mankind" (FAO, 2019a:3) The epistemic communities, apparently with the support of FAO has set Latin America to work, thus, showing a strong political will to tackle hunger- a human creation. Accordingly, the 2015 World report on the state of food security affirmed that, since 1990, the first region to split its hungry population by half in Latin America. Such achievement also equates some targets set by the United Nations MDGs/SDGs Agenda. That is, the figures of the Latin American and Caribbean population suffering from hunger declined from 14.7% in 1990 to 5.5% in 2014.

Having established the goal of eradicating hunger by the epistemic networks of global governance, the goal was collectively approved by all the states in the "Free Latin America and the Caribbean Initiative" in 2005 and also renewed their commitments during the 3rd Summit of the CELAC in Costa Rica. During the Summit, all Heads of State strengthened their commitments by way of "supporting the organization's Intergovernmental Plan for Food and Nutrition Security and the Eradication of Hunger by 2025" (FAO Reports, 2017). At the Summit, the Director-General(DG) of the FAO, José Graziano da Silva reiterated the significance of political responsibilities, cohesion and availability of the productive apparatus to achieve the anticipated outcomes.

Within these agreements by states, the FAO DG included viable cooperation within the South-South as the primary means in guaranteeing regional standpoint and accountability to overcoming hunger and promote food security. This plan, initiated by FAO, was supported by the Latin American Integration Association and the United Nations Economic Commission for Latin America and the Caribbean (ECLAC) after an extensive assessment. Over the years, the project tends to have improved the value of life all over the region by eradicating extreme poverty, guaranteeing food sufficiency and nutritional security.

In 2013, African presidents and Heads of government met in Ethiopia to sign a declaration that would end hunger by 2025. Following the tenets of the epistemic knowledge base, policy actors from different international organizations, civil society, and the private sectors, such as agriculturalists, researchers, scholars and several other partners were also present at the meeting (FAO,

2015). In the aftermath, the agreed declaration constituted a well-framed set of strategies that seeks to stimulate sustainable agricultural growth, increased food production, social security, protecting the vulnerable, and it also underlined the significance of non-state actors and stakeholders towards ensuring food security.

Consequently, the Addis Ababa Declaration also reiterated different African countries to domesticate and implement the "2003 Maputo Declaration on Agriculture and Food Security in Africa," which was signed under the Comprehensive Africa Agriculture Development Programme. Also, at the assembly in Ethiopia, African leaders strengthened their unshaken commitments to terminating hunger and germinating public investment and deals on agricultural development. In full acknowledgement of Africa's potential in agricultural expansion, its rising young population, the massive land area, water reservoirs and abundant natural resources, the avalanche of professionals and experts at the gathering in Ethiopia, pledged their technical support and resources to foster resilient cooperation between each African country and the development partners for food security in the Third World.

3.7　CONCLUSION

In conclusion, the epistemic community represents the most efficient composition of policy frameworks and recommendations for global governance. While the FAO has been the presiding body over the green world in the past 70 years, the enjoinment of the epistemic community since the 1990s has added more greenlight (towards ensuring food security), such that, the challenges created by greenhouse gases, even become less obstructive to achieving the goal of being food secured.

Drawing from extensive literature scoping, this study renews the existing perception that any agenda on food security is instituted on four solid pillars, which are; food availability, easily accessible, nutritional utility and stability of production. As strongly argued, the epistemic community is central to the coordination of food security policies in the Third World, at both regional and national levels. As expected, the collaboration between the relevant platforms of global governance and FAO has strengthened the institutional and legal frameworks, which has evidently improved food production and supply programs, and reduced wastage or loss, mostly in the Caribbean and Latin America. However, ensuring timely access to adequate, safe, nutritious, sufficient and sustainable food for everyone remains relatively impossible in almost all the Third World countries.

Although most African countries seem to be in the preface of those four pillars of food security, only a few African countries have a clear roadmap towards achieving food sufficiency and access to food. But, then, due to the systemic institutional decay in African nations, achieving food security will require the preparedness of the poor masses to suffer the consequences -of varying degrees- in the short and medium-term scale. For example, Nigeria currently shuts its land borders to ignite local rice production within the country. While that comes with a hike in the prices of available rice –even, despite the stones forming its bones-, the argument that the Nigerian government should have pre-informed its large population of such closure plans could as well be tagged the smugglers' argument- which would have enabled them to devise alternative routes, should the Nigerian government shuts its land borders. Also, the local production of rice enhances its nutritional value (the third pillar of food security), as a result of the organic relationship between our soil composition and the Negroid texture of Nigerians – unlike long-imported Asia's chemically-enhanced rice production which is organically suitable for the Mongoloid Asians alone (Brevit and Burgess, 2014).

Additionally, the efforts of FAO and the epistemic communities have facilitated stable production of food and rapid response to natural disasters in many Third World countries. Then, a lot of re-strategizing needs to be done to maximize production potentials, provide and manage food stocks, while also ensuring easy access to (and stability of) adequate food and water for the people. For instance, Mexico has launched its National Crusade against Hunger while Chile has also commenced its Choose Healthy Living program. While Grenada has implemented the Zero Hunger Challenge, it was recently added to the Venezuelan national strategy for food production and supply.

Peru, on the other hand, established an Inter-Sectoral Commission on Food and Nutritional Security as part of its commitment, whereas in Brazil, the prioritisation and development of renewed operational strategies for its Zero Hunger initiative are implemented. This programme in Brazil has lifted over nineteen million people out of severe poverty and also reduced undernutrition by almost 26%, all within 5 years of implementation. Consequently, we can realistically conclude that the realities of food security in the Third World are largely structured on the unrelenting efforts of FAO and the epistemic community, which serves as a rich source of fact-induced policies and testable recommendations for food security agenda in the Third World.

REFERENCES

Adato, M. & Besset, L. (2009). Social Protection to Support Vulnerable Children and Families: The Potential of Cash Transfers to Protect, Health and Nutrition. *AIDS Care Report*, 21(S1), 60–75. https://doi.org/10.1080/09540120903112351.

Djelic, M. & Quack, S. (2010). *Transnational Communities and Governance: Shaping Global Economic Governance*. Cambridge: Cambridge University Press.

Duncan, J. & Claeys, P. (2018). Politicizing Food Security Governance through Participation: Opportunities and Opposition. *Food Security*, 10, 1411–1424. https://doi.org/10.1007/s12571-018-0852-x.

Faleg, G. (2012). Between Knowledge and Power: Epistemic Communities and the Emergence of Security Sector Reform in the EU Security Architecture. *European Security*, 21(2), 161–184. https://doi.org/10.1080/09662839.2012.665882.

FAO (1996). Rome Declaration on World Food Security. http://www.fao.org/docrep/003/w3613e/w3613e00.HTM. Accessed 16 July 2019.

FAO (2003). Encyclopedia of Food and Culture. https://www.encyclopedia.com/food/encyclopedias-almanacs-transcripts-and-maps/fao-food-and-agriculture-organization-united-nations. Accessed 10 June 2019.

FAO (2006). Food Security. *Policy Brief* 2. http://www.fao.org/fileadmin/templates/faoitaly/documents/pdf/. Accessed 17 August July 2019.

FAO (2009a). Declaration of the World Summit on Food Security. *World Summit on Food Security*, Rome, 16–18 November 2009. WSFS 2009/2. http://www.fao.org/tempref/docrep/fao/Meeting/018/k6050e.pdf. Accessed 16 July 2019.

FAO (2009b). The State of Food Security in the World. Published in Rome. www.fao.org. Accessed 10 June 2019.

FAO. (2016). The State of Food and Agriculture. Climate Change, Agriculture and Food Security. Rome. https://www.fao.org/3/i6030e/i6030e.pdf Accessed 17 July 2019.

FAO (2017). The Future of Food and Agriculture – Trends and Challenges. http://www.fao.org/3/a-i6583e.pdf. Accessed 10 June 2019.

FAO (2019a). FAO Framework on Rural Extreme Poverty: Towards Reaching Target 1.1 of the Sustainable Development Goals. http://www.fao.org/3/ca4811en/ca4811en.pdf. Accessed 16 July 2019.

FAO (2019b). The State of the World's Biodiversity for Food and Agriculture, in Bélanger, J. and Pilling, D. (eds.). *FAO Commission on Genetic Resources for Food and AgricultureAssessments*. http://www.fao.org/3/CA3129EN/ca3129en.pdf. Accessed 16 July 2019.

FAO Reports (2015). Regional Overview of Food Insecurity Latin America and the Caribbean. Reports from FAO, United Nations. https://reliefweb.int/report/world/regionaloverview-food-insecurity-latin-america-ancaribbean. Accessed 15 May 2019.

FAO Reports (2017). Latin America and the Caribbean could be First Developing Region to Eradicate Hunger, Reports. https://reliefweb.int/report/nicaragua/latin-america-and-caribbean-could-be-firstdeveloping-region-eradicate-hunger. Accessed 17 July 2019.

FHI (2016). FAO and USAID's Food and Nutrition Technical Assistance III Project (FANTA), Managed by 360. https://www.fhi360.org. Accessed 16 July 2019.

Gitz, V. & Meybeck, A. (2011). The Establishment of the High-Level Panel of Experts on Food Security and Nutrition (HLPE). Shared, Independent and ComprehensiveKnowledgefor International Policy Coherence in Food Security and Nutrition. https://hal.archives-ouvertes.fr/hal-00866427. Accessed 10 June 2019.

Haas, P. M. (1989). Do Regimes Matter? Epistemic Communities and Mediterranean Pollution Control. *International Organization*, 43(3), 377–403. https://doi.org/10.1017/s0020818300032975.

Haas, P.M. (1990). *Saving the Mediterranean*. New York: Columbia University Press.

Haas, P. M. (1992). Introduction: Epistemic Communities and International Policy Coordination. *International Organization*, 46(1), 1–35. https://doi.org/10.1017/s0020818300001442.

HLPE (2010). Rules and Procedures for the Work of the High-Level Panel of Experts on Food Security and Nutrition. http://www.fao.org/cfs/cfs-hlpe/en/. Accessed 10 June 2019.

Hoffmann, U. (2011). Assuring Food Security in Developing Countries under the Challenges of Climate Change: Key Trade and Development Issues of a Fundamental Transformation of Agriculture. *United Nations Conference on Trade and Development (UNCTAD)*. https://unctad.org/en/Docs/osgdp20111_en.pdf. Accessed 10 June 2019.

Lang, T., Barling, D. & Caraher, M. (2009). *Food Policy: Integrating Health Environment and Society*. Oxford: Oxford University Press.

Morisse-Schilbach, M. (2015). Changing the World: Epistemic Communities, and the Democratizing Power of Science. *Innovation: The European Journal of Social Science Research*, 28(1), 18–26. https://doi.org/10.1080/13511610.2014.943163

OECD/FAO (2016). International Regulatory Co-operation and International Organisations: The Case of the Food and Agriculture Organization of the United Nations (FAO). https://www.oecd.org/gov/regulatory-policy/FAO_Full-Report.pdf. Accessed 10 June 2019.

Postnote (2006). Food Security in Developing Countries. www.parliament.uk/post. Accessed 10 June 2019.

Ringler, C. (2010). Climate Change Implications for Water Resources in the Limpopo Basin. Washington, DC: IFPRI Discussion paper. http://ebrary.ifpri.org/utils/getfile/collection/p15738coll2/id/761/filename/762.pdf. Accessed 10 June 2019.

Ruggie, J. G. (1975). International responses to technology: concepts and trends. *International Organization*, 29(03), s. 557–583.

Shiferaw, F., Kilmer, R.L., Gladwin, C. (2003). Determinants of food security in Southern Ethiopia. A selected paper presented at the 2003 American Agricultural Economics Association Meetings in Montreal, Canada.

Thondhlana, G. & Muchapondwa, E. (2014). Dependence on environmental resources and implications for household welfare: Evidence from the Kalahari drylands, South Africa. *Ecological Economics*, 108, 59–67. https://doi.org/10.1016/j.ecolecon.2014.10.003.

UK Parliamentary Office of Science and Technology (2006). Food Security in Developing Countries. https://www.parliament.uk/documents/post/postpn274.pdf. Accessed 10 June 2019.

United Nations Department of Economic and Social Affairs (DESA) (2019). The World Population Prospects 2019. https://www.un.org/development/desa/en/news/population/world-population-prospects2019.html. Accessed 10 June 2019.

USDA (2003). *Food Security Assessment, Agriculture and Trade Reports*. Washington, DC: USDA.

World Bank (2012). *Funding for Developing Countries*. Washington D.C.: World Bank Publications.

World Health Organisation (2010). *Making A Difference:World Health Report*. Geneva: World Health Organisation.

4 Recent Advances in Application of Biostimulants Derived from Beneficial Microorganisms
Agriculture and Environmental Perspective

Chioma Bertha Ehis-Eriakha and Charles Oluwaseun Adetunji
Edo State University Uzairue

CONTENTS

4.1 INTRODUCTION

The daily increase in the global population has necessitated the high demand for production of quality and request for improved food in terms of quality and quantity. The application of biological agents as fertilization reagents that are derived from natural sources could serve as a sustainable and environmentally friendly approach that could lead to increase in agricultural production.

Biostimulants could be referred to as products that pose the capability to act on plants' enzymatic and metabolic processes which could lead to improvement in the quality and production of crop quality. The application of biostimulants has been established to play a crucial role in assisting plant growth. The European Biostimulants Industry Council refers to biostimulants as microorganisms or substances that portends the ability to play a crucial role in stimulating the natural process of the plant when applied to the rhizosphere of the plant. These biostimulants possess the potential to improve the plat nutrients, increase crop quality nutrient efficiency and enhance the tolerance to abiotic stress. It has been stated that the global market for biostimulants will increase to $ 2,241 million by 2018 because they possess an active compound with an annual growth rate of 12.5% which

DOI: 10.1201/9781003268468-4

varies from 2013 to 2018 (Calvo et al., 2014). The economic significance of such products has been discovered not to be inconsequential when compared to the global market for biostimulants.

Moreover, some merits of biostimulants when applied to plants include enhancement of crop in term of quality and yield, decrease in the prevention of diseases and stress and enhancement of plant production. There are numerous types of biostimulants that could be grouped based on their mechanism of action, source of material, and some other crucial parameters (Yakhin et al., 2017). The biostimulant could be classified into seven groups which entail important seaweed extracts, bacteria and fungi, humic acid, inorganic compounds, fulvic acid, chitosan and protein hydrolysates (du Jardin, 2015).

The major sources of biostimulants also show numerous physiological features and the origin from where they are derived. A typical example includes the generation of biostimulants majorly from the extract of macroalgae (McHugh, 2013). The major constituents available in some biomass have capability to serve as biostimulants and could be grouped into a group of various molecules which entails phytohormones such as abscisic acids, cytokinin, ethylene, auxins, brassinosteroids, and gibberellins (Pacifici et al., 2015), polyamine (Fuell et al., 2010), amino acids (Colla et al., 2017).

Moreover, it has been observed that phytohormones derived from seaweed extracts have been shown to contain putative bioactive components which could be grouped as biostimulants (Stirk et al., 2013). Furthermore, it has been stated that algae entail hormones together with different types of carbohydrates which include betaines, minerals, alginate, proteins, fucoidan which could enhance the raid development of plant (Sharma et al., 2014).

Therefore, this chapter intends to provide detailed information about the application of biostimulants derived from beneficial microorganism and their utilization for an increase in agricultural productivity. The merits and demerits of the biostimulants derived from microorganism were highlighted. The modes of action that biostimulants utilized in executing their role were also stated.

4.2   MODES OF ACTION OF BIOSTIMULANTS

It has been observed that the application of biostimulants could lead to the improvement of agricultural crops, germination, and increase in the development of seedlings. This might be linked to the action of the various types of signaling bioactive molecules that play a crucial role in secondary and primary metabolism (Calvo et al., 2014). Ertani et al. (2013b) examined the influence of 6 seaweed extracts obtained from five extracts of *Ascophyllum nodosum* and one extract derived from *Laminaria therough* by supply 2 days at 0.5 mL/L. The authors tested the effectiveness of the biostimulant using biochemical, morphological and chemical respectively. It was established that the extract derived from *Ascophyllum nodosum* was the most effective in stimulating root morphological traits. This might be linked to the high level of stimulating root morphological traits. Their study indicated the application of vigorous chemical depiction of commercial seaweed extracts which shows that metabolic targets of seaweed extract are associated with the activity of the biostimulant.

Furthermore, the application of *Cladosporium sphaerospermum* enhances the rapid development of two pepper cultivars when exposed to the plant seedlings. Moreover, it was stated that *Cladosporium sphaerospermum* led to an improvement in the growth of tobacco plant. The result obtained shows an improvement in plant development which could be linked to numerous putative physiology and molecular modes of action which entail defense responses, cell expansion and cycle, phytohormone, homeostasis, and photosynthesis,

Additionally, it has been stated that soil conditions and unfavorable environment most especially extreme, temperature, drought and salinity could be linked to 70% of the variation displayed by the high rate of instability in global climatic alteration (Wang et al., 2003). Also, the high rate of fluctuation in climate changes has been observed to have a negative influence on the level of food security and crop production globally (Rouphael et al., 2018b). Therefore, in order to save this situation, the utilization of microbial biostimulants could lead to the increase in the production of agricultural productivity of crops (Rouphael et al., 2018a).

The utilization of microbial stimulants most especially arbuscular mycorhiza fungi poses the capability to prevent abiotic stress such as drought stress in tomato plants. Volpe et al. performed the influence of two arbuscular mycorhiza fungi strains which contain of *Rhizophagus intraradices* and *Funneliformis mosseae* on tomato plant by evaluating the molecular and physiological effect. The result obtained showed that *Funneliformis mosseae* produced more active volatile organic compounds generation when compared to *Rhizophagus intraradices* that led to enhanced traits which led to substantial higher water use effectiveness when subjected to austere drought stress. Moreover, it was discovered that *Rhizophagus intraradices* portends the capability to prevent biotic most especially against aphids natural enemies and abiotic stress. Mycorrhizal plants demonstrated more enhanced water extraction rates from the root length as well as the biomass as a result of the arbuscular mycorhiza fungi-mediated substrate hydraulic features. It was observed that the pots treated with arbuscular mycorhiza fungi indicated enhanced root extraction rates and led to the preservation of transpiration when subjected to progressive drought under limiting transpiration rates of soil water flow to root systems.

4.3 CONCEPT OF BIOSTIMULATION

A promising and ecologically sustainable development would be the utilization of plant biostimulants (PBs) obtained from natural sources that enhance plant growth and development, natural product set, crop profitability, enhanced nutrient efficiency, and are also capable of improving plant to tolerate a large scope of abiotic stressors (Colla and Rouphael, 2015). Biostimulants from plant origin also referred to as "agricultural biostimulant" consists of different compounds and microorganisms that boost plant growth (Calvo et al., 2014). The description of perception of plant biostimulants is still developing, which is attributed to the diversity of contributions and research before a material can be considered to be biostimulants (Calvo et al., 2014). Initially, biostimulants were applied only to enhance plant growth and as such tagged plant biostimulants which were originally defined by excluding some of its potentials in terms of plant protection and as fertilizer. However, in 1997, two scientists Zhang and Schmidt redefined PBs as "materials that, in minute quantities, promote plant growth". This definition did not in totality capture the functionalities of PBs because the term "minute" automatically excludes biostimulants as nutrients and soil amendments which are the major components that encourage the growth of plants and are usually applied in considerable amounts. More recently another review of this definition was required in 2012 by the European Commission to investigate the materials involved in the formulation of the PBs and this was documented by du Jardin (2012) in an article titled "The Science of Plant Biostimulants - A bibliographic Analysis" which birthed a new definition, stating that "Plant biostimulants are substances and materials, with the exception of nutrients and pesticides, which, when applied to plant, seeds or growing substrates in specific formulations, have the capacity to modify physiological processes of plants in a way that provides potential benefits to growth, development and/or stress responses". Based on this definition, the author went further to draw up a proposal assigning PBs into eight categories of biostimulation elements which include, organic materials from different sources (agro-, industrial- and urban waste materials, composts, animal manure and sewage sludge extracts), essential chemical elements (Aluminum, sodium, Selenium, and Silicon), chitosan derivates seaweed extracts (red, green and brown, macroalgae), chitin and, antitranspirants (kaolin and polyacrylamide), inorganic salts, protein-based compounds such as amino acids, peptides, polyamines and humic substances without the inclusion of any biostimulant from microbial origin. Furthermore, In 2015, du Jardin proposed an updated definition of PBs based on advanced research and an improved understanding of the mode of action effects in agriculture and other related practices that "A plant biostimulant is any substance or microorganism applied to plants with the aim to enhance nutrition efficiency, abiotic stress tolerance and/or crop quality traits, regardless of its nutrient content" which could include microorganisms. This definition was bore out from an investigation on "Biostimulation in Horticulture". The research findings demonstrated more recent views and reflected the effectiveness

of beneficial microrganisms in horticulture which prompted the creation of new PB categories (six non-microbial and three microbial categories of PBs): (i) chitosan, (ii) humic and fulvic acids, (iii) protein hydrolysates (iv) phosphites, (v) seaweed extracts, (vi) silicon, (vii) arbuscular mycorrhizal fungi (AMF), (viii) plant growth-promoting rhizobacteria (PGPR), and (ix) Trichoderma spp. (Pichyangkura and Chadchawan, 2015; Battacharyya et al., 2015; Savvas and Ntatsi, 2015; Rouphael et al., 2015; Ruzzi and Aroca, 2015; López-Bucio et al., 2015).

Over the last decade, a myriad of researches have been conducted by various researchers on plant biostimulants, establishing that microbial and biostimulants from non-microbial sources are very effective in enhancing plant productivity, physiological and metabolic and biochemical plant responses, enhancing nutrient use efficiency and as agents of biological control to protect plants from pathogens (Calvo et al., 2014; Halpern et al., 2015; Yakhin et al., 2017; Rouphael et al., 2018a). Apart from the role of biostimulants in agriculture and horticulture, the stimulation of microbial growth is also an essential aspect of this topic. Most essential microorganisms in the environment (soil, water, sediment, ground water, etc) require some level of rate-limiting nutrients for proliferation and activation of microbial metabolic. The roles of microorganisms in the environment are too numerous and important to be avoided which includes among others plant growth and development. For example, in bioremediation, hydrocarbonoclastic and oleophilic microbial groups are involved in the mineralization of hydrocarbons which serve as pollutants to environmental compartments. These groups of organisms can function optimally in most cases through the process of biostimulation by addition of nutrients and several other growth factors to support the growth of the microbes and activate their degradative potentials. In this study, our focus will be on biostimulants as it affects both plants and microorganisms so the term plant biostimulant will not be applicable

4.4   MICROBIAL AND NON-MICROBIAL BIOSTIMULANTS

Till date, there is no standard definition bounded by law or regulatory status for biostimulants anywhere in the world. This also implies that there is no detailed or standard category of substances or microorganisms covered by this concept. However, researchers have given a general categorization of biostimulants as microbial and non-microbial biostimulants which best describes the composition of any known biostimulant (du Jardin, 2015).

4.4.1   NON-MICROBIAL BIOSTIMULANTS

4.4.1.1   Humic and Fluvic Acid

Humic substances (HS) are formed when plants, animals and microbial residues decompose to form a natural composition of organic acid and also from soil microbes using substrates to perform their metabolic activity. Humic substances have been documented to be among the most available organic material on the planet (Sutton and Sposito, 2005), and consist of above 50% of the organic matter present in soils. Initially, it was thought that humic substances are polymers of organic origin linked together however, research frontiers have discovered that they are numerous small organic molecules bonded together by hydrogen bonds and hydrophobic interactions (Sutton and Sposito, 2005; Halpern et al., 2015). HS are made up of heterogeneous compounds, initially grouped based on their solubility and molecular weights (MW) into three categories; humins, humic acids (HA) and fulvic acids (FA). (i) Humic acids, can be easily dissolved in basic media and so can be removed from soil by precipitation in acidic media and dilute alkali (ii) FA, can be dissolved in both acid media and alkali, and (iii) humins, cannot be extracted from soil (Stevenson, 1994; Berbara and García, 2014; Calvo et al., 2014). Another marked difference between both HA and FA is that the former are mostly high-MW, while the latter are low-MW (Nardi et al., 2009). Also it was proposed that humin should not be described as a HS but be described as a humic containing substance because it comprises both humic and non-humic substances (Nardi et al., 2009). The precise

structural features of HA and FA differ based on the time of its transformation and the source of organic material and the (Berbara and García, 2014).

These compounds form supra-molecules as a result of association/dissociation of complex dynamics that may occur and it is majorly influenced by the roots of the plant through the discharge of protons and exudates. Recently, several research works have revealed that root exudates containing amphiphilic substances can dissolve HA into high and low molecular sizes. This new finding supports the existing assumption that the behavioral structure of humus dissolved in the rhizosphere as well as the relationship of humic constituents with cells in plant roots could be influenced by root exudates or organic acids excreted by microbes present in the soil solution (Edorado et al., 2013)

Since HS and its complexes are derived from the relation that occurs between organic matter, plant roots and microbes colonizing the soil, application of HS for enhancing plant growth would require optimization of this interaction to achieve the maximum result. This could be the reason why the application of HS (soluble HA and FA) show varying but positive results globally on plant growth. Certain factors support the inconsistency in the effectiveness of HS such as i) source of the HS, ii) the surrounding conditions, iii) the benefitting plant and iv)mode of HS application (du Jardin, 2015; Rose et al., 2014). With regards to source, Humic substances can be extracted from diverse sources, which include composts and vermicomposts, various mineral deposits (Peat and leonardite), soil (du Jardin, 2015) and municipal waste (Schmidt et al., 2007; Nikbakht et al., 2008; Bulgari et al., 2015). Kelleher and Simpson (2006) reported thathumic substance (HS) removed from soils consists of some compounds which signify very essential compound classes in microbes and plants such as biopolymers, aliphatic, proteins, lignin and carbohydrates. Application of HS in plants can be carried out in several ways including applications of foliar, direct application to the soil and in irrigation water (Salman et al., 2005).

Studies have shown that HS from different sources improves total nitrogen (NO_3) uptake in addition to other essential minerals, Cu, Zn, Fe, P and Mn in barely throughout the period of a season (Quaggiotti, 2004; Çelik et al., 2010; Halpern et al., 2015). Halpern et al. (2015) highlighted the impact of HS on nutrient uptake as: (i) soil structure improvement, (ii) enhancement of soil micronutrient solubility as well as impact on the plant's physiology which include: (iii) alterations in plant root morphology, (iv) an improvement in the activity of NO_3^- assimilation enzymes and (v) increase in root activity of HþATPase,

4.4.1.2 Protein-Based Biostimulant

Protein-based products have been categorized into two distinct groups: protein hydrolysates which comprise a mixture of amino acids and peptides from animal/plant source and pure amino acids (AA) which include glutamine, glutamate, glycine betaine and proline (Calvo et al., 2014). The mode of preparation of protein hydrolysates could be either by enzymatic, thermal, hydrolysis or chemical of different animal and plant deposits, such as connective tissues or epithelial, animal collagen and elastin, etc. (Grabowska et al., 2012; Cavani et al., 2006; Apone et al., 2010; De Lucia and Vecchietti, 2012; Ertani et al., 2013a). Also, it has been documented that there are also non-protein constituents in the hydrolysates to support the stimulatory effect this biostimulant possesses on plants. For example, carob germ extract hydrolysate contained other non-protein components such as carbohydrates, fats, macro and micronutrient elements as well as six (6) phytohormones in addition to free amino acids, proteins and peptides. Other examples include alfalfa hydrolysate, high in free AA also contained macro and micronutrients, gibberellin and auxin-related activities established by a bioassay study. In a similar study, IAA and triacontanol which are plant growth regulators were observed in alfalfa hydrolysate (Ertani et al., 2013a). Individual AA is the second group of protein-based components which comprises of twenty (20) structural AA responsible for synthesizing proteins and another 250 non-protein AA present in large quantities in selected plant varieties (Vranova et al., 2011). These two groups of AA have been considered effective in exogenous applications evidenced by protecting plants from abiotic/enviromental stresses, active in metabolic

signaling and N storage and chelation as phytosiderophores (Vranova et al., 2011; Liang et al., 2013; Halpern et al., 2015)

Protein hydrolysates (PHs) comprises signaling peptides and free AA that have gained a lot of attention as non-microbial biostimulant because of their abilities in enhancing agricultural activities such as to boost crop yield, promote germination, fruit and vegetable quality, plant development and seedling growth under abiotic stress circumstances (Colla et al., 2015; Rouphael and Colla, 2020). Animal- and plant-based PHs participate through physiological and molecular biostimulation mechanisms in agricultural and horticultural activities. It has been shown that the application of PHs modified plant microbiome to enhance the quantitative and qualitative activity and composition of the microbial community. The plant growth stimulatory effect of PHs appears to be distinct from the conventional addition of nutrient sources which extends its benefits to protect against environmental stresses (Ertani et al., 2009).

In a review by du Jardin (2015), stated that protein hydrolysate compounds have been shown to be involved in multiple roles of biostimulation of plant development (Halpern et al., 2015) through different mechanisms. They include modulation of nutrient uptake and assimilation controlled by structural genes and enzymes as well as by interring with the signaling nitrogen acquisition pathway in the root of plants. Other mechanisms include regulation of enzyme of the TCA cycle which contributes to tissue hydrolysates, interaction between C and N metabolic processes and hormonal functionalities in complex protein (du Jardin, 2015) and chelating effects observed in some amino acids which are involved in plant protection against heavy metals. Scavenging some of the nitrogenous compounds, such as proline glycine and betaine, contributes to the reduction of abiotic stress exhibit antioxidant activity by scavenging free radicals. In addition, PH have been identified to enhance microbial richness and functions, respiration in soil and soil biomass which are essential effects of plant growth promotion in agricultural practices. Santi et al. (2017) reported increased root and length surface area of a maize plant followed by a concomitant increase in K, Zn, Cu, and Mn when treated with protein hydrolysate compared to the same plant treated with inorganic fertilizers. The utilization of these protein-based compounds as biostimulants has shown positive and promising results in the promotion of plant development and agricultural activities in general (Subbarao et al., 2015; Verma et al., 2017). Also very importantly, these compounds serve as a nutrient source for soil microorganisms consequently improving biodiversity and biogeochemical cycling (Santi et al., 2017).

4.4.1.3 Seaweed Extracts

Seaweed extracts (SW) is another essential class of organic non-microbial biostimulant; however the SW is classified based on pigmentation an assigned into three main groups; Chlorophyta Phaeophyta and Rhodophyta, and for red, green, and brown macroalgae, respectively. They act as the most common SW applied in agriculture practices amongst over 9,000 other species with different commercial brands presently existing in the market (Rafiee et al., 2016). du Jardin (2015) stated in a review that the application of fresh seaweeds organic matter source and biofertiliser has been ongoing for a long time however, the biostimulation effects have only been recorded recently. Research has shown that SW contains several constituents that supports its biostimulation activity such as, alginates, laminarin, carrageenans, polysaccharides and their breakdown products. There are other components actively involved in the plant growth enhancement and development which include, hormones, micro- and macronutrients, N-containing compounds (betaines) and sterols (Craigie, 2011; Rouphael and Colla, 2018). According to Halpern et al. (2015), Based on the literature, plant hormones have been observed in SW such as abscisic acid auxins, cytokinins as well as amino acids which all contribute to improve plant yield/growth and biological activity (Battacharyya et al., 2015).

Application of SW on hydroponic solutions, soil, as foliar or on plants as treatments have been documented (Craigie, 2011; Khan et al., 2009). The polysaccharides produced by SW support soil water retention, formation of gel, and soil aeration which are essential requirements for agricultural

and microbiological activities. SW have shown very promising attributes in agricultural studies such as plant growth promotion, increased chlorophyll levels, flowering and yield and improved germination of seeds (Kumar and Sahoo, 2011). They enhance the achievement of *in vitro* proliferation (Vinoth et al., 2012) and promote biocontrol activities against pests and pathogens (Halpern et al., 2015). The seed germination and plant growth development occur due to hormonal effects which is the main reason for the biostimulation effects. SW enhances plant nutrition by interfering with some soil processes via diverse mechanisms such as: (i) improvement of ability of micronutrient to solubilize in the soil (ii) improvement of soil structure. Others include: (iii) increased root colonization by arbuscular mycorrhizal fungi and (iv) alterations in root morphological system which has a direct effect on plant's physiology (Kuwada et al., 2006; Khan et al., 2009; Spinelli et al., 2010).

4.4.2 MICRIOBIAL BISOTIMULANTS

The application of microbial biostimulants (MB) to improve plant growth and other microbial activities has gained global attention within the last few decades (Hayat et al., 2010; Calvo et al., 2014; Aamir et al., 2020). The soil ecosystem is associated with a diverse array of microbial population carrying out diverse metabolic functions and more importantly act as a major source of inoculum for the rhizosphere of plants which is the soil region attached to and influenced by roots of plants. An important aspect to MB is the level of interaction on a molecular basis between microbes and plants, microbes and other beneficial microbes and how these interactions support plant growth, development and microbial activity. Also, characterizing the key players of the biological activities of the essential plant-microbe associates within the soil microbiome is a key requirement for application (Aamir et al., 2020).

Microbial-based inoculants have been typically characterized by different trade names including biocontrol agents, biopesticides, bioinoculants, biostimulants, biofertilizers, bioformulations which have been actively involved in contributing to maintaining a sustainable ecological system and improved crop productivity under eco-friendly conditions (Singh et al., 2016). In this review, microbial-based inoculants will be referred to as microbial biostimulants (MB). According to Calvo et al. (2014), MB are biological products comprising viable microbes when applied to plant surface, seeds, soil or a nutrient-deficient environment enhances growth by diverse mechanisms which include, increasing nutrient uptake capacity, root biomass increase and area, increases availability and supply of nutrients (Feng et al., 2017). Microbial biostimulants include plant growth-promoting rhizobacteria and rhizospheric fungi (PGPR/PGPF), Arbuscular mycorrhizal fungi (AMF) and endophytic microbes are largely perceived as efficient and sustainable measures to effectively secure plant stability and yield under nutrient deficient or low-input conditions, especially nitrogen and phosphorus deficiency, environmental bioremediation (Ekwuabu et al., 2016) and also as an innovative technology to enhance plant ability to tolerate environmnetal stressors in cases of salinity, extreme temperature and drought (Berg, 2009; Calvo et al., 2014; du Jardin, 2015).

4.4.2.1 Plant Growth Promoting Microorganisms

The plant system is a chemical environment that supports interaction between plants and diverse groups of microorganisms (bacteria, fungi, actinobacteria) termed plant-microbe interaction. This interaction can be mutualistic or antagonistic depending on the functionalities and metabolic processes involved. The plant root system which is the main point of entry for most microorganisms is called the rhizosphere. In the rhizosphere, plants release root exudates in the form of chemicals that acts as signaling agents to attract diverse microbial communities to the rhizosphere and other parts of the plant. The interaction between plants and the beneficial microorganisms such as mycorrhizas, endophytes (which inhabit the internal part of the plant such as stems, roots, seeds and leaves without causing any negative effect on associated plants) and rhizospheric (those that colonize the plant rhizosphere and rhizoplane) is mutualistic because they provide plants with essential growth properties, protection from phytopathogens and tolerance to environmental stress (Dubey et al., 2020).

Hence, these microorganisms have been tagged plant growth-promoting microorganisms or plant growth stimulating microorganisms. Recently, these groups of microorganisms have been implicated in the remediation of toxic pollutants from the environment through several mechanisms. Plant growth-promoting microbes have been reportedly obtained from different plants in stressed and various natural soils (de Bashan et al., 2012; Nadeem et al., 2014; Akinsemolu, 2018).

4.4.2.1.1 Plant Growth Promoting Rhizobacteria (PGPR)

Kloepper and Schroth (1978) first used the term PGPR for soil-borne related PGP activity by colonizing the roots of plants. The PGPR is made up of heterogenous, non-pathogenic, bacteria colonizing the root of plants called the rhizosphere. The rhizosphere comprises a large quantity of soil surrounded by the roots of plants and a well-characterized ecological niche for diverse microbial population influenced by chemical exudates released by the plants (Goswami et al., 2016). This microenvironment is highly competitive with a diverse array of microorganisms with different functionalities mostly to support plant growth and guide plants from phytopathogens. PGPR has a chain of positive effects on plants ranging from improved plant yield and growth (Alam et al., 2011), biocontrol activity (Bashan and de Bashan, 2005), enhanced ability to tolerate salt (Alavi et al., 2013), enhanced heavy metal and several other pollutant resistance (Lucy et al., 2004) and improved plant nutrition (Richardson et al., 2009). Generally, PGPR conducts this range of activities on plants through a variety of mechanisms which could be direct or indirect. The former involves the activities that promote plant growth and yield by promoting the acquisition of essential nutrients and minerals from the immediate environment or by stimulating the availability of synthesized compounds some of which include Nitrogen fixation, phosphate solubilization, iron sequestration, zinc solubilization, phytohormone production and catalase activity. On the other hand, the latter are related to reducing the negative effects inflicted on plants by phytopathogens through lytic enzyme production (cellulases, 1,3-glucanases, chitinases, proteases and lipases), synthesis of antibiotics, siderophore production, HCN production, induced systemic resistance (ISR) and systemic acquired resistance (SAR). PGPR are classified as biofertilizer (ensures the presence of essential nutrients in the plant through phosphate solubilization, nitrogen fixation and iron acquisition), biopesticide or biocontrol agents (release of antifungal metabolites and production of antibiotics to control or suppress diseases), phytostimulators (production of phytohormones like IAA, gibberellins, cytokinins, etc.), and phytoremediators (mineralization of organic pollutants and reduction of metal toxicity).

PGPR has been intensively investigated as an effective alternative to chemical fertilizers in agricultural and environmental studies (Table 4.1), with successful field application of some commercialized products. Although, in developing countries like Nigeria, the commercialization of this technology has not been achieved compared to the developed countries however extensive research has been conducted, and some is still ongoing (Taiwo et al., 2017). This group of organisms has been obtained from several environments (agricultural fields, coastline, forests, plants) (Chowdhury et al., 2016; Uzair et al., 2018; Andreolli et al., 2019) and plants roots which include wheat (Fouzia et al., 2015), chickpea (Saini et al., 2013), tomato, aloe vera (Thakur et al., 2017), Mung bean (Kumari et al., 2019), soyabean (Subramanian et al., 2015) and corn (Ikeda et al., 2013). Some commonly isolated PGPR include, include *Enterobacter, Klebsiella, Serratia, Pseudomonas, Azospirillum, Acinetobacter, Azotobacter, Arthrobacter, Bacillus, Burkholderia*, and *Alcaligenes* and their PGP attributes demonstrated in several crops (Islam et al., 2016; Adesemoye et al., 2017).

Another important classification of PGPR is based on interaction level with root of plants and the groups include; (i), those living inside specialized cells within the root structures or nodules (ii) those living within the root environment, (iii) residing inside the root tissue and among spaces connecting cortical cells) and (iv) those that reside on the surface of root plant (bacteria colonizing rhizoplane). Broadly, based on these association, PGPR have been categorized into two (2) major categories (Hameed et al., 2014). The first category is the extracellular PGPR (ePGPR-symbiotic) that has a symbiotic interaction with the host plant existing within the rhizosphere and rhizosphere associated environment (rhizoplane and root cortex cells). ePGPR-symbiotic is known to be producers of

TABLE 4.1

PGPR/PGPF Triats Responsible for Agicultural and Environmental Related Activities

PGPR/PGPF	Host Plant	PGPR/PGPF Traits	Application	Reference
Bacillus sp.RZ2MS9, RZ2MS16 (Burkholderia ambifaria)	Guarana (Paullinia cupana)	IAA production, nitrogen fixaton, phosphate solubilization, siderophone production	Agriculture	Batista et al. (2018)
Burkholderia contaminans KNU17B11	Maize	phosphate solubilization, zinc solubilization, indole acetic acid (IAA) production, ammonia production, nitrogen fixation, 1-aminocyclopropane-1-carboxylate (ACC) deaminase activity, siderophore production and hydrolytic enzyme activity	Agriculture	Tagele et al. (2018)
Pseudomonas spp.	Durum wheat and barley	Antifungal activities (production of proteases, amylases, lipases)	Agriculture	Bakli and Zenasni (2019)
Bacillus licheniformis	Halotolerant plant, Sueada fruticosa	Phosphate solubilization, IAA production, tolerance to 1 M NaCl, Siderophore production, ammonia production, Antifungal activity against F. oxysporum	Agriculture	Goswami et al. (2014)
Pseudomonas putida BTCC1046 and Pseudomonas sp. 1S4	Rhizospheric soil of a plant	P solubilization, proteolytic and chitinolytic activity, antibacterial and anti-fungal activity, siderophore production and production of phosphatases	Agriculture	Georgieva et al. (2018)
Streptomyces rochei, WZS1-1 and WZS2-1 Streptomyces sundarbansensis.	Mikaniami crantha	Biosynthesis and production of secondary metabolites (aliphatic ketones, carboxylic acids, and esters), IAA production, phosphorus and potassium solubilization, nitrogen fixation	Agriculture	Han et al. (2018)
Burkholderia sp. 7016	Soybean	Promote activities of soil enzymes, antimicrobial activity, IAA production, nitrogen fixation, phosphate solubilization	Agriculture	Gao et al. (2015)
Alialigenes faecalis RZS2 and RZS3 Pseudomonas aeruginosa	Agricultural field	Siderophore production (heavy metals chelation), enhance seed germination, promote plant growth and chlorophyll content	Bioremediation, Agriculture	Patel et al. (2016)
Pseudomonas sp. ITRH25, Pantoea sp. BTRH79 and Burkholderia sp. PsJN	Carpet grass (Axonopus affinis)	Enhanced plant biomass production and hydrocarbon degradation	phytoremediation	Tara et al. (2014)
Enterobacter intermedius MH8b	Sinapis alba L.	ACC deaminase, IAA, and HCN, production, phosphate solubilization, increased plant biomass, Cd and Zn tolerant	Agriculture, phytoremediation	Plociniczak et al. (2013)
Acinetobacter sp.,	Barnyard grass (Echinochloa crusgalli Beauv.)	Plant growth-promoting properties, presence of enzyme ACC deaminase, IAA production and phosphate solubilization	Agriculture	Xun et al. (2015)

(Continued)

TABLE 4.1 (Continued)

PGPR/PGPF Triats Responsible for Agicultural and Environmental Related Activities

PGPR/PGPF	Host Plant	PGPR/PGPF Traits	Application	Reference
Pseudomonas protegens MP124	Rhizospheric soil in a temperate deciduous forest	Inhibitory effect against multiple fungal phytopathogens, synthesis antifungal compounds (2,4-diacetylphloroglucinol (2,4-DAPG), pyoluteorin and pyrrolnitrin), produces siderophore, ammonia, IAA and P solubilization	Agriculture	Andreolli et al. (2019)
Ralstonia eutropha (B1) and *Chrysiobacterium humi* (B2)	*Helianthus annuus*	Enhance short-term stabilization potential of plant in metal-contaminated land, reduce losses in plant biomass, reduces aboveground tissue contamination, Zn uptake/accumulation	Phytoremediatio/agriculture	Marques et al. (2013)
Pseudomonas sp., UW3 and UW4 *Pseudomonas corrugata* CMH3	Barley and oats	Salt tolerant, increase in plant biomass, (ACC) deaminase production	Phytoremediation/agriculture	Chang et al. (2014)
Klebsiella sp. D5A	Tall fescue	ACC deaminase activity, phosphate solubilization and IAA production, growth and enhanced remediation efficiency in petroleum-contaminated saline–alkaline soil	Agriculture/bioremediation	Liu et al. (2014)
Pseudomonas sp. AJ15 *Klebsiella* sp. strain CPSB4 (MH266218) and *Enterobacter* sp. strain CPSB49 (MH532567),	*Withania somnifera* Tomatoes	Phosphate solibulization, IAA and siderophone production Chromium resistant, enhanced plant biomass, root length, shoot length, fresh and dry weight, nutrient uptake, enhanced superoxide dismutase, catalase, peroxidase, total phenolic, and ascorbic acid production	Agriculture Agriculture/bioremediation,	Das and Kumar (2016) Gupta et al. (2020a)
Penicillium commune PS2	Rice	Phosphate solubilization, IAA, HCN, ammonia, siderophore production, enhanced root, shoot length, biomass production and photosynthetic efficacy	Agriculture	Banerjee and Dutta (2019)
Rhizoglomus iregulare *Trichoderma* spp. *T. koningii* IM 0956, *T. citrinoviride* IM 6325, *T. harzianum* KKP 534, *T. viride* KKP 792 and *T. virens* DSM 1963	Maize plants Rapeseed seedlings	Salt tolerant, improve crop productivity Increased length of roots and shoots, siderophore production, ACC deaminase activity and phosphate solubilization,	Agriculture Agriculture, biotranformation/bioremediation	Moreira et al. (2020) Nykiel-Szymańska et al. (2020)
Humicola sp. 2WS1,	*Bacopa monnieri* plant	Arsenic biomethylation property, ACC deaminase production, and IAA production	Bioremediation	Tripathi et al. (2020)
Purpureocillium lilacinum LPSC #876.	Agricultural field	IAA, ammonia production seed germination and antifungal activity	Agriculture	Cavello et al. (2015)

(continued)

TABLE 4.1 (Continued)
PGPR/PGPF Triats Responsible for Agicultural and Environmental Related Activities

PGPR/PGPF	Host Plant	PGPR/PGPF Traits	Application	Reference
Pseudomonas fluorescens UM16	*Medico truncatula*	Biofilm production, siderophore, IAA, protease, ACC deaminase production, antagonistic against phytopathogens	Agriculture	Hernández-León et al. (2015)
Talaromyces sp. NBP-61, *Penicillium* sp. NBP-45, *Trichoderma* sp. NBP-67	Chilli (*Capsicum annuum* L.)	Disease resistant, root colonization, phosphate solubilization IAA, siderophore cellulose, chitinase and HCN production.	Agriculture	Naziya et al. (2019)
Penicillium pinophilum SB-4	*Hosta undulata*	Petroleum hydrocarbon tolerant	bioremediation	Chang et al. (2020)
Hypocrea rufa PS2	*Arachis hypogaea L.*	IAA production, phosphate solubilization, chitinase production, increased plant biomass, phytopathogen tolerant	Agriculture	Thakor et al. (2016)
Lasiodiplodia sp. JAS12	Cotton and chili	Endosulfan tolerant	bioremediation	Abraham and Silambarasan (2014)
Fusarium equiseti GF18-3 and GF19-1	Turf grass rhizosphere	growth enhancement, disease tolerant	Agriculture	Saldajeno and Hyakumachi (2011)
Pseudomonas fluorescens (PF01) and *Bacillus subtilis* (BS01)	Rhizospheric-heavy metal polluted soil	Plant growth promotion, photosynthetic pigments production, heavy metal uptake	Phytoremediation/Agriculture	Rajendran and Sundaram (2020)
Bacillus sp. *Pseudomonas Fluorescens, Pseudomonas putida, Azospirillum brasieliense, Bacillus subtilis*	Fescue grass in heavy metal polluted soil	Protease production, Phosphorus solubilization, Nitrogen fixation, oxidase reaction, siderophores IAA and ACC deaminase, production, enhanced soil recovery of polluted soil	Bioremediation/Agriculture	Grobelak et al. (2018)
Pseudomonas sp, Bacillus sp and Acinetobacter sp	Mung Bean	IAA, phosphate solubilization, ammonia, catalase and siderophore production, antifungal activity	Agriculture	Kumari et al. (2019)

metabolites which work as efficient PGP agents. Some examples include *Flavobacterium*, *Micrococcus*, *Pseudomonas, Bacillus, Serratia, Agrobacterium, Azotobacter, Azospirillum, Burkholderia, Cellulomonas, Erwinia, Arthrobacter, Caulobacter, Chromobacterium*, (Hassan et al., 2019) while the second category is the intracellular PGPR, free-living and exist within the nodular structures of the plant cells (iPGPR-free living). Some examples of iPGPR members include *Frankia* species, Rhizobiaceae family (*Mesorhizobium, Bradyrhizobium, Rhizobium, Allorhizobium*,) and endophytes (Bhattacharyya and Jha 2012).

Two genus amongst the ePGPR *Bacillus* and *Pseudomonas* have been documented as the most studied and abundant as a result of their remarkable root-colonizing, biocontrol abilities and PGP acitivities (Verma et al., 2019; Aloo et al., 2019). The isolation of *Bacillus* and *Pseudomonas* as the only bacteria obtained from the rhizosphere of plants has been documented by some researchers (Caulier et al., 2018; Gamez et al., 2019; Chenniappan et al., 2019). Reports show that bacilli in plant rhizosphere make up to 95% of the Gram-positive rhizobacterial populations (Prashar et al., 2013). These bacterial species can survive in extreme environmental conditions with several physiological characteristics and are very versatile (Shafi et al., 2017). They have potentials to cause stress resistance, form multi-layered cell wall, produce endospore, and secrete different secondary metabolites (signal molecules, peptides, extracellular enzymes, antibiotics) which are essential in agriculture and the environment in general (Gutiérrez-Mañero et al., 2001; Kumar et al., 2013). Aloo et al. (2019) conducted an extensive review of the effect of Bacilli rhizobacteria on sustainable agriculture. The report highlights the benefits of this bacteria group as not only eco-friendly but also a resourceful technology (Shafi et al., 2017), for plant growth enhancement through diverse mechanisms which include nutrient solubilization plant bio-protection, phytohormone production, production of antibiotic and antifungal metabolites such as siderophore and lytic enzymes. For example, Zhou et al. (2016) discovered that *B. polymyxa* BFKC01 can enhance the availability of nutrients to plants, encourage the production of phytohormones and promote plant host ability to withstand environmental stresses. Other related research studies showcasing *Bacillus* as an outstanding microbial biostimulant have been documented (Chakraborty et al., 2011; Sokolova et al., 2011; Sharma et al., 2013; Khan et al., 2017; Park et al., 2017). Gowtham et al. (2018) reported that *B. amyloliquefaciens* possess immense potential to suppress anthracnose disease in Capiscum annum (Chilli) and also enhance the seed germination of the plant by 84.75%. A novel *Bacillus sonorensis* amongst other PGPR (*Paenibacillus polymyxa* and *Pantoea dispera*) significantly improved the growth nutrition and fruit yield of Chilly. In addition, the *B. sonorensis* possessed the following traits, phosphate solubilization, siderophore production, IAA, chitinase, HCN and biofilm formation (Thilagar et al., 2018). A new discovery has been reported by Xia et al. (2020). The report stated that a multi-stress *Bacillus xiamenensis* strain (PM14) isolated from rhizosphere of sugarcane is the first rhizospheric bacterium to possess biocontrol against twelve phytopathogens and positive outcome for all *in vitro* PGP traits excluding HCN production. Other related researches showcasing PGP potentials of *Bacillus* sp. have also been documented (Kumari et al., 2018; Guo et al., 2020; Lee et al., 2020)

Pseudomonads are also ubiquitous in soil and rhizosphere of diverse plants. *Pseudomonas* spp is among the most studied PGPR with multifarious PGP traits across all categories (biofertilizer, biocontrol agent, rhizoremediators and phytostimulants) (Tewari and Arora, 2016; Caulier et al., 2018; Mishra and Arora, 2018; Georgieva et al., 2018; Hassen et al., 2020). Chellilah (2018) isolated a metal tolerant PGPR *Pseudomonas* spp. with plant growth-promoting potentials and heavy metal accumulation by plants. Hernández-León et al. (2015) reported the *P. fluorescens* produced a volatile organic compound that aided the biocontrol of *Botrytis cinerea* in *Medicago truncatula* (tomato). Furthermore, the organism increased the plant's biomass content and chlorophyll. In another study, the toxicity level of cadmium exposed *Solanum lycopersicum* seedling was reduced by supplementation with *P. aeruginosa* and *Burkholderia gladioli*. This was achieved by the upregulation of anti-oxidative defense mechanism which increased the levels of different secondary metabolites such as phenolic compounds and osmolytes (Khanna et al., 2019). The most common members of

this group implicated in PGP activities are *P. putida*, *P. protengs*, *P. fluorescens*, *P. syringe andP. aeruginosa* (Dorjey et al., 2017; Andreolli et al., 2019).

Since the microbial community and activity within the root region are greatly influenced by the plant root exudates, it suggests that the species and plant variety can occasionally determine the benefits or plant promoting activity using a particular PGPR (Remans et al., 2008). This implies that different root exudates will be produced by different species which support the function of the applied microorganism and also acts as a base for the formulation of bio-active substance by the microorganism (Khalid et al., 2004; Hassan et al., 2019). PGPR promotes plant growth through increased uptake of essential nutrients from a pool of nutrients not readily available in the rhizosphere (Prasad et al., 2019). Several reports have shown that PGPR promotes plant growth via different mechanisms which include; biological fixation of N_2, biological solubilization of insoluble nutrients such as phosphorus, potassium and some micronutrients, production of volatile organic compounds (VOCs) and iron sequestration by siderophore production. Also, PGPR can elicit an "induced systemic tolerance" to ecological stresses (drought and salinity) which provides plant health and protection. A relatively new method of plant growth stimulation is zinc solubilization. Zinc (Zn) is an essential requirement by plants for their growth although in lesser quantities (micronutrients) compared to N and P. However, out, only a little quantity of the total Zn is present in soil solution and available for plant uptake, most part of the zinc occurs as minerals and insoluble complexes (Kamran et al., 2017; Rouphael and Colla 2020) which leads to zinc deficiencies in plant. Various researchers have identified PGPR as Zinc solubilizing agent. Kamran et al. (2017) analyzed the effect of zinc solubilizing rhizobacteria on wheat plant after isolation from wheat and sugarcane. The authors observed enhanced growth of wheat as well as zinc content and concluded that *Enterobacter cloacae*, *Pantoea*, and particularly *Pseudomonas fragi* could be as applied as a microbial-based biostimulant to conquer zinc deficiency under low input circumstances. Similarly, Eshaghi et al. (2019) reported *Pseudomonas japonica* as zinc-solubilizing rhizobacteria which significantly enhanced the plants' height, fresh and dry weight of corn. Various other PGPR such as *Bacillus* sp. *Pseudomonas*, *Azospirillum*, *Rhizobium* and *Bacillus aryabhattai* have been implicated to stimulate the growth of plant and enhance zinc availability and content when incorporated into plants. (Joshi et al., 2013; Hussain et al., 2015 Naz et al., 2016). Microorganisms that are metal-tolerant are known to remediate heavy metals by immobilization and reduced metal-plant availability through acidification, phosphate solubilization, release of chelating agents and redox changes (Abou-Shanab et al., 2003; Marwa et al., 2020). In phytoremediation, the rhizobacteria and host plant forms specific mutual association whereby the plant makes available carbon source which stimulates the bacteria to reduce phytoxicity of the polluted soil (Khan and Bano, 2018). This mechanism enhances the potential of microorganisms in close relationship with plant root to enhance remediation abilities. Other mechanisms include mobilization or transformation of metals into innocuous/inactive forms to permit the uptake of heavy metal ions by plants. Additionally, PGPR species have the ability to acidify the surrounding environment through the release of different chelating substances which results in an alteration in the redox potential and increase in heavy metal remediation (Khan and Bano, 2016). Other mechanisms involved in PGPR heavy metal remediation include, production of exopolysaccharides and lipopolysaccharides which removes the heavy metals through biosorption, Secretion of extracellular enzymes which reacts with heavy metals to enhance metabolization and assimilation by plants, production of siderophore which facilitates absorption of iron and reduction of free radicals formulation and lastly, sequestration of heavy metals through the process of biomineralization (Khan et al., 2017). In recent times, phytoremediation strategies have been categorized into five types phytoextraction/phytoaccumulation, (plants through roots and shoots accumulate metals from contaminated sites), phytostabilization; involved in the decreased mobility of metal by plants and decrease in bioavailability of a metal to avoid either entry into the plants or its leaching to groundwater, rhizofiltration; ability to absorb or ensure contaminant concentration by roots in with a continuous effluent flow in mostly hydroponic systems and phytovolatilization; ability to convert a heavy metal into volatile forms released it into the atmosphere

through stomata (small pores) (Khan and Bano, 2018; Pandey and Bajpai, 2019; Zubair et al., 2016). However, phytodegradation (mineralization of contaminants by enzymatic action into more simple or less-toxic forms by plants either in the root after their uptake or rhizosphere before uptake) does not apply to heavy metals, but toxic pollutants like hydrocarbons that can be degraded into simpler and innocuous substances.

The use of *Chenopodium album has been validated to be* stimulated by two consortia of growth-promoting rhizobacterial isolates belonging to the *Pseudomonas* genus for the phytoextraction of lead. The experimental result revealed the bioaccumulation efficiency of lead by the plants increased three times higher upon stimulation by the rhizospheric consortia (Abdelkrim et al., 2020).

4.4.2.1.2 Mechanisms for Plant Growth Promotion

Rhizobacteria are able to promote the growth of plants through different mechanisms (direct and indirect). Direct mechanisms occur externally and impact on the regulators of plant growth equilibrium. This is achieved due to (i) microbes harbour a plethora of plant-released hormones, (ii) induced plant metabolism leads to an improved adaptive capacity, (iii) micro-organisms produce growth regulators that are incorporated into the plant (Glick, 2014; Govindasamy et al., 2011). These mechanisms include solubilization of phosphate, atmospheric nitrogen fixation phytohormone production (auxins, ethylene, cytokinins, and gibberellins) and ammonia production (Siddikee et al., 2010) while the indirect mechanisms occur as an internal process in the plant host which allows the stimulation of the plants defensive response stimulated by bacteria colonizing the plants (Aeron et al., 2011). The indirect mechanisms include Hydrogen cyanide production, siderophores production, catalase activity and the synthesis of several other growth-promoting compounds (Pérez-Montaño et al., 2014).

4.4.2.1.3 Nutrient Uptake

Nitrogen fixation by microorganisms occurs through the biological nitrogen fixation (BNF) process especially the Diazotrophic microorganisms belong to archeae and bacteria groups and this occurs during nutrient cycling. Nitrogen fixation bacteria are two types: non-symbiotic (Cyanobacteria, Azospirillum), symbiotic (Rhizobia and Frankia) and the most widely studied. Rhizobia is capable of fixing 180×10^6 tonnes of N_2 with the BNF practice at a global scale (Verma et al., 2019) which significantly increases the utilizable nitrate availability for plant use and consequently increases productivity. *Rhizobium phaseoli* have been reported to effect an increase on the growth and yield of *Phaseolus vulgaris* as a result of increase in nitrogen fixation process. It is documented that *Azospirillum* is the most studied amongst other N fixing bacteria and reported in several crops, including sugarcane, wheat, cotton and maize, (Bashan and de- Bashan, 2010; Calvo et al., 2014).

Khan et al. (2016) isolated four N_2 nitrogen fixation, phosphate solubilization rhizobacteria, Amongst which *Bacillus pumilis* had the highest nitrogen fixing capacity (30.5%) of the total requirement of maize plant. Habibi et al. (2014) identified *Rhizobium daejeonense* strain JR5, *B. altitudinus* strain, JR 19, *Pseudomonas monteilii* and *Enterobactter cloacae* strain JW69 as N-fixation microorganisms from rice plants.

4.4.2.1.4 Phosphate Solubilization

Phosphorous is the second most essential nutritional requirement for plants after nitrogen. Though, unavailable in the form suitable for plant uptake despite its abundance in nature, plants have the capacity to uptake the soluble (utilizable) forms of phosphate (monobasic and dibasic) (Jha and Saraf, 2015). It is known that plants can only uptake the available quantity (<5%) of the total phosphorus present in soil. Calvo et al. 2014 reported the mineralization of organic phosphorus in soil by producing organic acids which solubilize aluminum, tricalcium and rock phosphates, (complex-structured phosphates) etc. to soluble forms through different groups: hydroxyl and carboxyl groups by cation chelation bond to the phosphate which are readily available for plants use (Sharma et al., 2013;

Goswami et al., 2014). Excretion of organic acids accompanied by a decrease in pH results in acidification of the surroundings and microbial cells, which consequently releases P ions by substitution of H^+ and CA^{2+}. Different organic acids are released by different species of microbes (strain-specific phenomenon) (Seshachala and Tallapragada, 2012). However, gluconic acid is the most common produced organic acid by bacteria such as *Azospirillum* sp., *Pseudomonas* sp., *Erwinia* sp., and *Burkholderia* sp. (Rodriguez et al., 2004) while *Bacillus* sp. produces a mixture of lactic, isovaleic, isobutyric and acetic acids (REFS). Also according to the work by Goswami et al. (2014), some soil bacteria which include ecto-rhizospheric (reside on roots and in rhizosohere of plants) strains from *Bacilli* and *Pseudomonas* were reported as effective phosphate solubilisers. Other examples include, *Achromobacter, Streptomyces, Rhizobium, Rastonia, Rhodococcus, Serratia, Burkholderia, Agrobacterium, Micrococcus, Azotobacter, Erwinia, Flavobacterium, Aereobacter* and *Enterobacter* (Gamalero and Glick, 2011; Halpern et al., 2015; David et al., 2014,; Alori et al., 2017). Examples of active fungal species involved in phosphate solubilization include; *Penicillium* (Wakelin et al., 2007) *Aspergillus, Trichoderma* (Sibi, 2011), *Cunninghamella, Curvularia, Fusarium, Achrothcium, Alternaria, Arthrobotrys Pythium, Rhizoctonia, YarrowiaRhizopus, Sclerotium, Cephalosporium, Cladosporium, Torula, Saccharomyces (*Srinivasan et al., 2012; Sharma et al., 2013; Alori et al., 2017) and *Talaromyces* (Kanse et al., 2015). Alori et al. (2017) documented the ability of fungi to effortlessly penetrate through and move farther more easily within the soil than bacteria which could be a relevant factor in phosphate solubilization. Also, Fungi are known for the secretion of several acids such as gluconic, oxalic, lactic, acetic, tartaric, gluconic and 2 ketogluconic acid (Sharma et al., 2013).

4.4.2.1.5 *Volatile Organic Compounds (VOC)*

Microorganisms release VOCs as an essential PGP mechanism to enhance plant growth and protect plants from phytopathogens. There are different types VOCs which promote microbe-microbe interaction and plant-microbe interaction (Peñuelas et al., 2014). Examples of some VOCs produced by microbes include compounds such as acetoin, 2–3-butane-diol, jasmonates and terpenes. When produced in sufficient concentration during interactions between PGPR and their host plant, VOCs act signaling agents (Verma et al., 2019). Volatile organic compounds are regarded as low molecular weight (LMW) compounds (< 300 mg/mol) with high vapour pressure to evaporate into the atmosphere. These compounds been linked severally with PGP by modulating plant proliferation and stimulating induced system resistance against phytopathogens (Shafi et al., 2017; Aloo et al., 2019). VOCs adopts a strain-specic phenomenon whereby each microbial strain produces specific type of volatile compounds that could be different from other strains. Park et al. (2015) reported the physiological function produced by a novel VOC from *P. fluorescens* SS101 (*Pf*SS101). The effect of this VOC *in planta* and *in vitro* produced a significant plant growth. Also, the VOCs emitted by this strain characterized by solid-phase micro extraction (SPME) and gas chromatography-mass spectrophotometry (GCMS) revealed 11 different compounds with molecular weight ranging between 27 and 240 mg/mol. The identities and distribution of the compounds is as follows: five (5) alcohol (2-methyl-1-propanol, 3,-methyl-butanol, 1,3,-tetradicadien-1-ol, 1-cyclohexyl-1-pentanol and trans-9-hexadecen-1-ol), a ketone (2-butanone), two (2) hydrocarbons (1,-nonene and 2-methyl-n-tridecene), and an ester (methyl-thioacetate) and an ether (decyl-oxirane) were observed in *Pf*SS101. Vidhyasri et al. (2019) obtained 12 VOCs from *Ochrobactrium* spp which elicited enhanced plant growth on tobacco plants while two of the produced VOCs (2, 3.-butanediol and acetoin) induced resistance on the plant.

4.4.2.1.6 *Phytohormone Production*

Production of phytohormones of different classes (IAA, gibberellins, ethylene, abscisic acid, cytokinins) is among the direct processes of PGP. Different groups of soil microorganisms (bacteria and fungi) have the capacity to produce phytohormones, especially those associated with plants. *Azospirillum brasilense* is the most identified rhizospheric microorganism among others to produce

phytohormones in soil (Goswami et al., 2014). Plants have a positive response to the availability of any phytohormone in the rhizosphere produced by microorganisms. These hormones have the capacity to control processes like cell plant extension, enlargement and division in the roots of plants (Glick, 2014).

4.4.2.1.7 Indole-3-Acetic Acid (IAA)

Plant growth promoting microorganisms produces an auxin called indole acetic acid. Auxins help to manage many phases of plant development *via* cell division, stimulate seed and tuber germination, mediate vegetative growth processees, pigment formation, apical dominance, initiate root formation (lateral and adventitious), cell elongation, increased the rate of root and xylem development, tissue differentiation, influence photosynthesis, biosynthesis of various metabolites, (Kravchenko et al., 2004; Gupta et al., 2015). Primarily, IAA has the potentials to stimulate long term and rapid plant responses by increasing cell division, elongation and differentiation. Researchers have revealed that about 80% of rhizospheric bacteria can synthesize IAA (REF). The ability to enhance nutrient and mineral uptake by host plant is displayed microbes which colonize the seed or root of plants together with endogenous IAA present in plant (Gupta et al., 2015). An important benefit of IAA after long-term application is exhibited in the highly developed roots which support and enhance plant uptake of nutrients for better growth and yield of plants (Aeron et al., 2011). Indole acetic acid has been observed in the different categories of plant-associated microbes such as PGPR/PGPF (Tagele et al., 2018; Batista et al., 2018; Ozimek et al., 2018) and endophytic bacteria and fungi (Passari et al., 2016; Haidar et al., 2018; Ramos-Solano et al., 2008).

4.4.2.1.8 Cytokinins

The cytokinins represent another essential class of phytohormones produced by some plant-associated microbes. They are involved in enhancing plant general development. Analogous to IAA, the response of plants to exogenous application cytokinins results in enhanced root development, improved formation of root hairs, enhanced cell division, inhibition of root elongation shoot initiation, including other physiological and metabolical responses (Amara et al., 2015).

4.4.2.1.9 Hydrogen Cyanide (HCN) Production

Hydrogen cyanide (HCN) is among the chemical compounds produced by rhizospheric microbes with biocontrol activities against plant diseases and PGP properties (Rijavec and Lapanje, 2016). Research has revealed that HCN-producing *Pseudomonas fluorescent* has the ability to suppress of phytopathogens in the root zone of the plant by inhibiting the cytochrome-oxidase pathway. Hydrogen cyanide has also been found to be antagonistic against fungi (Akintokun et al., 2016). The pathogenic bacterium *Pseudomonas entomophila* produces HCN as a secondary metabolite and displays biocontrol properties against pathogenicity induced by other bacteria (Ryall et al., 2009). Hydrogen cyanide (HCN) produced by a apsychrotolerant bacterium *Pseudomonas fragi* CS11RH1 (MTCC 8984) significantly increased the percentage germination, plant biomass, nutrient uptake and rate of germination of wheat seedlings (Selvakumar et al., 2009). Ahmad et al. (2009) also reported that the production of HCN has been identified to be a regular trait of *Pseudomonas* (88.89%) and *Bacillus* (50%) in the rhizosphere of the host plant

4.4.2.1.10 Catalase Activity

Plant cells produce antioxidant enzymes such as peroxidase and catalase. This enzyme is responsible for protecting plants against toxic free radicals generated mainly during environmental stresses (Kravchenko et al., 2004). Sem and Chandrasekhar (2015) reported that catalase activity in selected rhizobacteria strains increased under increased salt stress condition, helping them to perform their functions. Organisms that produce catalase therefore enhance plant tolerance to environmental stresses.

4.4.2.1.11 Lytic Enzyme Production

The PGPR/PGPF has potentials in suppressing the development and activities of plant pathogens through the secretion of lytic enzymes. These rhizospheric microbes secrete several enzymes such as cellulases, ACC-deaminase, proteases, lipases, and chitinases which initiate lysis of fungal cell wall. These cell wall mainly comprise polysaccharides, glucan and chitin, hence the corresponding enzyme (e.g. glucanase and chitinase) producing microbes is considered potent in inhibiting their growth and activity. These activities empower plant-associated microbes to be significant in environmental and agricultural activities. Several rhizospheric microorganisms have been reported to express the lytic enzymes which among other benefits protect plants from pathogens.

Saraf et al. (2010) reported that ACC deaminase-expressing rhizospheric microorganisms exert several positive benefits on the plant such as protection of plants against effects associated with extreme temperatures, pollutants, pathogens, drought and high salt concentration. Also, AAC enhance the production of VOCs, phytoremediation of impacted soils and promotion of root development

4.4.2.1.12 Plant Growth Promoting Fungi (PGPF)

Rhizospheric fungi can be categorized into different groups based on their ability to colonize different environments and their functional roles such as endophytic, mycorrhizal saprotrophic, epiphytic and pathogenic. The rhizosphere acts as a metabolically active and essential environment for interaction between rhizosphere-associated microbial, plant and soil (plant-soil-microbe interaction) which promotes an exchange of substances and energy. Another important rhizosphere-associated microbial community is the PGPF, although not as explored as rhizospheric bacteria, they play similar roles in promoting plant development and sustained environment. PGPF also adopts the indirect and direct mechanisms for plant growth promotion or PGP and protection from environmental stresses such as phosphate solubilization, HCN production, phytohormones, siderophore, ISR, enzyme production.

The dominant phyla of rhizospheric fungi *Ascomycota*, *Zygomycota*, *Basidiomycota*, and *Glomeromycota* (Song et al., 2018; Pattnaik and Busi, 2019). Most rhizospheric fungi are saprophytes that are responsible for adequately enhancing plant productivity, preventing phytopathogens, producing ISR and supporting environmental remediation. Some examples include *Trichoderma, Phoma, Aspergillus, Fusarium, Penicillium, Rhizoctonia*, and *Piriformospora*, which stimulate diverse PGP traits to promote plant yield (Jaber and Enkerli, 2017). This group of organisms modulate plant growth, enhance mineral uptake by plants, and increase biomass and crop yield. PGPF has been classified as endophytes, ectomycorrhizas, arbuscular mycorrhizae and saprophytic (Waghunde et al., 2017). These beneficial fungal groups have been applied in agriculture (biofertilizer, biopesticides and biocontrol agents) and bioremediation of environmental pollutants (Shelake et al., 2018).

Arbuscular mycorrhizal (AM) among the PGPF groups constitutes the most studied and important based on its extensive applications and occurrence in over 80% of plant species (Cervantes-Gamez et al., 2015; Badri et al., 2009). AM fungi function in various capacities which include, maintaining and enhancing soil fertility, elevated nutrient uptake and availability, stimulating plant systemic resistance by antioxidant induction during conditions of disease and stress, translocation of mineral nutrients essential for promoting plant growth/yield and enhancing community succession (Cui et al., 2018). Interestingly, AM has been isolated from various environments under different prevailing conditions such as aquatic environments, tropical rainforests, sodic or gypsum soils and strong saline conditions. The symbiotic relationship between plant and AM is the most widespread in nature, with beneficial contributions from both sides. The AMF promotes the potential of water and nutrient absorption of the plants while carbon is made available to the AMF by the plant soil (Smith and Read, 2008). An important feature of AMF is the ability to translocate and absorb nutrients in plants further than "rhizospheric depletion zones" of plants were secondary metabolism is altered (Rouphael et al., 2015). Thus, AMF contributes significantly to the growth development of plant and enhances plant resistant to environmental stresses (Song et al., 2015). More importantly, the AM

fungal possess an outer membrane that absorbs a number of heavy metals preventing their entry into the cell of the host plant. This mechanism aids in the detoxification of heavy metals and other xenobiotics from damaging root tissues. It also helps to prevent the entry of phytopathogens that could lead to poor crop yield. Several reports have been documented on the potential application of AMF in the detoxification of heavy metals from the environment (Abu-Elsaoud et al., 2017; Khan and Bano, 2018). Some heavy metals; Zn, Ni, Cr, and Cu are critical for the growth of plants, animals and the microbial community in concentrations that are required (Rajkumar et al., 2009). However, some heavy metals (e.g., Hg, Cd, and Pb) do not play any biotic or physiological role, hence possess risk to both plants and the environment (Soetan et al., 2010; Khan and Bano, 2018).

Kumar and Saxena (2019) in a review demonstrated the potential of mechanism of AMF remediation of heavy metals polluted soil. Plants colonized by AMF have the ability to grow under heavy metal contaminated soils. The review highlighted that AMF can immobilize heavy metals within soil by translocating metals into the hyphae and roots. The mechanisms involved in the adoption of AMF to environmental stresses include action of cell wall chitin, extraradical hyphae, release of certain proteins such as phytochelatins, metallothioneins and siderophore.

Pellegrino and Bedini (2014), conducted a field study which involved inoculation of chickpea plant with AMF. An increase in chickpea AMF root colonization, increased plant biomass/yield, improved nutritional value of grain by protein, Fe and Zn biofortification were observed.

Khan (2020) conducted an extensive review on the potential of bioenergy crop associated with mycorrhiza to remediate (mycorrhiza-remediation) uranium-contaminated mine sites. Based on his findings, the author proposed that mycorrhizo-remediation of U-contamination is a sustainable and efficient approach in addition to the production of bioenergy biomass during the process. Other related research on effect of AMF as a sustainable tool in agriculture and environment has be documented (Srivastava et al., 2017; Atakan et al., 2018; Chenchouni et al., 2020; Singh et al., 2020).

The benefits associated with the use of an integrated/co-inoculation system of applying both PGPR and PGPF (AMF or other rhizopsheric fungi) is considered a sustainable tool for ensuring environment management, crop stability and enhanced yield in low-input conditions and nutrient-deficient soils. Also, this innovative technology aids in the improvement of plant tolerance under abiotic and biotic stress conditions like salinity, drought and extreme temperatures (Maharshi et al., 2019; Rouphael and Colla, 2020). Khan (2002) and (2004), stated that the microbe rhizospheric population can be effective by utilizing an inoculum containing PGPR consortium, AMF as allied colonizers, nitrogen fixing rhizobacteria, and mycorrhiza-helping bacteriateria (MHB) as biostimulants which could provide plants with benefits decisive for eco-restoration of heavy metal contaminated soil, etc.

Rahimzadeh and Pirzad (2017) conducted a study on the possible effects of co-inoculation of AMF in association with phosphate solubilizing bacteria (PSB) and single inoculations of AMF (*Glomus mossae* and *G. intradadices*) and PSB (*Pseudomonas putida*) on linseed in drought stressed environment. The mycorrhizal symbiosis compensated for the drought stress-induced yield reduction observed in the PSB inoculated plant while a significant rise in crop productivity was detected in mycorrhizal plants and higher yield in the co-inoculated plants. It has been stated that a great success could be achive in the phytoremediation of iron polluted agricultural soil cultivated with *Sorghum bicolor* and *Pennisetum glaucum* using 3 AMF (*Scutellospora, Acaulospora,* and *Glomus*) and 4 PGPR genera (*Pseudomonas, Azotobacter, Streptomyces* and *Paenibacillus*). Co-inoculation of PGPR and AMF was more effective than single inoculation of AMF and PGPR under the same conditions. The co-inoculants produced siderophore which enhanced the absorption of iron. Mohamed et al. (2019) studied the potentials of combined or single inoculations of mycorrhizae and PGPR (*Bacillus* subtilis and Pseudomonas fluorescence) as a biocontrol agent of *Sclerotium rolfsii* in bean plant and as bio-fertilizer for promoting plants nutrient status. As compared to non-inoculated treatments, single inoculations enhanced green pod and straw growth and uptake of P and Fe by plants. However, co-inoculation further enhanced these tested parameters as compared with other inoculations and fungicide-treated plants.

Co-inoculation of AMF and PGPR also has environmental significance in terms of pollution control and supporting a sustainable environment. Dong et al. (2014) also documented that co-inoculation of PGPR *Serratiamarcenscens* BC3 and AMF *Glomus intraradices* enhanced plant biomass, antioxidant enzyme activities, and increased microbial populations in the rhizosphere of petroleum-contaminated soil. The report further stated that the breakdown percentage of total petroleum hydrocarbons with the co-inoculants was 72.24% compared to the single inoculation of either AMF or PGPR. Xun et al. (2015) investigated the phytoremediation effect of PGPR and AMF petroleum impacted saline-alkali soil. Results demonstrated that the petroleum inhibited plant growth was of malondialdehyde (MDA accumulation), free proline and the activities superoxide dismutase(antioxidant enzyme), peroxidase and catalase. However, the co-inoculants augmented the activities of the antioxidant enzymes, decreased MDA and free proline contents. Also, soil quality was improved evidenced by increased soil enzyme activities (dehydrogenase, sucrase, and urease). Additionally, a high hydrocarbon degradation rate was recorded indicating AMF and PGPR co-inoculation could enable plants to tolerate harmful hydrocarbon contaminants, enhance soil structure as well as remediate saline-alkali soil contaminated with petroleum hydrocarbon. Other more recent related research highlighting the prospects of AMF and PGPR in pollution control have been documented (Fecih and Baoune, 2019; Guarino et al., 2020)

4.4.2.1.13 Endophytic Plant Growth Promoting Microorganisms

Endophytes denote free living microorganism that resides in the intercellular and intracellular part of plants such as the root, stem, flowers, leaves and seeds depend upon the site of colonization without imposing any harmful effect on the host plant (Gaiero et al., 2013; Compant et al., 2016; Gupta et al., 2020a). Plant species harbor a large population of microorganisms within their tissues (Ray et al., 2018) during at least part of their life cycle without displaying any visible signs on the plant host (Jimtha et al., 2014). Endophytic colonization of plants involves two processes: chemotaxis and establishment of the rhizosphere. Plant exudation through the root induce chemical signaling molecules such as malic acid, fumaric acids, amino acids, carbohydrate, citric acid and other allelochemicals (Yuan et al., 2015) that allow the recognition, penetration and establishment of soil microbes into the root of plants (the rhizospheric region). The rhizosphere is very attractive to a large array of micro-organisms as a result of the presence of chemical exudates which contains several essential compounds such as amino acids, carbohydrate, and lipids (Kumar et al., 2015). These morphological diverse microorganisms enter into the plant through different routes but most often through the root surface into the plant tissues as endophytes (Kumar et al., 2016). This endophyte-plant interaction is symbiotic as both benefits from the interaction. The microbes get the essential nutrients and a conducive habitat while the plants benefit from the growth-promoting substances properties, bioactive compounds that aid in plant promotion, and protection from diseases environmental stresses to plants (Gupta et al., 2020a). Gupta et al. (2020b) in their report also revealed that endophytes are more effective stress-tolerant than rhizospheric and phyllospheric microbial strains due to their ability to survive inside the host plant.

Reports have highlighted the potential biotechnological application of endophytic microbes in environmental pollution control and agriculture (Santos e Silva et al., 2016; Liotti et al., 2018; Gupta et al., 2020; Guarino et al., 2020; Fu et al., 2020). Endophytes possess plant growth promoting attributes that assist plant development growth through direct and indirect processes such as acquisition of nutrient and water uptake which promotes hardness and a reduction in oxidative stress enzymes in plants present in polluted soil (Santoyo et al., 2016; Naveed et al., 2014), nitrogen fixation (Banik et al., 2016), solubilization of phosphorus (Zhu et al., 2011; Sarbadhikary and Mandal, 2018), inhibition of phytopathogens by stimulating induced system resistance (Carvalhais *et al.*, 2013), production and regulation of phytohormones such as gibberellins, cytokinnins, ethylene (Strader et al., 2010; Dudeja et al., 2012; Vanstraelen and Benková, 2012; Bhattacharyya and Jha, 2012), siderophore production (Kannan et al., 2018), Nitrogen fixation (Firdous et al., 2019), tolerance to stresses (Khan et al., 2011; Theocharis et al., 2012). Passari et al. (2016) reported that

an endophytic *Bacillus* sp BPSA16had the most efficient biofilm formation potential amongst other microorganisms tested. In addition, the bacterial sp. showed further ability to resist various abiotic and biotic stresses and also produced three phytohormones (IAA, Kinetin and 1,6-Benzyladenine). Wu et al. (2016) observed that *Pseudomonas saponiphilia* isolated from a medicinal plant produced IAA, solubilized phosphate and also possessed some antagonistic effects against phytopathogens through the production of siderophore, HCN and 2, 4- diacetylphloroglucinol. In a recent study by Vurukonda et al. (2018), *Streptomyces* sp. was reported as one of the most important endophytic microbes owing to the multifarious PGP traits it exhibits. They revealed that *Streptomyces* sp. has the potential to inhibit bacterial and fungal phytopathogens which makes this species a remarkable biocontrol agent. In addition, it produces antibiotics, volatile organic compounds and lytic enzymes such as chitinase, amylase, invertase, cellulase, peroxidase, xylanase, keratinase, pectinase, phytase, protease, and lipase which aids the degradation of complex nutrients into simple mineral forms. Some other examples of endophytic bacteria include: *Azotobacter chroococcum Avi2.* (Banik et al., 2016), *Bacillus* sp. SBER3 (Bisht et al., 2014), *Paenibacillus* spp (Díaz Herrera et al., 2016), *Burkholderia phytofirmans* PsJN (Naveed et al., 2014), *Cladosporium velox* (Singh et al., 2016). These endophytes have been found to be associated with several plants which include, Chilli, Potato;, Pea, Common Bean, Citrus, Sunflower, Cotton, Chickpea, Pearl millet, Soybean, wheat, Rice, Mustard, Chilli, Sugarcane, Tomato, Maize, and Strawberry (Yadav, 2018; Verma et al., 2017). The endophytic micro-organism has gained global attention of the scientific community owing to multifarious abilities (PGP, tolerance to extreme environmental conditions, bioremediation potential) they possess (Verma et al., 2017). The interactions between rhizobacterial and endophytes of a specific host plant could contribute to defining the success of bioinoculation technology proposed for promoting plant growth and yield (Thokchom et al., 2018)

Diverse microbial communities inhabit the plant environment including the endopsphere. These microbes sustain interaction with the host plants and conduct essential functions in plant growth and detoxification of polluted environments. Several endophytes are known to harbor pollutant-degrading genes for the mineralization of contaminated soil *in planta* (Feng et al., 2017). In light of this, several researchers have revealed that endophytes significantly contribute to pollution control directly (harboring hydrocarbon-degrading genes) or indirectly (phytoremediation). Endophytic bacteria have the ability to induce adaptation to abiotic and biotic stress tolerance as well as biomass increase in plants involved in phytoremediation (Ijaz et al., 2016). Additionally, endophytes have great potential abilities to facilitate phytoremediation properties of the plant since they possess degradative potentials which can be transferred to different parts of the plant enhancing the metabolic activity. The unique compatibility and interaction between plants and endophytes is a critical factor in expediting the phytoremediation of polluted sites. Pollutant degrading endophytes have the ability to directly detoxify or mineralize pollutants because they possess catabolic genes targeting specific components of the pollutant. For example, Crude oil pollutant comprises a plethora of hydrocarbons ranging from aliphatics to low and high molecular weight hydrocarbons. Barman et al. (2014) identified organophosphorus hydroxylase gene ophB, harbored in an endophyte Pseudomonas sp. BF 1-3, which effectively hydrolyze chlorpyrifos. It becomes more interesting when plant-associated microbes possess not only plant-growing traits, but also pollutant-degrading catabolic genes. This approach will facilitate phytoremediation efficiency of contaminated environment. Baoune et al. (2018), investigated the petroleum-degrading and PGP potentials of endophytic Actinobacteria. All petroleum-degrading isolates obtained were identified as *Streptomyces* sp. and achieved 98% removal after seven days of incubation. Also, all strains showed multifarious PGP traits which include siderophore, phosphate solubilization, ACC deaminase, nitrogen fixation and IAA production and also biosurfactant production. Wu et al. (2020) identified another strain with capacity to mineralize hydrocarbons and enhance plant growth. The investigation revealed that *Pseudomonas aeruginosa* L10 is a biosurfactant producing, petroleum-degrading and PGP endophytic bacterium which aided the breakdown of $C_{10} - C_{26}$ n-alkanes in diesel oil and also LMW PAHs (pyrene, phenanthrene and naphthalene). In addition, *P. aeruginosa* L10 possesses PGP traits siderophore, ACC

deaminase activity and IAA. The authors further screened the entire genome of the species and identified catabolic genes associated with hydrocarbon mineralization which include genes encoding monooxygenases, alcohol dehydrogenase, dioxygenase, and aldehyde dehydrogenase.

Mitter et al. (2019) explored the potential application of hydrocarbon-degrading endophytic strains (*Stenotrophomonas* sp., *Flavobacterium* sp., *Pantoea* sp. and *Pseudomonas* sp.) initially obtained from an oil rich sand and subsequently inoculated on white clover plant growing on diesel polluted soil. Results showed that the strains enhanced plant growth in the face of the pollutant with high biomass generated. Also, GCMS analysis revealed the soil was remediated by the inoculated plants. Diverse microbial groups have been reported as endophytes which include, bacteria, fungi, and archaea.

Endophytic fungi (EF) present certain features that are unique and promising in sustainable agriculture and environmental remediation (Table 4.2), however, endophytic bacteria have gained more attention over the years (Nandy et al., 2020; Mishra and Venkateswara, 2017). Endophytic fungi contribute significantly to plant development by enhancing nutrition (bidirectional nutrient transfer) production of phytohormones and biocontrol of plants against phytopathogens to boost plant health. Additional, EF and plant interaction provides plant protection against adverse abiotic conditions which include the ability to tolerate salinity, heavy metal, hydrocarbon and drought and increased resistance to other stressors (Mitter et al., 2019).

Endophytic fungi are ubiquitous, reportedly isolated from different host plants owing to their capacity to synthesize different biologically active compounds and a host of extracellular enzymes (Rana et al., 2019). Plants inoculated with EF result in significantly increased biomass production and enhanced commercial plant production. Relatively, EF have shown great prospects in biotechnological applications due to its diverse potentials of being used as source of secondary metabolites, bioremediation agents, biological control agents and plant growth promoting agents. A diverse array of EF species assigned to different genera including *Cryptococcus, Rhizoctonia, Rhodotorula, Talaromyces Curvularia, Fusarium, Geomyces, Alternaria, Aspergillus, Chaetomium, Collectotrichum, Glomus, Penicillium, Rhizopus, Phaeomoniella, Trichoderma,* and *Xylaria* were isolated from several selected host plants (Yadav et al., 2017). Reviews on the fungal diversity of endophytes from various crops revealed that the predominant genera include, *Penicillium Aspergillus, Piriformospora* and *Fusarium.* Endophytic fungi play significant functions in the breakdown of debris and other organic compounds which include plant residues (lignin, keratin, oligosaccharides, glucose, pectin, hemicelluloses and cellulose,) lipids and proteins, through its extracellular enzyme system to release nutrients, carbon and energy (Kudanga and Mwenje 2005). These organic compounds are present in litter, wood and leaf which consequently implies that endophytic fungi can readily degrade these materials and so it is of great importance to environmental management. These enzymes of interest include, phenol oxidase, proteinase cellulase, lipoidase, pectinase and lignin-degrading enzymes. For example, Correa et al. (2014) reported that fungal enzymes initiated the degradation of lignocellulosic materials in an oxidative system (ligninases and peoxidases) and hydrolytic system (xylanases and cellulases). Also, amylases and its derivatives; glucoamylases, α amylase and β-amylases are reportedly responsible for the conversion of starch to several sugar solutions (Krishnamurthy and Naik, 2017). These degradation processes mediated by the extracellular enzymes is very critical in promoting nutrient availability in soil and ecosystem balance because the end product of the decomposition increases biomass, organic matter and nutrient in the soil.

Endophytic fungi also impact on enhancing plant growth and protection by producing bioactive secondary metabolites, stimulating host plants to withstand both biotic and environmental stresses, resistance to diseases and other very desirable crop traits that will ensure sustainable agriculture. There is a growing attention in the exploration and potential applications of EF in agricultural practice. Potshangbam et al. (2017) investigated the factors influencing plant growth and biocontrol properties of endophytic fungi associated with maize and rice plants. The report stated that *Penicillium simplicisssum* (ENF22) and *Acremonium* sp. (ENF 31) could potentially prevent the growth of all four tested phytopathogens by producing defensive enzymes. ENF31 was observed to

TABLE 4.2

Endophytic Microorganisms with Agricultural and Environmental Applications

Endophytic Microbes	Host Plant/Source	Endophyte Potential Trait	Application	Reference
Bacillus sp.BPSAC6	*Clerodendrum colebrookianum* Walp.	Phytohormone production (IAA, Kinetin and 6-Benzyladenine), phosphate solubilization, ammonia production, HCN production, antifungal activity	Agriculture	Passari et al. (2016)
Paracoccus halophilus G062	Potato plant	nitrogen fixation, siderophore, phytohormone production (IAA, Gibberellin, and zeatin), phosphate solubilization	Agriculture	Akhidiya et al. (2014)
Streptomyces spp	Sorghum stems.	Antifungal activity and enhanced plant growth (increase in wet weight of roots and shoots)	Agriculture	Patel et al. (2018)
Streptomyces olivaceoviridis, S. rochei	Wheat	Auxin, gibberellin, and cytokinin synthesis	Agriculture	Aldesuquy et al. (1998)
Streptomyces spp.	*Terfezia leonis* Tul.	Siderophore production, IAA, and gibberellic acid production	Agriculture	Goudjal et al. (2015)
Bacillus sp. strain MBL_B17 *Pseudomonas psychrotolerans* strain MBL_B23 *Staphylococcus arlettae* strain MBL_B2 *Kocuria* sp. strain MBL_B19 *Micrococcus luteus* strain MBL_B18	Jute plant (*Corchorus olitorius*)	IAA, P solubilization, N_2 fixation, siderophore production	Agriculture	Haidar et al. (2018)
Bacillus sp strain B1920, B1923, B2084, B2088	Maize	P solubilization, IAA, siderophore, production, enhanced plant biomass, shoot and root dry weight	Agriculture	Ribeiro et al. (2018)
Pantoea sp.	Wheat	IAA, siderophore production, P solubilization, N_2 fixation	Agriculture	Díaz Herrera et al. (2016)
Bacillus amyloliquefaciens (MPE20) and Pseudomonas fluorescens (MPE115)	Medicinal plant	modulate withanolide biosynthetic pathway, antifungal tolerance, improvement in photochemical efficiency, upregulation of biosynthetic pathway genes	Agriculture	Mishra et al. (2018)
Bacillus sp, *Ochrobactrum*, sp, *Stenotrophomonas maltophilia, Paenibacillus* sp.	Guarana (*Paullinia cupana* var. sorbilis)	Growth pathogen inhibition, amylase, protease, cellulose and pectate-lyase enzyme production	Agriculture	Silva et al. (2016)
Pantoea dispera, Bacillus amyloliquefaciens, Staphylococcus hominis, Bacillus sp.	Maize	Potassium, Zinc and Phosphorus solubilization, IAA, HCN, and siderophore production, antifungal activity, BNF activity	Agriculture	Marag and Suman (2018)

(Continued)

TABLE 4.2 (Continued)

Endophytic Microorganisms with Agricultural and Environmental Applications

Endophytic Microbes	Host Plant/Source	Endophyte Potential Trait	Application	Reference
Bacillus subtilis strain EDR4	Wheat	Antifungal protein (E2) against numerous fungal pathogens. E2 exhibition of ribonuclease, hemagglutinating activities and trifle protease activity.	Agriculture	Liu et al. (2010)
Streptomyces pluricolorescen	Tomato (*Lycopersicon esculentum*)	Antimicrobial activity against numerous phytopathogens, IAA, siderophore production and phosphate solubilization	Agriculture	Fialho et al. (2010)
Pseudomonas aeruginosa PM389	pearl millet (*Pennisetum glaucum*)	Nitrogen fixation, mineral phosphate solubilization, siderophore production and antagonistic properties	Agriculture	Gupta et al. (2013)
Fusarium tricinctum RSF-4L *Alternaria alternata* RSF-6L	*Solanum nigrum*	IAA, promote plant growth (chlorophyll content, root-shoot length, and biomass production)	Agriculture	Khan et al. (2015)
Aspergillus fumigatus TS1 and *Fusarium proliferatum* BRL1	*Oxalis corniculata*	Promote plant growth (chlorophyll content, root-shoot length, and biomass production), Gibberellic acid and IAA production, synthesis of bioactive compounds	Agriculture	Bilal et al. (2018)
Curvularia geniculata	*Parthenium hysterophorus* L.,	P solubilization, IAA production, plant growth promotion	Agriculture	Priyadharsini and Muthukumar (2017)
Alternaria sp. A13	*Salvia miltiorrhiza*	enhanced the dry root biomass and secondary metabolite, stimulation of root growth	Agriculture	Zhou et al. (2018)
Aspergillus awamori W11	Drought-stressed *Withenia somnifera*	secondary metabolite production, including IAA, phenols and sugars, plant growth promotion,	Agriculture	Mehmood et al. (2019)
Trichoderma aureoviride Trichoderma harzianum	Wheat (*Triticum aestivum* L.)	salt, heavy metals, and drought tolerance, IAA, ammonia catalase and siderophore production, phosphate solubilization,	Bioremediation/Agriculture	Ripa et al. (2019)
Aspergillus sp. A31, *C. geniculata* P1, Lindgomycetaceae P87 and *Westerdykella* sp. P71	*Aeschynomene fluminensis* and *Polygonum acuminatum*	Removed 100% mercury from medium and enhanced plant growth,	Bioremediation	Pietro-Souza et al. (2019)
Mucor sp MHR-7	*Parthenium* sp	Heavy metal tolerant, IAA and ACC deaminase production phosphate making it excellent phytostimulant fungus.	Bioremediation	Zahoor et al. (2017)
Sphingomonas SaMR12	*Sedum alfredii*	promote zinc extraction/uptake, increased plant biomass, increased the exudation of oxalic acid	Bioremediation	Chen et al. (2014)

(Continued)

TABLE 4.2 (Continued)

Endophytic Microorganisms with Agricultural and Environmental Applications

Endophytic Microbes	Host Plant/Source	Endophyte Potential Trait	Application	Reference
Pseudomonas sp *Rhodococcus fascians* strain. *Microbacterium sp.*	Ryegrass (*Lolium perenne* L.)	multiple plant growth-promoting abilities, IAA production phosphate solubilization cellulolytic enzyme production, capacity motility production of siderophore, ammonium and hydrogen cyanide production, biosurfactants production and harbors hydrocarbon-degrading genes	Bioremediation/ Agriculture	Kukla et al. (2014)
Streptococcus and *Staphylococcus strains*	*Psidium guajava* (Guava) and *Mangifera indica* (Mango)	Tolerant to different heavy metals	Phytoremediation	Riskuwa-Shehu et al. (2019)
Enterobacter cloacae ATCC 13047.	*Ficus septica*	High tolerance to Chromium, production of IAA, exopolysaccaride and P solubilization	Phytoremediation	Rohma et al. (2020)
Pseudomonas stutzeri Z11	*Phragmites australis*	Salt-tolerant with diesel oil degradation abilities,	Bioremediation	Wu et al. (2019)
Serratia sp PW7	*Plantago asiatica*	Pyrene degradation, reduced pyrene uptake in roots and shoot of plant	Bioremediation	Zhu et al. (2017)

tolerate a large range of pH from 2 to 12, an essential requirement for studying plant growth in several soil types, most importantly acidic, as a significant factor that renders land unusable for agricultural activities while ENF22 grew within pH range 3–12, with 10% salt tolerance ability. Similarly, Bilal et al. (2018) isolated two endophytic fungi from the roots of oxalis corniculata, *Fusarium proliferatum* BRL$_1$ and*Aspergillus fumigatus* TS$_1$. The fungal species after *in vitro* screening produced siderophore, phosphate solubilization activity, IAA and gibberellins. Endophytic fungal strain of *Aspergillus* isolated from the leaf of *Schima wallichii* was also found to possess good phosphate solubilizing and IAA producing ability (Sarbadhikary and Mandal, 2018).

The availability of contaminants in soil distorts the endophytic community structure, population and functions by destroying the microbe-rhizosphere interface which consequently affects colonization and microbe-host plant interactions. In light of this, the availability of xenobiotic-resistant strains involved in remediation activities positively impacts plant development and growth (Rajkumar et al., 2010). Endophytic fungal poses to be very essential in phytoremediation because their hyphal structure contributes to higher surface area to absorb toxic pollutants. The plant-microbe interaction permits the exploration of the different phytoremediation strategies (phytoextraction, phytoaccumulation and biotransformation) to detoxify contaminated environments in the most sustainable way (Stępniewska and Kuźniar, 2013; Nandy et al., 2020). The application of EF in the mineralization of noxious pollutants such as hydrocarbons, polychlorinated biphenyls, polycyclic aromatic hydrocarbons and metals is a novel and essential resource because of its diverse enzyme system. It has been established that some endophytic hydrocarbon degrading bacteria obtained from moss plant *Macromitrium* sp could help in the bioremediation of polluted environment. These organisms have ability to grow in a mineral salt medium (MSM) amended with naphthalene and phenanthrene. HPLC analysis revealed that each of the isolates degraded the PAH amendments more than 85%. Endophytic fungi can be a candidate for environmental sustainability through the degradation of xenobiotics by a wide range of catabolic enzymes. These enzymes are also responsible for mineralization of complex (macromolecules) compounds into simpler molecules or transformation of noxious substances to innocuous substances to enhance adaptability. Furthermore, the report highlighted that endophytic fungi are actively involved in phytoremediation since they promote plant growth and interact directly with plants owing to their unique type of plant-microbe interaction, are heavy metal resistant and also can modulate metal translocation and accumulation plants. In a more recent study, four selected EF, *Aspergillus* sp A31, *Curvularia geniculata* P1, *Westerdykella* sp. P71 and. *Lindgomycetaceae* P87, were investigated in an *in vitro* system for the bioaccumulation and bioremediation of mercuryand also growth enhancement of *Aeschynomene fluminensis* and *Zea mays*. Results revealed that up to 100% of mercury was removed from the culture medium by all endophytes and enhanced plant growth of both plants in substrate with or without metal was recorded. The authors observed increase in host plant biomass was associated with a decrease in the concentration of mercury in soil due to bioaccumulation of metal or possible volatilization (Pietro-Souza et al., 2020). Marín et al. (2018) also established the fact that endophytic fungi possess degradation capabilities. The *in vitro* study revealed *Verticullum* sp. and *Xylaria* sp.1 showed momentous potentials as hydrocarbon degraders by removing 99.6% of crude oil in a mineral salt based medium amended with crude oil. It has been established that plants have the ability to bioaccumulate heavy metals or hydrocarbon compounds from the soil which is referred to as bioremediation. However, such pollutants especially the hydrocarbon-related pollutants cannot be effectively degraded inside the plant which is greatly detrimental to food security and public health. Research have shown that some endophytic fungi have the ability to degrade pollutants *in planta*.

4.5   CONCLUSION AND FUTURE RECOMMENDATION

This chapter has provided detailed information on the application of biostimulants derived from beneficial microorganism and their utilization for increase in agricultural productivity. The merits

and demerits of the biostimulants derived from microorganisms were highlighted. The modes of action that biostimulants utilized in executing their role was also stated. The application of genetic engineering and mutation could lead to enhancement in the production of important strains that could lead to the production of highly effective biostimulant strains that could lead to increase in agricultural production. This biostimulant could be mass produced on cheap agricultural substrates when compared to synthetic media. The application of bioinformatics, genomics and proteomics will play a crucial role toward the identification of a gene that could enhance the agronomical trail of plant as well as led to increase in nutritional attributes of agricultural crops. Moreover, the application of metabolomics will reveal the presence of necessary metabolites of interests that could lead to increase in agricultural production through the prevention of abiotic and biotic stress such as agricultural pest and pathogens.

REFERENCES

Aamir, M., Rai, K. K., & Zehra, A. (2020). Microbial bioformulation-based plant biostimulants: A plausible approach toward next generation of sustainable agriculture Multi-stress tolerant PGPR *Bacillus xiamenensis* PM14 activating sugarcane (*Saccharum officinarum* L.) red rot disease resistance, microbial endophytes, functional biology and applications, pp. 195–225.

Abdelkrim, S., Jebara, S. H., Saadani, O., Abid, G., Taamalli, W., Zemni, H., Mannai, K., Louati, F., & Jebara, M. (2020). In situ effects of lathyrus sativus- PGPR to remediate and restore quality and fertility of PB and cd polluted soils. *Ecotoxicology and Environmental Safety*, *192*, 110260. https://doi.org/10.1016/j.ecoenv.2020.110260.

Abou-Shanab, R., Angle, J., Delorme, T., Chaney, R., Van Berkum, P., Moawad, H., Ghanem, K., & Ghozlan, H. (2003). Rhizobacterial effects on nickel extraction from soil and uptake by alyssum murale. *New Phytologist*, *158*(1), 219–224. https://doi.org/10.1046/j.1469-8137.2003.00721.x.

Abraham, J. & Silambarasan, S. (2014). Role of novel fungus Lasiodiplodia sp. JAS12 and plant growth promoting bacteria klebsiella pneumoniae JAS8 in mineralization of endosulfan and its metabolites. *Ecological Engineering*, *70*, 235–240. https://doi.org/10.1016/j.ecoleng.2014.05.029.

Abu-Elsaoud, A. M., Nafady, N. A., & Abdel-Azeem, A. M. (2017). Arbuscular mycorrhizal strategy for zinc mycoremediation and diminished translocation to shoots and grains in wheat. *PLoS One*, *12*(11), e0188220. https://doi.org/10.1371/journal.pone.0188220.

Adesemoye, A. O., Yuen, G., & Watts, D. B. (2017). Microbial Inoculants for optimized plant nutrient use in integrated pest and input management systems. *Probiotics and Plant Health*, 21–40. https://doi.org/10.1007/978-981-10-3473-2_2.

Aeron, A., Kumar, S., Pandey, P., & Maheshwari, D. K. (2011). Emerging role of plant growth promoting rhizobacteria in agrobiology. In: Maheshwari, D. K. (Ed.), *Bacteria in Agrobiology: Crop Ecosystems*, pp. 1–36. https://doi.org/10.1007/978-3-642-18357-7_1.

Ajuzieogu, C., Ibiene, A., Okechukwu, H. (2015). Laboratory study on influence of plant growth promoting rhizobacteria (PGPR) on growth response and tolerance of *Zea mays* to petroleum hydrocarbon. *African Journal of Biotechnology*, 14, 2949–2956. https://doi.org/10.5897/AJB2015.14549.

Akhdiya, A., Wahyudi, A. T., Munif, A., & Darusman, L. K. (2014). Characterization of an endophytic bacterium G062 isolate with beneficial traits. *HAYATI Journal of Biosciences*, *21*(4), 187–196. https://doi.org/10.4308/hjb.21.4.187.

Akinsemolu, A. A. (2018). The role of microorganisms in achieving the sustainable development goals. *Journal of Cleaner Production*, 182, 139–155. https://doi.org/10.1016/j.jclepro.2018.02.081.

Akintokun, A.K., Kofoworola, Taiwo, M.O., Oluwambe (2016). Biocontrol potentials of individual specie of rhizobacteria and their consortium against phytopathogenic *Fusarium oxysporum* and *Rhizoctonia solani*. *International Journal of Scientific Research in Environmental Sciences*, 4, 219–227. https://doi.org/10.12983/ijsres-2016-p0219-0227.

Alam, M., Khaliq, A., Sattar, A., Shukla, R. S., Anwar, M., & Dharni, S. (2011). Synergistic effect of arbuscular mycorrhizal fungi and bacillus subtilis on the biomass and essential oil yield of rose-scented geranium (Pelargonium graveolens). *Archives of Agronomy and Soil Science*, *57*(8), 889–898. https://doi.org/10.1080/03650340.2010.498013.

Alavi, P., Starcher, M. R., Zachow, C., Müller, H., & Berg, G. (2013). Root-microbe systems: The effect and mode of interaction of stress protecting agent (SPA) Stenotrophomonas rhizophila DSM14405T. *Frontiers in Plant Science*, 4. https://doi.org/10.3389/fpls.2013.00141.

Aldesuquy, H. S., Mansour, F. A., & Abo-Hamed, S. A. (1998). Effect of the culture filtrates of Streptomyces on growth and productivity of wheat plants. *Folia Microbiologica*, *43*(5), 465–470. https://doi.org/10.1007/bf02820792.

Aloo, B., Makumba, B., & Mbega, E. (2019). The potential of bacilli rhizobacteria for sustainable crop production and environmental sustainability. *Microbiological Research*, *219*, 26–39. https://doi.org/10.1016/j.micres.2018.10.011.

Alori, E. T., Glick, B. R., & Babalola, O. O. (2017). Microbial phosphorus solubilization and its potential for use in sustainable agriculture. *Frontiers in Microbiology*, *8*. https://doi.org/10.3389/fmicb.2017.00971.

Amara, U., Khalid, R., & Hayat, R. (2015). Soil bacteria and phytohormones for sustainable crop production. *Bacterial Metabolites in Sustainable Agroecosystem*, 87–103. https://doi.org/10.1007/978-3-319-24654-3_5.

https://doi.org/10.1016/j.plaphy.2020.04.016.

Andreolli, M., Zapparoli, G., Angelini, E., Lucchetta, G., Lampis, S., & Vallini, G. (2019). Pseudomonas protegens MP12: A plant growth-promoting endophytic bacterium with broad-spectrum antifungal activity against grapevine phytopathogens. *Microbiological Research*, *219*, 123–131. https://doi.org/10.1016/j.micres.2018.11.003.

Apone, F., Tito, A., Carola, A., Arciello, S., Tortora, A., Filippini, L., Monoli, I., Cucchiara, M., Gibertoni, S., Chrispeels, M. J., & Colucci, G. (2010). A mixture of peptides and sugars derived from plant cell walls increases plant defense responses to stress and attenuates ageing-associated molecular changes in cultured skin cells. *Journal of Biotechnology*, *145*(4), 367–376. https://doi.org/10.1016/j.jbiotec.2009.11.021.

Atakan, A., Özgönen Özkaya, H., & Erdoğan, O. (2018). Effects of Arbuscular mycorrhizal fungi (AMF) on heavy metal and salt stress. *Turkish Journal of Agriculture - Food Science and Technology*, *6*(11), 1569. https://doi.org/10.24925/turjaf.v6i11.1569-1574.1992.

Ayomide, F., & Olubukola, B. (2020). Elucidating mechanisms of endophytes used in plant protection and other bioactivities with multifunctional prospects. *Frontiers in Bioengineering and Biotechnology*, *8*, 467. https://doi.org/10.3389/fbioe.2020.00467.

Bakli, M. & Zenasni, A. (2019). Isolation of fluorescent *pseudomonas* spp. Strains from rhizosphere agricultural soils and assessment of their role in plant growth and phytopathogen bio control. *Research Journal of Agricultural Science*, *51*(1), 20–29.

Banerjee, S. & Dutta, S. (2019). Plant growth promoting activities of a fungal strain *Penicillium commune* MCC 1720 and it's effect on growth of black gram. *The Pharma Innovation Journal*, *8*(12), 121–127.

Banik, A., Mukhopadhaya, S. K., & Dangar, T. K. (2016). Characterization of N2-fixing plant growth promoting endophytic and epiphytic bacterial community of Indian cultivated and wild rice (*Oryza* spp.) genotypes. *Planta*, *243*(3), 799–812.

Baoune, H., Ould El Hadj-Khelil, A., Pucci, G., Sineli, P., Loucif, L., & Polti, M. A. (2018). Petroleum degradation by endophytic Streptomyces spp. isolated from plants grown in contaminated soil of southern Algeria. *Ecotoxicology and Environmental Safety*, *147*, 602–609. https://doi.org/10.1016/j.ecoenv.2017.09.013.

Barman, M., Shukla, L. M., Datta, S. P., & Rattan, R. K. (2014). Effect of applied lime and boron on the availability of nutrients in an acid soil. *Journal of Plant Nutrition*, *37*(3), 357–373.

Bashan, Y. & de Bashan, L. E. (2005). Bacteria: Plant growth-promoting. In: Hillel, D. (Ed.), *Encyclopedia of Soils in the Environment*, vol. 1. Elsevier, Oxford, pp. 103–115.

Bashan, Y., & de Bashan, L. E. (2010). How the plant growth-promoting bacterium Azospirillum promotes plant growth—a critical assessment. *Advances in Agronomy*, *108*, 77–136.

Batista, B. D., Lacava, P. T., Ferrari, A., Teixeira-Silva, N. S., Bonatelli, M. L., Tsui, S., Mondin, M., Kitajima, E. W., Pereira, J. O., Azevedo, J. L., & Quecine, M. C. (2018). Screening of tropically derived, multi-trait plant growth- promoting rhizobacteria and evaluation of corn and soybean colonization ability. *Microbiological Research*, *206*, 33–42. https://doi.org/10.1016/j.micres.2017.09.007.

Battacharyya, D., Babgohari, M. Z., Rathor, P., & Prithiviraj, B. (2015). Seaweed extracts as biostimulants in horticulture. *Scientia Horticulturae*, *196*, 39–48. https://doi.org/10.1016/j.scienta.2015.09.012.

Berbara, R. L. L. & García, A.C. (2014). Humic substances and plant defense metabolism. In: Ahmad, P. & Wani, M. R. (Eds.), *Physiological Mechanisms and Adaptation Strategies in Plants under Changing Environment*, vol. 1. Springer Science+Business Media, New York, pp. 297–319.

Berg, G. (2009). Plant–microbe interactions promoting plant growth and health: Perspectives for controlled use of microorganisms in agriculture. *Applied Microbiology and Biotechnology*, *84*(1), 11–18. https://doi.org/10.1007/s00253-009-2092-7.

Bhardwaj, D., Ansari, M. W., & Sahoo, R.K. (2014). Biofertilizers function as key player in sustainable agriculture by improving soil fertility, plant tolerance and crop productivity. *Microbial Cell Factories*, *13*, 66. https://doi.org/10.1186/1475-2859-13-66.

Bhattacharyya, P. N. & Jha, D. K. (2012). Plant growth-promoting rhizobacteria (PGPR): Emergence in agriculture. *World Journal of Microbiology and Biotechnology*, *28*(4), 1327–1350. https://doi.org/10.1007/s11274-011-0979-9.

Bilal, L., Asaf, S., Hamayun, M., Gul, H., Iqbal, A., Ullah, I., Lee, I., & Hussain, A. (2018). Plant growth promoting endophytic fungi Asprgillus fumigatus TS1 and Fusarium proliferatum BRL1 produce gibberellins and regulates plant endogenous hormones. *Symbiosis*, *76*(2), 117–127. https://doi.org/10.1007/s13199-018-0545-4.

Bisht, S., Pandey, P., Kaur, G., Aggarwal, H., Sood, A., Sharma, S., Kumar, V., & Bisht, N. (2014). Utilization of endophytic strain bacillus sp. SBER3 for biodegradation of polyaromatic hydrocarbons (PAH) in soil model system. *European Journal of Soil Biology*, *60*, 67–76. https://doi.org/10.1016/j.ejsobi.2013.10.009.

Bulgari, R., Cocetta, G., Trivellin, A., Vernieri, P., & Ferrante, A. (2015). Biostimulants and crop responses: A review. *Biological Agriculture and Horticulture*, 31, 1–17. https://doi.org/10.1080/01448765.2014.964649.

Calvo, P., Nelson, L., & Kloepper, J. W. (2014). Agricultural uses of plant biostimulants. *Plant and Soil*, *383*(1–2), 3–41. https://doi.org/10.1007/s11104-014-2131-8.

Carvalhais, L. C., Dennis, P. G., Badri, D. V., Tyson, G. W., Vivanco, J. M., & Schenk, P. M. (2013). Activation of the Jasmonic acid plant defence pathway alters the composition of rhizosphere bacterial communities. *PLoS One*, *8*(2), e56457. https://doi.org/10.1371/journal.pone.0056457.

Caulier, S., Gillis, A., Colau, G., Licciardi, F., Liépin, M., Desoignies, N., Modrie, P., Legrève, A., Mahillon, J., & Bragard, C. (2018). Versatile antagonistic activities of soil-borne bacillus spp. and pseudomonas spp. against Phytophthora infestans and other potato pathogens. *Frontiers in Microbiology*, 9. https://doi.org/10.3389/fmicb.2018.00143.

Cavani, L., Ter Halle, A., Richard, C., & Ciavatta, C. (2006). Photosensitizing properties of protein hydrolysate-based fertilizers. *Journal of Agricultural and Food Chemistry*, *54*(24), 9160–9167. https://doi.org/10.1021/jf0624953.

Cavello, I. A., Crespo, J. M., García, S. S., Zapiola, J. M., Luna, M. F., & Cavalitto, S. F. (2015). Plant growth promotion activity of Keratinolytic fungi growing on a recalcitrant waste known as "Hair waste". *Biotechnology Research International*, *2015*, 1–10. https://doi.org/10.1155/2015/952921.

Çelik, H., Katkat, A. V., Aşık, B. B., & Turan, M. A. (2010). Effect of foliar-applied humic acid to dry weight and mineral nutrient uptake of maize under calcareous soil conditions. *Communications in Soil Science and Plant Analysis*, *42*(1), 29–38. https://doi.org/10.1080/00103624.2011.528490.

Cervantes-Gamez, R. G., Bueno-Ibarra, M.A., Cruz-Mendivil, A., Calderon-Vazquez, C.L., Ramirez-Douriet, C.M., Maldonado-Mendoza, I.E., et al. (2015). Arbuscular mycorrhizal symbiosis-induced expression changes in Solanum lycopersicum leaves revealed by RNA-seqanalysis. *Plant Molecular Biology Reporter*, *23*, 1–14.

Chakraborty, A. P., Dey, P., Chakraborty, B., Roy, S., & Chakraborty, U. (2011). Plant growth promotion and amelioration of salinity stress in crop plants by a salt-tolerant bacterium. *Recent Research Science Technology*, *3*, 61–70.

Chang, P., Gerhardt, K. E., Huang, X., Yu, X., Glick, B. R., Gerwing, P. D., & Greenberg, B. M. (2014). Plant growth-promoting bacteria facilitate the growth of Barley and oats in salt-impacted soil: Implications for Phytoremediation of saline soils. *International Journal of Phytoremediation*, *16*(11), 1133–1147. https://doi.org/10.1080/15226514.2013.821447.

Chang, Y., Reddy, M. V., Umemoto, H., Kondo, S., & Choi, D. (2020). Biodegradation of alkylphenols by rhizosphere microorganisms isolated from the roots of hosta undulata. *Journal of Environmental Chemical Engineering*, *8*(3), 103771. https://doi.org/10.1016/j.jece.2020.103771.

Chellaiah, E. R. (2018). Cadmium (heavy metals) bioremediation by pseudomonas aeruginosa: A minireview. *Applied Water Science*, *8*(6). https://doi.org/10.1007/s13201-018-0796-5.

Chen, B., Shen, J., Zhang, X., Pan, F., Yang, X., & Feng, Y. (2014). The endophytic bacterium, Sphingomonas SaMR12, improves the potential for zinc Phytoremediation by its host, sedum alfredii. *PLoS One*, *9*(9), e106826. https://doi.org/10.1371/journal.pone.0106826.

Chenchouni, H., Mekahlia, M. N., & Beddiar, A. (2020). Effect of inoculation with native and commercial arbuscular mycorrhizal fungi on growth and mycorrhizal colonization of olive (Olea europaea L.). *Scientia Horticulturae*, *261*, 108969. https://doi.org/10.1016/j.scienta.2019.108969.

Chenniappan, C., Narayanasamy, M., Daniel, G., Ramaraj, G., Ponnusamy, P., Sekar, J., & Vaiyapuri Ramalingam, P. (2019). Biocontrol efficiency of native plant growth promoting rhizobacteria against rhizome rot disease of turmeric. *Biological Control*, *129*, 55–64. https://doi.org/10.1016/j.biocontrol.2018.07.002.

Chowdhury, R.A., Bagchi, A., & Chandan, S. (2016). Isolation and characterization of plant growth promoting rhizobacteria (PGPR) from agricultural field and their potential role on germination and growth of spinach (*Spinacia oleracea* L.) plants. *International Journal of Current Agricultural Sciences*, *6*(10), 128–131.

Chuks, O., Chibuzor, K., Akpi, N., Unah, K., & Unah, U. (2019). Plant Growth Promoting Rhizobacteria (PGPR): A Novel Agent for Sustainable Food Production, *American Journal of Agricultural and Biological Science, 14,* 35–54. 10.3844/ajabssp.2019.35.54.

Colla, G., Hoagland, L., Ruzzi, M., Cardarelli, M., Bonini, P., Canaguier, R., & Rouphael, Y. (2017). Biostimulant action of protein hydrolysates: Unraveling their effects on plant physiology and microbiome. *Frontiers in Plant Science, 8.* https://doi.org/10.3389/fpls.2017.02202.

Colla, G., Nardi, S., Cardarelli, M., Ertani, A., Lucini, L., Canaguier, R., & Rouphael, Y. (2015). Protein hydrolysates as biostimulants in horticulture. *Scientia Horticulturae, 196,* 28–38. https://doi.org/10.1016/j.scienta.2015.08.037.

Colla, G., & Rouphael, Y. (2015). Biostimulants in horticulture. *Scientia Horticulturae, 196,* 1–134.

Compant, S., Saikkonen, K., Mitter, B., Campisano, A., & Mercado-Blanco, J. (2016). Editorial special issue: Soil, plants and endophytes. *Plant and Soil, 405*(1–2), 1–11. https://doi.org/10.1007/s11104-016-2927-9.

Correa, A., Pacheco, S., Mechaly, A. E., Obal, G., Béhar, G., Mouratou, B., Oppezzo, P., Alzari, P. M., & Pecorari, F. (2014). Potent and specific inhibition of glycosidases by small artificial binding proteins (Affitins). *PLoS One, 9*(5), e97438. https://doi.org/10.1371/journal.pone.0097438.

Craigie, J. S. (2011). Seaweed extract stimuli in plant science and agriculture. *Journal of Applied Phycology, 23*(3), 371–393. https://doi.org/10.1007/s10811-010-9560-4.

Cui, J., Bai, L., Liu, X., Jie, W., & Cai, B., (2018). Arbuscular mycorrhizal fungal communities in the rhizosphere of a continuous cropping soybean system at the seedling stage. *Brazilian Journal of Microbiology, 49*(2), 240–247, https://doi.org/10.1016/j.bjm.2017.03.017.

Das, A. J. & Kumar, R. (2016). Bioremediation of petroleum contaminated soil to combat toxicity on Withania somnifera through seed priming with biosurfactant producing plant growth promoting rhizobacteria. *Journal of Environmental Management, 174,* 79–86. https://doi.org/10.1016/j.jenvman.2016.01.031.

David, P., Raj, R. S., Linda, R., & Rhema, S. B. (2014). Molecular characterization of phosphate solubilizing bacteria (PSB) and plant growth promoting rhizobacteria (PGPR) from pristine soils. *International Journal of Innovative Science Engineering and Technology, 1,* 317–324.

De Lucia, B. & Vecchietti, L. (2012). Type of bio-stimulant and application method effects on stem quality and root system growth in L.A. Lily. *European Journal of Horticultural Science, 77,* 10–15.

de-Bashan, L. E., Hernandez, J., & Bashan, Y. (2012). The potential contribution of plant growth-promoting bacteria to reduce environmental degradation – A comprehensive evaluation. *Applied Soil Ecology, 61,* 171–189. https://doi.org/10.1016/j.apsoil.2011.09.003.

Díaz Herrera, S., Grossi, C., Zawoznik, M., & Groppa, M. D. (2016). Wheat seeds harbour bacterial endophytes with potential as plant growth promoters and biocontrol agents of Fusarium graminearum. *Microbiological Research, 186–187,* 37–43. https://doi.org/10.1016/j.micres.2016.03.002.

Dong, R., Gu, L., Guo, C., Xun, F., & Liu, J. (2014). Effect of PGPR Serratia marcescens BC-3 and AMF Glomus intraradices on phytoremediation of petroleum contaminated soil. *Ecotoxicology, 23*(4), 674–680. https://doi.org/10.1007/s10646-014-1200-3.

Dorjey, S., Dolkar, D., & Sharma, R. (2017). Plant growth promoting rhizobacteria pseudomonas: A review. *International Journal of Current Microbiology and Applied Sciences, 6,* 1335–1344. https://doi.org/10.20546/ijcmas.2017.607.160.

du Jardin, P. (2012). The science of plant biostimulants-a bibliographic analysis. Contract 30-CE0455515/00-96, ad hoc study on bio-stimulants products.

du Jardin, P. (2015). Plant biostimulants: Definition, concept, main categories and regulation. *Scientia Horticulturae, 196,* 3–14. https://doi.org/10.1016/j.scienta.2015.09.021.

Dubey, R. K., Tripathi, V., Prabha, R., Chaurasia, R., Singh, D. P., Rao, C. S., El-Keblawy, A., & Abhilash, P. C. (2020). Belowground microbial communities: Key players for soil and environmental sustainability. In: *Unravelling the Soil Microbiome.* SpringerBriefs in Environmental Science. Springer, Cham, pp. 5–22. https://doi.org/10..1007/978-3-030-15516-2_2.

Dudeja, S. S., Giri, R., Saini, R., Suneja-Madan, P., & Kothe, E. (2012). Interaction of endophytic microbes with legumes. *Journal of Basic Microbiology, 52*(3), 248–260. https://doi.org/10.1002/jobm.201100063.

Edoardo, P., Silvia, P., Nicolet, S., Ilenia, C., Gabriella, F., d, & Riccardo, S. (2013). Rhizosphere microbial diversity as influenced by humic substance amendments and chemical composition of rhizodeposits. *Journal of Geochemical Exploration, 129,* 82–94. https://doi.org/10.1016/j.gexplo.2012.10.006.

Ekwuabu, C. B., Chikere, C. B., & Akaranta, O. (2016). Effect of different nutrient amendments on eco-restoration of a crude oil polluted soil. *SPE African Health, Safety, Security, Environment, and Social Responsibility Conference and Exhibition,* 1–17. https://doi.org/10.2118/183608-ms.

Ertani, A., Cavani, L., Pizzeghello, D., Brandellero, E., Altissimo, A., Ciavatta, C., & Nardi, S. (2009). Biostimulant activity of two protein hydrolyzates in the growth and nitrogen metabolism of maize seedlings. *Journal of Plant Nutrition and Soil Science*, *172*(2), 237–244. https://doi.org/10.1002/jpln.200800174.

Ertani, A., Pizzeghello, D., Altissimo, A., & Nardi, S. (2013a). Use of meat hydrolyzate derived from tanning residues as plant biostimulant for hydroponically grown maize. *Journal of Plant Nutrition and Soil Science*, *176*(2), 287–295. https://doi.org/10.1002/jpln.201200020.

Ertani, A., Schiavon, M., Muscolo, A., & Nardi, S. (2013b). Alfalfa plant-derived biostimulant stimulate short-term growth of salt stressed zea mays L. plants. *Plant and Soil*, *364*(1–2), 145–158. https://doi.org/10.1007/s11104-012-1335-z.

Eshaghi, E., Nosrati, R., Owlia, P., Malboobi, M. A., Ghaseminejad, P., & Ganjali, M. R. (2019). Zinc solubilization characteristics of efficient siderophore-producing soil bacteria. *Iranian Journal of Microbiology*, *11*(5), 419–430. https://doi.org/10.18502/ijm.v11i5.1961.

e Silva, M. C. S, Polonio, J. C., Quecine, M. C., de Almeida, T. T., Bogas, A. C., Pamphile, J. A, Pereira, J. O., et al. (2016). Endophytic cultivable bacterial community obtained from the *Paullinia cupana* seed in Amazonas and Bahia regions and its antagonistic effects against *Colletotrichum gloeosporioides*. *Microbial Pathogenesis*, 98, 16–22.

Fecih, T. & Baoune, H. (2019). Arbuscular mycorrhizal fungi remediation potential of organic and inorganic compounds. In: Arora, P. (Ed.), *Microbial Technology for the Welfare of Society*. Microorganisms for Sustainability, vol. 17. Springer, Singapore, pp. 247–257.

Feng, N., Yu, J., Zhao, H., Cheng, Y., Mo, C., Cai, Q., Li, Y., Li, H., & Wong, M. (2017). Efficient phytoremediation of organic contaminants in soils using plant–endophyte partnerships. *Science of the Total Environment*, *583*, 352–368. https://doi.org/10.1016/j.scitotenv.2017.01.075.

Fialho de Oliveira, M., Germano da Silva, M., & Van Der Sand, S. T. (2010). Anti-phytopathogen potential of endophytic actinobacteria isolated from tomato plants (Lycopersicon esculentum) in southern Brazil, and characterization of Streptomyces sp. R18(6), a potential biocontrol agent. *Research in Microbiology*, *161*(7), 565–572. https://doi.org/10.1016/j.resmic.2010.05.008.

Firdous, J., Lathif, N. A., Mona, R. & Muhamad, N. (2019). Endophytic bacteria and their potential application in agriculture: A review. *Indian Journal of Agricultural Research*, *53*(1), 1–7.

Fouzia, A., Allaoua, S., Hafsa, C. S., & Mostef, G. (2015). Plant growth promoting and antagonistic traits of indigenous fluorescent *Pseudomonas* spp. isolated from Wheat rhizosphere and *A. Halimus* endosphere. *European Scientific Journal*, *11*(24), 129–148.

Fu, W., Xu, M., Sun, K., Chen, X., Dai, C., & Jia, Y. (2020). Remediation mechanism of endophytic fungus Phomopsis liquidambaris on phenanthrene in vivo. *Chemosphere*, *243*, 125305. https://doi.org/10.1016/j.chemosphere.2019.125305.

Fuell, C., Elliott, K. A., Hanfrey, C. C., Franceschetti, M., & Michael, A. J. (2010). Polyamine biosynthetic diversity in plants and algae. *Plant Physiology and Biochemistry*, *48*(7), 513–520. https://doi.org/10.1016/j.plaphy.2010.02.008.

Gaiero, J.R., McCall, C.A., Thompson, K.A., Day, N.J., Best, A.S., & Dunfield, K.E. (2013). Inside the root microbiome: bacterial root endophytes and plant growth promotion. *American Journal of Botany*, *100*(9), 1738–1750.

Gamalero, E., & Glick, B. R. (2011). Mechanisms used by plant growth-promoting bacteria. In: Maheshwari, D. K. (Ed.), *Bacteria in Agrobiology: Plant Nutrient Management*. Springer, Berlin Heidelberg, pp. 17–46.

Gamez, R., Cardinale, M., Montes, M., Ramirez, S., Schnell, S., & Rodriguez, F. (2019). Screening, plant growth promotion and root colonization pattern of two rhizobacteria (Pseudomonas fluorescens Ps006 and bacillus amyloliquefaciens Bs006) on banana CV. Williams (Musa acuminata Colla). *Microbiological Research*, *220*, 12–20. https://doi.org/10.1016/j.micres.2018.11.006.

Gao, M., Zhou, J., Wang, E., Chen, Q., Xu, J., & Sun, J. (2015). Multiphasic characterization of a plant growth promoting bacterial strain, Burkholderia sp. 7016 and its effect on tomato growth in the field. *Journal of Integrative Agriculture*, *14*(9), 1855–1863. https://doi.org/10.1016/s2095-3119(14)60932-1.

Georgieva, T., Evstatieva, Y., Savov, V., Bratkova, S., & Nikolova, D. (2018). Assessment of plant growth promoting activities of five rhizospheric pseudomonas strains. *Biocatalysis and Agricultural Biotechnology*, *16*, 285–292. https://doi.org/10.1016/j.bcab.2018.08.015.

Glick, B. R. (2014). Bacteria with ACC deaminase can promote plant growth and help to feed the world. *Microbiological Research*, *169*(1), 30–39. https://doi.org/10.1016/j.micres.2013.09.009.

Goswami, D., Dhandhukia, P., Patel, P., & Thakker, J. N. (2014). Screening of PGPR from saline desert of Kutch: Growth promotion in arachis hypogea by bacillus licheniformis A2. *Microbiological Research*, *169*(1), 66–75. https://doi.org/10.1016/j.micres.2013.07.004.

Goswami, D., Thakker, J. N., & Dhandhukia, P. C. (2016). Portraying mechanics of plant growth promoting rhizobacteria (PGPR): A review. *Cogent Food & Agriculture*, 2(1). https://doi.org/10.1080/23311932.2015.1127500.

Goudjal, Y., Zamoum, M., Meklat, A., Sabaou, N., Mathieu, F., & Zitouni, A. (2015). Plant-growth-promoting potential of endosymbiotic actinobacteria isolated from sand truffles (Terfezia leonis Tul.) of the Algerian Sahara. *Annals of Microbiology*, 66(1), 91–100. https://doi.org/10.1007/s13213-015-1085-2.

Govindasamy, V., Senthilkumar, M., Magheshwaran, V., Kumar, U., Bose, P., Sharma, V., & Annapurna, K. (2011). *Bacillus* and *Paenibacillus* spp.: Potential PGPR for sustainable agriculture. In: Maheshwari, D. K. (Ed.), *Plant Growth and Health Promoting Bacteria*. Springer-Verlag, Berlin, pp. 333–364.

Gowtham, H., Murali, M., Singh, S. B., Lakshmeesha, T., Narasimha Murthy, K., Amruthesh, K., & Niranjana, S. (2018). Plant growth promoting rhizobacteria- Bacillus amyloliquefaciens improves plant growth and induces resistance in chilli against anthracnose disease. *Biological Control*, 126, 209–217. https://doi.org/10.1016/j.biocontrol.2018.05.022.

Grabowska, A., Kunicki, E., Sękara, A., Kalisz, A., & Wojciechowska, R. (2012). The effect of cultivar and biostimulant treatment on the carrot yield and its quality. *Vegetable Crops Research Bulletin*, 77(1), 37–48. https://doi.org/10.2478/v10032-012-0014-1.

Grobelak, A., Kokot, P., Hutchison, D., Grosser, A., & Kacprzak, M. (2018). Plant growth-promoting rhizobacteria as an alternative to mineral fertilizers in assisted bioremediation - Sustainable land and waste management. *Journal of Environmental Management*, 227, 1–9. https://doi.org/10.1016/j.jenvman.2018.08.075.

Guarino, C., Marziano, M., Tartaglia, M., Prigioniero, A., Postiglione, A., Scarano, P., & Sciarrillo, R. (2020). Poaceae with PGPR bacteria and Arbuscular Mycorrhizae partnerships as a model system for plant microbiome manipulation for Phytoremediation of petroleum hydrocarbons contaminated agricultural soils. *Agronomy*, 10(4), 547. https://doi.org/10.3390/agronomy10040547.

Guo, D., Yuan, C., Luo, Y., Chen, Y., Lu, M., Chen, G., Ren, G., Cui, C., Zhang, J., & An, D. (2020). Biocontrol of tobacco Black shank disease (Phytophthora nicotianae) by bacillus velezensis Ba168. *Pesticide Biochemistry and Physiology*, 165, 104523. https://doi.org/10.1016/j.pestbp.2020.01.004.

Gupta, A., Singh, S. K., Singh, V. K., Singh, M. K., Modi, A., Zhimo, V. Y., Singh, A. V., & Kumar, A. (2020a). Endophytic microbe approaches in bioremediation of organic pollutants. *Microbial Endophytes*, 157–174. https://doi.org/10.1016/b978-0-12-818734-0.00007-3.

Gupta, G., Panwar, J., & Jha, P. N. (2013). Natural occurrence of pseudomonas aeruginosa, a dominant cultivable diazotrophic endophytic bacterium colonizing pennisetum glaucum (L.) R. BR. *Applied Soil Ecology*, 64, 252–261. https://doi.org/10.1016/j.apsoil.2012.12.016.

Gupta, G., Parihar, S.S., Kumar, A.N., Kumar, S.S., & Singh, V. (2015). Plant growth promoting rhizobacteria (PGPR): Current and future prospects for development of sustainable agriculture. *Journal of Microbiology and Biochemistry Technology*, 7, 2.

Gupta, P., Kumar, V., Usmani, Z., Rani, R., Chandra, A., & Gupta, V. K. (2020b). Implications of plant growth promoting klebsiella sp. CPSB4 and Enterobacter sp. CPSB49 in luxuriant growth of tomato plants under chromium stress. *Chemosphere*, 240, 124944. https://doi.org/10.1016/j.chemosphere.2019.124944.

Gutiérrez-Mañero, F. J., Ramos-Solano, B., Probanza, A., Mehouachi, J., R. Tadeo, F., & Talon, M. (2001). The plant-growth-promoting rhizobacteria bacillus pumilus and bacillus licheniformis produce high amounts of physiologically active gibberellins. *Physiologia Plantarum*, 111(2), 206–211. https://doi.org/10.1034/j.1399-3054.2001.1110211.x.

Habibi, S., Djedidi, S., Prongjunthuek, K., Mortuza, M. F., Ohkama-Ohtsu, N., Sekimoto, H., & Yokoyoma, T. (2014). Physiological and genetic characterization of rice nitrogen fixer PGPR isolated from rhizosphere soils of different crops. *Plant and Soil*, 379(1–2), 51–66. https://doi.org/10.1007/s11104-014-2035-7.

Haidar, B., Ferdous, M., Fatema, B., Ferdous, A. S., Islam, M. R., & Khan, H. (2018). Population diversity of bacterial endophytes from jute (Corchorus olitorius) and evaluation of their potential role as bioinoculants. *Microbiological Research*, 208, 43–53. https://doi.org/10.1016/j.micres.2018.01.008.

Halpern, M., Bar-Tal, A., Ofek, M., Minz, D., Muller, T., & Yermiyahu, U. (2015). The use of biostimulants for enhancing nutrient uptake. *Advances in Agronomy*, 141–174. https://doi.org/10.1016/bs.agron.2014.10.001.

Hameed, A., Egamberdieva, D., Abd-Allah, E.F., Hashem, A., Kumar, A., & Ahmad, P. (2014). Salinity stress and arbuscular mycorrhizal symbiosis in plants. In: Miransari, M. (Ed.), *Use of Microbes for the Alleviation of Soil Stresses*, vol. 1. Springer, New York, pp. 139–159.

Han, D., Wang, L., & Luo, Y. (2018). Isolation, identification, and the growth promoting effects of two antagonistic actinomycete strains from the rhizosphere of Mikania micrantha Kunth. *Microbiological Research*, 208, 1–11. https://doi.org/10.1016/j.micres.2018.01.003.

Hassan, M., McInroy, J., & Kloepper, J. (2019). The interactions of Rhizodeposits with plant growth-promoting Rhizobacteria in the rhizosphere: A review. *Agriculture*, *9*(7), 142. https://doi.org/10.3390/agriculture 9070142.

Hassen, W., Neifar, M., Cherif, H., Najjari, A., Chouchane, H., Driouich, R. C., Salah, A., Naili, F., Mosbah, A., Souissi, Y., Raddadi, N., Ouzari, H. I., Fava, F., & Cherif, A. (2020). Pseudomonas rhizophila S211, a new plant growth-promoting Rhizobacterium with potential in pesticide-bioremediation. *Frontiers in Microbiology*, *9*. https://doi.org/10.3389/fmicb.2018.00034.

Hayat, R., Ali, S., Amara, U., Khalid, R., & Ahmed, I. (2010). Soil beneficial bacteria and their role in plant growth promotion: A review. *Annals of Microbiology*, *60*(4), 579–598. https://doi.org/10.1007/s13213-010-0117-1.

Hernández-León, R., Rojas-Solís, D., Contreras-Pérez, M., Orozco-Mosqueda, M. D., Macías-Rodríguez, L. I., Reyes-de la Cruz, H., Valencia-Cantero, E., & Santoyo, G. (2015). Characterization of the antifungal and plant growth-promoting effects of diffusible and volatile organic compounds produced by pseudomonas fluorescens strains. *Biological Control*, *81*, 83–92. https://doi.org/10.1016/j.biocontrol.2014.11.011.

Hussain, A., Arshad, M., Zahir, Z.A. & Asghar, M. (2015). Prospects of zinc solubilizing bacteria for enhancing growth of maize. *Pakistan Journal of Agricultural Sciences*, *52*, 915–922.

Ijaz, A., Imran, A., Anwar ul Haq, M., Khan, Q. M., & Afzal, M. (2016). Phytoremediation: recent advances in plant-endophytic synergistic interactions. *Plant and Soil*, *405*(1), 179–195.

Ikeda, A. C., Bassani, L. L., Adamoski, D., Stringari, D., Cordeiro, V. K., Glienke, C., Steffens, M. B., Hungria, M., & Galli-Terasawa, L. V. (2013). Morphological and genetic characterization of endophytic bacteria isolated from roots of different maize genotypes. *Microbial Ecology*, *65*(1), 154–160. https://doi.org/10.1007/s00248-012-0104-0.

Islam, S., Akanda, A. M., Prova, A., Islam, M. T., & Hossain, M. M. (2016). Isolation and identification of plant growth promoting Rhizobacteria from cucumber rhizosphere and their effect on plant growth promotion and disease suppression. *Frontiers in Microbiology*, *6*, 1–12. https://doi.org/10.3389/fmicb.2015.01360.

Jaber, L.R., & Enkerli, J. (2017). Fungal entomopathogens as endophytes: can they promote plant growth?. *Biocontrol Science and Technology*, *27*(1), 28–41.

Jha, C. K., & Saraf, M. (2015). Plant growth promoting rhizobacteria (PGPR): A review. *E3 Journal of Agricultural Research and Development*, *5*, 108–119.

Jimtha, J. C., Smitha, P. V., Anisha, C., Deepthi, T., Meekha, G., Radhakrishnan, E. K., Gayatri, G. P., & Remakanthan, A. (2014). Isolation of endophytic bacteria from embryogenic suspension culture of banana and assessment of their plant growth promoting properties. *Plant Cell, Tissue and Organ Culture (PCTOC)*, *118*(1), 57–66. https://doi.org/10.1007/s11240-014-0461-0.

Joshi, D., Negi, G., Vaid, S., & Sharma, A. (2013). Enhancement of wheat growth and Zn content in grains by zinc solubilizing bacteria. *International Journal of Agriculture, Environment and Biotechnology*, *6*(3), 363. https://doi.org/10.5958/j.2230-732x.6.3.004.

Kamran, S., Shahid, I., Baig, D. N., Rizwan, M., Malik, K. A., & Mehnaz, S. (2017). Contribution of zinc solubilizing bacteria in growth promotion and zinc content of wheat. *Frontiers in Microbiology*, *8*. https://doi.org/10.3389/fmicb.2017.02593.

Kannan, R., Damodaran, T., Nagaraja, A., & Umamaheswari, S. (2018). Salt tolerant polyembryonic mango rootstock (ML-2 and GPL-1): A putative role of endophytic bacteria by using BOX-PCR. *Indian Journal of Agricultural Research*, *52*, 419–423. https://doi.org/10.18805/ijare.a-5001.

Kannangara, S., Ambadeniya, P., Undugoda, L., & Abeywickrama, K. (2016). Polyaromatic hydrocarbon degradation of moss endophytic fungi isolated from *Macromitrium* sp. in Sri Lanka. *Journal of Agricultural Science and Technology A*, *6*, 171–182. https://doi.org/10.17265/2161-6256/2016.03.004

Kanse, O.S., Whitelaw-Weckert, M., Kadam, T.A., Bhosale, H.J. (2015). Phosphate solubilization by stress-tolerant soil fungus Talaromyces funiculosus SLS8 isolated from the Neem rhizosphere. *Annals of Microbiology*, *65*(1), 85–93.

Kelleher, B.P., Simpson, A.J. (2006). Humic substances in soils: Are they really chemically distinct? *Environmental Science & Technology*, *40*(15), 4605–4611. https://doi.org/10.1021/es0608085. PMID: 16913113.

Khalid, A., Arshad, M., & Zahir, Z. (2004). Screening plant growth-promoting rhizobacteria for improving growth and yield of wheat. *Journal of Applied Microbiology*, *96*(3), 473–480. https://doi.org/10.1046/j.1365-2672.2003.02161.x.

Khan, A. G. (2002). Chap 8: The significance of microbes. In: Wong, M. H. & Bradshaw, A. D. (Eds.), *The Restoration and Management of Derelict Land: Modern Approaches*. World Scientific Publishing, Singapore, pp. 80–92.

Khan, A.G. (2004). Co-inoculation of vesicular arbuscular mycorrhizal fungi (AMF), mycorrhiza helping-bacteria (MBF) and plant-growth promoting rhizobacteria (PGPR) for phytoremediation of heavy metal

contaminated soil. In: Lou, Y. (Ed.), *Proceedings of the 5th International Conference on Environmental Geochem in the Tropics (GEOTROP)*, Haiko, China, March 21–26, p. 68.

Khan, A.G. (2020), In situ phytoremediation of uranium contaminated soils. In: Shmaefsky, B. (Ed.), *Phytoremediation*. Concepts and Strategies in Plant Sciences. Springer, Cham. https://doi.org/10.1007/978-3-030-00099-8_5.

Khan, A. L., Hamayun, M., Kim, Y., Kang, S., Lee, J., & Lee, I. (2011). Gibberellins producing endophytic aspergillus fumigatus sp. LH02 influenced endogenous phytohormonal levels, isoflavonoids production and plant growth in salinity stress. *Process Biochemistry, 46*(2), 440–447. https://doi.org/10.1016/j.procbio.2010.09.013.

Khan, A. R., Ullah, I., Waqas, M., Shahzad, R., Hong, S., Park, G., Jung, B. K., Lee, I., & Shin, J. (2015). Plant growth-promoting potential of endophytic fungi isolated from solanum nigrum leaves. *World Journal of Microbiology and Biotechnology, 31*(9), 1461–1466. https://doi.org/10.1007/s11274-015-1888-0.

Khan, N. & Bano, A. (2016). Modulation of phytoremediation and plant growth by the treatment with PGPR, Ag nanoparticle and untreated municipal wastewater. *International Journal of Phytoremediation, 18*(-12), 1258–1269. https://doi.org/10.1080/15226514.2016.1203287.

Khan, N. & Bano, A. (2018). Role of PGPR in the phytoremediation of heavy metals and crop growth under municipal wastewater irrigation. In: Ansari, A. A. et al. (Eds.), *Phytoremediation*. Springer Nature Switzerland AG, pp. 135–149. https://doi.org/10.1007/978-3-319-99651-6_5.

Khan, N., Bano, A., & Babar, M. A. (2016). The root growth of wheat plants, the water conservation and fertility status of sandy soils influenced by plant growth promoting rhizobacteria. *Symbiosis, 72*(3), 195–205. https://doi.org/10.1007/s13199-016-0457-0.

Khan, W., Rayirath, U. P., Subramanian, S., Jithesh, M. N., Rayorath, P., Hodges, D. M., Critchley, A. T., Craigie, J. S., Norrie, J., & Prithiviraj, B. (2009). Seaweed extracts as biostimulants of plant growth and development. *Journal of Plant Growth Regulation, 28*(4), 386–399. https://doi.org/10.1007/s00344-009-9103-x.

Khan, W. U., Ahmad, S. R., Yasin, N. A., Ali, A., Ahmad, A., & Akram, W. (2017). Application ofBacillus megateriummcr-8 improved phytoextraction and stress alleviation of nickel inVinca rosea. *International Journal of Phytoremediation, 19*(9), 813–824. https://doi.org/10.1080/15226514.2017.1290580.

Khanna, K., Jamwal, V. L., Sharma, A., Gandhi, S. G., Ohri, P., Bhardwaj, R., Al-Huqail, A. A., Siddiqui, M. H., Ali, H. M., & Ahmad, P. (2019). Supplementation with plant growth promoting rhizobacteria (PGPR) alleviates cadmium toxicity in solanum lycopersicum by modulating the expression of secondary metabolites. *Chemosphere, 230*, 628–639. https://doi.org/10.1016/j.chemosphere.2019.05.072.

Kloepper, J. W. & Schroth, M. N. (1978). Plant growth-promoting rhizobacteria on radishes. In: *Proceedings of the 4th International Conference on Plant Pathogenic Bacteria, Station de Pathologie Végétale et Phyto-Bactériologie*, Tours, France, pp. 879–882.

Kravchenko, L. V., Azarova, T. S., Makarova, N. M., & Tikhonovich, I. A. (2004). The effect of tryptophan present in plant root exudates on the Phytostimulating activity of Rhizobacteria. *Microbiology, 73*(2), 156–158. https://doi.org/10.1023/b:mici.0000023982.76684.9d.

Krishnamurthy, Y. L. & Naik, B. S. (2017). Endophytic fungi bioremediation. *Endophytes: Crop Productivity and Protection*, 47–60. https://doi.org/10.1007/978-3-319-66544-3_3.

Kudanga, T. & Mwenje, E. (2005). Extracellular cellulase production by tropical isolates of Aureobasidium pullulans. *Canadian Journal of Microbiology, 51*(9), 773–776. https://doi.org/10.1139/w05-053.

Kukla, M., Płociniczak, T., & Piotrowska-Seget, Z. (2014). Diversity of endophytic bacteria in lolium perenne and their potential to degrade petroleum hydrocarbons and promote plant growth. *Chemosphere, 117*, 40–46. https://doi.org/10.1016/j.chemosphere.2014.05.055.

Kumar, A., Bahadur, I., Maurya, B.R., Raghuwanshi, R., Meena, V.S., Singh, D.K., & Dixit, J. (2015). Does a plant growth-promoting rhizobacteria enhance agricultural sustainability. *Journal of Pure and Applied Microbiology, 9*(1), 715–724.

Kumar, G. & Sahoo, D. (2011). Effect of seaweed liquid extract on growth and yield of triticum aestivum Var. Pusa gold. *Journal of Applied Phycology, 23*(2), 251–255. https://doi.org/10.1007/s10811-011-9660-9.

Kumar, S. & Saxena, S. (2019). Arbuscular mycorrhizal fungi (AMF) from heavy metal-contaminated soils: Molecular approach and application in Phytoremediation. *Biofertilizers for Sustainable Agriculture and Environment*, 489–500. https://doi.org/10.1007/978-3-030-18933-4_22.

Kumar, S.S., Ram, K.R., Kumar, D.R., Panwar, S., & Prasad, C.S. (2013). Biocontrol by plant growth promoting rhizobacteria against black scurf and stem canker disease of potato caused by R. Solani. *Archives of Phytopathology and Plant Protection, 46*, 487–502.

Kumari, P., Meena, M., & Upadhyay, R.S. (2019). Characterization of plant growth promoting rhizobacteria (PGPR) isolated from the rhizosphere of *Vigna radiata* (mung bean). *Biocatalysis and Agricultural Biotechnology, 16*, 155–162.

Kuwada, K., Wamocho, L. S., Utamura, M., Matsushita, I., & Ishii, T. (2006). Effect of red and green algal extracts on hyphal growth of Arbuscular mycorrhizal fungi, and on mycorrhizal development and growth of papaya and Passionfruit. *Agronomy Journal*, *98*(5), 1340–1344. https://doi.org/10.2134/agronj2005.0354.

Lee, S., Trịnh, C. S., Lee, W. J., Jeong, C. Y., Truong, H. A., Chung, N., Kang, C., & Lee, H. (2020). Bacillus subtilis strain L1 promotes nitrate reductase activity in arabidopsis and elicits enhanced growth performance in arabidopsis, lettuce, and wheat. *Journal of Plant Research*, *133*(2), 231–244. https://doi.org/10.1007/s10265-019-01160-4.

Liang, X., Zhang, L., Natarajan, S. K., & Becker, D. F. (2013). Proline mechanisms of stress survival. *Antioxidants & Redox Signaling*, *19*(9), 998–1011. https://doi.org/10.1089/ars.2012.5074.

Liotti, R. G., Da Silva Figueiredo, M. I., Da Silva, G. F., De Mendonça, E. A., & Soares, M. A. (2018). Diversity of cultivable bacterial endophytes in Paullinia cupana and their potential for plant growth promotion and phytopathogen control. *Microbiological Research*, *207*, 8–18. https://doi.org/10.1016/j.micres.2017.10.011.

Liu, B., Huang, L., Buchenauer, H., & Kang, Z. (2010). Isolation and partial characterization of an antifungal protein from the endophytic bacillus subtilis strain EDR4. *Pesticide Biochemistry and Physiology*, *98*(2), 305–311. https://doi.org/10.1016/j.pestbp.2010.07.001.

Liu, W., Hou, J., Wang, Q., Ding, L., & Luo, Y. (2014). Isolation and characterization of plant growth-promoting rhizobacteria and their effects on phytoremediation of petroleum-contaminated saline-alkali soil. *Chemosphere*, *117*, 303–308. https://doi.org/10.1016/j.chemosphere.2014.07.026.

López-Bucio, J., Pelagio-Flores, R., & Herrera-Estrella, A. (2015). Trichoderma as biostimulant: Exploiting the multilevel properties of a plant beneficial fungus. *Scientia Horticulturae*, *196*, 109–123. https://doi.org/10.1016/j.scienta.2015.08.043.

Lucy, M., Reed, E., & Glick, B. R. (2004). Applications of free living plant growth-promoting rhizobacteria. *Antonie van Leeuwenhoek*, *86*(1), 1–25. https://doi.org/10.1023/b:anto.0000024903.10757.6e.

Maharshi, A., Kumar, G., Mukherjee, A., Raghuwanshi, R., Singh, H.B., Kumar, B. & Arbuscular, S. (2019). Mycorrhizal colonization and activation of plant defense responses against phytopathogens. In: Singh, D. P. et al. (Eds.), *Microbial Interventions in Agriculture and Environment*. Springer Nature Singapore Pvt Ltd, pp. 219– doi.org/10.1007/978-981-13-8391-5_8

Marag, P. S. & Suman, A. (2018). Growth stage and tissue specific colonization of endophytic bacteria having plant growth promoting traits in hybrid and composite maize (Zea mays L.). *Microbiological Research*, *214*, 101–113. https://doi.org/10.1016/j.micres.2018.05.016.

Marín, F., Navarrete, H., & Narvaez-Trujillo, A. (2018). Total petroleum hydrocarbon degradation by endophytic fungi from the ecuadorian Amazon. *Advances in Microbiology*, *08*(12), 1029–1053. https://doi.org/10.4236/aim.2018.812070.

Marques, A. P., Moreira, H., Franco, A. R., Rangel, A. O., & Castro, P. M. (2013). Inoculating helianthus annuus (sunflower) grown in zinc and cadmium contaminated soils with plant growth promoting bacteria – Effects on phytoremediation strategies. *Chemosphere*, *92*(1), 74–83. https://doi.org/10.1016/j.chemosphere.2013.02.055.

Marwa, N., Mishra, N., Singh, N., Mishra, A., Saxena, G., Pandey, V., & Singh, N. (2020). Effect of rhizospheric inoculation of isolated arsenic (As) tolerant strains on growth, as-uptake and bacterial communities in association with adiantum capillus-veneris. *Ecotoxicology and Environmental Safety*, *196*, 110498. https://doi.org/10.1016/j.ecoenv.2020.110498.

McHugh, D. J. (2013). *A Guide to the Seaweed Industry*. Food and Agriculture Organization of the United Nations, Rome.

Mehmood, A., Hussain, A., Irshad, M., Hamayun, M., Iqbal, A., & Khan, N. (2019). In vitro production of IAA by endophytic fungus aspergillus awamori and its growth promoting activities in zea mays. *Symbiosis*, *77*(3), 225–235. https://doi.org/10.1007/s13199-018-0583-y.

Mishra, A., Singh, S. P., Mahfooz, S., Bhattacharya, A., Mishra, N., Shirke, P. A., & Nautiyal, C. (2018). Bacterial endophytes modulates the withanolide biosynthetic pathway and physiological performance in Withania somnifera under biotic stress. *Microbiological Research*, *212–213*, 17–28. https://doi.org/10.1016/j.micres.2018.04.006.

Mishra, J. & Arora, N. K. (2018). Secondary metabolites of fluorescent pseudomonads in biocontrol of phytopathogens for sustainable agriculture. *Applied Soil Ecology*, *125*, 35–45. https://doi.org/10.1016/j.apsoil.2017.12.004.

Mishra, R. & Venkateswara, S.V. (2017). Mycoremediation of heavy metal and hydrocarbon pollutants by endophytic fungi. In: Prasad, R. (Eds), *Mycoremediation and Environmental Sustainability*. Fungal Biology. Springer, Cham, pp. 133–151. https://doi.org/10.1007/978-3-319-68957-9_8.

Mishra, R., & Venkateswara, S.V. (2017). Mycoremediation of heavy metal and hydrocarbon pollutants by endophytic fungi. In: Prasad R. (eds) *Mycoremediation and Environmental Sustainability. Fungal Biology*. Springer, Cham, pp. 133–151. https://doi.org/10.1007/978-3-319-68957-9_8.

Mishra, V., Gupta, A., Kaur, P., Singh, S., Singh, N., Gehlot, P., & Singh, J. (2016). Synergistic effects of Arbuscular mycorrhizal fungi and plant growth promoting rhizobacteria in bioremediation of iron contaminated soils, *International Journal of Phytoremediation*, 18(7), 697–703.

Mitter, E. K., Kataoka, R., De Freitas, J. R., & Germida, J. J. (2019). Potential use of endophytic root bacteria and host plants to degrade hydrocarbons. *International Journal of Phytoremediation*, 21(9), 928–938. https://doi.org/10.1080/15226514.2019.1583637.

Mohamed, I., Eid, K. E., Abbas, M. H., Salem, A. A., Ahmed, N., Ali, M., Shah, G. M., & Fang, C. (2019). Use of plant growth promoting Rhizobacteria (PGPR) and mycorrhizae to improve the growth and nutrient utilization of common bean in a soil infected with white rot fungi. *Ecotoxicology and Environmental Safety*, 171, 539–548. https://doi.org/10.1016/j.ecoenv.2018.12.100.

Moreira, H., Pereira, S. I., Vega, A., Castro, P. M., & Marques, A. P. (2020). Synergistic effects of arbuscular mycorrhizal fungi and plant growth-promoting bacteria benefit maize growth under increasing soil salinity. *Journal of Environmental Management*, 257, 109982. https://doi.org/10.1016/j.jenvman.2019.109982.

Nadeem, S. M., Ahmad, M., Zahir, Z. A., Javaid, A., & Ashraf, M. (2014). The role of mycorrhizae and plant growth promoting rhizobacteria (PGPR) in improving crop productivity under stressful environments. *Biotechnology Advances*, 32(2), 429–448. https://doi.org/10.1016/j.biotechadv.2013.12.005.

Nandy, S., Das, T., Tudu, C. K., Pandey, D. K., Dey, A., & Ray, P. (2020). Fungal endophytes: Futuristic tool in recent research area of phytoremediation. *South African Journal of Botany*, 134, 285–295. https://doi.org/10.1016/j.sajb.2020.02.015.

Nardi, S., Carletti, P., Pizzeghello, D., & Muscolo, A. (2009). Biological activities of humic substances. *Biophysico-Chemical Processes Involving Natural Nonliving Organic Matter in Environmental Systems*, 305–339. https://doi.org/10.1002/9780470494950.ch8.

Naveed, M., Mitter, B., Reichenauer, T. G., Wieczorek, K., & Sessitsch, A. (2014). Increased drought stress resilience of maize through endophytic colonization by Burkholderia phytofirmans PsJN and Enterobacter sp. FD17. *Environmental and Experimental Botany*, 97, 30–39. https://doi.org/10.1016/j.envexpbot.2013.09.014.

Naz, I., Ahmad, H., Khokhar, S. N., Khan, K., & Shah, A. H. (2016). Impact of zinc solubilizing bacteria on zinc contents of wheat. *American-Eurasian Journal of Agricultural and Environmental Sciences*, 16, 449–454. https://doi.org/10.5829/idosi.aejaes.2016.16.3.12886.

Naziya, B., Murali, M., & Amruthesh, K. N. (2019). Plant growth-promoting fungi (PGPF) instigate plant growth and induce disease resistance in capsicum annuum L. upon infection with Colletotrichum capsici (SYD.) butler & Bisby. *Biomolecules*, 10(1), 41. https://doi.org/10.3390/biom10010041.

Nikbakht, A., Kafi, M., Babalar, M., Xia, Y. P., Luo, A., & Etemadi, N. (2008). Effect of humic acid on plant growth, nutrient uptake, and Postharvest life of gerbera. *Journal of Plant Nutrition*, 31(12), 2155–2167. https://doi.org/10.1080/01904160802462819.

Rohmah, N. S.R. N. S., Jatmiko, Y. D., Siswanto, D., & Mustafa, I. (2020). The potency of endophytic bacteria isolated from ficus septica as phytoremediation promoting agent of CR (VI) contaminated soil. *Biodiversitas Journal of Biological Diversity*, 21(5), 1920–1927. https://doi.org/10.13057/biodiv/d210519.

Nykiel-Szymańska, J., Bernat, P., & Słaba, M. (2020). Biotransformation and detoxification of chloroacetanilide herbicides by Trichoderma spp. with plant growth-promoting activities. *Pesticide Biochemistry and Physiology*, 163, 216–226. https://doi.org/10.1016/j.pestbp.2019.11.018.

Ortíz-Castro, R., Contreras-Cornejo, H. A., Macías-Rodríguez, L., López-Bucio, J. (2009). The role of microbial signals in plant growth and development. *Plant Signaling & Behavior*, 4(8), 701–712.

Ozimek, E., Jaroszuk-Ściseł, J., Bohacz, J., Korniłłowicz-Kowalska, T., Tyśkiewicz, R., Słomka, A., Nowak, A., & Hanaka, A. (2018). Synthesis of indoleacetic acid, gibberellic acid and ACC-deaminase by Mortierella strains promote winter wheat seedlings growth under different conditions. *International Journal of Molecular Sciences*, 19(10), 3218. https://doi.org/10.3390/ijms19103218.

Pacifici, E., Polverari, L., & Sabatini, S. (2015). Plant hormone cross-talk: The pivot of root growth. *Journal of Experimental Botany*, 66(4), 1113–1121. https://doi.org/10.1093/jxb/eru534.

Pandey, V. C. & Bajpai, O. (2019). Phytoremediation: From theory towards practice. *Phytomanagement of Polluted Sites*, 1–49. https://doi.org/10.1016/b978-0-12-813912-7.00001-6.

Park, C., Yeo, H., Park, Y., Morgan, A., Valan Arasu, M., Al-Dhabi, N., & Park, S. (2017). Influence of indole-3-Acetic acid and gibberellic acid on Phenylpropanoid accumulation in common buckwheat (Fagopyrum esculentum Moench) sprouts. *Molecules*, 22(3), 374. https://doi.org/10.3390/molecules22030374.

Park, Y., Dutta, S., Ann, M., Raaijmakers, J. M., & Park, K. (2015). Promotion of plant growth by pseudomonas fluorescens strain SS101 via novel volatile organic compounds. *Biochemical and Biophysical Research Communications, 461*(2), 361–365. https://doi.org/10.1016/j.bbrc.2015.04.039.

Passari, A. K., Mishra, V. K., Leo, V. V., Gupta, V. K., & Singh, B. P. (2016). Phytohormone production endowed with antagonistic potential and plant growth promoting abilities of culturable endophytic bacteria isolated from Clerodendrum colebrookianum Walp. *Microbiological Research, 193*, 57–73. https://doi.org/10.1016/j.micres.2016.09.006.

Patel, T.M. & Minocheherhomji, F.P. (2018). Isolation and characterization of several siderophore producing bacteria from cotton plant, *Environmental Science.* Corpus ID: 212574224.

Patel, J. K., Madaan, S., & Archana, G. (2018). Antibiotic producing endophytic Streptomyces spp. colonize above-ground plant parts and promote shoot growth in multiple healthy and pathogen-challenged cereal crops. *Microbiological Research, 215*, 36–45. https://doi.org/10.1016/j.micres.2018.06.003.

Patel, P. R., Shaikh, S. S., & Sayyed, R. Z. (2016). Dynamism of PGRP in bioremediation and plant growth promotion in heavy metals and contaminated soil. *Indian Journal of Experimental Biology, 54*, 286–290.

Pattnaik, S. S. & Busi, S. (2019). Rhizospheric Fungi: Diversity and potential biotechnological applications. In: Yadav, N. et al. (Eds.), *Recent Advancement in White Biotechnology through Fungi.* Fungal Biology. https://doi.org/10.1007/978-3-030-10480-1_2.

Pellegrino, E. & Bedini, S. (2014). Corrigendum to "Enhancing ecosystem services in sustainable agriculture: Biofertilization and biofortification of chickpea (Cicer arietinum L.) by arbuscular mycorrhizal fungi.". *Soil Biology and Biochemistry, 75*, 314–315. https://doi.org/10.1016/j.soilbio.2014.03.018.

Peñuelas, J., Asensio, D., Tholl, D., Wenke, K., Rosenkranz, M., Piechulla, B., Schnitzler, J. (2014). Biogenic volatile emissions from the soil. *Plant, Cell & Environment, 37*(8), 1866–1891.

Pérez-Montaño, F., Alías-Villegas, C., Bellogín, R., Del Cerro, P., Espuny, M., Jiménez-Guerrero, I., López-Baena, F., Ollero, F., & Cubo, T. (2014). Plant growth promotion in cereal and leguminous agricultural important plants: From microorganism capacities to crop production. *Microbiological Research, 169*(5–6), 325–336. https://doi.org/10.1016/j.micres.2013.09.011.

Pichyangkura, R. & Chadchawan, S. (2015). Biostimulant activity of chitosan in horticulture. *Scientia Horticulturae, 196*, 49–65. https://doi.org/10.1016/j.scienta.2015.09.031.

Pietro-Souza, W., de Campos Pereira, F., Mello, I. S., Stachack, F. F. F, Terezo, A. J. et al. (2019). Mercury resistance and bioremediation mediated by endophytic fungi. *Chemosphere*, 124874. https://doi.org/10.1016/j.chemosphere.2019.124874.

Pietro-Souza, W., De Campos Pereira, F., Mello, I. S., Stachack, F. F., Terezo, A. J., Cunha, C. N., White, J. F., Li, H., & Soares, M. A. (2020). Mercury resistance and bioremediation mediated by endophytic fungi. *Chemosphere, 240*, 124874. https://doi.org/10.1016/j.chemosphere.2019.124874.

Płociniczak, T., Sinkkonen, A., Romantschuk, M., & Piotrowska-Seget, Z. (2013). Characterization of Enterobacter intermedius MH8b and its use for the enhancement of heavy metals uptake by Sinapis alba L. *Applied Soil Ecology, 63*, 1–7. https://doi.org/10.1016/j.apsoil.2012.09.009.

Potshangbam, M., Devi, S. I., Sahoo, D., & Strobel, G. A. (2017). Functional characterization of endophytic fungal community associated with oryza sativa L. and zea mays L. *Frontiers in Microbiology, 8*, 325. https://doi.org/10.3389/fmicb.2017.00325.

Prasad, M., Srinivasan, R., Chaudhary, M., Choudhary, M., & Jat, L. K. (2019). Plant growth promoting Rhizobacteria (PGPR) for sustainable agriculture. *PGPR Amelioration in Sustainable Agriculture*, 129–157. https://doi.org/10.1016/b978-0-12-815879-1.00007-0.

Prashar, P., Kapoor, N., & Sachdeva, S. (2013). Rhizosphere: Its structure, bacterial diversity and significance. *Reviews in Environmental Science and Bio/Technology, 13*(1), 63–77. https://doi.org/10.1007/s11157-013-9317-z.

Priyadharsini, P. & Muthukumar, T. (2017). The root endophytic fungus Curvularia geniculata from Parthenium hysterophorus roots improves plant growth through phosphate solubilization and phytohormone production. *Fungal Ecology, 27*, 69–77. https://doi.org/10.1016/j.funeco.2017.02.007.

Quaggiotti, S. (2004). Effect of low molecular size humic substances on nitrate uptake and expression of genes involved in nitrate transport in maize (Zea mays L.). *Journal of Experimental Botany, 55*(398), 803–813. https://doi.org/10.1093/jxb/erh085.

Rafiee, H., Naghdi, B. H., Mehrafarin, A., Qaderi, A., Zarinpanjeh, N., Sekara, A. & Zand, E. (2016). Application of plant biostimulants as new approach to improve the biological responses of medicinal plants- A critical review. *Journal of Medicinal Plants, 15*, 59.

Rahimzadeh, S. & Pirzad, A.R. (2017). Microorganisms (AMF and PSB) interaction on linseed productivity under water-deficit condition. *International Journal of Plant Production, 11*(2), 259–274.

Ramos-Solano, B., Barriuso-Maicas, J., Pereyra De La Iglesia, M. T., Domenech, J., Gutiérrez & Mañero, F.J. (2008). Systemic disease protection elicited by plant growth promoting rhizobacteria strains: relationship

between metabolic responses, systemic disease protection, and biotic elicitors. *Phytopathology*, *98*(4), 451–457.

Rais, A., Jabeen, Z., Shair, F., Hafeez, F. Y., & Hassan, M. N. (2017). *Bacillus* spp., a bio-control agent enhances the activity of antioxidant defense enzymes in rice against *Pyricularia oryzae*. *PLoS One*, *21*, 12(11), e0187412. https://doi.org/10.1371/journal.pone.0187412. PMID: 29161274; PMCID: PMC5697883.

Rajendran, S. K. & Sundaram, L. (2020). Degradation of heavy metal contaminated soil using plant growth promoting rhizobacteria (PGPR): Assess their remediation potential and growth influence of *Vigna radiata*. L. *International Journal of Agricultural Technology*, *16*(2), 365–376.

Rajkumar, M., Ae, N., Prasad, M. N., & Freitas, H. (2010). Potential of siderophore-producing bacteria for improving heavy metal phytoextraction. *Trends in Biotechnology*, *28*(3), 142–149. https://doi.org/10.1016/j.tibtech.2009.12.002.

Rajkumar, M., Vara Prasad, M. N., Freitas, H., & Ae, N. (2009). Biotechnological applications of Serpentine soil bacteria for phytoremediation of trace metals. *Critical Reviews in Biotechnology*, *29*(2), 120–130. https://doi.org/10.1080/07388550902913772.

Rana, K. L., Kour, D., Sheikh, I., Yadav, N., Yadav, A. N., Kumar, V., Singh, B. P., Dhaliwal, H. S. & Saxena, A. K. (2019). Biodiversity of endophytic fungi from diverse niches and their biotechnological applications. In: Singh, B. (Eds.), *Advances in Endophytic Fungal Research*. Fungal Biology. Springer, Cham, pp. 105–144. https://doi.org/10.1007/978-3-030-03589-1_6.

Ray, S., Mishra, S., Bisen, K., Singh, S., Sarma, B. K., & Singh, H. B. (2018). Modulation in phenolic root exudate profile of Abelmoschus esculentus expressing activation of defense pathway. *Microbiological Research*, *207*, 100–107. https://doi.org/10.1016/j.micres.2017.11.011.

Remans, R., Beebe, S., Blair, M., Manrique, G., Tovar, E., Rao, I., Croonenborghs, A., Torres-Gutierrez, R., El-Howeity, M., Michiels, J., & Vanderleyden, J. (2008). Physiological and genetic analysis of root responsiveness to auxin-producing plant growth-promoting bacteria in common bean (Phaseolus vulgaris L.). *Plant and Soil*, *302*(1–2), 149–161. https://doi.org/10.1007/s11104-007-9462-7.

Ribeiro, V. P., Marriel, I. E., Sousa, S. M., Lana, U. G., Mattos, B. B., Oliveira, C. A., & Gomes, E. A. (2018). Endophytic bacillus strains enhance pearl millet growth and nutrient uptake under low-P. *Brazilian Journal of Microbiology*, *49*, 40–46. https://doi.org/10.1016/j.bjm.2018.06.005.

Richardson, A. E., Barea, J., McNeill, A. M., & Prigent-Combaret, C. (2009). Acquisition of phosphorus and nitrogen in the rhizosphere and plant growth promotion by microorganisms. *Plant and Soil*, *321*(1–2), 305–339. https://doi.org/10.1007/s11104-009-9895-2.

Rijavec, T. & Lapanje, A. (2016). Hydrogen cyanide in the rhizosphere: Not suppressing plant pathogens, but rather regulating availability of phosphate. *Frontiers in Microbiology*, *18*(7). https://doi.org/10.3389/fmicb.2016.01785.

Ripa, F. A., Cao, W., Tong, S., & Sun, J. (2019). Assessment of plant growth promoting and abiotic stress tolerance properties of wheat endophytic fungi. *BioMed Research International*, *2019*, 1–12. https://doi.org/10.1155/2019/6105865.

Rodriguez, H., González, T, Goire, I, & Bashan, Y. (2004). Gluconic acid production and phosphate solubilization by the plant growth-promoting bacterium *Azospirillum* spp. *Die Naturwissenschaften*, 91, 552–555. https://doi.org/10.1007/s00114-004-0566-0. 865, 1–12. https://doi.org/10.1155/2019/6105865

Riskuwa-Shehu, M. L., Ismail, H. Y., & Ijah, U. J. (2019). Heavy metal resistance by endophytic bacteria isolated from guava (Psidium Guajava) and mango (Mangifera Indica) leaves. *International Annals of Science*, *9*(1), 16–23. https://doi.org/10.21467/ias.9.1.16-23.

Rose, M. T., Patti, A. F., Little, K. R., Brown, A. L., Jackson, W. R., & Cavagnaro, T. R. (2014). A metaanalysis and review of plant-growth response to humic substances: Practical implications for agriculture. *Advances in Agronomy*, *124*, 37–89.

Rouphael, Y. & Colla, G. (2020). Editorial: Biostimulants in agriculture. *Frontiers in Plant Science*, *11*(40), 1–7. https://doi.org/10.3389/fpls.2020.00040.

Rouphael, Y., & Colla, G. (2018). Synergistic biostimulatory action: Designing the next generation of plant biostimulants for sustainable agriculture. *Frontiers in Plant Science*, *9*. https://doi.org/10.3389/fpls.2018.01655.

Rouphael, Y., Franken, P., Schneider, C., Schwarz, D., Giovannetti, M., Agnolucci, M., Pascale, S. D., Bonini, P., & Colla, G. (2015). Arbuscular mycorrhizal fungi act as biostimulants in horticultural crops. *Scientia Horticulturae*, *196*, 91–108. https://doi.org/10.1016/j.scienta.2015.09.002.

Rouphael, Y., Kyriacou, M. C., & Colla, G. (2018a). Vegetable grafting: A toolbox for securing yield stability under multiple stress conditions. *Frontiers in Plant Science*, *8*. https://doi.org/10.3389/fpls.2017.02255.

Rouphael, Y., Kyriacou, M. C., Petropoulos, S. A., De Pascale, S., & Colla, G. (2018b). Improving vegetable quality in controlled environments. *Scientia Horticulturae*, *234*, 275–289. https://doi.org/10.1016/j.scienta.2018.02.033.

Ruzzi, M. & Aroca, R. (2015). Plant growth-promoting rhizobacteria act as biostimulants in horticulture. *Scientia Horticulturae*, *196*, 124–134. https://doi.org/10.1016/j.scienta.2015.08.042.

Ryall, B., Mitchell, H., Mossialos, D., & Williams, H. (2009). Cyanogenesis by the entomopathogenic bacteriumPseudomonas entomophila. *Letters in Applied Microbiology*, *49*(1), 131–135. https://doi.org/10.1111/j.1472-765x.2009.02632.x.

Saini, R., Dudeja, S. S., Giri, R., & Kumar, V. (2013). Isolation, characterization, and evaluation of bacterial root and nodule endophytes from chickpea cultivated in northern India. *Journal of Basic Microbiology*, *55*(1), 74–81. https://doi.org/10.1002/jobm.201300173.

Saldajeno, M. & Hyakumachi, M. (2011). The plant growth-promoting fungus Fusarium equiseti and the arbuscular mycorrhizal fungus Glomus mosseae stimulate plant growth and reduce severity of anthracnose and damping-off diseases in cucumber (Cucumis sativus) seedlings. *Annals of Applied Biology*, *159*(1), 28–40. https://doi.org/10.1111/j.1744-7348.2011.00471.x.

Salman, S., Abouhussein, S., Abdel-Mawgoud, A. M. R., & El-Nemr, M. A. (2005). Fruit yield and quality of watermelon as affected by hybrids and humic acid application. *Journal of Applied Sciences Research*, *1*, 51–58.

Santi, C., Zamboni, A., Varanini, Z., & Pandolfini, T. (2017). Growth stimulatory effects and genome-wide transcriptional changes produced by protein hydrolysates in maize seedlings. *Frontiers in Plant Science*, *8*, 433. https://doi.org/10.3389/fpls.2017.00433.

Santoyo, G., Moreno-Hagelsieb, G., Del Carmen Orozco-Mosqueda, M., & Glick, B. R. (2016). Plant growth-promoting bacterial endophytes. *Microbiological Research*, *183*, 92–99. https://doi.org/10.1016/j.micres.2015.11.008.

Saraf, M., Jha, C. K., & Patel, D. (2010). The role of ACC deaminase producing PGPR in sustainable agriculture. In: Maheshwari, D. (Ed.), *Plant Growth and Health Promoting Bacteria*. Microbiology Monographs, vol. 18. Springer, Berlin, pp. 365–385. https://doi.org/10.1007/978-3-642-13612-2_16.

Sarbadhikary, S. B. & Mandal, N. C. (2018). Elevation of plant growth parameters in two solanaceous crops with the application of endophytic fungus. *Indian Journal of Agricultural Research*, *52*, 424–428. https://doi.org/10.18805/ijare.a-4784.

Schiavon, M., Ertani, A., & Nardi, S. (2008). Effects of an alfaalfa protein hydrolysate on the gene expression and activity of enzymes of TCA cycle and N metabolism in Zea mays L. *Journal of Agricultural and Food Chemistry*, 56, 11800–11808. https://doi.org/10.1021/jf802362g

Savvas, D. & Ntatsi, G. (2015). Biostimulant activity of silicon in horticulture. *Scientia Horticulturae*, *196*, 66–81. https://doi.org/10.1016/j.scienta.2015.09.010.

Schmidt, W., Santi, S., Pinton, R., & Varanini, Z. (2007). Water-extractable humic substances alter root development and epidermal cell pattern in arabidopsis. *Plant and Soil*, *300*(1–2), 259–267. https://doi.org/10.1007/s11104-007-9411-5.

Selvakumar, G., Joshi, P., Nazim, S., Mishra, P. K., Bisht, J. K., & Gupta, H. S. (2009). Phosphate solubilization and growth promotion by pseudomonas fragi CS11RH1 (MTCC 8984), a psychrotolerant bacterium isolated from a high altitude himalayan rhizosphere. *Biologia*, *64*(2), 239–245. https://doi.org/10.2478/s11756-009-0041-7.

Sem, S. & Chandrasekhar, C. N. (2015). Effect of PGPR on enzymatic activities of rice (*Oryza sativa* L.) under salt stress. *Asian Journal of Plant Science and Research*, *5*(6), 44–48.

Seshachala, U. & Tallapragada, P. (2012). Phosphate Solubilizers from the Rhizospher of piper nigrum L. in Karnataka,India. *Chilean Journal of Agricultural Research*, 72(3), 397–403. https://doi.org/10.4067/s0718-58392012000300014.

Shafi,J.,Tian,H.,&Ji,M.(2017).Bacillusspeciesasversatileweaponsforplantpathogens:Areview.*Biotechnology & Biotechnological Equipment*, *31*(3), 446–459. https://doi.org/10.1080/13102818.2017.1286950.

Sharma, H. S., Fleming, C., Selby, C., Rao, J. R., & Martin, T. (2014). Plant biostimulants: A review on the processing of macroalgae and use of extracts for crop management to reduce abiotic and biotic stresses. *Journal of Applied Phycology*, *26*(1), 465–490. https://doi.org/10.1007/s10811-013-0101-9.

Sharma, S. B., Sayyed, R. Z., Trivedi, M. H., & Gobi, T. A. (2013). Phosphate solubilizing microbes: Sustainable approach for managing phosphorus deficiency in agricultural soils. *SpringerPlus*, *2*(1). https://doi.org/10.1186/2193-1801-2-587.

Shelake, R. M., Waghunde, R. R., Morita, E. H. et al. (2018). Plant-microbe-metal interactions: Basics, recent advances, and future trends. In: Egamberdieva, D. & Ahmad, P. (Eds.), *Plant Microbiome: Stress Response*. Microorganisms for Sustainability, vol 5. Springer, Singapore, pp. 1–5.

Sibi, G. (2011). Role of phosphate solubilizing fungi during phosphocompost production and their effect on the growth of tomato (*Lycopersicon esculentum* L) plants. *Journal of Applied and Natural Science*, *3*(2), 287–290.

Siddikee, M. A., Chauhan, P. S., Anandham, R., Han, G., & Sa, T. (2010). Isolation, characterization, and use for plant growth promotion under salt stress, of ACC deaminase-producing Halotolerant bacteria derived from coastal soil. *Journal of Microbiology and Biotechnology*, *20*(11), 1577–1584. https://doi.org/10.4014/jmb.1007.07011.

Silva, M. C., Polonio, J. C., Quecine, M. C., Almeida, T. T., Bogas, A. C., Pamphile, J. A., Pereira, J. O., Astolfi-Filho, S., & Azevedo, J. L. (2016). Endophytic cultivable bacterial community obtained from the Paullinia cupana seed in Amazonas and Bahia regions and its antagonistic effects against Colletotrichum gloeosporioides. *Microbial Pathogenesis*, *98*, 16–22. https://doi.org/10.1016/j.micpath.2016.06.023.

Singh, D. P., Singh, H. B., & Prabha, R. (Eds.) (2016). *Microbial Inoculants in Sustainable Agricultural Productivity*. Springer, New York.

Singh, R. K., Tripathi, R., Ranjan, A., & Srivastava, A. K. (2020). Fungi as potential candidates for bio-remediation. *Abatement of Environmental Pollutants Trends and Strategies*, 177–191. https://doi.org/10.1016/b978-0-12-818095-2.00009-6.

Smith, S.E., & Read, D. J. (2008). *Mycorrhizal Symbiosis*, Ed 3. Academic Press, New York.

Soetan, K. O., Olaiya, C. O., & Oyewole, O. E. (2010). The importance of mineral elements for humans, domestic animals and plants-a review. *African Journal of Food Science*, *4*(5), 200–222.

Sokolova, M. G., Akimova, G. P., & Vaishlya, O. B. (2011). Effect of phytohormones synthesized by rhizosphere bacteria on plants. *Applied Biochemistry and Microbiology*, *47*(3), 274–278. https://doi.org/10.1134/s0003683811030148.

Song, X. J., Kuroha, T., Ayano, M., Furuta, T., Nagai, K., Komeda, N., & Ashikari, M. (2015). Rare allele of a previously unidentified histone H4 acetyltransferase enhances grain weight, yield, and plant biomass in rice. *Proceedings of the National Academy of Sciences*, *112*(1), 76–81.

Song, X., Pan, Y., Li, L., Wu, X., & Wang, Y. (2018). Composition and diversity of rhizosphere fungal community in Coptis chinensis Franch. continuous cropping fields. *PLoS One*, *13*(3), e0193811. https://doi.org/10.1371/journal.pone.0193811.

Spinelli, F., Fiori, G., Noferini, M., Sprocatti, M., & Costa, G. (2010). A novel type of seaweed extract as a natural alternative to the use of iron chelates in strawberry production. *Scientia Horticulturae*, *125*(3), 263–269. https://doi.org/10.1016/j.scienta.2010.03.011.

Srinivasan, R., Yandigeri, M. S., Kashyap, S., & Alagawadi, A. R. (2012). Effect of salt on survival and P-solubilization potential of phosphate solubilizing microorganisms from salt affected soils. *Saudi Journal of Biological Sciences*, *19*(4), 427–434. https://doi.org/10.1016/j.sjbs.2012.05.004.

Srivastava, P., Saxena, B., & Giri, B. (2017). Arbuscular mycorrhizal fungi: Green approach/technology for sustainable agriculture and environment. In: Varma, A., Prasad, R., & Tuteja, N. (Eds.), *Mycorrhiza - Nutrient Uptake, Biocontrol, Ecorestoration*. Springer, Cham. http://doi.org/10.1007/978-3-319-68867-1_20.

Stępniewska, Z. & Kuźniar, A. (2013). Endophytic microorganisms—promising applications in bioremediation of greenhouse gases. *Applied Microbiology and Biotechnology*, *97*(22), 9589–9596. https://doi.org/10.1007/s00253-013-5235-9.

Stevenson, F. J. (1994). *Humus Chemistry: Genesis, Composition, Reactions*, 2nd ed. Wiley, New York, pp. 56–67.

Stirk, W. A., Tarkowská, D., Turečová, V., Strnad, M., & Van Staden, J. (2013). Abscisic acid, gibberellins and brassinosteroids in Kelpak®, a commercial seaweed extract made from Ecklonia maxima. *Journal of Applied Phycology*, *26*(1), 561–567. https://doi.org/10.1007/s10811-013-0062-z.

Strader, L. C., Chen, G. L., & Bartel, B. (2010). Ethylene directs auxin to control root cell expansion. *The Plant Journal*, *64*(5), 874–884. https://doi.org/10.1111/j.1365-313x.2010.04373.x.

Subbarao, S. B., Hussain, I. S. A., & Ganesh, P. T. (2015). Bio stimulant activity of protein hydrolysate: Influence on plant growth and yield. *Journal of Plant Science Research*, *2*, 1–6.

Subramanian, P., Kim, K., Krishnamoorthy, R., Sundaram, S., & Sa, T. (2015). Endophytic bacteria improve nodule function and plant nitrogen in soybean on Co-inoculation with Bradyrhizobium japonicum MN110. *Plant Growth Regulation*, *76*(3), 327–332. https://doi.org/10.1007/s10725-014-9993-x.

Sutton, R. & Sposito, G. (2005). Molecular structure in soil humic Substances: The new view. *Environmental Science & Technology*, *39*(23), 9009–9015. https://doi.org/10.1021/es050778q.

Tagele, S. B., Kim, S. W., Lee, H. G., Kim, H. S., & Lee, Y. S. (2018). Effectiveness of multi-trait Burkholderia contaminans KNU17BI1 in growth promotion and management of banded leaf and sheath blight in maize seedling. *Microbiological Research*, *214*, 8–18. https://doi.org/10.1016/j.micres.2018.05.004.

Taiwo, L., Ailenokhuoria, B., & Oyedele, A. (2017). Profiling rhizosphere microbes on the root of maize (Zea mays) planted in an Alfisol for selection as plant growth promoting Rhizobacteria (PGPR). *Microbiology Research Journal International*, *21*(5), 1–10. https://doi.org/10.9734/mrji/2017/36404.

Tara, N., Afzal, M., Ansari, T. M., Tahseen, R., Iqbal, S., & Khan, Q. M. (2014). Combined use of alkane-degrading and plant growth-promoting bacteria enhanced Phytoremediation of diesel contaminated soil.

International Journal of Phytoremediation, *16*(12), 1268–1277. https://doi.org/10.1080/15226514.2013. 828013.

Tewari, S. & Arora, N. K. (2016). Soybean production under flooding stress and its mitigation using plant growth-promoting microbes. In Miransari, M. (Ed.), *Environmental Stresses in Soybean Production*. Academic/Elsevier, New York, pp. 23–40.

Thakor, P., Goswami, D., Thakker, J., & Dhandhukia, P. (2016). Idiosyncrasy of local fungal isolate hypocrea rufa strain P2: Plant growth promotion and mycoparasitism. *Journal of Microbiology, Biotechnology and Food Sciences*, *05*(06), 593–598. https://doi.org/10.15414/jmbfs.2016.5.6.593-598.

Thakur, D., Kaur, M., & Mishra, A. (2017). Isolation and screening of plant growth promoting *Bacillus* spp. and *Pseudomonas* spp. and their effect on growth, rhizospheric population and phosphorous concentration of Aloe vera. *Journal of Medicinal Plants Studies*, *5*(1), 187–192.

Theocharis, A., Bordiec, S., Fernandez, O., Paquis, S., Dhondt-Cordelier, S., Baillieul, F., Clément, C., & Barka, E. A. (2012). Burkholderia phytofirmans PsJN primes vitis vinifera L. and confers a better tolerance to low Nonfreezing temperatures. *Molecular Plant-Microbe Interactions®*, *25*(2), 241–249. https://doi.org/10.1094/mpmi-05-11-0124.

Thilagar, G., Bagyaraj, D. J., Podile, A. R., & Vaikuntapu, P. R. (2018). Bacillus sonorensis, a novel plant growth promoting Rhizobacterium in improving growth, nutrition and yield of chilly (Capsicum annuum L.). *Proceedings of the National Academy of Sciences, India Section B: Biological Sciences*, *88*(2), 813–818. https://doi.org/10.1007/s40011-016-0822-z.

Thokchom, E., Thakuria, D., Kalita, M. C., Sharma, C. K., & Talukdar, N. C. (2018). Root colonization by host-specific rhizobacteria alters Indigenous root endophyte and rhizosphere soil bacterial communities and promotes the growth of Mandarin orange. *European Journal of Soil Biology*, *79*, 48–56. https://doi.org/10.1016/j.ejsobi.2017.02.003.

Tripathi, P., Khare, P., Barnawal, D., Shanker, K., Srivastava, P. K., Tripathi, R. D., & Kalra, A. (2020). Bioremediation of arsenic by soil methylating fungi: Role of Humicola sp. strain 2WS1 in amelioration of arsenic phytotoxicity in Bacopa monnieri L. *Science of the Total Environment*, *716*, 136758. https://doi.org/10.1016/j.scitotenv.2020.136758.

Uzair, B., Kausar, R., Bano, S. A., Fatima, S., Badshah, M., Habiba, U., & Fasim, F. (2018). Isolation and molecular characterization of a model antagonistic pseudomonas aeruginosa divulging in vitro plant growth promoting characteristics. *BioMed Research International*, *2018*, 1–7. https://doi.org/10.1155/2018/6147380.

Vanstraelen, M., & Benková, E. (2012). Hormonal interactions in the regulation of plant development. *Annual Review of Cell and Developmental Biology*, *28*(1), 463–487. https://doi.org/10.1146/annurev-cellbio-101011-155741.

Verma, P., Yadav, A. N., Kumar, V., Singh, D. P., & Saxena, A. K. (2017). Beneficial plant-microbes interactions: Biodiversity of microbes from diverse extreme environments and its impact for crop improvement. *Plant-Microbe Interactions in Agro-Ecological Perspectives*, 543–580. https://doi.org/10.1007/978-981-10-6593-4_22.

Verma, P. P., Shelake, R. M., Das, S., Sharma, P. & Kim, Y. M. (2019). Plant growth-promoting rhizobacteria (PGPR) and fungi (PGPF): Potential biological control agents of diseases and pests. In: *Microbial Interventions in Agriculture and Environment*, vol. 1: Research Trends, Priorities and Prospects. Springer, Singapore, pp. 281–311. https://doi.org/10.1007/978-981-13-8391-5_11.

Vidhyasri, M., Gomathi, V., & Kumar, U. S. (2019). Plant growth promotion of rice as influenced by Ochrobactrum sp. (MH685438) a Rhizospheric bacteria associated with Oryzae sativa. *International Journal of Current Microbiology and Applied Sciences*, *8*(05), 901–909. https://doi.org/10.20546/ijcmas.2019.805.105.

Vinoth, S., Gurusaravanan, P., & Jayabalan, N. (2012). Effect of seaweed extracts and plant growth regulators on high-frequency in vitro mass propagation of Lycopersicon esculentum L (tomato) through double cotyledonary nodal explant. *Journal of Applied Phycology*, *24*(5), 1329–1337. https://doi.org/10.1007/s10811-011-9717-9.

Vranova, V., Rejsek, K., Skene, K. R., & Formanek, P. (2011). Non-protein amino acids: Plant, soil and ecosystem interactions. *Plant and Soil*, *342*(1–2), 31–48. https://doi.org/10.1007/s11104-010-0673-y.

Vurukonda, S. S., Giovanardi, D., & Stefani, E. (2018). Plant growth promoting and biocontrol activity of Streptomyces spp. as endophytes. *International Journal of Molecular Sciences*, *19*(4), 952. https://doi.org/10.3390/ijms19040952.

Waghunde, R. R., Shelake, R. M., Shinde, M. S. et al. (2017). Endophyte microbes: A weapon for plant health management. In Deepak G. Panpatte, Yogeshvari K. Jhala, Rajababu V. Vyas, Harsha N. Shela (Eds.), *Microorganisms for Green Revolution*. Springer, Singapore, pp. 303–325.

Wakelin, S. A., Gupta, V. V., Harvey, P. R., Ryder, M. H. (2007). The effect of Penicillium fungi on plant growth and phosphorus mobilization in neutral to alkaline soils from southern Australia. *Canadian Journal of Microbiology*, *53*(1), 106–115.

Wang, W., Vinocur, B., & Altman, A. (2003). Plant responses to drought, salinity and extreme temperatures: Towards genetic engineering for stress tolerance. *Planta*, *218*(1), 1–14. https://doi.org/10.1007/s00425-003-1105-5.

Wu, L., Shang, H., Wang, Q., Gu, H., Liu, G., & Yang, S. (2016). Isolation and characterization of antagonistic endophytes from Dendrobium candidum Wall ex Lindl., and the biofertilizing potential of a novel Pseudomonas saponiphila strain. *Applied Soil Ecology*. 105, 101–108. https://doi.org/10.1016/j.apsoil.2016.04.008.

Wu, Y., Ma, L., Liu, Q., Vestergård, M., Topalovic, O., Wang, Q., & Feng, Y. (2020). The plant-growth promoting bacteria promote cadmium uptake by inducing a hormonal crosstalk and lateral root formation in a hyperaccumulator plant Sedum alfredii. *Journal of hazardous materials*, *395*, 122661.

Wu, T., Xu, J., Guo, W., Xia, J., Li, X., & Wang, R. (2019). Isolation and characterization of a Halotolerant, hydrocarbon-degrading endophytic bacterium from halobiotic reeds (Phragmites australis) growing in petroleum-contaminated soils. *Science of Advanced Materials*, *11*(2), 189–195. https://doi.org/10.1166/sam.2019.3438.

Wu, T., Xu, J., Xie, W., Yao, Z., Yang, H., Sun, C., & Li, X. (2018). Pseudomonas aeruginosa L10: A hydrocarbon-degrading, biosurfactant-producing, and plant-growth-Promoting endophytic bacterium isolated from a reed (Phragmites australis). *Frontiers in Microbiology*, 9. https://doi.org/10.3389/fmicb.2018.01087.

Xia, A. Y., Farooq, M. A., Javed, M. T., Kamran, M. A., Mukhtar, T., Ali, J., Tabassum, T., Rehman, S., Munis, M. F. H, Sultan, T., & Chaudhary, H. J. (2020). Multi-stress tolerant PGPR Bacillus xiamenensis PM14 activating sugarcane (Saccharum officinarum L.) red rot disease resistance. *Plant Physiology and Biochemistry*. https://doi.org/10.1016/j.plaphy.2020.04.016.

Xun, F., Xie, B., Liu, S., & Guo, C. (2015). Effect of plant growth-promoting bacteria (PGPR) and arbuscular mycorrhizal fungi (AMF) inoculation on oats in saline-alkali soil contaminated by petroleum to enhance phytoremediation. *Environmental Science and Pollution Research*, *22*(1), 598–608. https://doi.org/10.1007/s11356-014-3396-4.

Yadav, A. N. (2018). Biodiversity and biotechnological applications of host-specific endophytic fungi for sustainable agriculture and allied sectors. *Acta Scientific Microbiology*, *1*(5), 01–05.

Yadav, A. N., Kumar, R., Kumar, S., Kumar, V., Sugitha, T. C. K., Singh, B., Chauhan, V. S., Dhaliwal, H. S., & Saxena, A. K. (2017). Beneficial microbiomes: Biodiversity and potential biotechnological applications for sustainable agriculture and human health. *Journal of Applied Biology & Biotechnology*, *5*(6), 1–31. https://doi.org/10.7324/jabb.2017.50607.

Yadav, A. N., Kumar, V., Dhaliwal, H. S., Prasad, R., & Saxena, A. K. (2018). Microbiome in crops: Diversity, distribution, and potential role in crop improvement. *Crop Improvement through Microbial Biotechnology*, 305–332. https://doi.org/10.1016/b978-0-444-63987-5.00015-3.

Yaish, M. W., Antony, I., & Glick, B. R. (2015). Isolation and characterization of endophytic plant growth-promoting bacteria from date palm tree (Phoenix dactylifera L.) and their potential role in salinity tolerance. *Antonie van Leeuwenhoek*, *107*(6), 1519–1532. https://doi.org/10.1007/s10482-015-0445-z.

Yakhin, O. I., Lubyanov, A. A., Yakhin, I. A., & Brown, P. H. (2017). Biostimulants in plant science: A global perspective. *Frontiers in Plant Science*, 7. https://doi.org/10.3389/fpls.2016.02049.

Yaish, M. W., Antony, I. & Glick, B. R. (2015) Isolation and characterization of endophytic plant growth-promoting bacteria from date palm tree (Phoenix dactylifera L.) and their potential role in salinity tolerance. *Antonie Van Leeuwenhoek*, *107*(6), 1519–1532. https://doi.org/10.1007/s10482-015-0445-z. Epub 2015 Apr 10. PMID: 25860542.

Yuan, J., Zhang, N., Huang, Q., Raza, W., Li, R., Vivanco, J. M., & Shen, Q. (2015). Organic acids from root exudates of banana help root colonization of PGPR strain bacillus amyloliquefaciens NJN-6. *Scientific Reports*, *5*(1). https://doi.org/10.1038/srep13438.

Zahoor, M., Irshad, M., Rahman, H., Qasim, M., Afridi, S. G., Qadir, M., & Hussain, A. (2017). Alleviation of heavy metal toxicity and phytostimulation of brassica campestris L. by endophytic mucor sp. MHR-7. *Ecotoxicology and Environmental Safety*, *142*, 139–149. https://doi.org/10.1016/j.ecoenv.2017.04.005.

Zhou, C., Guo, J., Zhu, L., Xiao, X., Xie, Y., Zhu, J., Ma, Z., & Wang, J. (2016). Paenibacillus polymyxa BFKC01 enhances plant iron absorption via improved root systems and activated iron acquisition mechanisms. *Plant Physiology and Biochemistry*, *105*, 162–173. https://doi.org/10.1016/j.plaphy.2016.04.025.

Zhou, L., Tang, K., & Guo, S. (2018). The plant growth-promoting fungus (PGPF) Alternaria sp. A13 markedly enhances salvia miltiorrhiza root growth and active ingredient accumulation under

greenhouse and field conditions. *International Journal of Molecular Sciences*, *19*(1), *270*. https://doi.
    org/10.3390/ijms19010270.
Zhu, F., Qu, L., Hong, X., & Sun, X. (2011). Isolation and characterization of a phosphate-solubilizing Halophilic
    BacteriumKushneriasp. YCWA18 from Daqiao saltern on the coast of Yellow Sea of China. *Evidence-
    Based Complementary and Alternative Medicine*, *2011*, 1–6. https://doi.org/10.1155/2011/615032.
Zhu, X., Wang, W., Crowley, D. E., Sun, K., Hao, S., Waigi, M. G., & Gao, Y. (2017). The endophytic bacte-
    rium Serratia sp. PW7 degrades pyrene in wheat. *Environmental Science and Pollution Research*, *24*(7),
    6648–6656. https://doi.org/10.1007/s11356-016-8345-y.
Zubair, M., Shakir, M., Ali, Q., Rani, N., Fatima, N., Farooq, S., Shafiq, S., Kanwal, N., Ali, F., & Nasir, I. A.
    (2016). Rhizobacteria and phytoremediation of heavy metals. *Environmental Technology Reviews*, *5*(1),
    112–119. https://doi.org/10.1080/21622515.2016.1259358.

5 Genome Engineering in Agriculturally Beneficial Microorganisms Using CRISPR-Cas9 Technology

*Darshan T. Dharajiya, Yogesh R. Patel,
Kapil K. Tiwari and L. D. Parmar*
Sardarkrushinagar Dantiwada Agricultural University (SDAU)

CONTENTS

5.1 CRISPR–CAS9 TECHNOLOGY – A POWERFUL TOOL FOR GENOME ENGINEERING

5.1.1 INTRODUCTION TO GENOME ENGINEERING

The facility to modify and edit genetic information is very crucial for understanding gene function and revealing biological mechanisms. Scientists have been controlling and making use of prokaryotic molecules for genome engineering since 1971 when the production of particular DNA fragments using restriction enzymes was demonstrated for the first time. DNA-binding proteins which modify specific loci have extremely advanced science and applications. However, developing

modular DNA-binding proteins to bind at a defined target is a very complex process which often needs protein engineering expertise. This difficulty has been solved by the clustered regularly interspaced short palindromic repeats (CRISPR)-CRISPR-associated protein 9 (Cas9) technology as the target specificity of CRISPR–Cas9 depends on base pairing of nucleic acids rather than DNA-protein interaction (Pickar-Oliver and Gersbach 2019). In the past few years, highly versatile genome-editing technology, CRISPR–Cas9 has transformed genome engineering by providing investigators with the ability to introduce sequence-specific alterations into the genomes of a broad range of cell types and organisms (Gaj et al. 2016). The most common genome engineering/editing technologies are homing endonucleases (HENs) or meganucleases, zinc-finger nucleases (ZFNs), transcription activator-like effector nucleases (TALENS), and CRISPR-Cas9. Developments in genome editing technologies such as HENs, ZFNs and TALENS made it possible to more accurately target any gene of interest. However, these first-generation genome editing techniques involve difficult steps for protein engineering, which make them costly, laborious and time-consuming. Unlike those techniques, CRISPR-Cas9 technology involves simple designing and cloning procedures for the use of the same Cas9 with different guide RNAs (gRNAs) targeting multiple sites in the genome (Jaganathan et al. 2018).

5.1.2 Basics and History of CRISPR–Cas9 Technology

CRISPR-associated, RNA-guided endonuclease (RGENs) (Cas9) is the latest genome editing technology that has been discovered in bacteria as a unique defense mechanism to protect against invading bacteriophages (El-Sayed et al. 2017). The CRISPR concept was begun coincidentally when an uncommon structure was detected within *E. coli* genome while examining the *iap* gene (Ishino et al. 1987). CRISPR comprises repeats (short identical repeated sequences of DNA), separated by spacers (unusual short sequences originated from the invading phages or plasmid DNA) (Hsu et al. 2014). Jansen et al. (2002) named these unusual repeats as CRISPR. The history of CRISPR-Cas research is summarized in Table 5.1. The spacer sequences play a role as a genetic memory. The CRISPR system is a weapon for microorganisms to protect themselves from attack by bacteriophages and viruses (Bolotin et al. 2005; Pourcel et al. 2005).

5.1.3 Mechanism of CRISPR-Cas System

The CRISPR-mediated resistance mechanism involves three major stages (Figure 5.1).

1. Spacer (foreign DNA) acquisition
2. crRNA (CRISPR RNA) biogenesis/processing
3. RNA-guided target (viral element) interference

The first stage is spacer acquisition (also known as adaptation), which takes place in two steps: (i) recognition of 33 bp protospacer sequence through Cas1/Cas2 complex (two Cas1 dimers and one Cas2 dimer) by binding to foreign DNA (viruses or plasmids) in a sequence-independent way (Wiedenheft et al. 2009) and (ii) integration of these sequences into CRISPR locus by Cas1/Cas2 complex, which acts as integrases to perform two nucleophilic reactions (Nuñez et al. 2014). The second stage involves the transcription and processing of a CRISPR array into a mixture of small crRNAs, which contains a spacer (colored squares) and partial repeats (black squares), forming gRNA. The CRISPR array works as a memory of past invasions and aids in immunity on later exposure. An extended primary RNA molecule transcribed from the CRISPR locus known as pre-CRISPR RNA (pre-crRNA) is a transcript of partial repeats (black) and spacer sequences (colored). The pre-crRNA is processed by the host's RNaseIII, with the assistance of another RNA molecule known as the trans-activating CRISPR RNA (tracrRNA). The tracrRNA comprises the direct repeats so that it can attach to the pre-crRNA through sequence homology to

TABLE 5.1
Timeline of CRISPR-Cas Research

Year	Research Achievements	Reference
1987	First report of CRISPR clustered repeats	Ishino et al. (1987)
2000	CRISPR families exist in all prokaryotes	Mojica et al. (2000)
2002	Coined 'CRISPR' term and defined cas genes	Jansen et al. (2002)
2005	Identified foreign source of spacers, Proposed function of adaptive immunity	Mojica et al. (2005); Pourcel et al. (2005)
	Identified PAM sequence	Bolotin et al. (2005)
2007	First experimental proof of CRISPR adaptive immunity	Barrangou et al. (2007)
2008	CRISPR acts upon DNA targets	Marraffini and Sontheimer (2008)
	Spacers are converted into mature crRNAs that act as small gRNAs	Brouns et al. (2008)
2009	Type III-B Cmr CRISPR complexes cleave RNA	Hale et al. (2009)
2010	Cas9 (Type II) is guided by spacer sequences and cuts target DNA by DSBs	Garneau et al. (2010)
2011	tracrRNA forms a duplex structure with crRNA in association with Cas9	Deltcheva et al. (2011)
	Type II CRISPR systems are modular and can be heterologously expressed in other organisms	Sapranauskas et al. (2011)
2012	Cas9 is an RNA-guided DNA endonuclease	Jinek et al. (2012)
	In vitro characterization of DNA targeting by Cas9	Gasiunas et al. (2012)
2013	First experimentation of Cas9 genome editing in eukaryotic cells	Cong et al. (2013); Mali et al. (2013)
	CRISPR-Cas mediated plant genome engineering in rice, wheat, *Arabidopsis*, tobacco and *Sorghum*	Jiang et al. (2013)
2014	Genome-wide functional screening with Cas9	Shalem et al. (2014); Wang et al. (2014)
	Discovered crystal structure of apo-Cas9	Jinek et al. (2014)
	Discovered crystal structure of Cas9 in complex with gRNA and target DNA	Nishimasu et al. (2014)
2015	First report of genes edited in human embryos CRISPR-Cas9	Liang et al. (2015)
2016	Published research on upgraded type of CRISPR/Cas 9 with less risk of off-target DNA breaks	Kleinstiver et al. (2016)
2017	Published research on possibility of editing a gene defect in pre-implanted human embryos for preventing inherited heart disease, β-thalassemia and to study cause of infertility	Fogarty et al. (2017); Liang et al. (2017); Ma et al. (2017)

crRNAs: crispr RNAs, gRNAs: guide RNA, tracrRNA: trans-activating crispr RNA, PAM: protospacer adjacent motif, DSBs: Double-strand breaks.

form pre-crRNA:tracrRNA duplex. The host's RNaseIII cleaves dsRNA duplex to form mature crRNA which contains one spacer sequence and one repeat sequence. The spacer sequence delivers specificity for various exogenous target sequences due to complementarity with the target. The third stage comprises target interference by recognition of invading nucleic acid through sequence complementary to the spacer sequence of the crRNA and destruction by Cas nucleases (Barrangou et al. 2007). The crRNA:tracrRNA duplex binds to the Cas9 (DNA endonuclease) through the tracrRNA. The crRNA leads Cas9 to the target sequence through sequence homology provided by the spacer sequence and bind to the target through base-pairing to the complementary strand. Spacers get converted into mature crRNAs and act as gRNA to recognize and destroy the target foreign DNA (Brouns et al. 2008). A protospacer adjacent motif (PAM) sequence at downstream (3') of the target sequence is a short (3–5 nucleotide) sequence essential for nuclease binding. The PAM sequence for Cas9 is NGG, where the N can be any nucleotide. Ultimately, the nuclease (Cas9) cannot bind to the target sequence in the absence of the PAM sequence. The absence of the PAM sequence in the

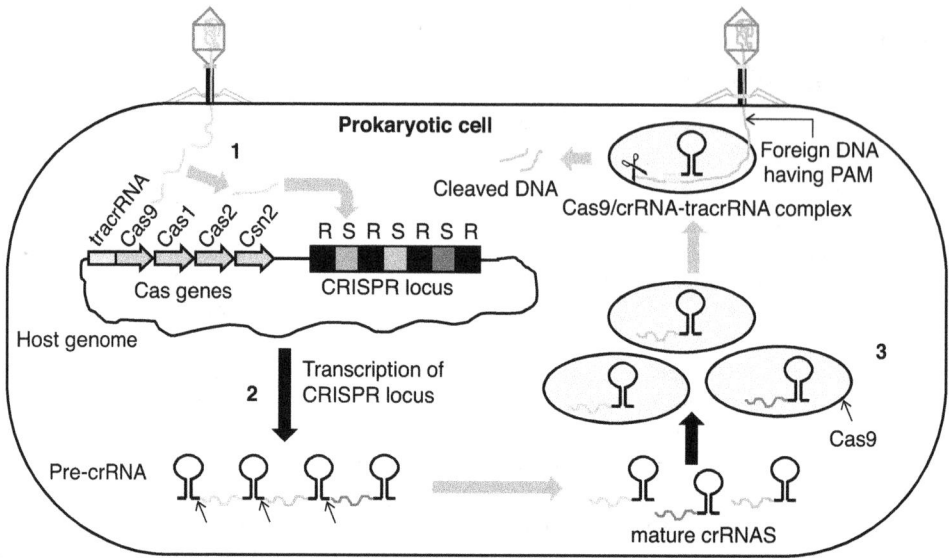

FIGURE 5.1   The mechanism of the CRISPR-Cas9 [1: stage 1 (spacer acquisition), 2: stage 2 (crRNA processing), 3: stage 3 (target interference), S: spacer sequence, R: repeat sequence].

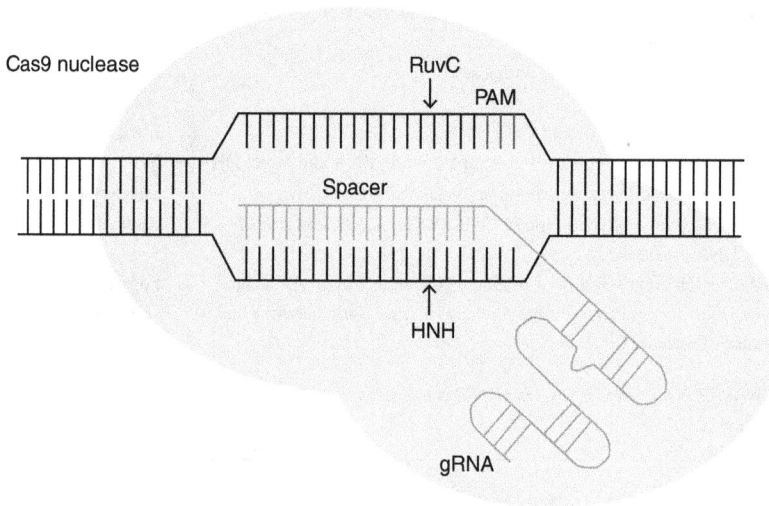

FIGURE 5.2   Schematic of the CRISPR/Cas9 system. (DNA and proteins are not drawn to scale).

spacer region of the endogenous CRISPR array inhibits the nuclease from cleaving and destroying its DNA.

The Cas9 endonuclease from *Streptococcus pyogenes* possesses two nuclease domains, which collectively create a double-stranded break (DSB) in the target DNA (Marraffini 2016). One domain is a RuvC-like nuclease located at the N-terminal and the second is an HNH-like domain located around the center of the protein (Shan et al. 2013). After binding at the PAM sequence/site, the Cas9 go through a conformational modification which directs the nuclease domains to bind with opposite strands. The DSB results from about three nucleotides upstream of the PAM sequence. The nuclease domain cleaves the strand non-complementary to the spacer in the crRNA, while the HNH domain breaks the complementary strand (Figure 5.2).

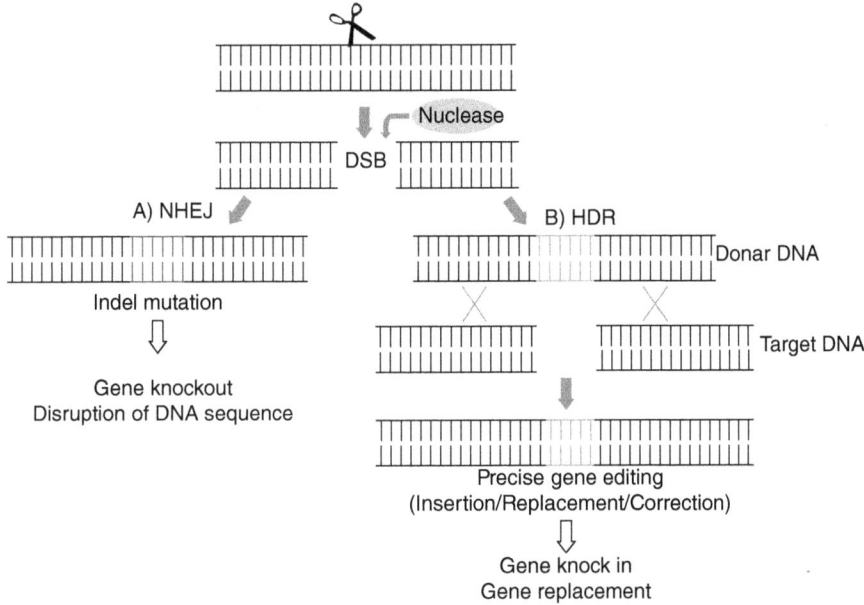

FIGURE 5.3 Repairing mechanisms of sequence-specific nuclease (SSN) (Cas9) induced DSBs of DNA by NHEJ and HDR pathways. [NHEJ (non-homologous end joining) -mediated repair can insert and/or delete a few nucleotides (indel mutation). HDR (homology-directed repair) - mediated repair can introduce precise point mutations, by corrections/insertions/replacement, based on the donor DNA template].

The Cas9-generated DSBs can be repaired by either error-prone non-homologous end joining (NHEJ) (Davis and Chen 2013) or homologous recombination (HR)/homology-directed repair (HDR) (Sung and Klein 2006) which stitch the two ends together (Figure 5.3). The DSB repaired by NHEJ contains small insertions or deletions (indels) at the target site that disrupt gene function (Lieber 2010; Chaudhary et al. 2018). Alternatively, if DSBs are repaired by HDR using donor template that mediates the exchange of the target sequences with the donor sequences, it leads to precise and perfect gene replacement and additions (Chiruvella et al. 2013; Kim and Kim 2014). The main characteristics of RGENs/CRISPR nucleases are listed in Table 5.2.

5.1.4 CLASSIFICATION OF CAS PROTEINS IN CRISPR-CAS SYSTEMS

Since CRISPR can be exploited for genome engineering in human cells (Cong et al. 2013; Mali et al. 2013), it has been extensively accepted to conduct various applications of genome editing. The CRISPR-Cas system has been classified into six types and furthermore divided into sixteen different subtypes (Makarova et al. 2015). There are two classes (class 1 and class 2) and six major types of CRISPR which includes (i) type I, III and IV defined by multi-subunit effector complexes (Cascade, Csm, Cmr) and (ii) type II, V and VI identified by single-subunit effector (Cpf1, Cas9) (Zetsche et al. 2015; Mougiakos et al. 2016; Ishino et al. 2018). The classification of Cas proteins in CRISPR-Cas systems according to their mode of action is illustrated in Figure 5.4.

5.1.5 ONLINE RESOURCES TO DESIGN CRISPR NUCLEASES

In recent times, genome editing with RGENs/CRISPR-associated nuclease has been shown to have great potential in a variety of applications in a broad spectrum of fields. However, their execution in genetic analysis generally depends on their specificity for the anticipated genomic target. Complex and large genomes often comprise very much repetitive/homologous sequences, which

TABLE 5.2

Main Characteristics of RGENs/CRISPR Nucleases (Cas9)

Characteristics	RGENs/CRISPR Nucleases
Nuclease	Cas9
Determinant of DNA targeting specificity	sgRNA or crRNA
Specificity-determining length of target site	22 bp (total length 23 bp)
Rate of success	High (approx. 90%)
Off-target effects	Variable
Average rate of mutation	High (approx. 20%)
Size	4.2 kb (Cas9 from *Streptococcus pyogenes*) + 0.1 kb (sgRNA)
Restriction in target site	End with an PAM sequence NGG or NAG (lower activity)
Design density	One per 8 bp (NGG PAM) or 4 bp (NGG and NAG PAM)
Cytotoxicity	Low

Class	Type	Subtypes	Adaptation	Pre-crRNA Transcript Processing	Target Binding	Target Cleavage
1	I	A-F, U	Cas1 Cas2 Cas4	Cas6	Cas7, Cas5, Cas8 (LS), SS, HD Cas3	Cas3
	III	A-D	Cas1 Cas2	Cas6	Cas7, Cas5, Cas10 (LS), SS,	Cas7 Cas10
	IV	A-B	Not known	Not known	Cas7, Cas5, Caf1 (LS), SS,	Not known
2	II	A-C	Cas1 Cas2 Cas4 Cas9	RNase III (Cas9)	Cas9	Cas9
	V	A-B, U	Cas1 Cas2 Cas4	Not known	Cas12	Cas12
	VI	A-C	Cas1 Cas2	Not known	Cas13	Cas13

FIGURE 5.4 Classification of Cas proteins in CRISPR-Cas systems according to their mode of action.

reduces the specificity of genome editing tools and could result in off-target activity (Periwal 2017). Recently, various computational approaches have been recognized to guide the design process and predict/reduce the off-target activity of CRISPR nucleases. Numerous databases, web servers, tools, and resources for genome editing could be competently used to guide the design of constructs for CRISPR nucleases and evaluate results after genome editing. Several computer programs are accessible that search for possible target sites of programmable nuclease in a particular DNA sequence

TABLE 5.3

Online Resources to Design RGENs/CRISPR Nucleases

Online Resources	Link	Applications	Reference
CRISPRs web server/ CRISPRcompar	http://crispr.u-psud.fr/	A gateway to publicly accessible CRISPRs database and software including CRISPRFinder, CRISPRdb and CRISPRcompar	Grissa et al. (2008)
CRISPI	http://crispi.genouest.org/	A web interface with graphical tools and functions allows users to find CRISPR in personal sequences.	Rousseau et al. (2009)
CRISPRTarget	http://bioanalysis.otago. ac.nz/CRISPRTarget	It predicts the most likely targets of CRISPR RNAs.	Biswas et al. (2013)
Zhang Lab Genome Engineering	http://www.genome-engineering.org/	CRISPR genome engineering resources website.	Cong et al. (2013)
CRISPRmap	http://rna.informatik.uni-freiburg.de/CRISPRmap	Web server provides an automated assignment of newly sequenced CRISPRs to standard classification system	Lange et al. (2013)
Crass: The CRISPR Assembler	http://ctskennerton.github. io/crass/	A program that searches through raw metagenomic reads for CRISPRs.	Skennerton et al. (2013)
CRISPRDetect	http://bioanalysis.otago. ac.nz/CRISPRDetect/	CRISPRDetect, in combination with CRISPRBank and CRISPRTarget, now provides an integrated resource for the detection and analysis of CRISPRs (CRISPRSuite)	Biswas et al. (2014)
E-CRISP	http://www.e-crisp.org/	It is a software tool to design and evaluate CRISPR target sites.	n/a
CRISPR RGEN Tools	http://www.rgenome.net/	Computational tools and libraries for RGENs.	n/a

and web-based computer programs are accessible for RGENs (Table 5.3). These computer algorithms and programs have different features and applications.

5.1.6 CRISPRi and CRISPRa for Precise Control of Gene Expression

When NHEJ-mediated repair occurs in CRISPR-based technology, Cas9 introduces short deletion(s) within a protein-coding open reading frame (ORF) which results in frame shifts that ultimately lead to a loss of function of the encoded protein. Numerous platforms for genetic screening have been applied and created on this strategy. CRISPR interference (CRISPRi) and CRISPR activation (CRISPRa) are approaches for reversible control of gene expression. In CRISPRi, nuclease-dead mutants of Cas9 (dCas9) retain sgRNA-directed binding of specific DNA sequences, which can stop the transcription of these genes in bacteria (Qi et al. 2013). CRISPRi using dCas9 can also perform transcription repression in mammalian cells. However, it is more effective when transcriptional repressor domains i.e. Krüppel-associated box (KRAB) are fused to dCas9 (Kampmann 2018). The fusion of dCas9 to the KRAB domain promotes the formation of heterochromatin and results in CRISPRi. The dCas9 can also be employed for the initiation of gene expression and this approach is known as CRISPRa (Gilbert et al. 2013). The CRISPRi is a loss-of-function technology while CRISPRa is a gain-of-function technology. Overall, CRISPRi and CRISPRa are useful in controlling transcript levels of endogenous genes (Kampmann 2018).

5.1.7 Applications of CRISPR-Cas9 Technology

Altered Cas9 versions (i.e., dCas9 and nCas9) have been employed for the development of programmable tools for genome editing using the CRISPR-Cas9 system. These CRISPR-based tools demonstrated their use in single base editing, RNA editing, gene regulation, genotyping, DNA barcoding,

epigenetic editing, gene tagging, chromatin engineering, imaging, gene targeting and many more. The next generation of CRISPR-based tools extended beyond DSB-based gene editing and imparted the competency to these tools to accurately target the DNA region (Shelake et al. 2019). The ease with which CRISPR-Cas9 can be constructed to identify novel genomic sequences has driven a revolution in genome editing that has enhanced scientific developments and discoveries in diverse disciplines such as synthetic biology, disease modeling, human gene therapy, neuroscience, drug discovery, and agricultural sciences (Gaj et al. 2016).

5.2 AGRICULTURALLY BENEFICIAL MICROORGANISMS

Agriculture mainly deals with the cultivation of plants and rearing of livestock. To enhance crop yield, extensive use of synthetic fertilizers, pesticides, heavy irrigation, monoculture pattern, high energy input, extensive tillage and concentrated animal feeding led to several negative impacts on agriculture. Therefore, it has become very crucial to address the major thrust area of agriculture like decreased soil integrity and fertility, ground water purity and environmental pollution nowadays (Wati et al. 2015). The theory of sustainable agriculture emphasises better usage of natural resources for higher agricultural productivity with no negative impact on climate and has been accepted by many people. The use of agriculturally beneficial microorganisms is very important for the production of healthy and safe agricultural products in an eco-friendly manner.

In agriculture, the relationship between plants and microorganisms is very complex. This relationship is either beneficial or harmful to the plant (Figure 5.5). Negative relationship results in plant diseases caused by pathogens. Positive relationship, on the other hand, offers many beneficial effects to the plant like growth and development, nutrient uptake, seed germination, plant defense against biotic and abiotic stresses, and improved crop yield which are due to plant-associated microbes, mostly rhizospheric. Such microorganisms have their unique structure, interaction, function, and habitat which perform various biotic activities in soil ecosystem (Kumar et al. 2018). Certain microorganisms have an obligate relationship and are limited to specific host plants, i.e. nodule forming rhizobia and leguminous plants, and certain endomycorrhizal fungi (Parniske 2008; Masson-Boivin et al. 2009). Some microorganisms display a non-specific relationship with plants and have the ability to colonize mostly on the surface of the plant root system and sometimes inside the root tissues to enhance plant growth and overall health of plant (Barea et al. 2005). In general, these microorganisms have beneficial effects on plants via direct or indirect activities. Direct mechanism involves nitrogen fixation, solubilization of phosphate, potassium, zinc, and other minerals, production of phytohormones like auxins, gibberellins, cytokinins, and abscisic acid. The indirect mechanism

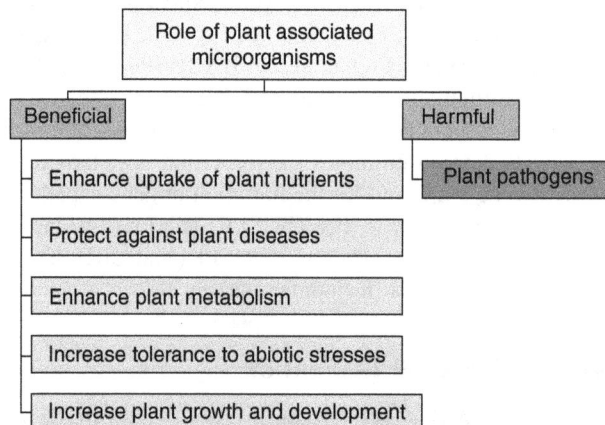

FIGURE 5.5 Effects of plant associated microorganisms on plant.

deals with the production of various chemical compounds (siderophores, cyanides, antibiotics, and lytic enzymes) to protect against plant pathogens (Stamenković et al. 2018). The microorganisms belonging to bacteria, fungi, actinomycetes and cyanobacteria groups are such agriculturally beneficial microorganisms.

5.2.1 IMPORTANT FUNCTIONS OF AGRICULTURALLY BENEFICIAL MICROORGANISMS

5.2.1.1 Microorganisms as Fertilizer

Soil and atmosphere provide nutrients to plants. The nutrients present in the soil are absorbed by root system in rhizosphere area. The phyllosphere on the other hand absorbs atmospheric nutrients (Bhattacharyya et al. 2016). The use of chemical fertilizers has several negative consequences like leaching out problem, surface and ground water contamination, reduction of soil fertility and integrity, eutrophication in water bodies, destruction of soil microorganisms and important insects, and increased pathogen susceptibility of plants (Wati et al. 2015). Plant root system releases certain compounds collectively termed as root exudates which influence different activities of rhizospheric microorganisms.

The enhanced activities of some beneficial microorganisms consequently lead to improved nutrient acquisition by plants. Biofertilizers are microbial preparations containing living or latent cells of beneficial microorganisms that interact with plants in rhizosphere and help in the uptake of nutrients from the soil. Nitrogen fixation is a prominent case in which different classes of bacteria are involved in the fixation of atmospheric nitrogen into ammonia or related nitrogenous compounds. The root-nodulating bacterial members of genera *Rhizobium, Sinorhizobium, Mesorhizobium, Bradyrhizobium,* and *Ensifer* make symbiotic associations with legume plants and facilitate the uptake of nitrogen by plants (Rathoure 2015). *Frankia* is one more group of nitrogen-fixing bacteria that live in symbiotic association with certain plants and induce nodulation in such plants. Free living nitrogen fixing bacteria belonging to *Azotobacter, Diazotrophicus, Gluconacetobacter, Azospirillum, Pseudomonas, Enterobacter, Acetobacter, Herbaspirillum, Bacillus, Azoarcus, and* Cyanobacteria *(Anabaena* and *Nostoc)* are involved in nitrogen fixation (Nath et al. 2018). The role of endophytic bacteria (*Gluconobacter, Azoarcus, Klebsiella, Herbaspirillum*, and *Burkholderia*) in nitrogen fixation in certain plant species is well documented (Sharma et al. 2018).

Phosphorus is the second most important macro-nutrient for plant growth. The occurrence of an insoluble form of phosphate in soil makes its utilization by plants more difficult. The conversion of insoluble to soluble form of phosphate is performed by number of phosphate solubilizing bacteria (*Pseudomonas, Rhizobium, Mycobacterium, Bacillus, Burkholderia, Rhodococcus, Enterobacter, Arthrobacter, Serratia, Beijerinckia, Flavobacterium*, and *Erwinia*), filamentous fungi (*Penicillium, Arthrobotrys,* and *Aspergillus*), and actomycorrhizal fungi (*Glomus* and *Hebeloma*) (Nath et al. 2018). These microorganisms have the capability of producing organic acids, which accelerates the mobilization of phosphate. Potassium has an essential role in ATP and nucleic acid synthesis, regulation of osmotic pressure, activation of enzymes, and many more. The solubilization of potassium from mica, illite, and other insoluble potassium sources is performed by a number of microorganisms like *Bacillus, Pseudomonas, Thiobacillus, Aspergillus, Micrococcus, Enterobacter,* and *Corynebacterium* (Sugumaran and Janarthanam 2007; Verma et al. 2016). The potential application of biofertilizer microorganisms is also explored for enhanced uptake of other nutrients like Zn, Ca, Fe, Mn, Si and Cu by plants. The usage of biofertilizer reduces input costs of chemical fertilizers and act as an important component of integrated nutrient management for sustainable agriculture.

5.2.1.2 Microorganisms as Plant Growth Promoter

Large numbers of microorganisms present in rhizosphere have plant growth promoting activities. They produce and secrete various phytostimulants in the vicinity of plant roots. Phytohormones like auxins, gibberellins, cytokinins, and abscisic acid are such plant growth stimulants (Gopalakrishnan et al. 2014). The effectual role of auxin in plant growth and development is well understood. It has

several functions like development of longer root with higher number of root hairs and lateral roots, stimulation of cellular division and differentiation, improvement of water and mineral uptake by increasing number of xylem cells, induction of flowering and fruiting, and inhibition or delaying abscission (Mohite 2013; Stamenković et al. 2018). Indole acetic acid (IAA) is an important form of physiologically active auxin. It is produced by several agriculturally important microorganisms like *Pseudomonas*, *Bacillus*, *Rhizobium*, *Azospirillum*, and many more.

Various developmental processes in plants like elongation of stem, breaking of seed dormancy, increased germination, development of larger sized fruits, higher number of buds, stimulation of parthenocarpy, and sex expression are performed by gibberellins (Escamilla 2000; Gelmi et al. 2000). Different genera of bacteria (*Bacillus*, *Pseudomonas*, *Rhizobium*, *Azospirillum*, and *Azotobacter*) and fungi (*Gibberella*, *Fusarium*, *Neurospora*, *Phaeosphaeria*, and *Sphaceloma*) have the ability to produce various forms of gibberellins (Desai 2017).

Cytokinin is third phytohormone which stimulates cellular division, formation of roots and root hairs, callus development, and shoot differentiation in certain plants (Gopalakrishnan et al. 2015; Vijayabharathi et al. 2016). Vast varieties of rhizospheric microorganisms (*Rhizobium* spp., *Azotobacter* spp., *Paenibacillus polymyxa*, *Pseudomonas fluorescens*, *Pantoea agglomerans*, *Bacillus subtilis* and *Rhodospirillum rubrum*) have been studied for their capacity to produce cytokinin (Glick 2012; García de Salamone et al. 2001).

Ethylene is a stress hormone that is up-regulated in plants in response to various abiotic stresses and biotic stress like fungal, viral and bacterial infections resulting in negative growth and development of plants. Rhizospheric microorganisms like bacteria (*Pseudomonas*, *Bacillus*, *Variovorax*, and *Achromobacter*) and fungus (*Trichoderma*) contain the enzyme 1-aminocyclopropane-1-carboxylate (ACC) deaminase which catalyses the conversion of ethylene precursor ACC into α-ketobutyrate and ammonia. This reaction reduces the level of secreted ACC and thereby ethylene level in plants. Therefore, ACC deaminase-producing microorganisms help plants to respond against a variety of stress conditions (Singh and Singh 2015).

Apart from phytohormones, rhizospheric microorganisms produce and release certain volatile compounds which have plant growth-promoting action. Volatile organic compounds (VOCs) such as acetoin, 2,3-butanediol, jasmonates, terpenes, etc. are produced by different species of rhizospheric bacteria (Ping and Boland 2004). Such VOCs act as signaling molecules for plant-microbe interactions and thereby stimulates certain plant responses (Ryu et al. 2003).

5.2.1.3 Microorganisms as Stress Defender

The overall plant health and development is highly influenced by various factors. These factors are broadly classified as abiotic and biotic stress. Abiotic stress is the certain nonliving environmental conditions which badly affect the plant. This includes pH, temperature, salts, water, and essential nutrients. The interaction between plants and microorganisms augments certain responses (local and/or systemic) which are capable to deal with changing abiotic stresses. The plant-associated microorganisms help in the regulation of various biochemical and physiological conditions in the plant. Regulation of phytohormone and antioxidant production, exopolysaccharide production, enhancing nutrient uptake, balancing water, organic solute, salt and mineral concentration, production of VOCs, over production of cold and heat shock proteins, siderophore production, production of various regulatory enzymes, microbes associated phytoremediation of pollutants, and many more functions are carried out by these microorganisms in favor to plant against drought, acidity, alkalinity, salinity, cold, heat, and pollutant stress (Shinwari et al. 2019).

The plant diseases caused by pathogens and damage caused by insects come under biotic stress conditions. Various species of bacteria, fungi, nematodes, viruses, insects, and parasitic plants negatively affect plants by causing disease or damage. The biocontrol strategy using different agriculturally beneficial and environmentally friendly microorganisms is a viable alternative for the control of such plant pathogens. The plant protection by biocontrol agents is performed by two mechanisms; direct and indirect antagonism. Direct mechanism deals with the physical interaction of antagonistic microorganisms or their metabolic products with the pathogens. Various extracellular hydrolytic

enzymes (protease, chitinase, lipases, and glucanase), siderophores, antibiotics, and HCN are produced by a number of microorganisms that have direct antagonistic effects (Bhatia et al. 2005; Dutta and Khurana 2015; Pundir and Jain 2015). In certain conditions, microorganisms attack plant pathogens and kill them called as hyperparasitism. In the competition mechanism, the microorganisms compete for colonization and nutrients against pathogens and thereby reducing the chances of host invasion and nutrient acquisition. The secondary metabolites having toxic effects are produced by microorganisms that serve as antibiotics and therefore the growth and metabolism of the pathogen are suppressed (Shinwari et al. 2019). Species of *Pseudomonas*, *Streptomyces*, *Bacillus*, and *Trichoderma* produce antibiotics, which induce systemic resistance in plants (Bakker et al. 2003). Chitinase is responsible for the degradation of the fungal cell wall and gut linings of insects, therefore, has wide application in fungal and insect pest management (Wati et al. 2015). *Pseudomonas fluorescens* is well known bacterial species for its antagonistic activity against several plant pathogens. The fungus, *Trichoderma* is widely used for the control of many fungal plant pathogens due to the production of mycolytic enzymes, antibiotics, nutrient competition, stress tolerance, pathogenic enzymes inactivation, solubilization of minerals, and induction of plant resistance (Monte 2001). The siderophores - low molecular weight iron chelating compounds produced by several microorganisms compete for iron availability with pathogens. *Rhizobium radiobacter*, *Bacillus megaterium*, *Azotobacter vinelandii*, *Pantoea allii*, and *Bacillus subtilis* are examples of siderophores producing microorganism (Ferreira et al. 2019). Various plant growth promoting activities along with ACC deaminase activity are reported in *Bacillus* species. Application of these microorganisms has suppressed the growth of plant pathogenic fungi such as *Rhizoctonia solani*, *Fusarium solani*, *Sclerotinia sclerotiotum*, and *Fusarium oxysporum* (Kumar et al. 2012).

Indirect antagonistic mechanism of pathogen control deals with the induction of plant defense mechanism by nonpathogenic microorganisms. Plants develop resistance against pathogens by certain defense mechanisms like induced systematic resistance (ISR), systematic acquired resistance (SAR), activation of certain chemicals, and physical defense (Singh and Pathak 2015). During a pathogen attack, the accumulation of salicylic acid-mediated pathogen-related proteins (PRPs) takes place in plants, which induces ISR in plants. The PRPs protect plants by various mechanisms such as the production of hydrolytic enzymes (chitinase and glucanase), cell wall strengthening and confined cell fatality (Waghunde et al. 2017). Certain nonpathogenic bacteria (*Bacillus*, *Pseudomonas*, and *Serratia*) activate ISR in plants by phytohormones like jasmonic acid, ethylene, and salicylic acid which ultimately protect plants against pathogens (Pieterse et al. 2012; Pieterse et al. 2014; Waghunde et al. 2017).

5.2.2 NECESSITY OF GENOME EDITING IN AGRICULTURALLY BENEFICIAL MICROORGANISMS

Species of bacteria, fungi and other microbes determine whether the plant-microbe's interaction is agriculturally good or bad. Various mutants have been generated to study plant microbe's interaction by using traditional genetic engineering methods, which produce random mutations in the genome. Hence, different mutants of important microorganisms frequently produce agriculturally beneficial/desired product(s) with very low content. Among all the approaches for genome engineering, the CRISPR-Cas9 technology has many advantages over other approaches. Hence, the use of CRISPR-Cas9 technology for genome editing of agriculturally beneficial microorganisms is significantly able to improve the strains of bacteria, fungi and other microbes due to its efficiency, accuracy and easiness (Alok et al. 2020).

5.3 CRISPR-CAS9 ASSISTED GENOME ENGINEERING IN AGRICULTURALLY BENEFICIAL MICROBES

Application of the CRISPR-Cas9 approach in agriculture has been mainly focused on plant genome modification for crop improvement by increasing crop yield, resistance to biotic stresses and tolerance to abiotic stresses. However, plant beneficial microbiota has direct or indirect effects

on growth, development and yield of the plant. Therefore, nowadays the concept of genome engineering of agriculturally beneficial microorganisms is attracting researchers. There are several studies of genome editing in microorganisms of agricultural importance among which few studies have been discussed below. There are publications available for utilization of the CRISPR-Cas9 system in various groups of agriculturally important microorganisms i.e. nitrogen-fixing cyanobacteria (*Synechocystis* sp. PCC 6803 (Yao et al. 2015), *Synechococcus* sp. strain PCC 7002 (Gordon et al. 2016), and *Anabaena* sp. PCC 7120 (Higo et al. 2017)), biofertilizer and biocontrol agent bacteria (*Paenibacillus polymyxa* DSM365 (Rütering et al. 2017)), rhizosphere-associated bacteria (*Pseudomonas putida* KT2440 (Aparicio et al. 2018; Sun et al. 2018), *Bacillus subtilis* HS3, and *Bacillus mycoides* EC18 (Yi et al. 2018)), biocontrol agent fungi (*Trichoderma reesei* (Liu et al. 2015), *Beauveria bassiana* (Chen et al. 2017), and *Purpureocillium lilacinum* (Jiao et al. 2019)), and endophytic fungus (*Phomopsis liquidambaris* (Huang et al. 2020)). For further discussion, the research in this field has been categorized into three main applications of CRISPR-Cas9 in agriculturally beneficial microorganisms i.e. enhancing plant growth and nutrient availability, understanding the basics of the plant-microbes interactions, and enhancing plant biotic stress resistance.

5.3.1 Enhancing Plant Growth and Nutrient Availability

Cyanobacteria as autotrophic prokaryotes are suitable for sustainable production of numerous beneficial compounds (Sharma et al. 2011). The first work on genome editing in cyanobacteria by a CRISPR/Cas9 system has been done using *Synechococcus elongatus* UTEX 2973 by Wendt et al. (2016). The system has been standardized to develop a genetic toolkit for wide-ranging genome editing of *S. elongatus* UTEX 2973 by targeting the *nblA* gene due to its major role in biological response to nitrogen deficit environments. An introduction of CRISPR-Cas9 system on a plasmid leads to temporary cas9 expression which allows effective markerless genome editing in a wild-type genetic background (Wendt et al. 2016).

Anabaena is a genus of filamentous cyanobacteria known for their nitrogen-fixing abilities and symbiotic relationships with some plants (Franche et al. 2009). *Anabaena* sp. PCC 7120 is a multicellular cyanobacterium heterocyst that executes nitrogen fixation and a well-studied model for multicellularity in prokaryotic cells. In recent times, CRISPRi has been employed in *Anabaena* sp. PCC 7120 for the photosynthetic production of ammonium through repression of an essential nitrogen assimilation gene (*glnA* that encodes glutamine synthetase) (Higo et al. 2017). It has been suggested that CRISPRi allows temporal control of preferred products and can be used as a tool for further research. Similarly, CRISPRi approach has been also applied to other cyanobacteria namely, *Synechococcus* sp. strain PCC 7002 (Gordon et al. 2016) and *Synechocystis* sp. PCC 6803 (Yao et al. 2015) to study essential genes and regulate metabolic pathways.

Paenibacillus polymyxa is a gram-positive bacterium found in soil, plant tissues and marine sediments (Lal and Tabacchioni 2009). It has potential future applications as a biofertilizer and biocontrol agent in agriculture (Tang et al. 2017). In the recent past, a CRISPR-Cas9 tool has been developed for *P. polymyxa* for the first time by Rütering et al. (2017). By using this developed system, study relevant to biosynthetic pathways, CRISPRi-mediated repression in *P. polymyxa* and related species can be conducted.

The CRISPR-Cas9 system has been implemented to execute gene knockout and chromosomal insertion in *B. subtilis* HS3 and *B. mycoides* EC18 strains to study different characters in plant-microbe interactions. These two strains are phylogenetically distant species of *Bacillus* and it has been suggested that due to the high efficiency of CRISPR-Cas9 system, it can be of great use in genome editing of rhizosphere Bacilli. In *B. subtilis* HS3, it has been reported that 2,3-butanediol is not the key VOC produced by *B. subtilis* HS3 to stimulate growth of grass by selectively colonizing on root hairs. In *B. mycoides* EC18, two siderophore biosynthesis genes, namely, *asbA* (encoding a petrobactin biosynthesis protein AsbA) and *dhbB* (encoding isochorismatase) has been disrupted by

CRISPR-Cas9 system and it has been revealed that petrobactin plays the important role in the plant growth promoting and root colonization activities of *B. mycoides* EC18 (Yi et al. 2018).

5.3.2 UNDERSTANDING THE BASICS OF THE PM INTERACTIONS

Phomopsis liquidambaris is an endophytic fungus that efficiently encourages the nitrogen metabolism and growth of host plants (Dheeman et al. 2017; Huang et al. 2020). The CRISPR-Cas9 system has been developed for this important fungus which may enhance detailed study for the understanding of interactions among the fungus and its host plants. The system has been developed for the targeted disruption of the *PmkkA* gene encoding a mitogen-activated protein kinase kinase (MAPKK). As compared with the wild-type, the mutant strain has an enhanced production of reactive oxygen species (ROS), glucanase activity and chitinase activity in rice seedlings which resulted in strong resistance. Hence, it has been proposed that the *PmkkA* gene plays an important role during the interaction with rice and inhibition of the immune system of host plants. The knock-out efficiency of the system has been achieved to over 60% by NHEJ based gene disruption. By this example, it has been suggested that the developed system will be of great use for the research related to the interaction between *P. liquidambaris* and its host plants (Huang et al. 2020).

Pseudomonas putida KT2440 is a gram-negative soil bacterium which is recognized as "safe"-certified strain with applications in bioremediation, synthetic biology, agriculture and biotechnology (Loeschcke and Thies 2015). The genome editing of *P. putida* KT2440 using CRISPR-Cas9 technology has been first time exploited by Aparicio et al. (2018). Successful multiplexing has been demonstrated along with the simultaneous deletion of *endA-1* (deoxyribonucleases I) and *flgM* (belongs to a cluster of genes responsible for the production of flagella) (Aparicio et al. 2018). A different group of researchers has established a rapid and convenient system of CRISPR-Cas9 in *P. putida* KT2440 with more than 70% mutation rate for gene insertion, gene deletion, and gene replacement (Sun et al. 2018). These recently developed methods will improve the understanding regarding the role of this strain and/or species in plant-microbe interaction, enhancement of plant's yield and pathogen resistance. This can also be the foundation to genome editing systems into other important *Pseudomonas* strains.

5.3.3 ENHANCING PLANT BIOTIC STRESS RESISTANCE

Beauveria bassiana is an ecologically friendly fungal substitute to chemical insecticides against many agricultural insect pests and vectors of human diseases (Shah and Pell 2003; Wang et al. 2004). Due to its sensitivity against abiotic stresses and slow killing of target insects, it has limited applications in agriculture (Rangel et al. 2008; Chen et al. 2017). Therefore, understanding of its physiological features and molecular mechanism of pathogenesis can facilitate the enhancement of its insecticidal activity in various environments. Recently, the researchers combined the use of blastospore-mediated transformation and uridine auxotrophy/*ura5* complementation for genome modification of *B. bassiana*. The CRISPR/Cas9 technology can be a potent tool for high-efficiency targeted gene knock-in and/or knock-out in *B. bassiana* in a single gene disruption and possesses the significant ability for advancing understanding of its pathogenesis. It has been further confirmed that the developed system permitted simultaneous disruption of multiple genes via homology-directed repair in a single transformation. This system for genome editing and functional genomics studies in *B. bassiana* may have broad applications to other entomopathogenic fungi (Chen et al. 2017).

Filamentous fungi possess remarkable applications in agriculture and biotechnology (Meyer et al. 2016). *Trichoderma* spp. are universally found filamentous fungi which have been extensively utilized for the production of numerous enzymes and proteins or as biocontrol agents in agriculture (Mukherjee et al. 2013). They are a great source of natural proteins that may be beneficial for the plants to survive in biotic as well as abiotic stress conditions (Hermosa et al. 2012; Lorito et al. 2010). *Trichoderma reesei* is a mesophilic filamentous fungus with good industrial applications

TABLE 5.4
Summary of Genome Editing in Agriculturally Important Microorganisms

Microorganism	Application Perspectives	Targeted Gene(s)	Main Strategy/ Mode of Action	Reference
Cyanobacteria				
Synechocystis sp. PCC 6803 (nitrogen-fixing cyanobacterium)	To repress formation of carbon storage compounds PHB and glycogen during nitrogen starvation	*phaE* and *glgC*	CRISPRi	Yao et al. (2015)
Synechococcus sp. strain PCC 7002 (nitrogen-fixing cyanobacterium)	Improve lactate production by repressing glutamine synthetase without reducing autotrophic growth rates or mutating chromosomal genes.	*glnA*	CRISPRi	Gordon et al. (2016)
Anabaena sp. PCC 7120 (symbiotic nitrogen-fixer cyanobacterium)	Conditional photosynthetic production of ammonium	*glnA*	CRISPRi	Higo et al. (2017)
Bacteria				
Paenibacillus polymyxa DSM365 (biofertilizer and biocontrol agent)	Biosynthesis of tailor-made EPS	*pepF*, *pepJ*, *pepC*, *ugdH1*, *manC*	Gene deletion	Rütering et al. (2017)
Pseudomonas putida KT2440 (rhizospheric bacteria)	To test functional efficiency of CRISPR–Cas9 system	*endA-1* and *flgM*	Gene deletion	Aparicio et al. (2018)
Pseudomonas putida KT2440 (rhizospheric bacteria)	Establishment of a two-plasmid CRISPR–Cas9 system	*nicC* and *rhla*	Gene deletion and gene replacement	Sun et al. (2018)
Bacillus subtilis HS3 (rhizosphere-associated bacteria)	To study various aspects of plantmicrobe interaction mechanisms	*sfp*, *alsD*, and *bdhA*	Gene disruption	Yi et al. (2018)
Bacillus mycoides EC18 (rhizosphere-associated bacteria)	To study various aspects of plantmicrobe interaction mechanisms	*asbA* and *dhbB*	Gene disruption	Yi et al. (2018)
Fungi				
Trichoderma reesei (biocontrol agent)	Reduction of sporulation and cellulolytic capability	*lae1* and *vib1*	Site-specific mutations	Liu et al. (2015)
Beauveria bassiana (entomopathogenic fungus)	To develop highly efficient, low false-positive background CRISPR-Cas9 system for disruption of single gene and multiple genes simultaneously	*ura5*	Gene disruption	Chen et al. (2017)
Purpureocillium lilacinum (biocontrol agent against plant nematodes and pathogens)	To enhance the effectiveness of homologous recombination	*lcsL*	Gene disruption and overexpression	Jiao et al. (2019)
Phomopsis liquidambaris (endophytic fungus)	Higher production of ROS, chitinase and glucanase activity in rice seedlings	*PmkkA*	Gene disruption	Huang et al. (2020)

CRISPRi: CRISPR interference, ROS: reactive oxygen species, EPS: exopolysaccharides, PHB: Polyhydroxybutyrate.

(Martinez et al. 2008). The establishment of a controllable and conditional CRISPR-Cas9 system in *T. reesei* has been demonstrated by specific codon optimization and *in vitro* RNA transcription that produced target gene mutation and multiplex genome editing (Liu et al. 2015). These methods of CRISPR-Cas9 can be utilized in other filamentous fungal species, especially *Trichoderma* species which may speed up research on functional genomics and strain improvement in this important group of fungi.

Purpureocillium lilacinum is a promising commercial biocontrol agent against plant pathogens and plant-parasitic nematodes (Kiriga et al. 2018). Recently, the CRISPR-Cas9 system has been developed to enhance the effectiveness of homologous recombination for the disruption of *lcsL*, a gene located in the *lcs* (leucinostatins - a family of lipopeptides) cluster of *P. lilacinum*. The production of leucinostatins has been decreased to an undetectable level by disruption *lcsL* gene and increased by overexpression of *lcsL* gene (Jiao et al. 2019). This system can be advantageous for understanding mechanism of action of *P. lilacinum* and other related fungi.

Recently, the *sfp* gene (encoding 4'-phosphopantetheinyl transferase) has been disrupted in *B. subtilis* HS3, and it has been revealed that the surfactin and fengycin family lipopeptides are responsible for the antagonistic inhibitory activity of *B. subtilis* HS3 against two fungal pathogens (Yi et al. 2018). Few works have been published related to the application of the CRISPR-Cas9 technology for genome engineering in agriculturally important microorganisms which are summarized in Table 5.4. Most of the recent research in this field is in the initial phase for the CRISPR-Cas9 system development that can be very important in targeted genome engineering in agriculturally beneficial microorganisms.

5.4 CONCLUSION

In the past few years, highly versatile genome-editing technology, CRISPR–Cas9 has renovated genome engineering techniques by providing an introduction of sequence-specific alterations into the genomes of a broad range of cell types and organisms. Although there are several strategies for genome engineering, including HENs, ZFNs, and TALENs, the CRISPR-Cas9 technology is the most advantageous and continuously evolving. CRISPR-Cas9 has been applied in humans, plants, animals, insects, and microorganisms. The application of this technology in microorganisms is mainly focused on industrially important microbes. Its utility in agriculturally important microbes is in the primary stage with little success. Therefore, in agriculture, genome engineering technology through CRISPR-Cas9 is being developed for microbes along with plants. Different bacteria, cyanobacteria, fungi and other microbes possess different plant-microbe's interactions which make them agriculturally beneficial or harmful. Various mutants have been generated to study plant microbe's interactions by CRISPR-Cas9 technology. Genome engineering of agriculturally beneficial microorganisms by CRISPR-Cas9 significantly improved the strains of bacteria, fungi and other microbes due to its efficiency, accuracy and ease.

REFERENCES

Alok, A., Tiwari, S., & Kaur, J. (2020). CRISPR/Cas9 mediated genome engineering in microbes and its application in plant beneficial effects. *Molecular Aspects of Plant Beneficial Microbes in Agriculture*, 351–359. https://doi.org/10.1016/b978-0-12-818469-1.00028-6.

Aparicio, T., De Lorenzo, V., & Martínez-García, E. (2018). CRISPR/cas9-based counterselection boosts recombineering efficiency in Pseudomonas putida. *Biotechnology Journal*, 13(5), 1700161. https://doi.org/10.1002/biot.201700161.

Bakker, P. A., Ran, L. X., Pieterse, C. M., & Van Loon, L. C. (2003). Understanding the involvement of rhizobacteria-mediated induction of systemic resistance in biocontrol of plant diseases. *Canadian Journal of Plant Pathology*, 25(1), 5–9. https://doi.org/10.1080/07060660309507043.

Barea, J., Pozo, M. J., Azcón, R., & Azcón-Aguilar, C. (2005). Microbial co-operation in the rhizosphere. *Journal of Experimental Botany*, 56(417), 1761–1778. https://doi.org/10.1093/jxb/eri197.

Barrangou, R., Fremaux, C., Deveau, H., Richards, M., Boyaval, P., Moineau, S., Romero, D. A., & Horvath, P. (2007). CRISPR provides acquired resistance against viruses in prokaryotes. *Science*, *315*(5819), 1709–1712. https://doi.org/10.1126/science.1138140.

Bhatia, S., Dubey, R. C., & Maheshwari, D. K. (2005). Enhancement of plant growth and suppression of collar rot of sunflower caused by Sclerotium rolfsii through fluorescent Pseudomonas. *Indian Phytopathol*, *58*(1), 17–24.

Bhattacharyya, P. N., Goswami, M. P., & Bhattacharyya, L. H. (2016). Perspective of beneficial microbes in agriculture under changing climatic scenario: A review. *Journal of Phytology*, *8*, 26–41. https://doi.org/10.19071/jp.2016.v8.3022.

Biswas, A., Fineran, P. C., & Brown, C. M. (2014). Accurate computational prediction of the transcribed Strand of CRISPR non-coding RNAs. *Bioinformatics*, *30*(13), 1805–1813. https://doi.org/10.1093/bioinformatics/btu114.

Biswas, A., Gagnon, J. N., Brouns, S. J., Fineran, P. C., & Brown, C. M. (2013). CRISPRTarget. *RNA Biology*, *10*(5), 817–827. https://doi.org/10.4161/rna.24046.

Bolotin, A., Quinquis, B., Sorokin, A., & Ehrlich, S. D. (2005). Clustered regularly interspaced short palindrome repeats (CRISPRs) have spacers of extrachromosomal origin. *Microbiology*, *151*(8), 2551–2561. https://doi.org/10.1099/mic.0.28048-0.

Brouns, S. J., Jore, M. M., Lundgren, M., Westra, E. R., Slijkhuis, R. J., Snijders, A. P., Dickman, M. J., Makarova, K. S., Koonin, E. V., & Van der Oost, J. (2008). Small CRISPR RNAs guide antiviral defense in prokaryotes. *Science*, *321*(5891), 960–964. https://doi.org/10.1126/science.1159689.

Chaudhary, K., Chattopadhyay, A., & Pratap, D. (2018). The evolution of CRISPR/Cas9 and their cousins: Hope or hype? *Biotechnology Letters*, *40*(3), 465–477. https://doi.org/10.1007/s10529-018-2506-7.

Chen, J., Lai, Y., Wang, L., Zhai, S., Zou, G., Zhou, Z., Cui, C., & Wang, S. (2017). CRISPR/cas9-mediated efficient genome editing via blastospore-based transformation in entomopathogenic fungus Beauveria bassiana. *Scientific Reports*, *7*(1). https://doi.org/10.1038/srep45763

Chiruvella, K. K., Liang, Z., & Wilson, T. E. (2013). Repair of Double-Strand breaks by end joining. *Cold Spring Harbor Perspectives in Biology*, *5*(5), a012757–a012757. https://doi.org/10.1101/cshperspect.a012757.

Cong, L., Ran, F. A., Cox, D., Lin, S., Barretto, R., Habib, N., Hsu, P. D., Wu, X., Jiang, W., Marraffini, L. A., & Zhang, F. (2013). Multiplex genome engineering using CRISPR/Cas systems. *Science*, *339*(6121), 819–823. https://doi.org/10.1126/science.1231143.

Davis, A. J., & Chen, D. J. (2013). DNA double strand break repair via non-homologous end-joining. *Translational Cancer Research*, *2*(3), 130–143.

Deltcheva, E., Chylinski, K., Sharma, C. M., Gonzales, K., Chao, Y., Pirzada, Z. A., Eckert, M. R., Vogel, J., & Charpentier, E. (2011). CRISPR RNA maturation by trans-encoded small RNA and host factor RNase III. *Nature*, *471*(7340), 602–607. https://doi.org/10.1038/nature09886.

Desai, S. A. (2017). Isolation and characterization of gibberellic acid (GA3) producing rhizobacteria from sugarcane roots. *Bioscience Discoveries*, *8*(3), 488–494.

Dheeman, S., Maheshwari, D. K., & Baliyan, N. (2017). Bacterial endophytes for ecological intensification of agriculture. In: Maheshwari DK (ed) *Endophytes: Biology and Biotechnology, vol. 1. Sustainable Development and Biodiversity, vol. 15*. Springer, Cham, pp. 193–231.

Dutta, S., & Khurana, S. M. P. (2015). Plant growth-promoting rhizobacteria for alleviating abiotic stresses in medicinal plants. In: Egamberdieva D, Shrivastava S, Varma A (eds) *Plant-Growth-Promoting Rhizobacteria (PGPR) and Medicinal Plants*. Soil biology, vol. 42. Springer, Cham, pp. 167–200.

El-Sayed, A. S., Abdel-Ghany, S. E., & Ali, G. S. (2017). Genome editing approaches: Manipulating of lovastatin and taxol synthesis of filamentous fungi by CRISPR/Cas9 system. *Applied Microbiology and Biotechnology*, *101*(10), 3953–3976. https://doi.org/10.1007/s00253-017-8263-z.

Escamilla, S. E. (2000). Optimization of gibberellic acid production by immobilized gibberella fujikuroi mycelium in fluidized bioreactors. *Journal of Biotechnology*, *76*(2–3), 147–155. https://doi.org/10.1016/s0168-1656(99)00182-0.

Ferreira, C. M., Vilas-Boas, Â., Sousa, C. A., Soares, H. M., & Soares, E. V. (2019). Comparison of five bacterial strains producing siderophores with ability to chelate iron under alkaline conditions. *AMB Express*, *9*(1). https://doi.org/10.1186/s13568-019-0796-3.

Fogarty, N. M., McCarthy, A., Snijders, K. E., Powell, B. E., Kubikova, N., Blakeley, P., Lea, R., Elder, K., Wamaitha, S. E., Kim, D., Maciulyte, V., Kleinjung, J., Kim, J., Wells, D., Vallier, L., Bertero, A., Turner, J. M., & Niakan, K. K. (2017). Genome editing reveals a role for OCT4 in human embryogenesis. *Nature*, *550*(7674), 67–73. https://doi.org/10.1038/nature24033.

Franche, C., Lindström, K., & Elmerich, C. (2009). Nitrogen-fixing bacteria associated with leguminous and non-leguminous plants. *Plant and Soil*, *321*(1–2), 35–59. https://doi.org/10.1007/s11104-008-9833-8.

Gaj, T., Sirk, S. J., Shui, S., & Liu, J. (2016). Genome-editing technologies: Principles and applications. *Cold Spring Harbor Perspectives in Biology*, *8*(12), a023754. https://doi.org/10.1101/cshperspect.a023754.

García de Salamone, I. E., Hynes, R. K., & Nelson, L. M. (2001). Cytokinin production by plant growth promoting rhizobacteria and selected mutants. *Canadian Journal of Microbiology*, *47*(5), 404–411. https://doi.org/10.1139/w01-029.

Garneau, J. E., Dupuis, M., Villion, M., Romero, D. A., Barrangou, R., Boyaval, P., Fremaux, C., Horvath, P., Magadán, A. H., & Moineau, S. (2010). The CRISPR/Cas bacterial immune system cleaves bacteriophage and plasmid DNA. *Nature*, *468*(7320), 67–71. https://doi.org/10.1038/nature09523.

Gasiunas, G., Barrangou, R., Horvath, P., & Siksnys, V. (2012). Cas9-crrna ribonucleoprotein complex mediates specific DNA cleavage for adaptive immunity in bacteria. *Proceedings of the National Academy of Sciences*, *109*(39), E2579–E2586. https://doi.org/10.1073/pnas.1208507109.

Gelmi, C., Pérez-Correa, R., González, M., & Agosin, E. (2000). Solid substrate cultivation of gibberella fujikuroi on an inert support. *Process Biochemistry*, *35*(10), 1227–1233. https://doi.org/10.1016/s0032-9592(00)00161-8.

Gilbert, L., Larson, M., Morsut, L., Liu, Z., Brar, G., Torres, S., Stern-Ginossar, N., Brandman, O., Whitehead, E., Doudna, J., Lim, W., Weissman, J., & Qi, L. (2013). CRISPR-mediated modular RNA-guided regulation of transcription in eukaryotes. *Cell*, *154*(2), 442–451. https://doi.org/10.1016/j.cell.2013.06.044.

Glick, B. R. (2012). Plant growth-promoting bacteria: Mechanisms and applications. *Scientifica*, *2012*, 1–15. https://doi.org/10.6064/2012/963401.

Gopalakrishnan, S., Sathya, A., Vijayabharathi, R., Varshney, R. K., Gowda, C. L., & Krishnamurthy, L. (2015). Plant growth promoting rhizobia: Challenges and opportunities. *3 Biotech*, *5*(4), 355–377. https://doi.org/10.1007/s13205-014-0241-x.

Gordon, G. C., Korosh, T. C., Cameron, J. C., Markley, A. L., Begemann, M. B., & Pfleger, B. F. (2016). CRISPR interference as a titratable, trans-acting regulatory tool for metabolic engineering in the cyanobacterium Synechococcus sp. strain PCC 7002. *Metabolic Engineering*, *38*, 170–179. https://doi.org/10.1016/j.ymben.2016.07.007.

Grissa, I., Vergnaud, G., & Pourcel, C. (2008). CRISPRcompar: A website to compare clustered regularly interspaced short palindromic repeats. *Nucleic Acids Research*, *36*(Web Server), W145–W148. https://doi.org/10.1093/nar/gkn228.

Hale, C. R., Zhao, P., Olson, S., Duff, M. O., Graveley, B. R., Wells, L., Terns, R. M., & Terns, M. P. (2009). RNA-guided RNA cleavage by a CRISPR RNA-cas protein complex. *Cell*, *139*(5), 945–956. https://doi.org/10.1016/j.cell.2009.07.040.

Hermosa, R., Viterbo, A., Chet, I., & Monte, E. (2012). Plant-beneficial effects of Trichoderma and of its genes. *Microbiology*, *158*(1), 17–25. https://doi.org/10.1099/mic.0.052274-0.

Higo, A., Isu, A., Fukaya, Y., Ehira, S., & Hisabori, T. (2017). Application of CRISPR interference for metabolic engineering of the heterocyst-forming multicellular Cyanobacterium Anabaena sp. PCC 7120. *Plant and Cell Physiology*, *59*(1), 119–127. https://doi.org/10.1093/pcp/pcx166.

Hsu, P., Lander, E., & Zhang, F. (2014). Development and applications of CRISPR-cas9 for genome engineering. *Cell*, *157*(6), 1262–1278. https://doi.org/10.1016/j.cell.2014.05.010.

Huang, P., Yang, Q., Zhu, Y., Zhou, J., Sun, K., Mei, Y., & Dai, C. (2020). The construction of CRISPR-cas9 system for endophytic Phomopsis liquidambaris and its pmkka-deficient mutant revealing the effect on rice. *Fungal Genetics and Biology*, *136*, 103301. https://doi.org/10.1016/j.fgb.2019.103301.

Ishino, Y., Krupovic, M., & Forterre, P. (2018). History of CRISPR-Cas from encounter with a mysterious repeated sequence to genome editing technology. *Journal of Bacteriology*, *200*(7). https://doi.org/10.1128/jb.00580-17.

Ishino, Y., Shinagawa, H., Makino, K., Amemura, M., & Nakata, A. (1987). Nucleotide sequence of the iap gene, responsible for alkaline phosphatase isozyme conversion in escherichia coli, and identification of the gene product. *Journal of Bacteriology*, *169*(12), 5429–5433. https://doi.org/10.1128/jb.169.12.5429-5433.1987.

Jaganathan, D., Ramasamy, K., Sellamuthu, G., Jayabalan, S., & Venkataraman, G. (2018). CRISPR for crop improvement: An update review. *Frontiers in Plant Science*, *9*, 985. https://doi.org/10.3389/fpls.2018.00985.

Jansen, R., Embden, J. D., Gaastra, W., & Schouls, L. M. (2002). Identification of genes that are associated with DNA repeats in prokaryotes. *Molecular Microbiology*, *43*(6), 1565–1575. https://doi.org/10.1046/j.1365-2958.2002.02839.x.

Jiang, W., Zhou, H., Bi, H., Fromm, M., Yang, B., & Weeks, D. P. (2013). Demonstration of CRISPR/Cas9/sgrna-mediated targeted gene modification in arabidopsis, tobacco, sorghum and rice. *Nucleic Acids Research*, *41*(20), e188–e188. https://doi.org/10.1093/nar/gkt780.

Jiao, Y., Li, Y., Li, Y., Cao, H., Mao, Z., Ling, J., Yang, Y., & Xie, B. (2019). Functional genetic analysis of the leucinostatin biosynthesis transcription regulator lcsL in Purpureocillium lilacinum using

CRISPR-cas9 technology. *Applied Microbiology and Biotechnology*, *103*(15), 6187–6194. https://doi. org/10.1007/s00253-019-09945-2.

Jinek, M., Chylinski, K., Fonfara, I., Hauer, M., Doudna, J. A., & Charpentier, E. (2012). A programmable Dual-RNA-Guided DNA endonuclease in adaptive bacterial immunity. *Science*, *337*(6096), 816–821. https://doi.org/10.1126/science.1225829.

Jinek, M., Jiang, F., Taylor, D. W., Sternberg, S. H., Kaya, E., Ma, E., Anders, C., Hauer, M., Zhou, K., Lin, S., Kaplan, M., Iavarone, A. T., Charpentier, E., Nogales, E., & Doudna, J. A. (2014). Structures of Cas9 endonucleases reveal RNA-mediated conformational activation. *Science*, *343*(6176), 1247997–1247997. https://doi.org/10.1126/science.1247997.

Kampmann, M. (2018). CRISPRi and CRISPRa screens in mammalian cells for precision biology and medicine. *ACS Chemical Biology*, *13*(2), 406–416. https://doi.org/10.1021/acschembio.7b00657.

Kim, H., & Kim, J. (2014). A guide to genome engineering with programmable nucleases. *Nature Reviews Genetics*, *15*(5), 321–334. https://doi.org/10.1038/nrg3686.

Kiriga, A. W., Haukeland, S., Kariuki, G. M., Coyne, D. L., & Beek, N. V. (2018). Effect of Trichoderma spp. and Purpureocillium lilacinum on Meloidogyne javanica in commercial pineapple production in Kenya. *Biological Control*, *119*, 27–32. https://doi.org/10.1016/j.biocontrol.2018.01.005.

Kleinstiver, B. P., Pattanayak, V., Prew, M. S., Tsai, S. Q., Nguyen, N. T., Zheng, Z., & Joung, J. K. (2016). High-fidelity CRISPR–cas9 nucleases with no detectable genome-wide off-target effects. *Nature*, *529*(-7587), 490–495. https://doi.org/10.1038/nature16526.

Kumar, A., Singh, V. K., Tripathi, V., Singh, P. P., Singh, A. K. (2018). Plant growth-promoting rhizobacteria (PGPR): Perspective in agriculture under biotic and abiotic stress. In: Prasad R, Gill SS, Tuteja N (eds) *Crop Improvement through Microbial Biotechnology*. Elsevier, USA pp. 333–342.

Kumar, P., Dubey, R., & Maheshwari, D. (2012). Bacillus strains isolated from rhizosphere showed plant growth promoting and antagonistic activity against phytopathogens. *Microbiological Research*, *167*(8), 493–499. https://doi.org/10.1016/j.micres.2012.05.002.

Lal, S., & Tabacchioni, S. (2009). Ecology and biotechnological potential of Paenibacillus polymyxa: A minireview. *Indian Journal of Microbiology*, *49*(1), 2–10. https://doi.org/10.1007/s12088-009-0008-y.

Lange, S. J., Alkhnbashi, O. S., Rose, D., Will, S., & Backofen, R. (2013). CRISPRmap: An automated classification of repeat conservation in prokaryotic adaptive immune systems. *Nucleic Acids Research*, *41*(17), 8034–8044. https://doi.org/10.1093/nar/gkt606.

Liang, P., Ding, C., Sun, H., Xie, X., Xu, Y., Zhang, X., Sun, Y., Xiong, Y., Ma, W., Liu, Y., Wang, Y., Fang, J., Liu, D., Songyang, Z., Zhou, C., & Huang, J. (2017). Correction of β-thalassemia mutant by base editor in human embryos. *Protein & Cell*, *8*(11), 811–822. https://doi.org/10.1007/s13238-017-0475-6.

Liang, P., Xu, Y., Zhang, X., Ding, C., Huang, R., Zhang, Z., Lv, J., Xie, X., Chen, Y., Li, Y., Sun, Y., Bai, Y., Songyang, Z., Ma, W., Zhou, C., & Huang, J. (2015). CRISPR/cas9-mediated gene editing in human tripronuclear zygotes. *Protein & Cell*, *6*(5), 363–372. https://doi.org/10.1007/s13238-015-0153-5.

Lieber, M. R. (2010). The mechanism of Double-Strand DNA break repair by the Nonhomologous DNA end-joining pathway. *Annual Review of Biochemistry*, *79*(1), 181–211. https://doi.org/10.1146/annurev. biochem.052308.093131.

Liu, R., Chen, L., Jiang, Y., Zhou, Z., & Zou, G. (2015). Efficient genome editing in filamentous fungus Trichoderma reesei using the CRISPR/Cas9 system. *Cell Discovery*, *1*(1). https://doi.org/10.1038/celldisc.2015.7.

Loeschcke, A., & Thies, S. (2015). Pseudomonas putida—a versatile host for the production of natural products. *Applied Microbiology and Biotechnology*, *99*(15), 6197–6214. https://doi.org/10.1007/s00253-015-6745-4.

Lorito, M., Woo, S. L., Harman, G. E., & Monte, E. (2010). Translational research on Trichoderma: From 'Omics to the Field. *Annual Review of Phytopathology*, *48*(1), 395–417. https://doi.org/10.1146/annurev-phyto–073009–114314

Ma, H., Marti-Gutierrez, N., Park, S., Wu, J., Lee, Y., Suzuki, K., Koski, A., Ji, D., Hayama, T., Ahmed, R., Darby, H., Van Dyken, C., Li, Y., Kang, E., Park, A., Kim, D., Kim, S., Gong, J., Gu, Y., … Mitalipov, S. (2017). Correction of a pathogenic gene mutation in human embryos. *Nature*, *548*(7668), 413–419. https://doi.org/10.1038/nature23305.

Makarova, K. S., Wolf, Y. I., Alkhnbashi, O. S., Costa, F., Shah, S. A., Saunders, S. J., Barrangou, R., Brouns, S. J., Charpentier, E., Haft, D. H., Horvath, P., Moineau, S., Mojica, F. J., Terns, R. M., Terns, M. P., White, M. F., Yakunin, A. F., Garrett, R. A., Van der Oost, J., … Koonin, E. V. (2015). An updated evolutionary classification of CRISPR–cas systems. *Nature Reviews Microbiology*, *13*(11), 722–736. https://doi.org/10.1038/nrmicro3569.

Mali, P., Yang, L., Esvelt, K. M., Aach, J., Guell, M., DiCarlo, J. E., Norville, J. E., & Church, G. M. (2013). RNA-guided human genome engineering via Cas9. *Science*, *339*(6121), 823–826. https://doi. org/10.1126/science.1232033.

Marraffini, L. A. (2016). The CRISPR-Cas system of *Streptococcus pyogenes*: Function and applications. In: Ferretti JJ, Stevens DL, Fischetti VA (eds) *Streptococcus pyogenes:* Basic Biology to Clinical Manifestations. University of Oklahoma Health Sciences Center, Oklahoma City, OK, pp. 1–17.

Marraffini, L. A., & Sontheimer, E. J. (2008). CRISPR interference limits horizontal gene transfer in staphylococci by targeting DNA. *Science, 322*(5909), 1843–1845. https://doi.org/10.1126/science.1165771

Martinez, D., Berka, R. M., Henrissat, B., Saloheimo, M., Arvas, M., Baker, S. E., Chapman, J., Chertkov, O., Coutinho, P. M., Cullen, D., Danchin, E. G., Grigoriev, I. V., Harris, P., Jackson, M., Kubicek, C. P., Han, C. S., Ho, I., Larrondo, L. F., De Leon, A. L., ... Brettin, T. S. (2008). Genome sequencing and analysis of the biomass-degrading fungus Trichoderma reesei (syn. Hypocrea jecorina). *Nature Biotechnology, 26*(5), 553–560. https://doi.org/10.1038/nbt1403.

Masson-Boivin, C., Giraud, E., Perret, X., & Batut, J. (2009). Establishing nitrogen-fixing symbiosis with legumes: How many rhizobium recipes? *Trends in Microbiology, 17*(10), 458–466. https://doi.org/10.1016/j.tim.2009.07.004.

Meyer, V., Andersen, M. R., Brakhage, A. A., Braus, G. H., Caddick, M. X., Cairns, T. C., De Vries, R. P., Haarmann, T., Hansen, K., Hertz-Fowler, C., Krappmann, S., Mortensen, U. H., Peñalva, M. A., Ram, A. F., & Head, R. M. (2016). Current challenges of research on filamentous fungi in relation to human welfare and a sustainable bio-economy: A white paper. *Fungal Biology and Biotechnology, 3*(1). https://doi.org/10.1186/s40694-016-0024-8.

Mohite, B. (2013). Isolation and characterization of indole acetic acid (IAA) producing bacteria from rhizospheric soil and its effect on plant growth. *Journal of soil science and plant nutrition, 13*(3), 638–649. https://doi.org/10.4067/s0718-95162013005000051.

Mojica, F. J., Díez-Villaseñor, C., García-Martínez, J., & Soria, E. (2005). Intervening sequences of regularly spaced prokaryotic repeats derive from foreign genetic elements. *Journal of Molecular Evolution, 60*(-2), 174–182. https://doi.org/10.1007/s00239-004-0046-3.

Mojica, F. J., Diez-Villasenor, C., Soria, E., & Juez, G. (2000). Biological significance of a family of regularly spaced repeats in the genomes of Archaea, bacteria and mitochondria. *Molecular Microbiology, 36*(1), 244–246. https://doi.org/10.1046/j.1365-2958.2000.01838.x.

Monte, E. (2001). Understanding Trichoderma: Between biotechnology and microbial ecology. *International Microbiology, 4*(1), 1–4. https://doi.org/10.1111/mbt.2008.1.issue-4.

Mougiakos, I., Bosma, E. F., De Vos, W. M., Van Kranenburg, R., & Van der Oost, J. (2016). Next generation prokaryotic engineering: The CRISPR-Cas toolkit. *Trends in Biotechnology, 34*(7), 575–587. https://doi.org/10.1016/j.tibtech.2016.02.004.

Mukherjee, P. K., Horwitz, B. A., Herrera-Estrella, A., Schmoll, M., & Kenerley, C. M. (2013). Trichoderma research in the genome era. *Annual Review of Phytopathology, 51*(1), 105–129. https://doi.org/10.1146/annurev-phyto-082712-102353.

Nath, M., Bhatt, D., Bhatt, M. D., Prasad, R., Tuteja, N. (2018). Microbe-mediated enhancement of nitrogen and phosphorus content for crop improvement. In: Prasad R, Gill SS, Tuteja N (eds) *Crop Improvement through Microbial Biotechnology*, Elsevier, pp. 293–304.

Nishimasu, H., Ran, F., Hsu, P., Konermann, S., Shehata, S., Dohmae, N., Ishitani, R., Zhang, F., & Nureki, O. (2014). Crystal structure of Cas9 in complex with guide RNA and target DNA. *Cell, 156*(5), 935–949. https://doi.org/10.1016/j.cell.2014.02.001.

Nuñez, J. K., Kranzusch, P. J., Noeske, J., Wright, A. V., Davies, C. W., & Doudna, J. A. (2014). Cas1–cas2 complex formation mediates spacer acquisition during CRISPR–cas adaptive immunity. *Nature Structural & Molecular Biology, 21*(6), 528–534. https://doi.org/10.1038/nsmb.2820.

Parniske, M. (2008). Arbuscular mycorrhiza: The mother of plant root endosymbioses. *Nature Reviews Microbiology, 6*(10), 763–775. https://doi.org/10.1038/nrmicro1987.

Periwal, V. (2017). A comprehensive overview of computational resources to aid in precision genome editing with engineered nucleases. *Briefings in Bioinformatics, 18*(4), 698–711. https://doi.org/10.1093/bib/bbw052.

Pickar-Oliver, A., & Gersbach, C. A. (2019). The next generation of CRISPR–cas technologies and applications. *Nature Reviews Molecular Cell Biology, 20*(8), 490–507. https://doi.org/10.1038/s41580-019-0131-5.

Pieterse, C. M., Van der Does, D., Zamioudis, C., Leon-Reyes, A., & Van Wees, S. C. (2012). Hormonal modulation of plant immunity. *Annual Review of Cell and Developmental Biology, 28*(1), 489–521. https://doi.org/10.1146/annurev-cellbio-092910-154055.

Pieterse, C. M., Zamioudis, C., Berendsen, R. L., Weller, D. M., Van Wees, S. C., & Bakker, P. A. (2014). Induced systemic resistance by beneficial microbes. *Annual Review of Phytopathology, 52*(1), 347–375. https://doi.org/10.1146/annurev-phyto-082712-102340.

Ping, L., & Boland, W. (2004). Signals from the underground: Bacterial volatiles promote growth in arabidopsis. *Trends in Plant Science, 9*(6), 263–266. https://doi.org/10.1016/j.tplants.2004.04.008.

Pourcel, C., Salvignol, G., & Vergnaud, G. (2005). CRISPR elements in Yersinia pestis acquire new repeats by preferential uptake of bacteriophage DNA, and provide additional tools for evolutionary studies. *Microbiology*, *151*(3), 653–663. https://doi.org/10.1099/mic.0.27437-0.

Pundir, R. K., & Jain, P. (2015). Mechanism of prevention and control of medicinal plant-associated diseases. *Soil Biology*, 231–246. https://doi.org/10.1007/978-3-319-13401-7_11.

Qi, L., Larson, M., Gilbert, L., Doudna, J., Weissman, J., Arkin, A., & Lim, W. (2013). Repurposing CRISPR as an RNA-guided platform for sequence-specific control of gene expression. *Cell*, 152(5), 1173–1183. https://doi.org/10.1016/j.cell.2013.02.022.

Rangel, D. E., Anderson, A. J., & Roberts, D. W. (2008). Evaluating physical and nutritional stress during mycelial growth as inducers of tolerance to heat and UV-B radiation in Metarhizium anisopliae conidia. *Mycological Research*, 112(11), 1362–1372. https://doi.org/10.1016/j.mycres.2008.04.013.

Rathoure, A. K. (2015). Competent soil microorganisms for agricultural sustainability: New dimension. In: Singh J, Gehlot P (ed) *Microbes: In Action*, 1st edn. Agrobios, Jodhpur, pp. 1–18.

Rousseau, C., Gonnet, M., Le Romancer, M., & Nicolas, J. (2009). CRISPI: A CRISPR interactive database. *Bioinformatics*, *25*(24), 3317–3318. https://doi.org/10.1093/bioinformatics/btp586.

Rütering, M., Cress, B. F., Schilling, M., Rühmann, B., Koffas, M. A., Sieber, V., & Schmid, J. (2017). Tailor-made exopolysaccharides—CRISPR-cas9 mediated genome editing in Paenibacillus polymyxa. *Synthetic Biology*, 2(1). https://doi.org/10.1093/synbio/ysx007.

Ryu, C., Farag, M. A., Hu, C., Reddy, M. S., Wei, H., Pare, P. W., & Kloepper, J. W. (2003). Bacterial volatiles promote growth in arabidopsis. *Proceedings of the National Academy of Sciences*, 100(8), 4927–4932. https://doi.org/10.1073/pnas.0730845100.

Sapranauskas, R., Gasiunas, G., Fremaux, C., Barrangou, R., Horvath, P., & Siksnys, V. (2011). The streptococcus thermophilus CRISPR/Cas system provides immunity in escherichia coli. *Nucleic Acids Research*, *39*(21), 9275–9282. https://doi.org/10.1093/nar/gkr606.

Shah, P. A., & Pell, J. K. (2003). Entomopathogenic fungi as biological control agents. *Applied Microbiology and Biotechnology*, *61*(5–6), 413–423. https://doi.org/10.1007/s00253-003-1240-8.

Shalem, O., Sanjana, N. E., Hartenian, E., Shi, X., Scott, D. A., Mikkelsen, T. S., Heckl, D., Ebert, B. L., Root, D. E., Doench, J. G., & Zhang, F. (2014). Genome-scale CRISPR-cas9 knockout screening in human cells. *Science*, *343*(6166), 84–87. https://doi.org/10.1126/science.1247005.

Shan, Q., Wang, Y., Li, J., Zhang, Y., Chen, K., Liang, Z., Zhang, K., Liu, J., Xi, J. J., Qiu, J., & Gao, C. (2013). Targeted genome modification of crop plants using a CRISPR-Cas system. *Nature Biotechnology*, *31*(8), 686–688. https://doi.org/10.1038/nbt.2650.

Sharma, M., Kansal, R., & Singh, D. (2018). Endophytic microorganisms: Their role in plant growth and crop improvement. *Crop Improvement through Microbial Biotechnology*, 391–413. https://doi.org/10.1016/b978-0-444-63987-5.00020-7.

Sharma, N. K., Tiwari, S. P., Tripathi, K., & Rai, A. K. (2011). Sustainability and cyanobacteria (blue-green algae): Facts and challenges. *Journal of Applied Phycology*, *23*(6), 1059–1081. https://doi.org/10.1007/s10811-010-9626-3.

Shelake, R. M., Pramanik, D., & Kim, J. (2019). Exploration of plant-microbe interactions for sustainable agriculture in CRISPR era. *Microorganisms*, *7*(8), 269. https://doi.org/10.3390/microorganisms7080269.

Shinwari, Z. K., Tanveer, F., & Iqrar, I. (2019). Role of microbes in plant health, disease management, and abiotic stress management. *Microbiome in Plant Health and Disease*, 231–250. https://doi.org/10.1007/978-981-13-8495-0_11.

Singh, S., & Singh, J. (2015). Plant growth promotion by soil microorganisms through root interaction. In: Singh J, Gehlot P (ed) *Microbes: In Action*, 1st edn. Agrobios, Jodhpur, pp. 41–57.

Singh, S. K., & Pathak, R. (2015). Ecological manipulations of rhizobacteria for curbing medicinal plant diseases. In: Egamberdieva D, Shrivastava S, Varma A (eds) *Plant-Growth-Promoting Rhizobacteria (PGPR) and Medicinal Plants*. Soil Biology, vol 42. Springer, Cham, pp. 217–230.

Skennerton, C. T., Imelfort, M., & Tyson, G. W. (2013). Crass: Identification and reconstruction of CRISPR from unassembled metagenomic data. *Nucleic Acids Research*, *41*(10), e105. https://doi.org/10.1093/nar/gkt183.

Stamenković, S., Beškoski, V., Karabegović, I., Lazić, M., & Nikolić, N. (2018). Microbial fertilizers: A comprehensive review of current findings and future perspectives. *Spanish Journal of Agricultural Research*, *16*(1), e09R01. https://doi.org/10.5424/sjar/2018161-12117.

Sugumaran, P., & Janarthanam, B. (2007). Solubilization of potassium containing minerals by bacteria and their effect on plant growth. *World Journal of Agricultural Research*, *3*(3), 350–355.

Sun, J., Wang, Q., Jiang, Y., Wen, Z., Yang, L., Wu, J., & Yang, S. (2018). Genome editing and transcriptional repression in pseudomonas putida KT2440 via the type II CRISPR system. *Microbial Cell Factories*, *17*(1). https://doi.org/10.1186/s12934-018-0887-x.

Sung, P., & Klein, H. (2006). Mechanism of homologous recombination: Mediators and helicases take on regulatory functions. *Nature Reviews Molecular Cell Biology*, *7*(10), 739–750. https://doi.org/10.1038/nrm2008.

Tang, Q., Puri, A., Padda, K. P., & Chanway, C. P. (2017). Biological nitrogen fixation and plant growth promotion of lodgepole pine by an endophytic diazotroph Paenibacillus polymyxa and its GFP-tagged derivative. *Botany*, *95*(6), 611–619. https://doi.org/10.1139/cjb-2016-0300.

Verma, A., Patidar, Y., & Vaishampayan, A. (2016). Isolation and purification of potassium solubilizing bacteria from different regions of India and its effect on crop's yield. *Indian Journal of Microbiological Research*, *3*(4), 483–488.

Vijayabharathi, R., Sathya, A., & Gopalakrishnan, S. (2016). A renaissance in plant growth-promoting and biocontrol agents by endophytes. In: Singh D., Singh H., Prabha R. (eds) *Microbial Inoculants in Sustainable Agricultural Productivity*. Springer, New Delhi, pp. 37–60.

Waghunde, R. R., Shelake, R. M., Shinde, M. S., & Hayashi, H. (2017). Endophyte microbes: A weapon for plant health management. In: Panpatte D, Jhala Y, Vyas R, Shelat H (eds) *Microorganisms for Green Revolution*. Microorganisms for Sustainability, vol. 6. Springer, Singapore, pp. 303–325.

Wang, C., Fan, M., Li, Z., & Butt, T. (2004). Molecular monitoring and evaluation of the application of the insect-pathogenic fungus Beauveria bassiana in southeast China. *Journal of Applied Microbiology*, *96*(4), 861–870. https://doi.org/10.1111/j.1365-2672.2004.02215.x.

Wang, T., Wei, J. J., Sabatini, D. M., & Lander, E. S. (2014). Genetic screens in human cells using the CRISPR-cas9 system. *Science*, *343*(6166), 80–84. https://doi.org/10.1126/science.1246981.

Wati, L., Raj, K., & Goel, A. (2015). Role of microbes in agriculture. In: Singh J, Gehlot P (ed) *Microbes: In Action*, 1st edn. Agrobios, Jodhpur, pp. 19–40.

Wendt, K. E., Ungerer, J., Cobb, R. E., Zhao, H., & Pakrasi, H. B. (2016). CRISPR/Cas9 mediated targeted mutagenesis of the fast growing cyanobacterium Synechococcus elongatus UTEX 2973. *Microbial Cell Factories*, *15*(1). https://doi.org/10.1186/s12934-016-0514-7.

Wiedenheft, B., Zhou, K., Jinek, M., Coyle, S. M., Ma, W., & Doudna, J. A. (2009). Structural basis for DNase activity of a conserved protein implicated in CRISPR-mediated genome defense. *Structure*, *17*(6), 904–912. https://doi.org/10.1016/j.str.2009.03.019.

Yao, L., Cengic, I., Anfelt, J., & Hudson, E. P. (2015). Multiple gene repression in Cyanobacteria using CRISPRi. *ACS Synthetic Biology*, *5*(3), 207–212. https://doi.org/10.1021/acssynbio.5b00264.

Yi, Y., Li, Z., Song, C., & Kuipers, O. P. (2018). Exploring plant-microbe interactions of the rhizobacteriaBacillus subtilisandBacillus mycoidesby use of the CRISPR-cas9 system. *Environmental Microbiology*, *20*(12), 4245–4260. https://doi.org/10.1111/1462-2920.14305.

Zetsche, B., Gootenberg, J., Abudayyeh, O., Slaymaker, I., Makarova, K., Essletzbichler, P., Volz, S., Joung, J., Van der Oost, J., Regev, A., Koonin, E., & Zhang, F. (2015). Cpf1 is a single RNA-guided endonuclease of a class 2 CRISPR-Cas system. *Cell*, *163*(3), 759–771. https://doi.org/10.1016/j.cell.2015.09.038.

6 AI-Based Agricultural Knowledge System

Sherine F. Mansour
Agricultural Economics - Desert Research Center

Nabil Ibrahim Elsheery
Tanta University

CONTENTS

6.1 INTRODUCTION

For several developing countries, the agriculture sector faces great challenges if it is to achieve the food security and development targets set out in the action plan of the World Food Summit. Science and technology underpin agricultural production, but in many countries funding is insufficient for research and development at the national level. At the same time, frameworks for recording and disseminating science's technological outputs are low. As a consequence, research results are often insufficiently registered, and are not transmitted to producers, policymakers, and those who need to adapt and implement them. In science and technology, conventional national systems are augmented, and sometimes even substituted, by regional/sub-regional networks that cut across organizational structures and political boundaries, and in reality national initiatives in this field need to be fostered and strengthened. Such multiple systems have different methods to collect and disseminate research findings. In reality, the manner in which work is financed and results are disseminated is

changing in agricultural knowledge and information systems, and the type of information that farmers need is also changing (FAO-DFID, 2005).

6.2 IMPORTANCE OF AGRICULTURAL INFORMATION SYSTEMS AND THEIR NEEDS FOR AGRICULTURAL PRODUCTION

Agricultural information is an important factor which interacts with other factors in production.

With relevant, reliable and useful information, the productivity of these other factors, such as land, labor, capital, and managerial ability, can be improved. Information given by extension, research, education, and farm organizations helps farmers make better choices. Therefore, the operation of a specific agricultural information system needs to be understood to control and enhance it (Demiryürek et al., 2008).

According to the findings of Maningas et al. (2005), information within the hands of the farmers means empowerment through control over their resources and decision-making processes. They noted that being an effective and efficient delivery system of essential information and technology services facilitates the clients' critical role in decision-making towards improved agricultural production, processing, trading, and marketing. Food and Agriculture Organization points out, information is very important for rural development because improving the income of the farming community will depend crucially upon raising agricultural productivity. Achieving sustainable agricultural development is less based on material inputs (e.g., seeds and fertilizer) than on the people involved in their use. For achieving this there is a need to focus on human resources for increased knowledge and information sharing about agricultural production, as well as on appropriate communication methodologies, channels and tools. New agricultural technologies are generated by research institutes, universities, private companies, and by the farmers themselves. Agricultural information and knowledge delivery services (including extension, consultancy, business development and agricultural information services) are expected to disseminate new technologies amongst their clients (people who are involved in agriculture). The role of research and advisory services is to give highly accurate, specific and unbiased technical and management information and advice in direct response to the needs of their clients. Due to poor linkages between research and advisory services, the adoption of new agricultural technologies by farmers is often very slow and research is not focusing on the actual needs of farmers. In many countries, low agricultural production has been attributed, among other factors, to poor linkages between Research-Extension-Farmers and to ineffective technology delivery systems, including poor information packaging, inadequate communication systems and poor methodologies. Therefore, the information systems which integrate farmers, agricultural educators, researchers, extensionists and farmers should be introduced to the agriculture sector. They operate as facilitators and communicators helping farmers in their decision-making and ensuring that appropriate knowledge is implemented in order to obtain the best results in terms of sustainable production and general rural development) and the private sector (support and input services, traders) to harness knowledge and information from various sources for better farming and improved livelihoods (FAO, 2005). However, this integration among people and institutions, particularly in the research-extension farmer relationship, has not been successful in many parts of the developing (and developed for that matter) world. There is also a basic difference in the information needs between market-oriented, transitional and subsistence-based farming. In addition, recent experiences show that the human components of the system such as researchers, educators, extensionists and farmers are not connected together in information flow. Therefore, it is a current need to investigate the proper information delivery systems for the agriculture sector and people that are involved in agriculture. However, there have been limited studies about agricultural information systems.

Thus, there is a need for substantial information about these issues, including the mechanisms of the information systems, interactions between components in the system, and their activity. Specifically, the information requirements of farmers, the structure of the organizations involved in these activities are issues that need to be explored (Demiryürek et al., 2008).

Attention in recent years has been focused on the linkages between research and extension, and there are many interesting theoretical and practical considerations involved in the interface between these major components of the information system (Rolls et al., 1994). Each information system aims to regulate organizational processes and is expected to provide important contributions on the subject of farm management that are necessary for farm holdings. Researchers agree that an easy-to-use farm management system that integrates production information and assists in making sustainable agricultural management decisions would be valuable for farm managers (Brennan et al., 2003).

A major issue in the agricultural sector today concerns strengthening the linkage between research and extension, particularly technology transfer to rural areas (Swanson et al., 1990).

Agricultural research and the agricultural extension system are dependent on each other. The objectives of applied agricultural research are to produce knowledge to solve farmers' problems and disseminate the results of research to farmers. It should be pointed out that the suitability and effective application of agricultural research mainly depends on good agricultural extension. On the other hand, agricultural extension also depends on agricultural research. Extension people obtain knowledge from the agricultural research system to convey it to farmers (Talug, 1990).

It is very clear that agricultural research and extension systems should be integrated into the agricultural knowledge system in terms of producing knowledge and disseminating it to farmers in a good way. However, the strength of the linkage between research and extension is often not very strong in practice. The weak relationship between agricultural research and extension is an important problem in Turkey. Also, the over-fragmentation of research institutions and lack of effective communication between the agricultural research institutions has inhibited the generation and dissemination of information (Talug et al., 1990).

Agricultural extension models can take several forms. The most common approaches are Training/Visit (T&V), Farmer Field Schools (FFS) and fee-for-service. In Training and visit plan, specialists/field staff provides technical information and village visits to communities selected. In many cases, field agents train and work with contact farmers, or farmers who have successfully adopted new technologies and can train others. World Bank promoted T&V and applied it in more than 70 countries between 1975 and 1995 (Feder et al., 2006). Farmer field schools (FFS), designed specifically to replace integrated pest management (IPM) methods around Asia. FFS also utilize contact farmers, relies on participatory training methods and builds farmer capacities. Fee-for-service extension comprises both public and private initiatives and public funding. Farmer groups contract extension agents with accurate information and service requests.

6.3 DEFINING OF AGRICULTURAL INFORMATION SYSTEM

An agricultural information system can be defined as a system, a system is a group of interacting components, operating together for a common purpose (Spedding, 1988). According to Checkland (1981) a system is a model of an entity. It is characterized in terms of its hierarchical structure, emergent properties, communication and control. The term subsystem is equivalent to system, contained within a larger system. The system approach is a way of looking at an entity and dealing with problems in order to identify and improve the particular system. It can be applied to any subject (Spedding, 1988). The system approach has also shown a high potential for offering a conceptual framework to analyse, manage and improve a current system and to design a better one (Cavallo, 1982). Models of social system can be used as a tool for analysing the information requirements of actors involved in a system (Checkland and Holwell, 1998).

Information is structured data within a context that gives it meaning (Checkland and Holwell, 1998). Information can be processed, generated, transformed and shared (Röling, 1988), through complex processes of coding and decoding, generally known as communication. The communication of information is a major concern for agricultural extension services (Demiryürek, 2000).

An agricultural information system can be defined as a system, in which agricultural information is generated, transformed, transferred, consolidated, received and fed back in such a manner that

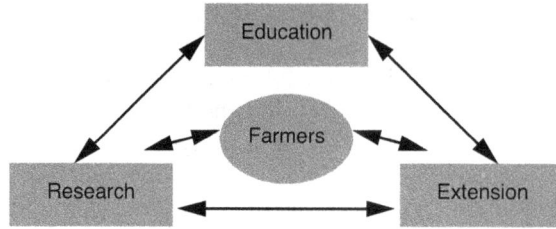

FIGURE 6.1 The FAO and The World Bank AKIS/RD model. (*Source*: FAO and The World Bank, 2000.)

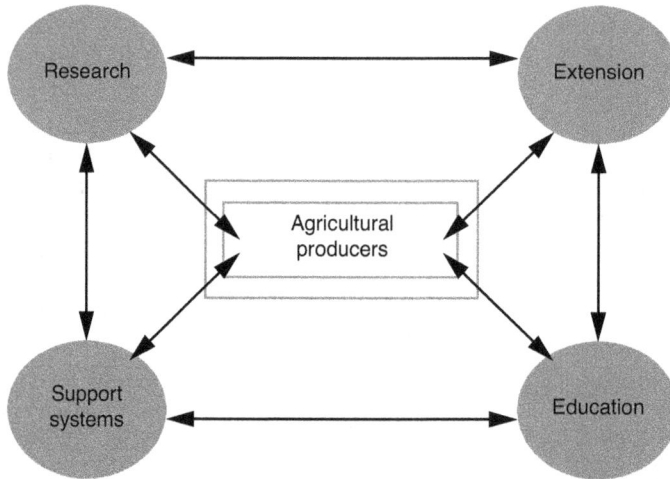

FIGURE 6.2 Idealised four pillar AKIS model. (*Source*: Rivera, Qamar and Mwandemere, 2005:6.)

these processes function synergistically to underpin knowledge utilization by agricultural producers. According to FAO and The World Bank (2000), "an Agricultural Knowledge and Information System links people and institutions to promote mutual learning and generate, share and utilize agriculture-related technology, knowledge and information. The system integrates farmers, agricultural educators, researchers and extensionists to harness knowledge and information from various sources for better farming and improved livelihoods." improved upon the agricultural knowledge triangle and brought to light the purpose of the system, which was——to serve the farmers or producers emphasizing the importance of agriculture in rural development (AKIS/RD) (Rivera et al., 2005:5) (Figure 6.1).

However, the FAO and the World Bank model failed to recognize the important role of agricultural innovation and the importance of markets. In addition, Rivera et al. (2005:5–6) noted that the FAO and The World Bank (2000) model did not incorporate other key factors such as government, the private sector, civil society, markets, support systems and knowledge and information. The shortcomings of the FAO and The World Bank model led to the development of the Pakistan four pillar model comprising the knowledge creation, knowledge diffusion, knowledge utilization and agricultural support sub-systems. The agricultural support included various functions such as credit, market and input. Rivera et al. (2005:5–6) improved upon the Pakistan model and placed agricultural producers at the core of the model, which was referred to as the idealized four pillar quadrangular model, comprising education, extension, research and support systems (Rivera et al., 2005:6) (Figure 6.2).

Accordingly, an agricultural information system consists of components (subsystems), information related processes (generation, transformation, storage, retrieval, integration, diffusion and

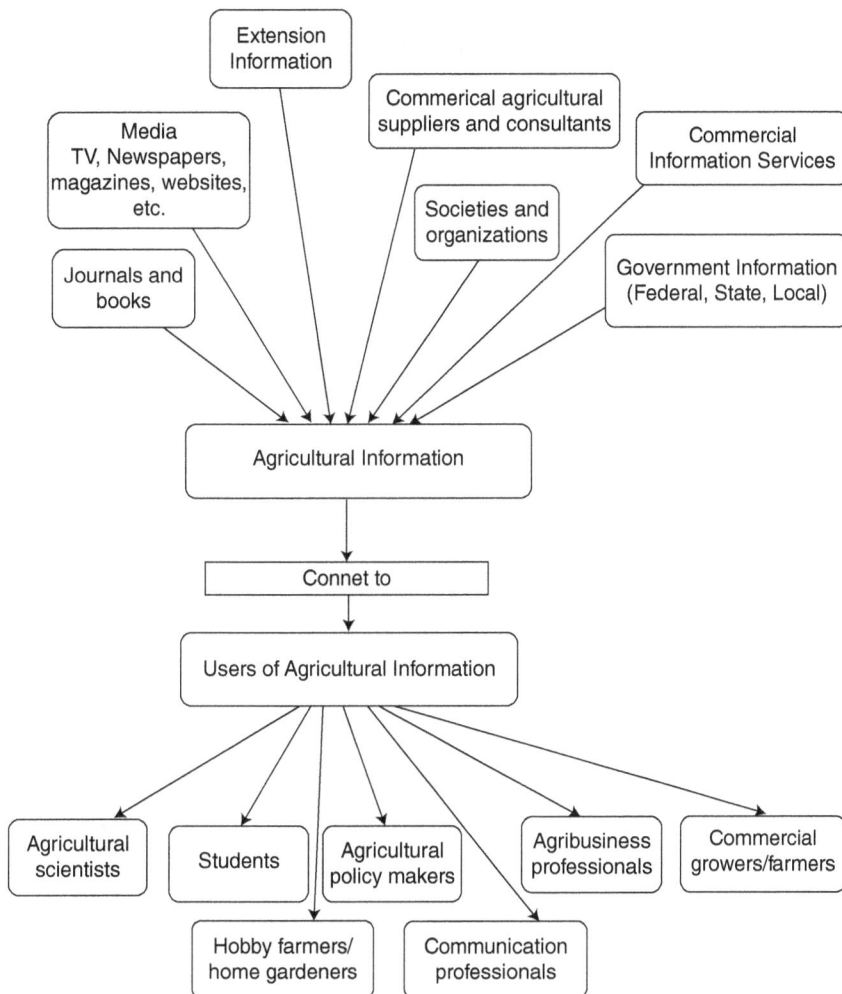

FIGURE 6.3   Concept map for agricultural information. (*Source*: McCue et al., 2005:5.)

utilization), system mechanisms (interfaces and networks) and system operations (control and management). Agricultural information is considered as an essential input to agricultural education, research and development and extension activities. Different kinds of information are required by different kinds of users for different purposes. The potential users of agricultural information include government decision-makers, policy-makers, planners, researchers, teachers and students, program managers, field workers and farmers (Zaman, 2002). Figure 6.3 gives an illustration of the flow of agricultural information.

The analysis of the agricultural information system in a specific farming system may provide the identification of basic components and structure of the system, the different sources of information used by different components in the system, the understanding of how successfully the system works and how to improve system performance (system management) (Demiryürek, 2000). This approach is also useful to identify possible defaults and improve the coordination between components (i.e., information management). Rogers (1995) emphasizes that the exchange of information (communication) and its diffusion take place within a social system.

Actors such as individuals, informal groups, organizations and subsystems are the members of the system and the structure of the social system and their actors or members' roles affect the

diffusion process. When considering the actors farmers and agricultural extension officers are key persons in between information flow. Some authors criticize the system approach to agricultural information system and especially knowledge dissemination and its ultimate utilization. They defend a different approach, namely an actor-oriented view (Leeuwis, 2004). They emphasize that knowledge and information are the elements of a single process in which information is internalized to become a part of knowledge.

Thus, it is difficult to distinguish between knowledge and information. The actor-oriented approach views knowledge processes as social processes which may lead to conflict among social groups or common perceptions and interests. Ramkumar (1995) developed an actor-oriented information system approach which considers the farmers' social, economic and cultural characteristics. This approach helps to understand the complexity of farmers' information systems and their relations with other systems.

6.3.1   GIS (Geographic Information System)

Geographic Information System is a specialized information system that builds to manage spatial information data (spatial reference) or a computer system that has the ability to build, store, manage and display geo-referenced information, such as identified data by its location in a database (Li et al., 2010; Kadir et al., 2015).

Geographic information system technology can be used for scientific investigation, resource management, development planning, cartography and route planning (Yasmin et al., 2016; Abdul et al., 2016).

6.3.2   Centre of Agriculture

The term agricultural centers can be defined as certain areas that are projected to produce certain agricultural products such as rice, corn, soybeans and other food crops and horticulture (Masters, 2009). In the Agricultural Development Master Plan 2015–2045, agricultural development in the next 5 years (2015–2019) refers to the agricultural for a development paradigm that positioned the agricultural sector as a driver for a balanced and comprehensive development transformation demographic, economic, intersectional, spatial, institutional, and agricultural development transformations (Christian Funch, 2008).

6.4   TYPES OF INFORMATION, PROVIDERS AND USERS

It is useful to distinguish between formal and informal information. Formal information is typically written and may be divided into data (numbers and other raw information) and processed information that is based on interpretation and analysis of the raw data. Informal information consists of information obtained through conversation and business transactions. Gossip is an important source of informal information. Sources of formal information include public agencies such as the USDA and Cooperative Extension, commodity groups, and a wide array of private providers including commercial vendors, agricultural and non-agricultural media and, in some cases, in-house analysis in which large farms hire professionals to interpret information. The information users can be divided into two groups: end-users of information (e.g., farmers) and intermediaries, for example, consultants, who serve as the main suppliers of information to the end-user. To understand information-use patterns, we conducted a national survey (Zilberman, 2004).

The study conducted by Rees et al. (2000), on agricultural knowledge and information systems (AKIS) undertaken by Kenya summarizes the types of information obtained by farmers. It is mentioned that technical information was reportedly received by 16%–33% of farmers. However, most end users felt that the information flow for this category was particularly deficient; the major knowledge gap expressed in the feedback meetings in all four districts was for technical information

(e.g. how to manage late blight in potatoes, where to get certified seed, the most appropriate varieties for a given location, housing and management of livestock, etc.). Other Information types it mentioned are marketing information and operational information.

The findings of Ozowa (1995) indicate that the information needs may be grouped into five headings: agricultural inputs; extension education; agricultural technology; agricultural credit; and marketing. Modern farm inputs are needed to raise small farm productivity. These inputs may include fertilizers, improved variety of seeds and seedlings, feeds, plant protection chemicals, agricultural machinery, and equipment and water. An examination of the factors influencing the adoption and continued use of these inputs will show that information dissemination is a very important factor. It is a factor that requires more attention than it now gets.

6.5 SOURCES OF AGRICULTURAL INFORMATION

Information source is an institution or individual that creates or brings about a message (Statrasts, 2004). The characteristics of a good information source are relevance, timelessness, accuracy, cost-effectiveness, reliability, usability, exhaustiveness and aggregation level (Statrasts, 2004). According to Oladele (1999), the efficiency of technologies generated and disseminated depends on effective communication which is the key process of information dissemination. The development of agricultural technologies requires among other inputs, a timely and systematic transmission of useful and relevant agricultural information (messages) through relatively well-educated technology dissemination (extension) from formal technology generation system (research) via various communication media (channels) to the intended audience – farmers (Oladele, 1999). It is expected that the message from the client (effect) be passed back to the source or research (feedback) for the communication process to be complete. Despite the attempt at technological innovation transfer, the wide gap between the levels of production that research contends are attainable and that which farmers achieve suggests a missing link (Oladele, 1999). Also, weak linkages between the farmer, extension and researcher mean that the farmers are not included in the planning of the innovation and hence do not know where to get their technologies despite the fact that they are the end users. Agricultural information disseminated by different information sources needs to be determined. It is imperative therefore to identify the sources of agricultural information utilized by farmers.

A study by Njuguna and Kooijman (1999) stated that neighbors, local meetings and the extension of government were the most frequently rated sources of agricultural knowledge in the region. Farmer training colleges and coordinated tours were least frequently listed. Half to three-quarters of respondents identified government agricultural extension workers as valuable sources of knowledge, and neighbors and relatives were classified as among the most important sources. In some branches, non- and churches represented essential sources of knowledge. Radio has been cited as an important agricultural knowledge tool. Opara's (2008) study studied the sources of agricultural knowledge accessible to farmers in Imo State, Nigeria, as well as favored sources from the farmers. The findings indicate that 88.1% of farmers identified agricultural extension agents as their information source, 71.2% indicated fellow farmers, 63.2% indicated radio, 43.3% indicated television, etc. The findings also revealed that the majority (70.0%) favored the extension agent to the other media (28.4% radio, 27.2% friends and family, 19.1% TV, etc.). The results emphasize the need for the extension agency to recognize certain sources of information that farmers use, or use most, on a regular basis as this will allow them to effectively deliver agricultural information to farmers. The Rees et al. (2000) research summarizes the different and varying sources of knowledge cited by the farmers. Community-based organizations and traders were identified as major sources in the sub-location input meetings, friends, families, neighbors, women's groups and school/youth groups. Most farmers considered the Ministry of Agriculture to be a major source of information, and most also reported receiving agricultural information from barazas (local meetings called by regional chiefs – presidential office appointees). Half to three-quarters of respondents identified government

extension workers as important sources of knowledge, and neighbors and relatives were classified as among the most valuable sources.

6.6 BRIEF EXPLANATION OF TRADITIONAL SYSTEM DESIGN AND IMPLEMENTATION PROCESSES

In system design, the system development life cycle (SDLC) methodology has been followed over the years. The primary cycle involves a five (5) step process consisting of Analysis to the implementation process (Joshua, 2015). Figure 6.4 show this cycle.

Using this cycle, we will discuss an example of how a simple fund transfer system can be designed and implemented. This is done in other to create an understanding of how this cycle works. We will focus on the analysis stage of the SDLC. **Example 1: design of a simple online account balance access system.** *This system will provide end users the functionality of accessing funds online.* **Phase 1**: Analysis: Under analysis, we seek to discover the needs of the users. That is, what users expect of the fund transfer system. These requirements are analyzed and translated into logical patterns as understood by the computer. User needs: Pay in money into a bank account – access funds balance online. Here the needs of the user are clearly defined. No complicated situations are referenced. Thus, a user case diagram can be easily created as shown in Figure 6.5 below (Joshua, 2015).

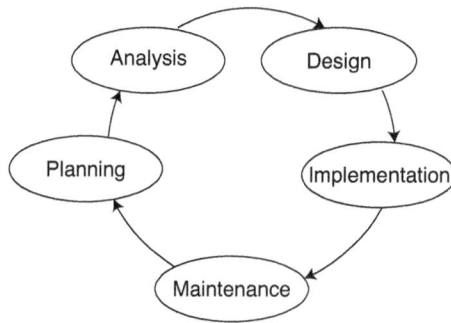

FIGURE 6.4   System development life cycle.

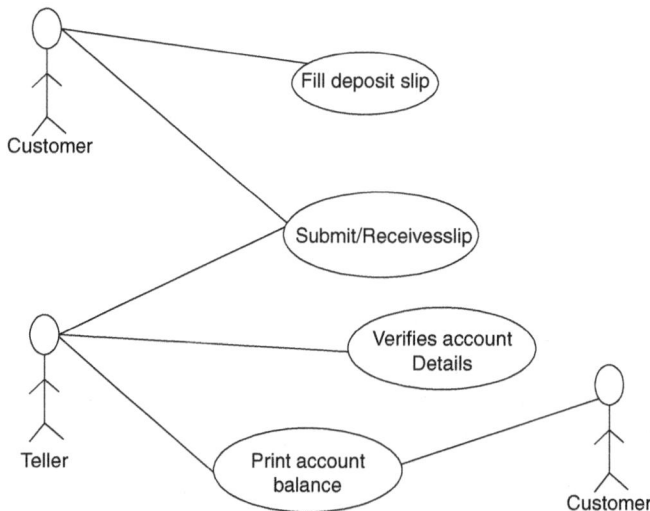

FIGURE 6.5   User case diagram.

Also, the user's abilities are also considered here. For example, user language, educational level, ethical variations, etc. When these factors are successfully considered, and analysis of system requirements is done based on these factors and many others, essential knowledge of system requirement has been developed. A fundamental process that is important in the analysis of system is referred to as the business process. The business process lists and links as simple as possible the internal and external logic behind the system development. In precise terms, a business process can be seen as steps to achieving a solution. For example, if a robot were to be programmed to pull out a bad electric bulb, the business process would be the steps needed to be followed by the robot to achieve its mission. Successfully defining the business process is a problem half solved. For our above example, the business process can be defined based on two (2) different headings; these headings are derived from the users' requirements (Masters, 2009):

- Pay in money into a bank account
- Access fund balance online pay in money into a bank account: In practice, the steps in paying money into the bank involve all or some of the following:
  1. Customer fills a deposit slip
  2. Submits the slip to a teller
  3. Teller collects and verifies the money and records the information on the slip against the stated account number.

6.6.1 Access Fund Balance Online

To access funds balance online, several things need to be considered. This includes the banks' processes of accessing funds balance (not online). Using this as a base, the designer can then create a business process on accessing funds balance online. First knowledge of how fund balances are accessed in banks:

1. Customer fills an account inquiry slip (balance)
2. Submits the slip to a teller
3. Teller accesses customer's account (using account number, name, etc.)
4. Presents account balance to the customer on a slip with the successful analysis of these business processes, the logic behind how the systems should work can now be quickly developed.

6.6.2 Process A

Access fund balance online (System Logic) Interpreting business process to enable coding:

Customer fills an account inquiry slip (balance): A digital mind slip is generated that will be filled online by customers.

Submits the slip to a teller: account slip on completion is presented for query Teller accesses customers account (using account number, name etc): Based on successful queries, account details are retrieved.

Presents account balance to the customer on a slip: System outputs the account balance to the customer on screen (Verma, 2012).

Phase 2–5: Based on the business process, a flowchart is designed, and subsequent coding is carried out to achieve the logic defined by the enterprise process. After these stages, the developed system is implemented with maintenance performed as required. As observed in the example above, generating a business process is the bedrock to the successful design of a system. However, this is not the only determinant to the successful implementation of a system. Other factors that include behavioral, educational, cultural, social, structural, etc. can determine to a great extent the success of a system,- especially in relation to different sectors. The Agricultural sector is one of these sectors. Developing a business process for the Agricultural sector is quite different from the process developed above. This

difference is great because of its unstructured pattern as we shall discuss below. With this knowledge, one can infer the reasons for implementation problems in Agriculture (Joshua, 2015).

6.7   DEVELOPING A FRAMEWORK FOR MARKET ACCESS

This study employs a pro-poor development approach; "the Making Markets work the Poor" (M4P) in designing the framework for accessing markets and market information. The M4P framework, mostly used by governments and agencies in the private sector, was introduced in the late 1990s and focuses on developing market institutions that benefit the poor (Denison and Vidra, 2013). M4P is a framework that focuses on inclusive growth, or pro-poor growth i.e. M4P is "based on recent thinking about how to use market systems to meet the needs of the poor and how to support the private sector through market mechanisms that bring about sustainable change" (Denison and Vidra, 2013; DFID, 2005; Wildt et al., 2006). M4P aims to enhance the poor's access to opportunities and their capacity to respond to opportunities within the economic mainstream, either as producers/entrepreneurs, workers or consumers (Tschumi and Hagan, 2008). According to the literature (Wildt et al., 2006), M4P places great emphasis on understanding the system in which the poor are located, the root cause of constraints that they face (rather than the symptoms of the problem) and ways in which the system might be changed to benefit them. M4P approach basically involves three basic steps outlined below (Ferrand et al., 2004; Tschumi and Hagan, 2008; Wildt et al., 2006): poverty reduction through market system development, developing a framework for understanding the market systems in which the poor exist and providing guidance for intervention practices. The M4P can leverage the market system by improving service delivery, changing practices, roles and important market players and functions, and changing the attitudes of market players (Tschumi and Hagan, 2008). The M4P framework has been implemented successfully in some countries like South Africa, Nigeria, Jamaica, and Bangladesh (Denison and Vidra, 2013). Using the M4P approach, information and communication technologies emerge as an intervention mechanism that can facilitate access to markets and marketing information by rural farmers. The goal of intervention is to ensure that farmers benefit from their produces and thus encourage more investment in farm production. Also, M4P acknowledges that access to markets and marketing information has a strong influence on poverty reduction (Jones, 2012). A framework consists of a set of standards, guidelines, policies and procedures which are implemented either manually, or where possible, automated through technology. A framework ensures data and information are managed in a secure, structured and consistent manner.

Figure 6.6 depicts a proposed framework for accessing agricultural market information. The management component in the framework specifies the issues that need attention in facilitating access to the agricultural market and agricultural market information. In doing so, the management component specifies the strategies for tackling the issues, opportunity for up-scaling and supporting policies. The framework also shows the services provided by the business systems and applications that accept *inputs* and produce *outputs*.

In the proposed framework, the strategies seek to transform the current situation of accessing agricultural markets and market information by unlocking the potential of utilizing ICTs. Overall, this can be achieved by improving the physical infrastructure like railways, roads and establishing ICT connectivity in rural areas. The service component of the framework shows some of the services that can be provided by the business systems and applications. The output of the system depends upon the given input. The system is expected to collect, store, share, analyze, organize etc. the received data and information. Different people can use the services provided by the system like farmers, traders, policymakers, government officials etc.

6.7.1   MANAGEMENT

The M4P approach requires different actors in the market to play some roles. The management is charged with the overall functions to oversee the market information provided to users and the

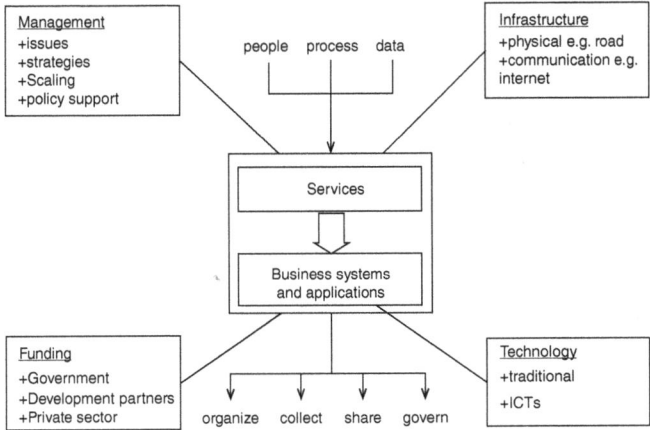

FIGURE 6.6 Building blocks of framework for accessing agricultural market information.

service is sustainable. The service can be provided by the government, private sector or state in collaboration with the private sector. Usually, the management provides supporting functions such as coordination, improving markets, market linkage and market infrastructure, creating awareness and providing information, developing required capacity and skills, and conducting research related to market development. The government as a management part is charged with establishing rules and regulations to guide the marketing activities, improving physical infrastructure like roads, and ensuring rural electrification. The management has the role of ensuring the rural areas are connected with the ICT infrastructure (e.g. providing cellular towers, Internet, etc.). The management has also the responsibility of promoting the use of standardized measures (e.g. weights, quality grades) and ensuring contracts engaged by market players are respected and enforced. Lastly, the government formulates and implements policies to guide agricultural marketing activities.

6.7.2 INFRASTRUCTURE

Access to markets and agricultural markets to a great extent depends on both established physical and ICT infrastructure. Lack of rural roads leads to insufficient participation of rural community in agricultural marketing and also discourages traders to travel to rural areas to collect produce. Also, lack of communication facilities denies important information to both traders and producers which can help them to make marketing decisions. Improvement in road infrastructure can significantly reduce both transport and transaction costs. Government, rural communities and government partners can collaborate in improving infrastructure in rural areas.

6.7.3 TECHNOLOGY

Technology is very important in delivering market information to users. The advantage of using recent technology of ICT includes delivery of up-to-date information and providing analysis of the information for different purposes. Respondents were concerned with the means of delivering information to them. For example, they preferred to use battery-powered radios and mobile phones due to their availability and accessibility. Respondents had concerns that radio programs are few and sometimes are not relevant as they are not targeted to them. Broadcasting time may be an issue as farmers may not be available during program time. The use of the Internet seems to be limited due to the absence of electrical power and lack of Internet infrastructure. This requires that when designing and implementing market information systems, it is very important to consider the technology that can be used to deliver information to recipients.

6.7.4 FUNDING

Access and use of agricultural market information may involve some costs. For example, radios broadcasting market information may decide to charge for such a service though recently they are broadcasting freely. For farmers to receive a package of market information intended for them (i.e. location specific for their produce), costs of collecting information and broadcasting or delivering are involved. Due to priorities and budget constraints, most African governments are yet to initiate the establishment of agricultural market information systems. Proving funds to assist in collecting, storing and disseminating agricultural market information may be relevant to rural communities. Funds may be provided by governments, development partners, NGOs etc. Funds may be used in training data collection staff and establishing means of delivering information. Creating awareness, promoting the program and introducing user fees are very important in ensuring its sustainability.

6.7.5 INPUTS AND OUTPUTS

The overall goal of the framework is to transform the inputs into useful output. Users, processes and data are considered the inputs to the system. Users can be farmers, traders, retailers, government officials etc. Different users have different information needs depending on their activities. The output of the system may help users to organize, manage, maintain, and govern their activities.

6.8 END USERS

To the small-scale producers, the M4P framework ensures that they have access to markets and that can overcome any form of market exclusion, can afford in making purchase, have good returns from selling their produce and labor, have choices, and risks are reduced in their agricultural marketing activities.

Linking farmers to markets and providing them with agricultural market information is anticipated with different benefits. Some of the benefits include greater interaction between farmers and traders, improvements in productivity, good returns from sales of produce, improvement in rural livelihood and emerging of strong institution representing farmers. Some benefits are provided by Tschumi and Hagan (2008) as outlined below:

- Improved delivery (such as increase in access or participation rates, improved quality or levels of satisfaction)
- Changes in practices, roles and performance of important system players and functions
- Changed attitudes of, and evident ownership by, market players
- Demonstrated dynamism of market players and functions (for example, responsiveness to changed conditions in the system)
- Independent and continuing activity in the system (i.e. the extent to which changes are maintained after direct intervention support has ceased)

6.9 CONCLUSION

Among other items, information support is also important for the carrying out of various activities by farmers and rural areas. As discussed in the chapter, majority of the rural farmers are not having access to most of the required agricultural information. Therefore, the implementation of information support systems for agriculture based on ICT is very important for the dissemination of agricultural knowledge and technical know- by rural farming community. To the betterment of information systems in agriculture, it is highly recommended to establish communication between farmers, coordinators, agricultural experts, research centers, and community by information technology. The information must be based on farmers' needs, the internet is used as a mode to transfer the advanced

agricultural information to the farming community. Farmers may be analphabets and speak a local language, and are not supposed to use the program directly. Thus these requirements should also be taken into account when introducing better information systems for agriculture and rural people. The analysis of the agricultural information systems can recognize the system's basic components and networks. It may be extended to any particular farming systems to examine how the information system functions. This approach is also useful in identifying potential defaults and improving the management of the details. Furthermore, the exchange of information (communication) among system components through networks is of critical importance for effective technology generation and information transfer. The information system analysis indicates that more interactive information sources are needed. It can inspire older farmers to turn to new farming approaches. More active experts working with selected local leaders could have accelerated these changes if they had established and improved ties with the public (especially extension and research) and private sources of knowledge. The sophistication of the farm information system leads to undervaluation among end-users. Lack of knowledge of agricultural information will weaken support for funding of public information as a major agricultural priority. An increase in public information funding would allow farmers to have more access to public information. For easy access and effective utilization of agricultural information in this digital age, there is a need for the establishment of information centers. Such information centers would be able to provide the rural farmers with the desired agricultural information in a format that would be comprehensible to them, taking into cognizance the prevailing high illiteracy rate, cultural differences and limited technology (Aina, 2007). For effective dissemination of agricultural information in rural communities by extension staff, research institutes and other responsible persons, there is a need for the construction of good access roads that would lead to all the remote rural communities in the country. Finally, with an effort to tackle the problem of information systems and improve production, different methods have been proposed and designed. It is observed that most systems designed for agriculture have not been fully implemented. With a critical review of works done in this area, we found that most development is centered on programming logic and not on the clients. For example, the structure used in the design of an IS system for a top business organization should not be the same structure that is used for industry full of uneducated employees. With this knowledge, it is appropriate that designers should focus on the human aspect of the design of an IS system for agriculture to aid in implementation feasibility. Human-computer Interaction (HCI) guides system analysts/designers in modeling systems that can interact appropriately with humans (Fakhreddine et al., 2008). Systems are designed for easy adaptation by man. Figure 6.7 describes HCI approaches that can be used in modeling an IS system. This system will focus as much as possible to satisfy the following HCI conditions:

- Safety
- Utility
- Effectiveness
- Efficiency
- Usability
- Appeal

This will as much as possible ensure the usability of IS systems. Usable for all classes of farmers: Ranging from the most educated to the least educated.

Using the HCI approach, developers are more concerned with the way users view and interact with the system. Cultural differences, languages, and social status/classes are all put into consideration in designing systems. Ease of learning and ease of use are major concerns for developers using this approach (Fakhreddine et al., 2008). Questions like: Can I use the basic functions of a new system without reading the manual? Does the software facilitate us to learn new features quickly? Etc. are significant questions that must be answered before software is considered Human-friendly. The poor attitude of farmers in using Information Systems could be enhanced by the complexity of IS

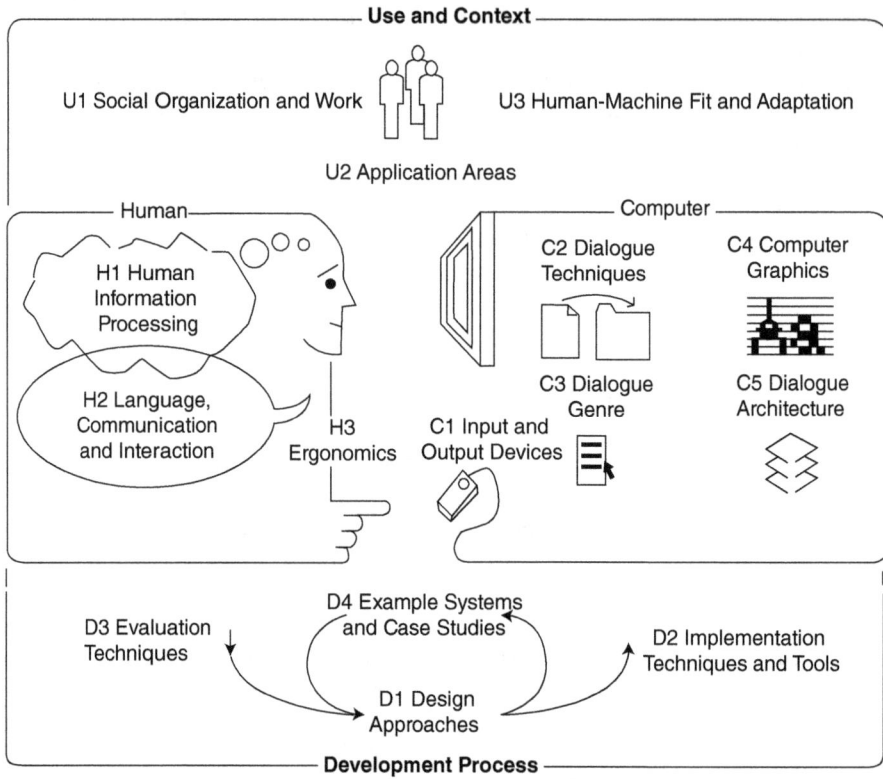

FIGURE 6.7   HCI approaches in modeling an IS system.

systems compared to the educational level or language of the end users. We believe that if systems are structured as simple as possible, many of the limitations to the implementation of Information Systems in agriculture in Africa will be handled (Joshua, 2015).

REFERENCES

Abdul F, Xu H, and Gao P (2016) *International Conference on Earth and Environmental Science*, 110, 45–51.
Aina LO (2007) Globalization and small- scale farming in Africa: What role for information Centres? *World Library and Information Congress: 73rd IFLA General Conference and Council*, Durban, South Africa.
Brennan MF, Pruss JA, and Brumfield RG (2003). A crop information management system to support and assess sustainable agricultural practices. *Proceedings of the 7th International Conference on Computers in Agriculture sponsored by the Society for Engineering in Agriculture*, Food and Biological Systems, ASAE. http://wcca.ifas.ufl.edu/archive/7thProc/BRENNAN/BRENNAN.htm#N_1.
Cavallo RE (1982) *Systems Methodology in Social Science Research: Recent Developments*. Boston, MA: Kluwer Publishing.
Checkland PB (1981) *Systems Thinking, Systems Practice*. Chichester, UK: John Wiley & Sons.
Checkland P and Holwell S (1998) *Information, Systems and Information Systems: Making Sense of the Field*. Chichester: John Wiley & Sons.
Christian F (2008) Information and communication technologies (ICTs) & society. University of Westminster, Communication and Media Research Institute, School of Media, Arts and Design.
Demiryürek K (2000) The analysis of information systems for organic and conventional hazelnut producers in three villages of the Black Sea region, Turkey. [PhD Thesis.] The University of Reading, Reading.
Demiryürek K, Erdem H, Ceyhan V, Atasever S, and Uysal O. (2008). Agricultural information system and communication networks: The case of dairy cattle farmers in Samsun province of Turkey. *Information Research*, 13, 343.

Denison M and Vidra RK (2013) Analytical Frameworks for Assessing ICT and Agribusiness Ecosystems. Accessed from: http://partnerplatform.org/?q7lvazvz.

DFID (2005) Making market systems work better for the poor (M4P): An introduction to the concept. Discussion paper prepared for the ADB-DFID 'learning event', ADB Headquarters, Manila.

Fakhreddine K, Milad A, Jamil A, and Monours A, (2008) Human-computer interaction: Overview on state of the art. *International Journal on Smart Sensing and Intelligent Systems*, 1(1).

FAO (2005) Agricultural Knowledge and Information Systems for Rural Development (AKIS/RD). Strategic Vision and Guiding Principles.

FAO and World Bank (2000) Agricultural Knowledge and Information Systems for Rural Development (AKIS/RD). Strategic Vision and Guiding Principles. Food and Agriculture Organization of the United Nations, and World Bank: Rome.

FAO-DFID (2005) Programme: Information systems in agricultural science and technology project GCP/INT/997/UK-2005-10.

Ferrand D, Gibson A, and Scott H (2004) Making markets work for the poor, an objective and an approach for Governments and Development Agencies, The ComMark Trust.

Jones L (2012) Discussion paper for an M4P WEE framework: How can the making markets work for the poor framework work for poor women and for poor men? The Springfield Centre for Business in Development, Durham.

Joshua CS (2015) Importance of the use of information systems by farmers in Africa. *International Journal of Scientific Research in Information Systems and Engineering (IJSRISE)* 1(1).

Kadir AA, Kaamin M, and Azizan NS (2015) *International Conference on Engineering*, 67, 4–10.

Leeuwis C (2004) *Communication for Rural Innovation: Rethinking Agricultural Extension*. Oxford: Blackwell Science.

Li G, Zhou K, Wang J, Sun L, Wang Q, and Qin Y (2010) *International Conference on Information Systems Engineering (ICISE)*, 41, 56–60.

Maningas RV, Perez VO, Macaraig AJ, Alesna WT and Villagonzalo J (2005) Electronic Information Dissemination through the Farmers' Information and Technology Services (FITS)/Techno Pinoy Program. Bringing Information and Technology within the Reach of the Farmers. Accessed from http://jsai.or.jp/afita/afita-conf/2000/part08/p231.pdf.

Masters WA (2009) *Africa's Turn: From Crisis to Opportunity in African Agriculture*. Sub-Saharan Africa: The Food and Financial Crisis.

McCue J, Craycraft C, Dunham T, Fretz T, McGeachin R, Wilson P and Young E (2005) Stocking the Shelves. Leadership Council for Agricultural Information and Outreach. Subcommittee on Content Report. Lexington, KY:5.

Njuguna E and Kooijman M (1999) Gender considerations in farm characterization and problem identification in the hill masses of Machakos and Makueni districts. In: Curry J, Kooijman M and Recke H (Eds.) *Institutionalizing Gender in Agricultural Research: Experiences from Kenya*. Nairobi: KARI, pp. 49–57.

Oladele OI (1999) Extension communication methods for reaching small-ruminant Farmers in South Western Nigeria. *In 26th Annual Nigerian Society of Animal Production Conference*, Kwara Hotels, Ilorin.

Opara UN (2008) Agricultural information sources used by Farmers in Imo State, Nigeria. *Information Development*, 24(4), 289–295.

Ozowa VN (1995) Information needs of small scale farmers in Africa: The Nigerian Example. Quarterly Bulletin of the International Association of Agricultural Information Specialists, IAALD/CABI. 40.

Ramkumar SN (1995) The analysis of farmer information systems for feeding of dairy cattle in two villages of Kerala State, India, The University of Reading, Reading.

Rees D, Momanyi M, Wekundah J, and Ndungu F (2000) Agricultural knowledge and information systems in Kenya–implications for technology dissemination and development. Agricultural Research & Extension Network, Network Paper 107.

Rivera WM, Qamar MK, and Mwandemere HK (2005) Enhancing coordination among AKIS/RD actors: An analytical and comparative review of country studies on agricultural knowledge and information systems for rural development. Rome: FAO. Accessed from http://www.fao.org/sd/dim_kn3/docs/kn3_050901d_en.pdf>.

Rogers EM (1995) *Diffusion of Innovations*. New York: The Free Press.

Röling NG (1988) *Extension Science: Information System in Agricultural Development*. Cambridge: Cambridge University Press.

Rolls MJ, Salleh M, Hassan H, Garforth CJ and Kamsah MF (1994) The agricultural information system for smallholder farmers in Peninsular Malaysia. Rural Extension and Education Research Report No: 1, AERDD, University of Reading, UK.

Spedding CRW (1988) *An Introduction to Agricultural Systems* (2nd ed). London: Elsevier Applied Science.

Statrasts AM (2004) Battling the Knowledge factor: A study of Farmers' Information Seeking Learning and Knowledge Process with an online environment in Queensland. Unpublished Ph.D Thesis.

Swanson BE, Farmer BJ, and Bahal R (1990) The current status of agricultural extension worldwide. Global Consultation on Agricultural Extension, 4–8 December 1989. Rome, Italy, pp. 43–76.

Talug C (1990) *Agricultural Research-Development and the Examination of Extension Services for Agricultural Productivity*. Ankara: MPM Publishing.

Talug C, Tatlıdil H, Kumuk T, and Ceylan C (1990) Agricultural extension services, problems and suggestions in Turkey. *In 3rd Technical Congress*, 8–12 January 1990. Maya Publishing, pp. 691–699.

Tschumi P and Hagan H (2008) A Synthesis of the Making Markets Work for the Poor (M4P) Approach. Swiss Agency for Development and Cooperation.

Verma R (2012) Role of Information Technology in Development of Rural Himachal, No.-172027.

Wildt MdRd, Elliott D, and Hitchins R (2006) Making Markets Work for the Poor: Comparative Approaches to Private Sector Development, The Springfield Centre.

Yasmin N, Ismail A, and Rothman, RA (2016) *International Journal of Engineering*, 76, 2614–2637.

Zaman MA (2002) Present status of agricultural information technology systems and services in Bangladesh. Accessed from http//jsai.or.jp/afita/afita-conf/2002/part 1/p075.pdf.

Zilberman D (2004) Giannini Foundation of Agricultural Economics, Information Systems in Agriculture.

7 Augmentation of Precision Agriculture by Application of Artificial Intelligence

Binny Naik and Ashir Mehta
Indus University

Manan Shah
Pandit Deendayal Petroleum University

CONTENTS

DOI: 10.1201/9781003268468-7

7.1   INTRODUCTION

The integration of technological advancements with bio-systems has profited well (Shah et al., 2020; Patel et al., 2020a; Ahir et al., 2020). One such example is in the management of agricultural production systems. Like industrial age brought mechanization and synthesized fertilizers to agriculture, and technology age brought genetic engineering and automation, the industrial age brought the potential for blending technological progress into Precision Agriculture (PA) (Zhang et al., 2002; Whelan et al., 1997). Initially known as Site-Specific Management (SSM), Precision Agriculture is the employment of a comprehensive management approach that includes the administration of farm information such as fertilizers, herbicides, seed, etc by executing the right practices at the right place and at the right time (Pierce and Nowak, 1999; Pierce et al., 1994). This is traditionally facilitated by utilizing sensor engineering and geo-spatial analysis systems. The key technologies that enable PA are: (i) Global Positioning System (GPS), (ii) Geographic Information System (GIS), (iii) Remote Sensing (RS), (iv) Variable rate technology, (v) Computers, and (vi) Information processing. These technologies are data-intense approaches that perform the task of acquisition and processing of spatial field data and also aid in the evolution of a novel type of farm tools with computer-aided controllers and sensors. The combination of these technologies has empowered farmers and their service providers to accomplish tasks not earlier possible, at levels of accuracy never previously attainable, and, when executed perfectly, at levels of quality never previously obtained (Fortin and Pierce, 1997). Other than these key technologies used in precision agriculture, the use of Artificial Intelligence is a novel approach to attain precision agriculture. The major domains managed under precision agriculture include crop management, pest management, soil and irrigation management, disease management, weed management, and agricultural product monitoring and storage control. The usual steps of the PA system include data collection, field variability mapping, decision making, and lastly management practice (Zhang and Kovacs, 2012). The adoption of precision agriculture imparts various benefits such as increased profitability, improved sustainability, increased crop quality, food safety associated with product traceability, lower management risk (Robert, 2002; Lowenberg-Deboer and Aghib, 1999), environmental protection, and rural development through new skills transferable to other activities (Robert, 2002).

The onset of technology in the digital world has encouraged humans to expand their thinking abilities and thereby strive to integrate the normal brain with an artificial one (Kakkad et al., 2019; Pandya et al., 2020; Sukhadia et al., 2020). This progressive investigation gave commencement to a completely innovative field of Artificial intelligence (Jha et al., 2019; Talaviya et al., 2020; Pathan et al., 2020). The advent of AI gave birth to many innovative logical ideas and hence different methods were identified and developed, which when implemented on machines, make the

process of problem-solving and decision-making more simple (Patel et al., 2020b; Kundalia et al., 2020; Jani et al., 2020). The machines are trained on normal behavior and are instructed for learning and self-improving upon their execution through abnormal behavior. Some of the AI methods include Machine Learning (e.g. SVM, KNN), Fuzzy logic, Neural Networks (e.g. Artificial Neural Networks (ANN) and Convolution Neural Networks (CNN)), Multi-Agent Systems, Expert systems, and Bio-inspired computing (e.g. swarm intelligence and evolutionary methods) (Parekh et al., 2020; Gandhi et al., 2020; Panchiwala and Shah, 2020). Both ANN and CNN are Deep Learning methods that involve processing any kind of stimuli using neurons in a layer-wise manner (Parekh et al., 2020; Shah et al., 2020; Patel et al., 2020b). In ANN, each neuron is connected to every other neuron, whereas in CNN only the last layer is entirely connected. Both SVM and KNN are supervised learning algorithms that classify the input data but KNN classifies data based on the distance metric whereas SVM requires a proper training phase. Both swarm intelligence and evolutionary methods are metaheuristic approaches and come under Nature Inspired Computing (NIC). Swarm intelligence works on the collective behavior of self-organizing entities and evolutionary methods are general population-based optimization procedures. With the accelerated technological improvement and the enormous area of implementation, AI is becoming ubiquitous very quickly because of its strong capability to resolve difficulties, especially problems that cannot be resolved accurately by humans and conventional computing (Bannerjee et al., 2018; Rich and Knight, 1991). The implementations of AI in various fields like agriculture, education, medical science, finance, security, etc prove to be beneficial.

The utilization of computers in agriculture was initially seen in 1983 (Bannerjee et al., 2018; Baker et al., 1983). The employment of AI techniques in agriculture can lead to a paragon transformation in agricultural practices as it is witnessing the adoption of such techniques very swiftly. Precision agriculture is information-intense and could not be accomplished without practicing immense improvements in networking and computer processing power (Stafford, 2000). Computer learning is especially proficient in becoming a groundbreaking technology in the field of agriculture (Bagchi, 2018). To resolve the current obstacles that persist in agriculture, various methods have been proposed beginning from the database (Martiniello, 1988) to decision support systems (Thorpe et al., 1992). But as far as accuracy and robustness are concerned, AI systems prove to be the best strategies among all the other methods. This is because AI techniques have allowed us to achieve the complex aspects of each circumstance and consequently give a solution that is optimally fit for that particular situation (Bannerjee et al., 2018). The advancement in AI-supported technology will empower farmers to work more efficiently and enhance the quality of the harvest. Some of the AI solutions available for the farmers include a chatbot. Chatbots are communicative virtual assistants which automate interactions with end-users. It helps farmers by providing answers to their queries, and giving guidance and suggestions on specific farm-related problems (Mokaya, 2019). Thus the farmers can be cognizant of technological advancements and can understand, learn and respond accordingly to issues related to farming. Other AI approaches employed in improving agriculture are: (i) Learning techniques- Machines are trained and developed based on traditional weather patterns, soil quality and the kind of crops that can be grown for increasing the crop yield, crop quality, and provide better water management techniques, (ii) Image processing- The implementation of AI techniques can help to interpret the drone captured images of field more efficiently and thus ensure rapid actions to the farmers, (iii) Prediction models- Cognitive AI solutions can easily understanding the data generated by Drones and cameras and can produce strong predictions on weather patterns, soil reports, and pest infestation to increase yield, (iv) AI aided sensors- The identification of insects that adversely affects the crop health can be done through the derivation of Vegetation Indices from reflectance spectral data and the damage severity of the crop can be calculated.

Precision agriculture is intuitively attractive as it provides a medium to enhance crop production and environmental quality in agriculture (Pierce and Nowak, 1999; Wolf and Nowak, 2015). Farmers have constantly been aware of the environmental impacts, but they were deprived of the

instruments to estimate, outline and control these fluctuations accurately. Thus, precision agriculture can have an impact on crop production facing the challenge of rising environmental variations. PA is viewed as a medium for the farmers to choose and implement the best inputs at the proper time and at the precise rate (Stafford, 2000). Crop generation is more specific and requires optimized inputs for better yield. This entails a reduction in production cost and amelioration of the impacts of the environment on crop management. Thus the need for proper implementation of Precision Agriculture is increasing. There are numerous perks of implementing precision farming in agriculture. Some of them are: (i) boosting agricultural fertility, (ii) obviating soil degradation, (iii) reducing chemical application on crops, (iv) better irrigation practices, (v) controlled use of herbicides and pesticides, and (vi) improved sustainability.

In this paper, the enabling technologies in Precision Agriculture, i.e., Geospatial technologies involving GPS, GIS, and Remote Sensing, have been overviewed. Additionally, the barriers pertaining to their application in precision agriculture have also been mentioned. The main aim of the paper is to discuss the Artificial Intelligence techniques that improve the management tasks associated with agriculture and hence aid to accomplish precision agriculture. The paper focuses to study the AI methods including Neural Networks- Artificial Neural Networks (ANN) and Convolution Neural Networks (CNN), Machine Learning classifiers- Support Vector Machine (SVM) and K-Nearest Neighbour (KNN), and Bio-inspired computing- Swarm Intelligence and Evolutionary methods for Crop management, Soil management, and Water management tasks. Finally, the future scope and challenges related to the use of AI in PA are presented and necessary conclusions are drawn.

7.2 GEOSPATIAL TECHNOLOGIES IN PRECISION AGRICULTURE

7.2.1 GLOBAL POSITIONING SYSTEM

Global Positioning System is a satellite-based radio-navigation system that provides highly accurate direction, time, and three-dimensional location data based on latitude, longitude, and elevation of anywhere in the world. The system has attained its complete operational ability with a whole set consisting of at least 24 satellites orbiting the earth in a prudently created pattern (Yousefi and Razdari, 2015). GPS has transformed the positioning notion, though it began originally as a navigation system. Depending on the selective tools used, GPS can provide a broad spectrum of accuracy from tens of meters to centimeters (Lange, 2015). There are two broad types of GPS receivers: (i) Code based, and (ii) Carrier phase based. The method with which a GPS receiver estimates distances to the satellites depends on the type of GPS receiver.

7.2.1.1 Role of GPS in PA

With the introduction of the Global Positioning System in agriculture, farmers have obtained the potential to take the consideration of spatial variability in the form of geographically encoded data. The fundamental task in precision agriculture includes the acquisition of data, such as soil samples, on some kind of spatial platform. The precise determination of the location of the acquisition points of such samples becomes a must to map the variability patterns. The use of GPS technology for such georeferencing procedures displays great accuracy (Borgelt et al., 1996). The GPS receivers aid to obtain location information for mapping farm borders, irrigation methods, streets, and problem domains in crops regarding weeds or disease.

Long et al. (1991) presented the use of GPS in a soil survey. The applied system gave adequately accurate outcomes in positioning and navigating the agricultural land. It also provided the digitalization of soil boundaries and showed greater in-field efficiencies than the traditional techniques.

McGovern et al. (2015) developed an innovative system that collected the field data straight into a digital database comprising soil, water, road, yield, and contour maps covered on remote sensing imagery. A GPS receiver was connected to a notebook computer that displayed relevant and

preloaded information panels. The arriving GPS signals were then merged with the displayed data using a software package. This helped the user to comprehend where they were relative to map components. Different panels of information could simply be edited and revised in the field and brand-new data could thus be joined. Additionally, the system also enabled the user to store GPS data to see and trace field movement at a succeeding stage.

Apart from enabling accurate soil sample collection from specific locations, GPS also allows the farmers to properly monitor crop health and conditions (Yousefi and Razdari, 2015; Qian and Zheng, 2006). Considering that the GPS technology in the area of agriculture has generated more features already available, thus, it contributes to a greater database for users (Tayari et al., 2015). In the future, the purpose of GPS systems in precision agriculture may help the farmers to harvest the results of edge technologies without settling for the quality of land and produce.

7.2.1.2 Barriers to Using GPS in PA

The implementation of PA using GPS requires proper correction signals. Though GPS receivers are easily available for agricultural use, the correction signals are not readily accessible. They require an expensive annual subscription. Thus PA farmers have to wait for greater precision signals, usually in centimeter ranges, in order to get a better organization of inputs, quicker applications, and night services (Robert, 2002). GPS is, however, a continuously evolving technology and thus acts as a barrier for precision agriculture adopters by imparting uncertain benefits (Boyer et al., 2014). Also, the primary barrier for PA adopters as well as non-adopters was that the GPS technology was too expensive to use.

Another limitation is the requirement of a clear 'line of sight' between the receiver's antenna and various orbiting satellites. Any obstruction lying between the antenna and satellite can block and weaken the GPS signals, thus, rendering unreliable positioning. Multi-path interference is another problem wherein the receiver is unable to differentiate between the original signal arriving directly from the satellite and the signal arriving indirectly after bouncing off from nearby objects (Lange, 2015). The proper geometric organization of the satellites is also a must for reliable position measurements.

7.2.2 Geographic Information Systems

Geographical Information System is a computer system that aids storage, manipulation, examination, and visualization of data. GIS is a distinctive type of information system that maintains a record of situations, actions, and objects in addition to where these situations, actions or objects occur or exist (Oshunsanya and Aliku, 2016; Longley et al., 2006). The ability of GIS to produce a quick adaptation to real circumstances and get answers in near-real-time is because GIS enables the updating of geographical information and its relevant attributes. Being a dynamic product rather than a static product, GIS offers to update, change, and generate maps easily (Coleman and Galbraith, 2000). Geographical Information System is a thematic mapping system, which permits the generation of maps based on subjects such as soils or hydrology (Oshunsanya and Aliku, 2016; Coleman and Galbraith, 2000).

7.2.2.1 Role of GIS in PA

Geographical information systems have been in occurrence for nearly three decades, but these applications have extensively been applied for agronomic and natural resource management only in the last 10 years (Burrough, 2015). GIS has been demonstrated to be very advantageous to those working in the farming sector because of its strength to examine and visualize agricultural settings and workflows (Andreo, 2013; Orellana et al., 2006). Data produced by PA are efficiently managed by GIS. However, the use of some additional software applications to treat the information collected in the GIS can enable appropriate data interpretation and management-taking decisions (McBratney and Whelan, 2001). ArcGIS, Manifold GIS, GeoMedia, MapInfo, etc are some of the examples of GIS

software. The GIS designed by a computing framework enables the generation of a complex view of the fields and offers accurate agrotechnological decisions (Neményi et al., 2003; Pecze, 2001).

Using GIS technology, various studies have been performed in the field of agriculture that proved to be beneficial. These include studies that focused on rendering the optimal or near-optimal nutrient (Neményi et al., 2003; Csizmazia, 1993) and chemical (Neményi et al., 2003; László, 1992; Mesterházi et al., 2001) content for crops and appropriate cultivation for the individual part of the field (Neményi et al., 2003; Jóri and Erbach, 1998). Consequently, the judicious use and application of chemicals and fertilizers on crops can help to conserve capital and prevent the environmental pollution that is created by the percolation of the nutrient and the overuse of chemicals (Neményi et al., 2003; Pecze et al., 2001a,b).

Sarantuya et al. (2011) studied GIS techniques in the livestock and crop sector to the improvement of the then existing agricultural conditions in Mongolia. They suggested the utilization of high resolution: Landsat ETM+ and SPOT and very high resolution: IKONOS and Quickbird satellite data sets obtained in the visible and near-infrared range of the electro-magnetic spectrum combined with different other ancillary and attribute information collected within a GIS for (i) reviewing of the development of different vegetation, (ii) monitoring, and management of other crops, and (iii) providing more certain spatial decision-making.

Another application of GIS in PA is seen for plantation crops. This can be seen as the development of GIS Anchored Integrated Plantation Management for tea in India. It included the production of a digital map by utilizing the existing map and high-resolution satellite image; the development of DEM (Digital Elevation Model); the formation of soil, land use, land cover, and drainage map; data accommodation in a centralized place; acquisition and storage of data into palmtop computers from the field instrumentation sensors, etc. (Mondal and Basu, 2009).

7.2.2.2 Barriers to Using GIS in PA

The essential GIS functionalities of integration, analysis, and modeling of the data obtained from diverse sources, cannot be entirely accomplished if the GIS database is incomplete, inaccurate, or obsolete. The data included in a GIS database are either spatial or thematic. Conventionally, these data are digitized from current topographic or land-use maps. These maps are secondary in nature and thus cannot display all the desired characteristics because of map generalization. Also, these maps may be outdated due to active alterations on the ground.

Though the use of different software for data collection, processing, and analysis in a GIS environment has made notable improvements in readiness of use, openness, spatial analysis, and 3D display, much more is yet required and has to be developed. Decision-making methods and expert systems are needed to be generated to optimize the utility of multi-layers and multi-years spatial data (Robert, 2002).

7.2.3 REMOTE SENSING

Remote sensing is the science of acquiring information about any object or any area of interest from a distance. It refers to non-contact measurements of radiation reflected or emitted from agricultural fields. Normally, remote sensing concerns the measurement of reflected radiation, rather than transmitted or absorbed radiation (Andreo, 2013). Remote sensing technique can be divided into (i) Active remote sensing: wherein the signal is emitted by an aircraft or satellite and the sensor detects its reflection by an object, and (ii) Passive remote sensing: wherein the sensor detects the reflection of sunlight. Examples of active remote sensing include RADAR and LiDAR, whereas passive remote sensors include film photography, radiometers, and infrared.

7.2.3.1 Role of RS in PA

The synergy of electromagnetic radiation with soil or plant matter makes the basis for remote sensing utilization in agriculture. Generally, there are three means of remote sensing data available, that are,

applied in agriculture: Proximal sensors, Airborne sensors, and Satellite sensors. Thus, Remote sensing applications in agriculture are naturally categorized according to the kind of platform for the sensor, including satellite, aerial, and ground-based platforms (Mulla, 2013). The application of a remote sensing system for farm management entails some unique factors. These include (i) Timeliness: to achieve optimal usefulness, data must be ready and accessible within minutes, and (ii) Frequency of coverage: the optimum usefulness is accomplished when continuous coverage is available. Plant-related properties such as canopy architecture, dry material, and leaf area index are best evaluated in the morning time, whereas stress-related characteristics are best accomplished within an hour following solar noon, and (iii) Spatial Resolution: these requirements are dependent upon the specific employment for the data (Jackson, 1984). Based on its features and capabilities, RS provides the possibility of mapping, monitoring and examining crop and soil variability and thus consequently renders an efficient approach to manage the impacts of any circumstances that affect crop health, yield, and quality.

The very first application of remote sensing in precision agriculture was reported when Bhatti et al. (1991) performed a remote sensing-based study to determine spatial patterns in soil organic matter content. They utilized Landsat imagery of bare soil to evaluate spatial patterns in soil organic matter content, which were then employed as auxiliary data simultaneously with ground-based measurements to determine spatial patterns in soil phosphorus and wheat grain yield. The spatial resolution of Landsat, SPOT and IRS satellites is somewhat coarse (20–30 m) for prevailing utilization in precision agriculture (Andreo, 2013).

Presently, Remote sensing applications in agriculture have contributed to a wide range of attempts. Hatfield and Pinter (1993) performed a comprehensive and practical review of crop protection by explaining relevant analysis into remote sensing of crop stress including weeds, water, diseases, frost, insects, and soil temperature. Other applications include the examination of crop nutrients and water stress (Clay et al., 2006; Moller et al., 2007; Tilling et al., 2007), crop leaf temperature and associations that could support the scheduling of irrigation followed (Moran, 1994; Garrot et al., 2003), infestations of weeds (Thorp and Tian, 2004). Baumgardner et al. (1986) studied soil properties such as particle size distribution, structure, surface roughness, Fe oxides, organic matter content, moisture content, and abundance of carbonate minerals that influenced reflectance, and hence, the remote sensing of soils. Other RS applications include the study of crop yield and biomass (Shanahan et al., 2001; Yang et al., 2000), and soil properties such as organic matter, moisture, clay content, pH (Christy, 2008), or salinity (Corwin and Lesch, 2003).

7.2.3.2 Barriers to Using RS in PA

The utilization of remote sensing, in general, has been depicted to be advantageous and profitable (Zhang and Kovacs, 2012; Godwin et al, 2003; Seelan et al, 2003; Tenkorang and Lowenberg-DeBoer, 2008), but existing applications in PA are however limited. These limitations include prompt acquisition and transfer of images, the scarcity of necessary spatial resolution data, image interpretation and data extraction problems; and, the synthesis of these data with agronomic data into expert systems (Zhang and Kovacs, 2012; Jackson, 1984; Du et al., 2008). Apart from these, some ambiguity may also arise regarding remote sensing terms. Remote sensing works based on a determination of energy that is reflected or emitted from the object of concern. An error can occur when the signal from the object is corrupted by energy arriving from the atmosphere, or some other objects in the vicinity, or is changed in processing when a film is utilized as the measuring equipment. The data that is not rectified for any of such intruding energy are called raw data, digital numbers, or just, DN. Until these raw DN are not calibrated upon some model in the view, they are not useful for quantitative evaluation or temporal analysis (Frazier et al., 2015). Considering the case of evaluation of crop conditions, remote sensing gives an inadequate image of the root region environment, because the data signify the reflectance of the surface substance, which might be bare topsoil, crop material, or a blend of both (Gebbers and Adamchuk, 2010).

Spry et al. (1996) performed a search for the implementation of digital imagery in California. They discovered that majority of the growers and agronomists were unfamiliar with remote sensing

and they were unable to easily associate images with obstacles and landscape characteristics found in the fields (Frazier et al., 2015). Also, TRW led a nationwide survey in 1994 to determine the extent of remote sensing applications in agribusiness (MacDonald et al., 1996) and obtained an insignificant presence of remote sensing in agricultural industries (Frazier et al., 2015).

Thus in sum, the use of remote sensing is limited due to the lack of knowledge and understanding of its applications and advantages, expenses, availability of imagery (cloud cover), quality and resolution of imagery, lag in preparing the imagery, and shortage of skills in rural regions for the interpretation of imagery (Robert, 2002).

7.3   AI-BASED CROP MANAGEMENT

Crop management refers to the group of precise agricultural practices that aims to optimize the crop development procedure by managing different crop properties and examining its surrounding environmental conditions. This management technique is a comprehensive approach that involves the administration of various agricultural aspects such as crop yield prediction, disease detection, pest infestation, weed detection, crop quality, and species recognition. Crop management tasks consist of a series of events that starts with the sowing of seeds, further proceeding with crop maintenance during the growth and development stage, and terminate with crop harvest, storage, and distribution (Madsen, 1995; Tivy, 1990). Such planning procedures can be optimized with the aid of different AI-driven technologies that can produce more favorable crop yields by determining the best hybrid seed choices, optimal crop selection, and adequate resource utilization.

7.3.1   Neural Nets

7.3.1.1   ANN

Robinson and Mort (1997) proposed a neural network system to foretell the rate of frost from the meteorological data. The neural network system had many feed-forward architectures that consisted of one or two hidden layers in it. The parameters on which the system works are Wind Speed and Direction, Humidity, Cloud Cover, Maximum Temperature, Precipitation, Maximum wind speed and direction and lastly Maximum temperature. From the parameters, the minimum temperature of the next night is taken as an indication which is then classified for preventive measures. As the parameters, cloud cover and wind direction are not in the numerical form. So the raw data is coded into the binary as a part of preprocessing. Feed-Forward network is used for classification whose weights are modified with the help of a back-propagation network. The outcomes of the study display the efficiency and advantage of using various parameters in comparison to a single parameter.

Cho et al. (2002) developed a machine vision system employing a device camera for the detection of weeds on the radish farm. The primary aim of the study was to generate an algorithm for differentiating weeds by employing ANN by recognizing the shape feature for the detection of the weeds. For the study, a total of 150 images were considered. Further, the images were pre-processed on the eight different parameters like Roundness, Elongation, Length to perimeter, Perimeter cube to area by length, Aspect, Compactness, Perimeter to broadness and length to width were considered. For the classification, the back-propagation network was built using 31 images of radish and 40 images of weed. All the eight shape features were taken as an input having one hidden layer. The shape features were examined along with the binary images which were taken from the color images of radish and weeds. The model was successful with a detection rate of 92% and 98% for radish and weeds respectively. Also, features like Perimeter, Aspect, and Elongation were elected as essential variables.

7.3.1.2   Convolutional Neural Network (CNN)

Barbosa et al. (2020) developed a Convolutional Neural Network to obtain the appropriate spatial structures of distinct attributes which were then blended into a yield model for the seed rate and

nutrient management for predictions. The data was taken from nine different cornfields in the US which were then tested and compared with the proposed model. For the testing of the data four architectures connecting the input at various layers of the network are estimated and then compared to the commonly used model. The attributes considered in the model were Elevation Map, Satellite Images, Nitrogen and seed rate and soil's electro-conductivity. In a supervised manner, the model was trained and the outcome was compared to Random Forest Regression, Linear Model, Support Vector Machine and Fully Connected Neural Network. This model displayed a decrease in the RMSE up to 68% in comparison to the linear regression models and up to 29% in comparison to the random forest. Also, the advantage of the model is the variability in the spatial structures.

Oppenheim and Shani (2017) exhibit a disease classification algorithm for the potato. A deep convolutional neural network is utilized in the algorithm which classifies the tubers into five different classes consisting of one healthy potato class and four disease classes. The images utilized in the study consist of potatoes of various sizes and shapes. Each visual indication of the disease was noted and labeled. The proposed method was trained with various sizes of the training set for the classification of the four diseases of the potatoes. For the complete study, the training and testing phase were repeated nine times over various sizes. A CNN consisting of eight layers from which the first five were convolutional layers and the remaining three were fully connected layers was used for the classification. The hyper parameters involved in the training were Stochastic Gradient Descent, Batch Size, Weight Decay, Learning Rate, and Momentum. The outcome of the study exhibited that the accuracy for eight of the nine training sets was above 90%. The best training set manifests a classification accuracy of 96%.

7.3.2 Machine Learning

7.3.2.1 Support Vector Machine (SVM)

Zheng et al. (2015) examined the power of the Support Vector Machine by distinguishing different crop types in the crop system. The training set for the model was created adopting the Intelligent Selection Approach and Stratified Random Approach with the help of local knowledge. The model was employed in the Landsat Time-Series NDVI (Normalized Difference Vegetation Index). Nine major crops were taken for the classification. There are a total of four functions in the SVM that is Polynomial, Sigmoid, Linear and Radial Basis Function (RBF) from which RBF was utilized in this study as it is capable of controlling the nonlinear relationship between the attribute and the classes. Also, the parameter of RBF like penalty parameter C and the gamma parameter was considered. The results displayed that the model was successful in classifying the nine major crops with an overall accuracy of 90%. Also, the selected approaches were able to give high classification accuracy compared to the stratified random approach.

Devadas et al. (2012) developed a technique based on the SVM classifier. The object-based data produced by the multi-temporal Landsat images were classified using the developed technique. For the study Landsat area of 352,456 hectares was taken. The pre-processing of the images includes two parts: Cloud Masking and Image Segmentation. Using the cloud-detection techniques every image of the data went through the process of cloud detection and masking. This technique also determines the irregularities in the time series and includes the region growing filters. The images were divided into objects by applying the multiresolution segmentation algorithm which works on the principle of the region merging technique. A total of 36 parameters were considered for the classification which consists of 10 spectral, 3 shape-based and 23 textual parameters. The outcome of the study was then compared with the traditional methods. The examination showed that the object-based SVM achieved an overall accuracy of 95% in comparison to the pixel-based technique. Also, the multi-temporal Landsat images were observed to be having a greater influence on the accuracy of the SVM model.

7.3.2.2 K Nearest Neighbor

Rangel et al. (2016) proposed a method to classify and diagnose the grapevine leaves with potassium deficiency with the help of image processing. A total of 50 images were taken from which 50% were used for testing the algorithm and rest for the data sampling. Healthy leaves, Leaves with potassium deficiency, Leaves having other disease or deficiency was the images included in the data. The images are then pre-processed; a white balance is applied to the images also further the gray scale image is used for the lighter pixel so that the background appears white. The analysis of potassium deficiency can be considered the base of the three classes. First, the absolute white background after the preprocessing, Second green tones define the healthy tissue and third the reddish tissues. In this study, the deficiency is measured as the ratio of the affected region to the entire region. Using the proposed KNN method rather than the histogram method proved to be beneficial as better results were obtained.

Vaishnnave et al. (2019) developed software for the identification and classification of groundnut leaf disease using the KNN classifier. The groundnut leaf images having the RGB color perfectly visible are taken into consideration. Further, the preprocessing is done to remove the unwanted noise from the images. The image quality is increased by smoothing the image followed by increasing the contrast. The image is thus converted into a binary masked image. The image is then segmented as the process of segmentation divides the image into many segments which are easy to study. Hence the binary mask is converted into an HSV image. The next step is feature extraction, in the study the features such as color, morphology, texture, and arrangement are utilized for disease detection. Out of 250 images 45 images were used to train the model and the remaining were used for testing the model. As a result, four types of groundnut disease: Early leaf spot, Rust, Late leaf spot, early and late spot Bud Necrosis were classified.

7.3.3 BIO-INSPIRED

7.3.3.1 Swarm

Vazquez and Garro (2016) offer an approach for classifying the crops on the bases of GLCM from the satellite images with the help of the Artificial Bee Colony algorithm. The satellite images were obtained from the internet which was segmented manually. Each of the segments of the image defines some class which makes the application of feature extraction easier. The features have been extracted using the GLCM method. Five different classes have been recognized and from the segmented images, a set of polygons has been drawn out. The dataset consists of 2,752 patterns along with 24 features. In the testing phase for the validation of the outcomes the proposed method was performed 30 times using the Manhattan and the Euclidean distance equation. The data set was divided into two parts, one for the training and another one for the testing model. Both the distance algorithms have exhibited better results for the few runs but the best result was achieved when the value of th was 0.5 and 0.9 for Euclidean and Manhattan respectively.

Wan et al. (2017) proposed the use of the Artificial Bee Colony algorithm for the classification of the images of rice based on remote sensing imagery. The study is divided into two parts the first part includes the collection of the ancillary information and the NDVI. The second part includes estimating the efficiency of the information that has been included. A total of 181 data samples including 46 data of paddy rice and 135 data of non-paddy samples. The approach used for the classification is clustering. The image fusion improves the image quality. The study examines the algorithm with one spectral indices, an algorithm with NDVI and texture indices. The factor that influences the classification is Entropy which depicts the absence of organization among the decision and the attributes. By adding the ancillary information the classification accuracy for the ABC classifier increased to 89.5% which outperformed the ACO algorithm.

7.3.3.2 Evolutionary

Singh et al. (2015) exhibited the algorithm for the image segmentation technique for the automatic detection and classification of plant leaf diseases. For the detection of the disease, various images

of the leaf are taken for the examination. For that, the image-processing technique is applied to obtain the features for the analysis of the leaf. Then the affected images are segmented using the genetic algorithm. Further, these segments are used to classify the disease. The study is carried out in MATLAB. The leaves of the rose, leaves of Beans, Lemon leaves, Banana leaves, leaves of Jackfruit, Mango leaves, potato leaves, Tomato leaves, and sapota leaves are the ten leaves on which the algorithm is tested. The study displayed an optimal result with fewer efforts. Also, the benefit of using the algorithm is that the disease is identified at an early stage.

Olakulehin and Omidiora (2014) displayed the potentiality of enhancing the production of the crops and to maintain the fertility of the soil using the Genetic Algorithm. The parameters considered for the crop yield model are Soil Fertility, Climatic Influence, Weed Competition, Varietal Choices of Crops, Water Usage, Pests and Diseases, Waste, and Land Use. While the parameters considered for Soil Fertility are Soil depth, Drainage System, pH, Organic Matter, Availability of Water, Aeration, Mineral Compositions, and Soil Organisms. Further for the optimized result, various operations are carried out on the parameter such as Genetic, Crossover, Mutation, and Local optimization. The GA parameters that were set for the optimized result are Soil Fertility = 100, Crossover Rate = 0.8, Penalty Factor = 0.5, Maximum number of crop yields = 200, and Mutation rate = 0.1. The algorithm generated 20 solutions having the average yield total production obtained was 1,315,945 while the range was between 1308954–1324541.13. The result displayed that for every change in the soil fertility the other linked factors are able to change the yield up to 42% (Table 7.1).

7.4   AI-BASED SOIL MANAGEMENT

Soil is a heterogeneous natural resource that follows complex processes and has various important properties. It is the power of such soil properties that enables the researchers to comprehend the dynamics of ecosystems and understand their impacts on agriculture. Soil management refers to the accurate determination and control of soil conditions such as soil temperature, organic matter, soil structure, soil color, moisture content, salinity, etc. Apart from depending on its internal potential, the crop vitality widely depends on the predominating ambient situations in the soil (Mishra et al., 2019). For example, soil moisture plays a crucial role in crop yield variability (Liakos et al., 2018). By applying various artificial intelligence techniques in the administration of soil properties it is possible to obtain valuable support and insights for appropriate decision-making actions that enable precision agriculture.

7.4.1   NEURAL NETS

7.4.1.1   Artificial Neural Network (ANN)

Dai et al. (2011) aimed to discover the relation between crop yield and soil moisture and also to see the behavior of crop yield to the various saline conditions and soil moisture. For the examination six soil samples were acquired from five different depths. The content of the soil was predicted with the help of Electrical Conductivity and Gravimetric Methods. A three-layered Back-Propagation Network was used in the study. The training of the network was executed by the forward propagation of the inputs while backpropagation of the errors. Sigmoidal Logistic function was used to determine the relation. A total of 108 samples were taken for the study in which the samples were divided into four parts which were used to train the ANN four times with four different samples. The remaining samples were used for the testing phase. The employed ANN model gave high accuracy compared to the Multi-Linear Regression. The model was successful in depicting the relationship between the soil salinity, moisture and sunflower yield.

Arif et al. (2012) developed an ANN model to predict the soil moisture from the meteorological data which was then compared to the observed value. The soil moisture was predicted with the help of the Precipitation and Eva-Transpiration (ETo) which was predicted with the help of the

TABLE 7.1

Crop Management Using AI Techniques

Crop	Functionality	Features	Domain	Method	Performance	References
Sugarcane	Automated detection of diseases	Color and shape	ML	SVM+KNN	Accuracy obtained was 96% over the sample of 200 images	–
Cotton	Performance analysis	Remote sensing imagery and vegetation index calculation	ML	SVM+Fuzzy clustering	SR outperforms other vegetation indices by achieving 88.72%, 88.71% and 89.15% accuracy values for image sets 1, 2 and 3 respectively	Kawarkhe and Musande (2014)
Vegetable crops	Detecting fungal disease	Local binary patterns	ML	Neuro-KNN	The classification accuracy is 91.54%	Pujari et al. (2014)
Paddy	Identification of blast and brown spot diseases	Image processing, pattern recognition, and geometrical features	ML	KNN	The proposed method obtained a result of 76.59%	Suresha et al. (2017)
Plant	Determine the rate of growth, detect diseases, classify and recognize classes of the plant images	Image processing and labeling	Neural nets	CNN	The results of the model were 99.58%	Abdullahi et al. (2017)
Seasonal crops	To detect crop diseases	Batch normalization, rectified linear unit activation function	Neural nets	CNN and autoencoders	Accuracy of 97.50% for 2×2 filter size in 100 epochs and accuracy of 100% for 3×3 filter size	Khamparia et al. (2019)
Seedlings	Detecting phalaenopsis seedling infected with BSR, BBS, and PBR	Color and texture features	Neural nets	ANN	The accuracy of phalaenopsis seedling is 89.6%	Huang (2007)
Corn	Corn yield and grain quality	7–13 important factors affecting spatial variability in corn yield and grain quality	Neural nets	ANN	The accuracy of the factors ranged from 68% to 92% for 7–12 factors of 68%–99% of the for 8–13	Miao et al. (2006)

(Continued)

TABLE 7.1 (*Continued*)
Crop Management Using AI Techniques

Crop	Functionality	Features	Domain	Method	Performance	References
Crop	Crop classification	High resolution satellite image	Bio inspired	PSO	If the training process tends to be multisource classification then the swarm intelligence models are best fit	Omkar et al. (2008)
Crop	Optimal field coverage	B-patterns approach and CVRP	Bio inspired	Ant colony optimization	The optimal plans with conventional plans show reductions in the in-field non-working distance in the range of 19.3%–42.1% whereas the savings in the total non-working distance were in the range of 18%–43.8%	Bakhtiari et al. (2013)
Tomato plant leaf	Automatic disease detection and classification approach of plant leaf	HSI conversion and texture feature analysis	Bio inspired	Genetic algorithm	Highest overall classification accuracy of 97% was achieved by genetic algorithm compared to PNN	Taohidul Islam and Mazumder (2019)
Rice	Nutrient management	Sensitivity analysis on different weight structures	Bio inspired	Fuzzy goal programming based on genetic algorithm	The suggested FGP based GA model is productive in choosing the action that will augment the yield of rice keeping the soil fertility maintained	Sharma and Jana (2009)

parameters like Minimum, Maximum, and Average value of the air temperature. For the study, Two three-layered ANN models were used in the study. The first model was used to predict the ETo. Then the predicted ETo and precipitation were given as input in the second model for the prediction of the soil moisture. The Back Propagation was taken as the training model which used the sigmoidal function. The data set was split into two parts. One part was to train the model while the second part was used to verify the model with the help of the Coefficient. The First model predicted the ETo precisely having the value of R^2 0.96 and 0.95 as the training and validation stage. The second model predicted the soil moisture with an R^2 value of more than 0.72.

7.4.1.2   Convolutional Neural Network

Sobayo et al. (2018) proposed a CNN-based regression model to estimate the Soil Moisture (SM) content using thermal infrared (TIR) images. The images were captured with the help of thermal cameras that were mounted on a drone. Each pixel of the TIR image represented the temperature estimates of the corresponding area. Soil moisture was assumed to be in association with the temperatures of the target crops and such temperatures were reflected on the TIR images through the colors. The model was trained by the TIR images and the soil moisture values corresponding to those TIR images obtained by in situ measurements. TIR images that were not utilized during the training phase were used for testing. The resulting SM estimates were close to manual measurements by the sensors. The model was compared with a DNN-based regression model. The CNN model avoided overfitting, required fewer resources than the DNN-based one and generated results much faster using the same hardware resources. SM estimated by CNN was far better than DNN.

Padarian et al. (2019) developed a CNN model for generating Digital Soil Maps (DSM) and simultaneously predicting soil organic carbon at multiple depths. As data, 485 soil profiles of Chilean soil were used with soil organic carbon (SOC) content (%) at depths 0–5, 5–15, 15–30, 30–60, and 60–100 cm. The input to the model was represented as a 3-D stack of covariates images that examined spatial contextual information by ascertaining non-linear local spatial relationships of neighboring pixels. Data augmentation was performed to reduce overfitting. The results revealed that compared to the conventional techniques that only employed point information of covariates, the CNN model decreased the error by 30% producing results with less prediction uncertainty. Also, the model predicted soil carbon at deeper soil layers more precisely. As the CNN model takes input as the covariate-represented images, it appears to be a simplistic and efficient framework for future DSM models.

7.4.2   MACHINE LEARNING

7.4.2.1   Support Vector Machine (SVM)

George and Kumar (2015) developed a model to map the different soil salinity classes with the help of hyperspectral indices that are produced with Hyperion data. Level 1 Radiometric was utilized having 242 bands with a resolution of 10 nm. Then a total of 196 bands from the total were calibrated which refer to the VNIR and SWIR. The data is processed by the EO-1 which is further corrected radiometrically level 1R. To distinguish the waterlogged areas from the soil NDWI is utilized. The SVM classifier was used to determine the hyperplane which divides the predefined number of classes. The division of the classes is done with the help of a decision surface which increases the boundary between the classes and will reduce misclassification. Different kernels have been applied in the model that can generate non-linear hyperplanes. The classification accuracy achieved was 78.13% along with a kappa statistic of 0.71. Also, the performance of the soil with high salinity outperformed the other classes.

Qiao and Zhang (2012) suggested a method for recommendation for the fertilizer by analyzing the P, K, and N nutrients of the soil based on Near Infrared Spectroscopy. The soil samples were acquired with the help of the DGPS for the soybean crop. After that, the sample of particle size

2 mm was obtained through the process like air-dry, and hand ground. As a part of pre-processing, the removal of the noise from the data was done by computing Smoothing Average. Out of 54 data samples, 14 were used for the testing phase while 40 were used for the training purpose. After pre-processing, PCA was used to obtain all the eight principle components. Further, Least Square Vector SVM was used for the estimation of the soil nutrients in which six Principle Components were taken as input. Also, hyper parameters like RBF and GAM were obtained. The proposed method is easy and the correlation between the estimated values and observed value were similar.

7.4.2.2   K Nearest Neighbor

Meng et al. (2014) proposed a novel k-nearest neighbor algorithm to predict soil moisture in the maize field. One hundred and eighty maize plants in sixteen areas, with different sample capacities in different areas, are taken as sample datasets. Morphological characteristics having non-negative values such as maize plant height, leaf area, stem diameter, dry weight, and fresh weight were used as the features in the estimation of soil moisture. The model was tested on datasets in six growth stages of maize, likewise seedling, jointing, heading, grain filling, milky, and mature stage. I-divergence was used as the distance metric. To evaluate the performance of the proposed model accuracy and macro-F1 measures were used as the evaluation indexes. The results showed that the proposed model was more effective than the traditional k-nearest neighbor algorithm.

Maniyath et al. (2018) suggested the use of the KNN classifier to detect soil color using digital image processing. Munsell soil chart images were utilized to create the database. The soil images for the dataset are collected from different regions. A median filter which is a non-linear spatial filter is used for filtering processes. Pre-filtering is done to remove noises such as salt, pepper, etc. Next, the image is first converted from RGB to HSV and the hue component is given an upper and a lower limit. Then the thresholding method of segmentation was performed on the hue and the image was converted back into RGB and displayed to see the thresholding results. Post-filtering was performed to define the boundaries properly. Euclidean was used as the distance metric. KNN classifier efficiently and accurately classifies the images based upon their RGB values and labels the images with Munsell soil notation.

7.4.3   BIO-INSPIRED

7.4.3.1   Swarm

Lasisi et al. (2015) examined the CLONALG (Clonal selection algorithm) along with the Artificial Immune System for the extraction of information about the data of agriculture for crop management. For the identification of the information which is the aim of CLONALG, the antibodies play a vital role that is produced by the B-cells. Also to augment the efficiency of the identification process two methods namely global search and local search are utilized. In the first stage for the presentation of the similar B-Cell, Artificial Recognition Ball is utilized in the AIS method. In the second stage, the antigens are trained. While the third step includes obtaining the memory cell. Further, the efficiency is enhanced by applying the fuzzy rough set for the selection of the feature. The suggested AIS method, when compared to the SMO and MLP for the Eucalyptus Soil Conservation data, was performed accurately. Also, the selection of features enhances the accuracy of the CLONALG.

Zhang et al. (2019) suggested the utilization of the Ant Colony Optimization algorithm for the estimation of the nitrogen content in the soil with the help of spectral features. Six hundred and thirty samples were chosen randomly to estimate the nitrogen content. Also to verify the result two experiments were carried out. In the soil samples, the NIR absorbance was estimated by the spectrometer. The soil was examined ten times and the overall average was taken as the relative absorbance. The MI method is used for the pre-processing of the data. It reduces unnecessary data from spectral information. The ACO algorithm with the help of the evaluation criteria features variables was selected. The selected criterion represents the relation between the variables and the

wavebands. The accuracy of the model was estimated based on R^2, RPD, and RMSE. The PLS model, MLR and SVM model were used for the verification of the outcome of the study. All the eight sensitive wavelengths that were chosen for the ACO-MI model manifest better estimation ability for the nitrogen content of the soil.

7.4.3.2 Evolutionary

Puente et al. (2009) aim to fabricate a method that produces vegetation indices that identify healthy, dead and dry vegetation. The study determines the best-fit indices which augment the association of the data. Genetic Programming is the Evolutionary Computation method that is used to determine the target function. The proposed method is capable of modifying a number of programs into better and new ones. The data in the study depicts the analysis of various parameters that determines the C factors. Afterward, pixels are determined on the LANDSAT Imagery for all 47 samples. The further terminal was formed using the analogous band information of each pixel. A total of six indices were found in the study namely GPVI17, GPVI26, NGPVI13, GPVI25, NGPVI4, and NGPVI17 having the correlation coefficients −0.74, 0.75, 0.73, 0.79, 0.88, and 0.86 respectively. The proposed methodology manifests that indices that are able to synthesize are rectified using the C factor by applying the Genetic Programming method.

Mansor et al. (2012) proposed a study for mainly three purposes, firstly to estimate the evaluation of land suitability for Wheat, Almond, and Potato. Secondly, presented water erosion geospatial pattern to discover the fitness function. Thirdly, to determine the best-fit region by optimizing the land use by applying the Genetic Algorithm. For the preparation of the data layers, satellite images were utilized. As a part of the pre-processing nearest neighbor algorithm was performed because it protects the spectral integrity. The accuracy was determined with the help of the error matrix. Once it is generated overall accuracy, Kappa Statistics, errors, and user's accuracy can be acquired. For the determination of the rate, the number of the parameters considered were Population, Mutation rate, Determination of each variable. Crossover Rate, The number of variables, and research domain. The outcome depicted that the implemented methodology executes better for resolving the multi-objective spatial optimization problem and gives precise decision-making for the optimization of the Land Use (Table 7.2).

7.5   AI-BASED WATER MANAGEMENT

Water management is the practice of managing and executing the optimum usage of water resources. In the agricultural domain, the management of water plays a crucial role in maintaining the balance between hydrological, climatological, and agronomical cycles (Liakos et al., 2018). The precise determination of Evapotranspiration, estimation of Daily dew point temperature, generation of automated irrigation systems, etc, are some of the strategies involved in the water management practices for agriculture. Such strategies are of high importance in improving crop productivity by identifying the expected water requirements for crop fecundity. The amount of water utilized for crop growth influences soil conditions which, in turn, impacts the crop development process. Crop yield is also affected by weather variations. Determining the rainfall patterns and precipitation scenarios can also contribute to precision agriculture. Applying Artificial Intelligence methods in such water management tasks helps in predicting the right quantity of water needed by the crop for its healthy growth as well as helps in conserving the water resources and enabling its judicious use.

7.5.1   NEURAL NETS

7.5.1.1   Artificial Neural Network (ANN)

Hinnell et al. (2010) presented a study in which ANN acts as an approximation for the numerical data of the infiltration from the drip emitters. In the Neuro-Drip approach, the water that has been

TABLE 7.2

Soil Management Using AI Techniques

Functionality	Features	Domain	Method	Performance	References
Soil water content forecasting	Divide-and-conquer principle	ML	SVM+ANN	The accuracy of the applied hybrid architecture is higher than the traditional method, also it doesn't require numerous training sets and so the computation time for the forecast is less compared to others	Liu et al. (2008)
Soil water simulation and daily long-term prediction	Nonlinear stochastic model and chaotic time series analysis	ML	LS-SVM	The outcomes manifested that the de-noising method and wavelet transformation had little effect on the delay time and embedding dimension obtained by the C–C method	Deng et al. (2011)
Analysis of soil behaviour	pH, organic carbon, nitrogen, phosphorus, potassium, sulphur, zinc, iron, copper, manganese	ML	KNN	The proposed method is beneficial to the farmers and the soil analysts for the decision making of sowing seed in which land that gives more efficient crop production	Paul et al. (2015)
Mapping of soil properties	Least absolute shrinkage and selection operator method	ML	KNN	The proposed method proved to be efficient and successful to standardize the soil data. As the approach gives a number of possibilities of mapping it has an advantage over previous soil models	Mansuy et al. (2014)
Soil moisture estimation	Stochastic behavior	Neural nets	MLP, RBF and ANFIS	The results suggested that the use of stochastic parameters along with the models is an appropriate fit for the estimation providing RMSE ranging from 1.27% to 1.30%	Johann et al. (2016)
Irrigation control	Irrigation closed-loop adaptive controller	Neural nets	ANN	The proposed method is capable of adapting the physical changes in the characteristics of the soil	Capraro et al. (2008)
Local soil moisture conditions	Linear approximation and discrete-time of the richards equation	Neural nets	CNN	The results of the CNN model suggested that it can reduce the water level consumption by 52% and also efficient to handle the irregularities in irrigation level	Tseng et al. (2018)
Predicting soil organic matter content	VIS-NIR spectroscopy	Neural nets	CNN	Deep learning gives an efficient method for predicting the Soil Organic Matter content by VIS-NIR Spectroscopy. Also, DenseNet provides an assuring method for diminishing the amount of data pre-processing	Xu et al. (2019)
Soil moisture prediction	Topology structure of BP, improved initial weights and threshold of BP	Bio inspired	IPSO-IBP	IPSO-BP manifests high prediction accuracy and it is quicker	Xiaoxia and Chengming (2016)
Estimation of soil mechanical resistance parameter	Bulk density, volumetric soil water content and soil mechanical resistance	Bio inspired	PSO	The employed PSO model gave higher R^2 and VAF values and low MAPE and RMSE value in comparison to the three models employed for comparison	Hosseini et al. (2016)
Soil moisture forecast model	Sunshine hours, humidity, temperature, evaporation, rainfall and other meteorological factors	Bio inspired	GA+BP	The GA-BP outperformed the BP neural network in terms of prediction accuracy, beneficial application prospect and application value	Huang et al. (2011)
Soil temperature modeling	Periodicity and different depth	Bio inspired	GP	The outcomes of the study indicates that the outperformed ANFIS and ANN in estimating soil temperature	Kisi et al. (2016)

added to the surface during the process of drip irrigation is estimated by ANN. For predicting the water known numerical data is taken as an input so that the ANN can predict the unknown data by reproducing infiltration with the help of numeric data for the values of the parameters of soil hydraulic. The efficiency of the proposed model is estimated using a part of the data that has been kept for the testing purpose but as there is a relation between the training and the testing phase the testing becomes a weak test. So completely new data is utilized for the testing which is known as the blind test. The proposed model gives precise decision-making predictions that are achieved without any assumptions made as it was done in the current methodology.

Umair and Usman (2010) suggested an intelligent control system for efficient irrigation planning based on the Artificial Neural Network. The controller used in the study has been developed with the help of MATLAB. Before the modeling of the data, the inputs are acquired from various parameters like Temperature, Radiation, Air Humidity, and Wind Speed. Then the Evapotranspiration Model is used to transform the parameters of the input into actual soil moisture. The main functionality of the controller is to maintain the level of soil moisture of the actual and the required soil so that accurate results can be obtained about the requirement of the amount of water for the optimization. The yellow and the red-colored signals in MATLAB depict the required soil moisture and actual soil moisture respectively while the green-colored signal depicts the outcome. The suggested approach is an efficient controller implementation. Also, it works by modifying according to the condition and not like the traditional methods that require prior information.

7.5.1.2 Convolutional Neural Network (CNN)

Zhang et al. (2018) presented a study that has been divided into two sections: first is employing the CNN classifier for the identification of the Centre Pivot Irrigation System (CPIS) and while the second section includes the variance-based method for precisely determining the center of the CPIS. For the collection of the data Landsat data and Crop Data Layer were utilized. The study has been As all the images were having circular shape so the images were clipped to a window size of 34×34. For the comparison of the proposed network, three CNN architectures were utilized. The inputs were given to each network and the training and testing set ratio was 9:1. Then the second section estimates the value of the variance for each pixel and then the pixel having the lowest variance value is considered as the center point. The study was efficient, reliable and gives a wide range of solutions for the long-term benefits. Also, the variance approach proved to be effective for determining the center of CPIS.

Abbas et al. (2019) proposed an autonomous canal traversal utilizing the Convolutional Neural Network. The images for training the model were acquired by the MAV controller. The data was split into two parts: A total of 26,282 images for the training phase and 3,000 images for the testing phase. The proposed network is built by the ResNet50. The image pixel having three color channels was taken as an input in the RGB order. After the inputs are given convolution is employed with a total of 64 filters. Also, the bias is added during each convolution execution. All the inputs were then normalized to maintain the activation at 0 the output is an activation at level 1. Further, this activation is applied in the second layer of the network and as an output activation at level 2 is produced. The whole process was repeated 12 times. Lastly, MAV was used to obtain the result. The network was successful to produce higher efficiency in comparison to the other state of art CNN models.

7.5.2 Machine Learning

7.5.2.1 Support Vector Machine (SVM)

Kisi and Cimen (2009) examine evapotranspiration with the help of the Support Vector Machine. For the examination, various parameters like Meteorological Data, Air Temperature, Wind Speed, Solar Radiation, and Relative Humidity were given as input to the SVM. For the measurement of the various parameters, various tools are utilized. Such as a pyranometer is used to determine the

solar radiation, Anemometers are used to determine the wind speed, and the thermistor is used to determine the relative humidity and air temperature. Evapotranspiration is calculated by the CIMIS Penman approach. It was found that solar radiation and air temperature had the best and second-best correlation with evapotranspiration. The inputs R_s, RH, T, U_2 given to the model are evaluated by the FAO-56 PM method which is further calibrated using the SVR model. RMSE (Root Mean Square Errors), MAE (Mean Absolute Errors), and R^2 (Determination Coefficient) are the statistical data that are considered based for the comparison. The outcomes when compared exhibit that the SVM is capable of modeling of the evapotranspiration process.

Suzuki et al. (2013) suggested an irrigation prediction control system based on the Support Vector Machine which automatically regulates the amount of water using the sensor data. Water for irrigation depends upon various factors. The proposed model has several functionalities like acquiring the environmental data using a wireless sensor, processing the sensor data, storing the sensor data, and controlling the sensor data. The data is processed in two parts which are Data Stream Processing and Data Mining. In the data stream processing, the data acquired from the sensor are transformed into DBMS. Further, the data is then converted into the format needed for Data Mining. One part of the data mining using the sensor data and soil moisture determines the quantity of water required for irrigation. These estimations are then visualized by monitoring them and further are given as functionalities to the users. The main benefit of the proposed method is that the user even with no knowledge of irrigation can irrigate properly and more efficiently.

7.5.2.2 K Nearest Neighbor

Sharif and Burn (2007) proposed an improved KNN weather generating model. The model allows the nearest neighbor to resample the historical data with perturbations to create realistic weather sequences while preserving the notable statistical features, such as the inter-station correlations. New values were achieved by adding a random component to the individual resample data points. Through this approach, the generation of unprecedented precipitation amounts is possible which is crucial for the realization of extreme cases. Daily weather variables likewise, maximum temperature, minimum temperature, and precipitation were simulated at various stations in and about the Upper Thames River Basin in Ontario. Results proved the model to be efficient. Among other benefits, Cross-correlation within the variables was maintained, which is especially significant for erosion, crop production, and rainfall-runoff models.

Rajagopalan and Lall (1999) developed a multivariate, nonparametric k nearest-neighbor simulator for daily precipitation and other weather variables which proves to be crucial for crop yield. The approach generated random sequences of daily weather variables that were in accordance with the statistical attributes of the historical data of the same weather variables at the site. Weather variables such as solar radiation, maximum temperature, minimum temperature, average dew point temperature, average wind speed, and precipitation on a particular day were resampled from the historical data by treating the feature vector on a preceding day. The resampling was done from the k nearest neighbors in the state scope of the feature vector by utilizing a weight function. The proposed model displayed to be better at maintaining the cross-dependence and frequency structure than earlier models. This is useful for crop modeling.

7.5.3 Bio-Inspired

7.5.3.1 Swarm

Noory et al. (2012) developed a linear and a mixed-integer linear (MIL) model for optimizing an irrigation water allocation and a multi-crop planning problem using Continuous Particle Swarm Optimization (CPSO) and Discrete Particle Swarm Optimization (DPSO) algorithms respectively. In the linear model, cultivated areas of crops and orchards, which were examined in continuous as well as discrete states, and monthly released volumes of irrigation water from reservoirs are

considered as the variables. Optimization was done with the Linear Programming (LP) method and CPSO. The optimal solution obtained by both of them was comparable but was not straightly suitable in real crop planning circumstances. However, the DPSO-based MIL model showed a notable effect on allotted areas and reservoir control policies. Both CPSO and DPSO algorithms limited the variations of annual net benefit within the range of 2%. DPSO gave 167,000 numbers of function evaluations for getting optimal annual net benefit and 0.81 standard deviation of the results whereas CPSO gave 200,000 and 1.09 respectively.

Reddy and Kumar (2007) presented a nonlinear reservoir model for multi-crop irrigation. The model is capable of managing non-linear relations. The main objective is to augment the relative crop yield that is constraint-dependent. Reservoir Level and Farm Level are the main constraints for the model. The Elitist-Mutated Particle Swarm Optimization technique is utilized for the optimization of the model. Firstly the inputs of the problem are taken then the population of the particles is initialized. Further, for each particle fitness values are evaluated and the GBest is calculated. The performance of the model is tested for crop yield sensitivity and various water shortage conditions. The outcome includes the storage, evaporation losses, Water Allocation, Overflows, Evapotranspiration, Soil Moisture for every crop for 10 days and thus making it more optimized for the decision-making for usability of the available water resources.

7.5.3.2   Evolutionary

Ines et al. (2006) proposed a Genetic algorithm and remote sensing-based approach to examine water management alternatives in irrigated agriculture. First, system characterization was done using a stochastic data assimilation scheme where the irrigation system properties and operational management practices were determined using RS data. Sowing dates, irrigation application criteria, soil hydraulic properties, depth to groundwater and water quality were chosen as stochastic variables (distributed data) and a modified-microGA was utilized to estimate the means and standard deviations for them. They served to be inputted to a soil–water–atmosphere–plant model (SWAP) for regional modeling. They were estimated by reducing the residuals among the distributions of field-scale evapotranspiration (ET) affected by the regional employment of SWAP, and by surface energy balance algorithm for land (SEBAL) utilizing two Landsat7 ETM+ images. Secondly, water management optimization was done under different levels of water availability. The inputs for the same were the derived distributed data. The genetic algorithm was applied in data assimilation and water management optimizations. Results revealed that crop productivity could increase when water and crop management occur simultaneously under limited water conditions.

Raju and Kumar (2004) suggested an application of genetic algorithms for irrigation planning. The use of GA for forming a cropping pattern that yields maximum benefits for an irrigation project in India includes continuity equation, land and water requirements, crop diversification and restrictions on storage as its constraints. The penalty function approach is applied to convert a constrained problem into an unconstrained one. Various combinations of population, generations, crossover and mutation probabilities were considered to fix GA parameters. The 200, 50, 0.6 and 0.01 respectively were selected as the values for the number of generations, population size, crossover probability, and mutation. GA results were compared with the Linear Programming solution and were found to be reasonably close. Thus GA showed to be an effective optimization tool for irrigation planning (Table 7.3).

7.6   FUTURE SCOPE AND CHALLENGES OF AI IN PRECISION AGRICULTURE

The pace at which Artificial Intelligence technologies are developing and expanding, it would seem that the farming sector is at the peak of a technological revolution with AI as its driving force. The use of cognitive technologies such as AI, in agriculture, could assist in ascertaining the best crop selection by analyzing seed types, soil types, and early prediction of pest infestation, disease detection, etc, for

TABLE 7.3

Water Management Using AI Techniques

Functionality	Features	Domain	Method	Performance	References
Sugarcane monitoring	Temperature, moisture, and humidity	ML	SVM	The accuracy achieved for 200 images is 96%	–
Water demand forecasting	Rough set theory	ML	SVM	The proposed method gave assuring forecasting efficiency, accelerated execution speed, and generalization ability	Xuemei et al. (2010)
Precipitation simulation	Gamma kernel	ML	KNN	The outcome of the proposed algorithm works efficiently to store historical statistics. Also, it provides a different amount of precipitation in comparison to the observed data	Goyal et al. (2013)
Daily potato crop evapotranspiration	Air temperature, solar radiation, wind speed and relative humidity	ML	KNN	When a limited amount of meteorological data is present the best fit for the estimation of the ETc is the KNN model	Yamaç and Todorovic (2020)
Estimating daily pan evaporation	Solar radiation, temperature, wind speed, relative humidity and rainfall	Neural nets	ANN	The proposed ANN method generated higher r^2 and lower RMSE values in comparison to multiple linear regression model or the priestley-Taylor equation	Bruton et al. (2000)
Portable temperature-moisture sensing unit	Moisture and temperature	Neural nets	ANN	The suggested methodology was applied on maize and rice filed over an area of 1 acre for a period of 3 weeks due to which 94% of plants were found to be alive after that	Tyagi et al. (2011)
Irrigation water demand forecasting	Soft-computing hybrid model	Neural nets	CNN	The proposed model is efficient as the requirements for the execution are less as the main aim is to plan water pumping and to reduce the execution cost for the forecasting of the irrigation water demand	Pulido-Calvo and Gutiérrez-Estrada (2009)
Optimal crop and irrigation water allocation	Domain knowledge through visibility factors	Bio inspired	ACO	The utilization of the VFs depicts its ability to recognize solutions at all stages especially for the small number of function evaluations	Nguyen et al. (2016)
Multiobjective irrigation planning	Hyperbolic and exponential membership functions	Bio inspired	PSO	From the outcome of the method it has been observed that the Hyperbolic Membership Function outperformed the Exponential Membership Function	Morankar et al. (2016)
To optimize the effective infiltration parameters	Q_m, T_c, L, W, DU, E_a, DPR, TWR, DR	Bio inspired	GA programming and visual basic	By implementing the method the parameters for the furrow irrigation has been optimized and maximum efficiency is obtained compared to previous studies	Valipour and Montazar (2012)
Calibration of crop model	IRI, BIR, ARMN, ARMX, WA, TB, TG, DLAI, RLAD, RBMD	Bio inspired	GA	The evaluated values for the parameters used were able to reproduce irrigation water use upto 13% with an accuracy of 19%	Bulatewicz et al. (2009)

varying climatic conditions and better suited to farmer's requirements (Mokaya, 2019; Blackmore et al., 2005). To serve such purposes, different promising AI technologies are available.

One such is the Machine Learning-based technologies of Drone and Unmanned Aerial Vehicles (UAV). UAVs can collect information 24 hours a day over the entire field. The amount of data that these sensors can generate is remarkable. Practical applications of such technologies are continually progressing and consequently, Drone-enabled solutions for monitoring crop growth and revealing agricultural-related issues such as irrigation problem, pest invasion, soil variation, etc will possibly be on the apex over the next few years. Another scope is in the area of Robotic agriculture. It is an envisioned future still to be fully realized in the next 10–15 years. Driverless vehicle technology is a nice example of autonomous robotic technology for improving agricultural practices. They have been utilized across a wide range of technological firms. For agriculture, there are driverless tractors that are implanted with sensors that can execute the expected exercises, monitor barriers and implement where to apply the farm inputs (Mokaya, 2019; Zhang and Kovacs, 2012). Apart from benefiting crop growth and maintenance, the use of AI technologies in irrigation systems has scope for managing environmental resources too. They can indirectly aid to save water resources. ML-enabled automated irrigation systems help to predict the desired amount of water needed by the soil to maintain optimal soil conditions for optimal crop yield. Such judicious use of water can help conserve water as well as can help to reduce production costs and labor.

Such sophistication in software and machinery is generating new possibilities for energizing farming. The pressure on workforce and labor will be reduced and farmers will make more land to be worked on for longer periods.

Although the use of AI is promising but when it comes to farming, the formation of AI algorithms can be challenging in an agricultural context. Precision Agriculture is developing but not as quickly as predicted 5 years ago. The construction of correct decision-support systems for executing precision decisions persists to be a significant barrier to its adoption (McBratney et al., 2005).

Another challenge lies, particularly in Indian agriculture, in the non-availability of data from remote regions and farmlands that don't satisfy minimum hectare criteria during surveys. As a result, such fragmented lands are left out from data collection and thus from precise farming.

Due to ever-changing unstable climatic conditions, sometimes the AI technologies are unable to ascertain the weather conditions in the soil texture. Also, the unprecedented entry of pests and diseases remains hidden even with adequate protection steps taken. The testing, validation and successful rise of such technologies become much more laborious than most other productions as these operations carried out in one environment cannot be accepted for other environmental conditions; the problem is that no two environments will be exactly similar anytime (Mokaya, 2019).

7.7   CONCLUSION

Precision farming enables the precise monitoring and management of crop production. The managing task generates a huge amount of data, and thus considerable work is required to interpret these data. The foundational enabling technologies in precision agriculture are GIS, GPS, and remote sensing. The use of these technologies in agriculture, serves to acquire location data of the agricultural field and consequently generate maps that can be analyzed and simulated to devise a decision support system and aid better resource utilization. But these technologies have limitations in the form of lack of understanding, orientation of satellites, map generalization, cost, etc. However, crop productivity can significantly be improved through the employment of Artificial Intelligence. It relies on a holistic approach from enhancing crop yield to increasing profitability for farmers. AI methods such as Neural Networks, Machine Learning, and Bio-Inspired computing are seen to efficiently achieve crop growth by managing soil conditions and water resources. Such tasks can be realized by employing AI-enabled weather prediction models, cameras, robots, drones, improved information extraction models from sensing images, etc. The Neural nets methods- ANN and CNN, ML methods- SVM and KNN, and Bio-inspired computing methods- Swarm Intelligence and

Evolutionary method were studied for crop yield prediction, nutrient management, disease detection, identifying crop classes, managing soil moisture content, determining soil type and soil salinity, generating improved, automated and efficient irrigation system. These AI methods have shown to be a great force that enhanced the tasks related to PA. AI-based precision agriculture proved to be a promising technique for growing and maintaining crops efficiently. Although some technical and environmental challenges lie in the utilization of AI in PA, Artificial Intelligence shows a great potential that can be utilized in enterprising farming and advantaging the farmers in the form of workforce as well as increased profitability.

ACKNOWLEDGMENTS

The authors are grateful to the Department of Computer Engineering, Indus University, Department of Chemical Engineering School of Technology, Pandit Deendayal Petroleum University for the permission to publish this research.

REFERENCES

Abbas, S.M., Ali, H., & Muhammad, A. (2019). Autonomous canal following by a micro-aerial vehicle using deep CNN. *IFAC-Papers on Line*, *52*(30), 243–250. doi: 10.1016/j.ifacol.2019.12.529.

Abdullahi, H. S., Sheriff, R. E., & Mahieddine, F. (2017). Convolution neural network in precision agriculture for plant image recognition and classification. *2017 Seventh International Conference on Innovative Computing Technology (INTECH)*. doi: 10.1109/intech.2017.8102436.

Ahir, K., Govani, K., Gajera, R., & Shah, M. (2020). Application on virtual reality for enhanced education learning, military training and sports. *Augmented Human Research*, *5*(1). doi: 10.1007/s41133-019-0025-2.

Andreo, V. (2013). Remote sensing and geographic information systems in precision farming. *Instituto de Altos Estudios Espaciales "Mario Gulich"-CONAE/UNC Facultad de Matematica*. Astronomia y Física–UNC.

Arif, C., Mizoguchi, M., Mizoguchi, M., & Doi, R. (2012). Estimation of soil moisture in Paddy field using artificial neural networks. *International Journal of Advanced Research in Artificial Intelligence*, *1*(1), 17–21. doi: 10.14569/ijarai.2012.010104.

Bagchi, A. (2018). *Artificial Intelligence in Agriculture*. Mindtree Semantic Scholar. https://pdfs.semantic-scholar.org/f16e/55c87af252e99c4010cf74ed6317f4da3604.pdf.

Baker, D. N., Lambert, J. R., & McKinion, J. M., (1983). GOSSYM: A simulator of cotton crop growth and yield. South Carolina, Agricultural Experiment Station, Technical bulletin (USA).

Bakhtiari, A., Navid, H., Mehri, J., Berruto, R., & Bochtis, D. D. (2013). Operations planning for agricultural harvesters using ant colony optimization. *Spanish Journal of Agricultural Research*, *11*(3), 652–660. doi: 10.5424/sjar/2013113-3865.

Bannerjee, G., Sarkar, U., Das, S., & Ghosh, I. (2018). Artificial intelligence in agriculture: A literature survey. *International Journal of Scientific Research in Computer Science Applications and Management Studies*, *7*(3), 1–6.

Barbosa, A., Trevisan, R., Hovakimyan, N., & Martin, N. F. (2020). Modeling yield response to crop management using convolutional neural networks. *Computers and Electronics in Agriculture*, *170*, 105197. doi: 10.1016/j.compag.2019.105197.

Baumgardner, M. F., Silva, L. F., Biehl, L. L., & Stoner, E. R., 1986. Reflectance properties of soils. In: Brady, N. C. (Eds.), *Advances in Agronomy*, Vol. 38. Academic Press: Cambridge, MA, pp. 1–44.

Bhatti, A., Mulla, D., & Frazier, B. (1991). Estimation of soil properties and wheat yields on complex eroded hills using geostatistics and thematic mapper images. *Remote Sensing of Environment*, *37*(3), 181–191. doi: 10.1016/0034-4257(91)90080-p.

Blackmore, S., Stout, B., Wang, M., & Runov, B. (2005). Robotic agriculture-the future of agricultural mechanisation. *In the 5th European Conference on Precision Agriculture (ECPA)*, Upsala (Sweden).

Borgelt, S. C., Harrison, J. D., Sudduth, K. A., & Birrell, S. J. (1996). Evaluation of GPS for applications in precision agriculture. *Applied Engineering in Agriculture*, *12*(6), 633–638. doi: 10.13031/2013.25692.

Boyer, C. N., English, B. C., Roberts, R., Larson, J., Lambert, D. M., Velandia, M. M., Zhou, V., Larkin, S. L., Marra, M. C., Rejesus, R. M., & Falconer, L. L. (2014). *Results from a Cotton Precision Farming Survey Across Fourteen Southern States*. National Cotton Council of America: New Orleans, LA.

Bruton, J. M., McClendon, R. W., & Hoogenboom, G. (2000). Estimating daily pan evaporation with artificial neural networks. *Transactions of the ASAE*, *43*(2), 491–496. doi: 10.13031/2013.2730.

Bulatewicz, T., Jin, W., Staggenborg, S., Lauwo, S., Miller, M., Das, S., Andresen, D., Peterson, J., Steward, D. R., & Welch, S. M. (2009). Calibration of a crop model to irrigated water use using a genetic algorithm. *Hydrology and Earth System Sciences*, *13*(8), 1467–1483. doi: 10.5194/hess-13-1467-2009.

Burrough, P. A., McDonnell, R., McDonnell, R. A., & Lloyd, C. D. (2015). *Principles of Geographical Information Systems*. Oxford University Press: Oxford.

Capraro, F., Patino, D., Tosetti, S., & Schugurensky, C. (2008). Neural network-based irrigation control for precision agriculture. *2008 IEEE International Conference on Networking, Sensing and Control*. doi: 10.1109/icnsc.2008.4525240.

Cho, S., Lee, D., & Jeong, J. (2002). AE: Automation and emerging technologies. *Biosystems Engineering*, *83*(3), 275–280. doi: 10.1006/bioe.2002.0117.

Christy, C. (2008). Real-time measurement of soil attributes using on-the-go near infrared reflectance spectroscopy. *Computers and Electronics in Agriculture*, *61*(1), 10–19. doi: 10.1016/j.compag.2007.02.010.

Clay, D. E., Kim, K., Chang, J., Clay, S. A., & Dalsted, K. (2006). Characterizing water and nitrogen stress in corn using remote sensing. *Agronomy Journal*, *98*(3), 579–587. doi: 10.2134/agronj2005.0204.

Coleman, A. L. & Galbraith, J. M. (2000). Using GIS as an agricultural land use planning tool. *Bulletin*, *2*, 1–93.

Corwin, D. L. & Lesch, S. M. (2003). Application of soil electrical conductivity to precision agriculture. *Agronomy Journal*, *95*(3), 455–471. doi: 10.2134/agronj2003.4550.

Csizmazia, Z. (1993). Technical conditions of equalised fertiliser application. *Hungarian Agricultural Research*, *4*, 16–22.

Dai, X., Huo, Z., & Wang, H. (2011). Simulation for response of crop yield to soil moisture and salinity with artificial neural network. *Field Crops Research*, *121*(3), 441–449. doi: 10.1016/j.fcr.2011.01.016.

Deng, J., Chen, X., Du, Z., & Zhang, Y. (2011). Soil water simulation and predication using stochastic models based on LS-SVM for red soil region of China. *Water Resources Management*, *25*(11), 2823–2836. doi: 10.1007/s11269-011-9840-z.

Devadas, R., Denham, R. J., & Pringle, M. (2012). Support vector machine classification of object-based data for crop mapping, using multi-temporal landsat imagery. *The International Archives of the Photogrammetry, Remote Sensing and Spatial Information Sciences*, *XXXIX-B7*, 185–190. doi: 10.519 4/isprsarchives-xxxix-b7-185-2012.

Du, Q., Chang, N., Yang, C., & Srilakshmi, K. R. (2008). Combination of multispectral remote sensing, variable rate technology and environmental modeling for citrus pest management. *Journal of Environmental Management*, *86*(1), 14–26. doi: 10.1016/j.jenvman.2006.11.019.

Fortin, M. C., & Pierce, F. (1997). *Toward an Agriculture Information System to Maximize Value in Agricultural Data*. CRC/St. Lucie Press: Boca Raton, FL.

Frazier, B., Walters, C., & Perry, E. (2015). Role of remote sensing in site-specific management. *The State of Site Specific Management for Agriculture*, *149*–160. doi: 10.2134/1997.stateofsitespecific.c8.

Gandhi, M., Kamdar, J., & Shah, M. (2020). Preprocessing of non-symmetrical images for edge detection. *Augmented Human Research*, *5*(1). doi: 10.1007/s41133-019-0030-5.

Garrot Jr, D. J., Pitcher, S. N., & Agrometrics Inc (2003). Aircraft based infrared mapping system for earth based resources. U.S. Patent 6,549,828.

Gebbers, R. & Adamchuk, V. I. (2010). Precision agriculture and food security. *Science*, *327*(5967), 828–831. doi: 10.1126/science.1183899.

George, J. & Kumar, S. (2015). Hyperspectral remote sensing in characterizing soil salinity severity using SVM technique - A case study of alluvial plains. *International Journal of Advanced Remote Sensing and GIS*, *4*(1), 1344–1360. doi: 10.23953/cloud.ijarsg.122.

Godwin, R., Richards, T., Wood, G., Welsh, J., & Knight, S. (2003). An economic analysis of the potential for precision farming in UK cereal production. *Biosystems Engineering*, *84*(4), 533–545. doi: 10.1016/s1537-5110(02)00282-9.

Goyal, M. K., Burn, D. H., & Ojha, C. S. (2013). Precipitation simulation based on k-nearest neighbor approach using gamma kernel. *Journal of Hydrologic Engineering*, *18*(5), 481–487. doi: 10.1061/(asce)he.1943-5584.0000615.

Hatfield, P., & Pinter, P. (1993). Remote sensing for crop protection. *Crop Protection*, *12*(6), 403–413. doi: 10.1016/0261-2194(93)90001-y.

Hinnell, A. C., Lazarovitch, N., Furman, A., Poulton, M., & Warrick, A. W. (2010). Neuro-drip: Estimation of subsurface wetting patterns for drip irrigation using neural networks. *Irrigation Science*, *28*(6), 535–544. doi: 10.1007/s00271-010-0214-8.

Hosseini, M., Movahedi Naeini, S. A., Dehghani, A. A., & Khaledian, Y. (2016). Estimation of soil mechanical resistance parameter by using particle swarm optimization, genetic algorithm and multiple regression methods. *Soil and Tillage Research*, *157*, 32–42. doi: 10.1016/j.still.2015.11.004.

Huang, C., Li, L., Ren, S., & Zhou, Z. (2011). Research of soil moisture content forecast model based on genetic algorithm BP neural network. *Computer and Computing Technologies in Agriculture*, *IV, 309*–316. doi: 10.1007/978-3-642-18336-2_37.

Huang, K. (2007). Application of artificial neural network for detecting phalaenopsis seedling diseases using color and texture features. *Computers and Electronics in Agriculture*, *57*(1), 3–11. doi: 10.1016/j.compag.2007.01.015.

Ines, A. V., Honda, K., Das Gupta, A., Droogers, P., & Clemente, R. S. (2006). Combining remote sensing-simulation modeling and genetic algorithm optimization to explore water management options in irrigated agriculture. *Agricultural Water Management*, *83*(3), 221–232. doi: 10.1016/j.agwat.2005.12.006.

Jackson, R. D. (1984). Remote sensing of vegetation characteristics for farm management. *Remote Sensing: Critical Review of Technology*, *475*, 81–97. doi: 10.1117/12.966243.

Jani, K., Chaudhuri, M., Patel, H., & Shah, M. (2020). Machine learning in films: An approach towards automation in film censoring. *Journal of Data, Information and Management*, *2*(1), 55–64. doi: 10.1007/s42488-019-00016-9.

Jha, K., Doshi, A., Patel, P., & Shah, M. (2019). A comprehensive review on automation in agriculture using artificial intelligence. *Artificial Intelligence in Agriculture*, *2*, 1–12. doi: 10.1016/j.aiia.2019.05.004.

Johann, A. L., De Araújo, A. G., Delalibera, H. C., & Hirakawa, A. R. (2016). Soil moisture modeling based on stochastic behavior of forces on a no-till chisel opener. *Computers and Electronics in Agriculture*, *121*, 420–428. doi: 10.1016/j.compag.2015.12.020.

Jóri, J. I. & Erbach, C. D. (1998). Annual international meeting sponsored by ASAE, Paper No. 981051.

Kakkad, V., Patel, M., & Shah, M. (2019). Biometric authentication and image encryption for image security in cloud framework. *Multiscale and Multidisciplinary Modeling, Experiments and Design*, *2*(4), 233–248. doi: 10.1007/s41939-019-00049-y.

Kawarkhe, M., & Musande, V. (2014). Performance analysis of possisblistic fuzzy clustering and support vector machine in cotton crop classification. *2014 International Conference on Advances in Computing, Communications and Informatics (ICACCI)*, pp. 961–967. doi: 10.1109/icacci.2014.6968584.

Khamparia, A., Saini, G., Gupta, D., Khanna, A., Tiwari, S., & De Albuquerque, V. H. (2019). Seasonal crops disease prediction and classification using deep convolutional encoder network. *Circuits, Systems, and Signal Processing*, *39*(2), 818–836. doi: 10.1007/s00034-019-01041-0.

Kişi, O., & Çimen, M. (2009). Evapotranspiration modelling using support vector machines/Modélisation de l'évapotranspiration a l'aide de 'support vector machines'. *Hydrological Sciences Journal*, *54*(5), 918–928. doi: 10.1623/hysj.54.5.918.

Kisi, O., Sanikhani, H., & Cobaner, M. (2016). Soil temperature modeling at different depths using neuro-fuzzy, neural network, and genetic programming techniques. *Theoretical and Applied Climatology*, *129*(3–4), 833–848. doi: 10.1007/s00704-016-1810-1.

Kundalia, K., Patel, Y., & Shah, M. (2020). Multi-label movie genre detection from a movie poster using knowledge transfer learning. *Augmented Human Research*, *5*(1), 11. doi: 10.1007/s41133-019-0029-y.

Lange, A. F. (2015). Centimeter accuracy differential GPS for precision agriculture applications. *Proceedings of the Third International Conference on Precision Agriculture*, 675–680. doi: 10.2134/1996.precisionagproc3.c81.

Lasisi, A., Ghazali, R., Herawan, T., Lasisi, F., & Deris, M. M. (2015). Knowledge extraction of agricultural data using artificial immune system. *2015 12th International Conference on Fuzzy Systems and Knowledge Discovery (FSKD)*, 1653–1658. doi: 10.1109/fskd.2015.7382193.

László, A. (1992). Automatizálás a vegyszerkijuttatásban (Automatisation in chemical application). *Mezogazdasági Technika (Agricultural Technique)*, *33*(4), 33.

Liakos, K., Busato, P., Moshou, D., Pearson, S., & Bochtis, D. (2018). Machine learning in agriculture: A review. *Sensors*, *18*(8), 2674. doi: 10.3390/s18082674.

Liu, H., Xie, D., & Wu, W. (2008). Soil water content forecasting by ANN and SVM hybrid architecture. *Environmental Monitoring and Assessment*, *143*(1–3), 187–193. doi: 10.1007/s10661-007-9967-9.

Long, D. S., DeGloria, S. D., & Galbraith, J. M. (1991). Use of the global positioning system in soil survey. *Journal of soil and water conservation*, *46*(4), 293–297.

Longley, P., Goodchild, M. F., Maguire, D. J., & Rhind, D. W. (2006). *Geographic Information Systems and Science*. John Wiley & Sons: Hoboken, NJ.

Lowenberg-DeBoer, J. & Aghib, A. (1999). Average returns and risk characteristics of site specific P and K management: Eastern corn belt on-farm trial results. *Journal of Production Agriculture*, *12*(2), 276–282. doi: 10.2134/jpa1999.0276.

MacDonald, M. C., Peterson, J. R., Prokop, R., Stampley, R., & Weinhaus, F. (1996). TRW Ag resource mapping: A commercial airborne agricultural remote sensing system. *Proceedings of the Second International Airborne Remote Sensing Conference and Exhibition-Technology, Measurement & Analysis*, San Francisco, CA.

Madsen, E. L. (1995). Impacts of agricultural practices on subsurface microbial ecology. *Advances in Agronomy*, 1–67. doi: 10.1016/s0065-2113(08)60897-4.

Maniyath, S. R., Hebbar, R., Akshatha, K. N., Architha, L. S., & Subramoniam, S. R. (2018). Soil color detection using Knn classifier. *2018 International Conference on Design Innovations for 3Cs Compute Communicate Control (ICDI3C)*. doi: 10.1109/icdi3c.2018.00019.

Mansor, S. B., Pormanafi, S., Mahmud, A. R., & Pirasteh, S. (2012). Optimization of land use suitability for agriculture using integrated geospatial model and genetic algorithms. *ISPRS Annals of the Photogrammetry, Remote Sensing and Spatial Information Sciences*, *I-2*, 229–234. doi: 10.5194/isprsannals-i-2-229-2012.

Mansuy, N., Thiffault, E., Paré, D., Bernier, P., Guindon, L., Villemaire, P., Poirier, V., & Beaudoin, A. (2014). Digital mapping of soil properties in Canadian managed forests at 250m of resolution using the k-nearest neighbor method. *Geoderma*, *235–236*, 59–73. doi: 10.1016/j.geoderma.2014.06.032.

Martiniello, P. (1988). Development of a database computer management system for retrieval on varietal field evaluation and plant breeding information in agriculture. *Computers and Electronics in Agriculture*, *2*(3), 183–192. doi: 10.1016/0168-1699(88)90023-3.

McBratney, A. & Whelan, B. (2001). Precision AG: OZ style. Australian Centre for Precision Agriculture, University of Sydney: Sydney, NSW. Available at: www.usyd.edu.au/su/agric/acpa/papers/PA-Oz.

McBratney, A., Whelan, B., Ancev, T., & Bouma, J. (2005). Future directions of precision agriculture. *Precision Agriculture*, *6*(1), 7–23. doi: 10.1007/s11119-005-0681-8.

McGovern, M. A., Hirose, T., Hopp, B. K., & Huffman, T. E. (2015). Agricultural ground Truthing GPS-GIS system. *Proceedings of the Fourth International Conference on Precision Agriculture*, 975. doi: 10.2134/1999.precisionagproc4.c100.

Meng, X., Zhang, Z., & Xu, X. (2014). A novel k-nearest neighbor algorithm based on I-divergence with application to soil moisture estimation in maize Field. *Journal of Software*, *9*(4), 841–846. doi: 10.4304/jsw.9.4.841-846.

Mesterházi, P. Á., Pecze, Z. S., & Neményi, M. (2001). A precíziós nővényvèdelmi eljárások müszakitèrinformatikai feltètelrendszere (the engineering and GIS background of the precision farming technology). *Nővényvèdelem (Plant Protection)*, *37*(6), 273–281.

Miao, Y., Mulla, D. J., & Robert, P. C. (2006). Identifying important factors influencing corn yield and grain quality variability using artificial neural networks. *Precision Agriculture*, *7*(2), 117–135. doi: 10.1007/s11119-006-9004-y.

Mishra, A., Pant, P. K., Bhatt, P., Singh, P., & Gangola, P. (2019). Management of soil system using precision agriculture technology. *Journal of Plant Development Sciences, 11*(2), 73–78.

Mokaya, V. (2019). Future of precision agriculture in India using machine learning and artificial intelligence. *International Journal of Computer Sciences and Engineering*, *7*(3), 422–425. doi: 10.26438/ijcse/v7i3.422425.

Moller, M., Alchanatis, V., Cohen, Y., Meron, M., Tsipris, J., Naor, A., Ostrovsky, V., Sprintsin, M., & Cohen, S. (2007). Use of thermal and visible imagery for estimating crop water status of irrigated grapevine. *Journal of Experimental Botany*, *58*(4), 827–838. doi: 10.1093/jxb/erl115.

Mondal, P., & Basu, M. (2009). Adoption of precision agriculture technologies in India and in some developing countries: Scope, present status and strategies. *Progress in Natural Science*, *19*(6), 659–666. doi: 10.1016/j.pnsc.2008.07.020.

Moran, M. (1994). Irrigation management in Arizona using satellites and airplanes. *Irrigation Science*, *15*(1), 35–44. doi: 10.1007/bf00187793.

Morankar, D. V., Srinivasa Raju, K., Vasan, A., & AshokaVardhan, L. (2016). Fuzzy multiobjective irrigation planning using particle swarm optimization. *Journal of Water Resources Planning and Management*, *142*(8), 05016004. doi: 10.1061/(asce)wr.1943-5452.0000657.

Mulla, D. J. (2013). Twenty five years of remote sensing in precision agriculture: Key advances and remaining knowledge gaps. *Biosystems Engineering*, *114*(4), 358–371. doi: 10.1016/j.biosystemseng.2012.08.009.

Neményi, M., Mesterházi, P., Pecze, Z., & Stépán, Z. (2003). The role of GIS and GPS in precision farming. *Computers and Electronics in Agriculture*, *40*(1–3), 45–55. doi: 10.1016/s0168-1699(03)00010-3.

Nguyen, D. C., Dandy, G. C., Maier, H. R., & Ascough, J. C. (2016). Improved ant colony optimization for optimal crop and irrigation water allocation by incorporating domain knowledge. *Journal of Water Resources Planning and Management, 142*(9), 04016025. doi: 10.1061/(asce)wr.1943-5452.0000662.

Noory, H., Liaghat, A. M., Parsinejad, M., & Haddad, O. B. (2012). Optimizing irrigation water allocation and multicrop planning using discrete PSO algorithm. *Journal of Irrigation and Drainage Engineering, 138*(5), 437–444. doi: 10.1061/(asce)ir.1943-4774.0000426.

Olakulehin, O. J. & Omidiora, E. O. (2014). A genetic algorithm approach to maximize crop yields and sustain soil fertility. *Net Journal of Agricultural Science, 2*(3), 94–103.

Omkar, S. N., Senthilnath, J., Mudigere, D., & Manoj Kumar, M. (2008). Crop classification using biologically-inspired techniques with high resolution satellite image. *Journal of the Indian Society of Remote Sensing, 36*(2), 175–182. doi: 10.1007/s12524-008-0018-y.

Oppenheim, D. & Shani, G. (2017). Potato disease classification using convolution neural networks. *Advances in Animal Biosciences, 8*(2), 244–249. doi: 10.1017/s2040470017001376.

Orellana, J., Best, S., & Claret, M. (2006). Sistemas de Información Geográfica (SIG). *Procisur, Agricultura de Presición: Integrando Conocimientos para una Agricultura Moderna y Sustentable.*

Oshunsanya, S. O. & Aliku, O. (2016). GIS applications in agronomy. *Geospatial Technology: Environmental and Social Applications.* doi: 10.5772/64528.

Padarian, J., Minasny, B., & McBratney, A. B. (2019). Using deep learning for digital soil mapping. *Soil, 5*(1), 79–89. doi: 10.5194/soil-5-79-2019.

Panchiwala, S., & Shah, M. (2020). A comprehensive study on critical security issues and challenges of the IoT world. *Journal of Data, Information and Management, 2*(4), 257–278. doi: 10.1007/s42488-020-00030-2.

Pandya, R., Nadiadwala, S., Shah, R., & Shah, M. (2020). Buildout of methodology for meticulous diagnosis of K-complex in EEG for aiding the detection of Alzheimer's by artificial intelligence. *Augmented Human Research, 5*(1). doi: 10.1007/s41133-019-0021-6.

Parekh, P., Patel, S., Patel, N., & Shah, M. (2020). Systematic review and meta-analysis of augmented reality in medicine, retail, and games. *Visual Computing for Industry, Biomedicine, and Art, 3*(1). doi: 10.1186/s42492-020-00057-7.

Parekh, V., Shah, D., & Shah, M. (2020). Fatigue detection using artificial intelligence framework. *Augmented Human Research, 5*(1). doi: 10.1007/s41133-019-0023-4.

Patel, D., Shah, D., & Shah, M. (2020a). The intertwine of brain and body: A quantitative analysis on how big data influences the system of sports. *Annals of Data Science, 7*(1), 1–16. doi: 10.1007/s40745-019-00239-y.

Patel, H., Prajapati, D., Mahida, D., & Shah, M. (2020b). Transforming petroleum downstream sector through big data: A holistic review. *Journal of Petroleum Exploration and Production Technology, 10*(6), 2601–2611. doi: 10.1007/s13202-020-00889-2.

Pathan, M., Patel, N., Yagnik, H., & Shah, M. (2020). Artificial cognition for applications in smart agriculture: A comprehensive review. *Artificial Intelligence in Agriculture, 4*, 81–95. doi: 10.1016/j.aiia.2020.06.001.

Paul, M., Vishwakarma, S. K., & Verma, A. (2015). Analysis of soil behaviour and prediction of crop yield using data mining approach. *2015 International Conference on Computational Intelligence and Communication Networks (CICN)*, pp. 766–771. doi: 10.1109/cicn.2015.156.

Pecze, Z. S. (2001). Case maps of the precision (site-specific) farming. Thesis of Ph.D. Dissertation. University of West-Hungary, Mosonmagyaróvár.

Pecze, Z. S., Nemènyi, M., Debreczeni, B., Csathő, P., & Árendás, T. (2001a). Helyspecifikus tápanyagvisszapótlás kukorica növènynèl (site-specific nutrient replacement in case of maize). *Növènytermelès (Plant Production), 50*(2/3), 269–284.

Pecze, Z. S., Nemènyi, M., & Mesterházi, P. Á. (2001b). A helyspecifikus tápanyagvisszapótlás müszaki háttere (the technical background of the site-specific nutrient replacement). *Mezögazdasági Technika (Agricultural Technique), 42*(2), 5–6.

Pierce, F. J. & Nowak, P. (1999). Aspects of precision agriculture. *Advances in Agronomy, 67*, 1–85. doi: 10.1016/s0065-2113(08)60513-1.

Pierce, F. J., Robert, P. C., & Mangold, G. (1994). Site specific management: The pros, the cons, and the realities. *Proceedings of the Integrated Crop Management Conference.* doi: 10.31274/icm-180809-454.

Puente, C., Olague, G., Smith, S. V., Bullock, S., Gonzalez, M. A., & Hinojosa, A. (2009). Genetic programming methodology that synthesize vegetation indices for the estimation of soil cover. *Proceedings of the 11th Annual Conference on Genetic and Evolutionary Computation: GECCO'09*, pp. 1593–1600. doi: 10.1145/1569901.1570114.

Pujari, J. D., Yakkundimath, R., & Byadgi, A. S. (2014). Neuro-kNN classification system for detecting fungal disease affected on vegetables using local binary patterns. *Agricultural Engineering International: The CIGR e-Journal, 16*(4), 299–308.

Pulido-Calvo, I. & Gutiérrez-Estrada, J. C. (2009). Improved irrigation water demand forecasting using a soft-computing hybrid model. *Biosystems Engineering*, *102*(2), 202–218. doi: 10.1016/j.biosystemseng.2008.09.032.

Qian, P. & Zheng, Y. (2006). *Study and Application of Agricultural Ontology*. China Agricultural Science and Technology Publishing House: Beijing.

Qiao, Y. & Zhang, S. (2012). Near-infrared spectroscopy technology for soil nutrients detection based on LS-SVM. *Computer and Computing Technologies in Agriculture*, 325–335. doi: 10.1007/978-3-642-27281-3_39.

Rajagopalan, B. & Lall, U. (1999). Ak-nearest-neighbor simulator for daily precipitation and other weather variables. *Water Resources Research*, *35*(10), 3089–3101. doi: 10.1029/1999wr900028.

Rangel, B. M., Fernandez, M. A., Murillo, J. C., Pedraza Ortega, J. C., & Arreguin, J. M. (2016). KNN-based image segmentation for grapevine potassium deficiency diagnosis. *2016 International Conference on Electronics, Communications and Computers (CONIELECOMP)*, pp. 48–53. doi: 10.1109/conielecomp.2016.7438551.

Reddy, M. J., & Kumar, D. N. (2007). Optimal reservoir operation for irrigation of multiple crops using elitist-mutated particle swarm optimization. *Hydrological Sciences Journal*, *52*(4), 686–701. doi: 10.1623/hysj.52.4.686.

Rich, E. & Knight, K. (1991). *Artificial Intelligence*. McGraw-Hill: New Delhi.

Robert, P. C. (2002). Precision agriculture: A challenge for crop nutrition management. *Progress in Plant Nutrition: Plenary Lectures of the XIV International Plant Nutrition Colloquium*, pp. 143–149. doi: 10.1007/978-94-017-2789-1_11.

Robinson, C. & Mort, N. (1997). A neural network system for the protection of citrus crops from frost damage. *Computers and Electronics in Agriculture*, *16*(3), 177–187. doi: 10.1016/s0168-1699(96)00037-3.

Sarantuya, G., Amarsaikhan, D., & Uuganbayar, D. (2011). The role of RS and GIS for agriculture in Mongolia. *Proceeding papers AGC2–5*. https://a-a-r-s.org/proceeding/ACRS2005/Papers/AGC2-5.pdf. Accessed, 2.

Seelan, S. K., Laguette, S., Casady, G. M., & Seielstad, G. A. (2003). Remote sensing applications for precision agriculture: A learning community approach. *Remote Sensing of Environment*, *88*(1–2), 157–169. doi: 10.1016/j.rse.2003.04.007.

Shah, D., Dixit, R., Shah, A., Shah, P., & Shah, M. (2020). A comprehensive analysis regarding several breakthroughs based on computer intelligence targeting various syndromes. *Augmented Human Research*, *5*(1). doi: 10.1007/s41133-020-00033-z.

Shanahan, J. F., Schepers, J. S., Francis, D. D., Varvel, G. E., Wilhelm, W. W., Tringe, J. M., Schlemmer, M. R., & Major, D. J. (2001). Use of remote-sensing imagery to estimate corn grain yield. *Agronomy Journal*, *93*(3), 583–589. doi: 10.2134/agronj2001.933583x.

Sharif, M., & Burn, D. H. (2007). Improved k-nearest neighbor weather generating model. *Journal of Hydrologic Engineering*, *12*(1), 42–51. doi: 10.1061/(asce)1084-0699(2007)12:1(42).

Sharma, D. K., & Jana, R. (2009). Fuzzy goal programming based genetic algorithm approach to nutrient management for rice crop planning. *International Journal of Production Economics*, *121*(1), 224–232. doi: 10.1016/j.ijpe.2009.05.009.

Singh, V., Varsha, & Misra, A. K. (2015). Detection of unhealthy region of plant leaves using image processing and genetic algorithm. *2015 International Conference on Advances in Computer Engineering and Applications*, pp. 1028–1032. doi: 10.1109/icacea.2015.7164858.

Sobayo, R., Wu, H., Ray, R., & Qian, L. (2018). Integration of convolutional neural network and thermal images into soil moisture estimation. *2018 1st International Conference on Data Intelligence and Security (ICDIS)*, pp. 207–210. doi: 10.1109/icdis.2018.00041.

Spry, K., Marsh, D. J., & Dana, J. (1996). The applications of digital aerial imagery to precision farming. *Proceedings of the Second International Airborne Remote Sensing Conference and Exhibition: Technology, Measurement, & Analysis*, San Francisco, CA.

Srinivasa Raju, K. & Nagesh Kumar, D. (2004). Irrigation planning using genetic algorithms. *Water Resources Management*, *18*(2), 163–176. doi: 10.1023/b:warm.0000024738.72486.b2.

Stafford, J. V. (2000). Implementing precision agriculture in the 21st century. *Journal of Agricultural Engineering Research*, *76*(3), 267–275. doi: 10.1006/jaer.2000.0577.

Sukhadia, A., Upadhyay, K., Gundeti, M., Shah, S., & Shah, M. (2020). Optimization of smart traffic governance system using artificial intelligence. *Augmented Human Research*, *5*(1). doi: 10.1007/s41133-020-00035-x.

Suresha, M., Shreekanth, K. N., & Thirumalesh, B. V. (2017). Recognition of diseases in Paddy leaves using knn classifier. *2017 2nd International Conference for Convergence in Technology (I2CT)*, pp. 663–666. doi: 10.1109/i2ct.2017.8226213.

Suzuki, Y., Nakamatsu, K., & Mineno, H. (2013). A proposal for an agricultural irrigation control system based on support vector machine. *2013 Second IIAI International Conference on Advanced Applied Informatics*, pp. 104–107. doi: 10.1109/iiai-aai.2013.65.

Talaviya, T., Shah, D., Patel, N., Yagnik, H., & Shah, M. (2020). Implementation of artificial intelligence in agriculture for optimisation of irrigation and application of pesticides and herbicides. *Artificial Intelligence in Agriculture, 4*, 58–73. doi: 10.1016/j.aiia.2020.04.002.

Taohidul Islam, S., & Mazumder, B. (2019). Wavelet based feature extraction for rice plant disease detection and classification. *2019 3rd International Conference on Electrical, Computer & Telecommunication Engineering (ICECTE)*, pp. 626–629. doi: 10.1109/icecte48615.2019.9303567.

Tayari, E., Jamshid, A. R., & Goodarzi, H. R. (2015). Role of GPS and GIS in precision agriculture. *Journal of Scientific Research and Development, 2*(3), 157–162.

Tenkorang, F. & Lowenberg-DeBoer, J. (2008). On-farm profitability of remote sensing in agriculture. *Journal of Terrestrial Observation, 1*(1), 6.

Thorp, K., & Tian, L. (2004). A review on remote sensing of weeds in agriculture. *Precision Agriculture, 5*(5), 477–508. doi: 10.1007/s11119-004-5321-1.

Thorpe, K., Ridgway, R., & Webb, R. (1992). A computerized data management and decision support system for Gypsy moth management in suburban parks. *Computers and Electronics in Agriculture, 6*(4), 333–345. doi: 10.1016/0168-1699(92)90004-7.

Tilling, A. K., O'Leary, G. J., Ferwerda, J. G., Jones, S. D., Fitzgerald, G. J., Rodriguez, D., & Belford, R. (2007). Remote sensing of nitrogen and water stress in wheat. *Field Crops Research, 104*(1–3), 77–85. doi: 10.1016/j.fcr.2007.03.023.

Tivy, J. 1990. *Agricultural Ecology.* Longman Group Ltd.: Essex, England Niley, New York.

Tseng, D., Wang, D., Chen, C., Miller, L., Song, W., Viers, J., Vougioukas, S., Carpin, S., Ojea, J. A., & Goldberg, K. (2018). Towards automating precision irrigation: Deep learning to infer local soil moisture conditions from synthetic aerial agricultural images. *2018 IEEE 14th International Conference on Automation Science and Engineering (CASE)*, pp. 284–291. doi: 10.1109/coase.2018.8560431.

Tyagi, A., Apoorv Reddy, A., Singh, J., & Roy Chowdhury, S. (2011). A low cost portable temperature-moisture sensing unit with artificial neural network based signal conditioning for smart irrigation applications. *International Journal on Smart Sensing and Intelligent Systems, 4*(1), 94–111. doi: 10.21307/ijssis-2017-428.

Umair, S. M., & Usman, R. (2010). Automation of irrigation system using ANN based controller. *International Journal of Electrical & Computer Sciences IJECS-IJENS, 10*(2), 41–47.

Vaishnnave, M., Devi, K. S., Srinivasan, P., & Jothi, G. A. (2019). Detection and classification of groundnut leaf diseases using KNN classifier. *2019 IEEE International Conference on System, Computation, Automation and Networking (ICSCAN)*. doi: 10.1109/icscan.2019.8878733.

Valipour, M. & Montazar, A. A. (2012). Optimize of all effective infiltration parameters in furrow irrigation using visual basic and genetic algorithm programming. *Australian Journal of Basic Appllied Sciences, 6*(6), 132–137.

Vazquez, R. A. & Garro, B. A. (2016). Crop classification using artificial bee colony (ABC) algorithm. *Lecture Notes in Computer Science*, 171–178. doi: 10.1007/978-3-319-41009-8_18.

Wan, S., Chang, S., Peng, C., & Chen, Y. (2017). A novel study of artificial bee colony with clustering technique on Paddy rice image classification. *Arabian Journal of Geosciences, 10*(9). doi: 10.1007/s12517-017-2992-2.

Whelan, B. M., McBratney, A. B., & Boydell, B. C. (1997). The impact of precision agriculture. proceedings of the ABARE outlook conference. The Future of Cropping in NW NSW, Moree, 5.

Wolf, S., & Nowak, P. (2015). The status of information-based agrichemical management services in Wisconsin's Agrichemical supply industry. *Site-Specific Management for Agricultural Systems*, 909–920. doi: 10.2134/1995.site-specificmanagement.c67.

Xiaoxia, Y., & Chengming, Z. (2016). A soil moisture prediction algorithm base on improved BP. *2016 Fifth International Conference on Agro-Geoinformatics (Agro-Geoinformatics)*. doi: 10.1109/agro-geoinformatics.2016.7577668.

Xu, Z., Zhao, X., Guo, X., & Guo, J. (2019). Deep learning application for predicting soil organic matter content by VIS-NIR spectroscopy. *Computational Intelligence and Neuroscience, 2019*, 1–11. doi: 10.1155/2019/3563761.

Xuemei, L., Lixing, D., & Jinhu, L. (2010). Agriculture irrigation water demand forecasting based on rough set theory and weighted LS-SVM. *2010 Second International Conference on Communication Systems, Networks and Applications, 2*, 371–374. doi: 10.1109/iccsna.2010.5588826.

Yamaç, S. S. & Todorovic, M. (2020). Estimation of daily potato crop evapotranspiration using three different machine learning algorithms and four scenarios of available meteorological data. *Agricultural Water Management, 228*, 105875. doi: 10.1016/j.agwat.2019.105875.

Yang, C., Everitt, J. H., Bradford, J. M., & Escobar, D. E. (2000). Mapping grain sorghum growth and yield variations using airborne multispectral digital imagery. *Transactions of the ASAE, 43*(6), 1927–1938. doi: 10.13031/2013.3098.

Yousefi, M. R., & Razdari, A. M. (2015). Application of GIS and GPS in precision agriculture (a review). *International Journal of Advanced Biological and Biomedical Research, 3*(1), 7–9. doi: 10.18869/ijabbr.

Zhang, C. & Kovacs, J. M. (2012). The application of small unmanned aerial systems for precision agriculture: A review. *Precision Agriculture, 13*(6), 693–712. doi: 10.1007/s11119-012-9274-5.

Zhang, N., Wang, M., & Wang, N. (2002). Precision agriculture: A worldwide overview. *Computers and Electronics in Agriculture, 36*(2–3), 113–132. doi: 10.1016/s0168-1699(02)00096-0.

Zhang, C., Yue, P., Di, L., & Wu, Z. (2018). Automatic identification of center pivot irrigation systems from Landsat images using convolutional neural networks. *Agriculture, 8*(10), 147. doi: 10.3390/agriculture8100147.

Zhang, Y., Li, M., Zheng, L., Qin, Q., & Lee, W. S. (2019). Spectral features extraction for estimation of soil total nitrogen content based on modified ant colony optimization algorithm. *Geoderma, 333*, 23–34. doi: 10.1016/j.geoderma.2018.07.004.

Zheng, B., Myint, S. W., Thenkabail, P. S., & Aggarwal, R. M. (2015). A support vector machine to identify irrigated crop types using time-series Landsat NDVI data. *International Journal of Applied Earth Observation and Geoinformation, 34*, 103–112. doi: 10.1016/j.jag.2014.07.002.

8 Modes of Action of Beneficial Microorganism as a Typical Example of Microbial Pesticides

O.P. Ikhimalo and A.M. Ugbenyen
Edo University Iyamho

CONTENTS

8.1   INTRODUCTION

Microorganisms are ubiquitous in all ecosystems, they are the most abundant of all groups on earth (Vitorino and Bessa, 2018; Panizzon, et al., 2015; Tsiamis et al., 2014). The advent of biotechnology has totally changed the way and manner these organisms, which were once thought to be nothing but pathogens are viewed today. They are being utilized much more today than was once thought with only a few found usefulness which is their utilization in the production of bread, alcohol and beverages (Linares et al., 2015) the prehistoric uses. Biotechnology modifies the original organisms by introducing/deleting new genes, enhancing/tone down already present genes for the benefit of the end user. Biotechnology heralded several fields including Industrial microbiology for the bulk production of microbial products.

The list of the application of microbes and microbial product is countless; it is amazing how these organisms are continuously shaping our world today. Their roles in the civilization of man and the growth of the world economy are of no small measure. Nowadays the utilization of microorganisms and their products is seen in several industries such as food, medicine, pharmacy, waste management, and agriculture. A more recent discovery of their use as cell factories has opened up a new dimension of producing several products of immense benefit. Advances in biotechnology and other molecular tools increasingly open up novel uses for these microbes. In this work, we take a cursory look at microbial pesticide and their mechanism of action.

DOI: 10.1201/9781003268468-8

8.2   BENEFICIAL MICROORGANISMS

The application of beneficial microbes is seen in virtually all industries. Great strides have been achieved with their utilization; moreover, it is becoming increasingly difficult to live without these microbes.

In the food industries, the age-long technique of fermentation has led to the production of several foods, food products and beverages. Microbial extract has industrial potential for the formulation of cosmetics and personal care products (Gupta et al., 2019); in addition, colouring and food additives have substantially been exploited from microbes for colourants and to impact therapeutic benefits (Sen et al., 2019; Dufossé, 2018; Yangilar and Yildiz, 2016).

In the pharmaceutical industries, the use of microbes in the production of drugs, vaccines, enzymes, antibiotics, vitamins and other therapeutic product is of immense cognizance to human health. Mimicking of beneficial host-microbe interaction in the gut microbiome has led to the discovery and production of some therapeutics such as probiotics and prebiotics which are now gaining widespread use in health, especially in areas pertaining to the gastrointestinal tract. Probiotics are live organism that imputes beneficial qualities when administered in appropriate quantity to the host (FAO/WHO, 2001) and has application in treating and managing inflammatory bowel disease and diabetes (Neel and Onkar, 2017; Wilkins and Sequoia, 2017). Examples include *L. rhamnosus*, *L. acidophilus* and *L. delbrueckii, Bifidobacteria longum, B. breve*. The use of microbes as cell factories has made easier the process of production of several drugs, hormones, vaccines and other metabolites which were aforetime impossible to obtain.

In the waste industry, microbes are increasingly being used for the bioconversion of waste like lignocellulose biomass, feathers and animal excreta to less harmful substances and the production of beneficial products.

In the agricultural sector, great strides have been achieved in the use of microbes for the production of limiting nutrients in the feed industries such as lysine, methionine, isoleucine and arginine (Mukhtar et al., 2017); likewise, products of microbial metabolism such as enzymes, vaccine and antibiotics have also enhanced sustainable agriculture.

Microbial colonization of the root zone known as the rhizosphere provides a shield against pathogenic organisms, aids the development of the plant and also supports plant growth in the presence of abiotic and biotic stress. Synthetic fertilizers overtime impact negatively on soil health and the environment; it also disrupts the microbial communities present in the rhizosphere and may predispose plants to disease thus jeopardizing the overall plant health. Microbiome present in the rhizosphere categorized as plant growth-promoting bacterial (PGPB), plant growth-promoting rhizobacterial (PGPR) and ambuscular mycorrhiza (AM) fungi enhances soil health and improves the bioavailability and assimilation of nutrients (Mosa et al., 2016). Nowadays, PGPB, PGPR and AMF are being exploited as biofertilizers; biofertilizers are active or latent formulations of microbes applied to induce soil microbial activity so as to enhance production and mobilization of nutrients (Singh et al., 2016). Biofertilizers are now being produced industrially as alternatives to chemical fertilizers.

Microbes are making a huge difference in the use of biological control agents. Biological control agents are living organisms such as plants, animals or microbes used to control pest population. They could be used as biopesticides, bioherbicides, biofungicides and bionematicides.

8.3   BIOPESTICIDES

Pesticides are chemical compounds with deleterious effects on pests; the adverse effects of these chemical pesticides on the environment, humans, plants and animals are enormous. They have a nonspecific mode of action and may leave toxic residues on crop plant and the environment which would also have an effect on humans and other living organisms in the long run. This calls for a need for other alternatives; moreover, the world is gradually moving away from synthetic to natural products. One of such is biopesticides.

Biopesticides are pesticides of natural origin such as plants, animals, bacteria and some minerals which have direct effects or whose product and byproduct are pathogenic to pest harmful to plants (EPA, 2016). Biopesticides are specific to target pest, nontoxic to the environment, humans and other living things. They are also cost-effective with their use being sustainable and when integrated with other insect pest management systems, results are comparable and could serve as replacements to the use of synthetic chemicals (Usta, 2013; Koul, 2011). An added advantage of edibility of crops immediately when harvested in that there is no pre-harvest time for such crops after the application of biopesticides. In addition, the multiple modes in which they are antagonistic reduces pest resistance common to synthetic pesticide (Hubbard et al., 2014).

The biopesticide market is currently valued at about 4 billion dollars and has been projected to reach $10.24 Billion by 2025 (Marrone, 2019; Dunham and Trimmer, 2018).

According to EPA (2016), there are three categories of biopesticides based on their active agent.

i. Biochemical pesticides: these are naturally occurring substances that control pests by non-toxic mechanisms. They do this by attracting insects to traps or by interfering with mating.
ii. Plant Incorporated Protectant: These are pesticides produced from genetically engineered plants.
iii. Microbial pesticides

8.3.1 MICROBIAL PESTICIDES

Microbial pesticides consist of a microorganism (e.g., a bacterium, fungus, virus or protozoan) as the active ingredient and there are over 100 microbial pesticides registered by the EPA.

Quite a lot of microbes and their metabolite have been reported to have active roles in the control of plant disease, some of which have been developed into microbial products for the control of plant pest. They are composed of bacteria, viruses, yeast and fungi.

8.3.1.1 Bacteria

Although there is a never-ending war between GMOs and non-GMO's, the era of biotechnology has been of huge benefit to agriculture, especially in pest and disease control. Sub-species and strains of *Bacillus thuringiensis* (Bt) have been the most widely used biopesticides as each strain produces a different blend of proteins that is specific to one or few related species of insect larvae (EPA, 2016). Other species of bacteria used includes *Bacillus sphaericus, B. substilis, B. velezensis, Serratia marcescens* and *Pseudomonas taiwanensis*. Investigations on the exploitation of bacterial floral present in insects for the production of biopesticides have also been achieved. A Turkish study showed high mortality of 98% was recorded for nymphs of *Palomena prasina* when treated with *Bacillus megaterium* isolated from dead male and female *P. prasina* (Aksoy et al., 2018).

During spore formation, *Bacillus thuringiensis* (Bt) produces crystals containing the cry proteins encoded by the cry genes. These cry proteins are endotoxins and upon ingestion by insect forms pores by lysing midgut epithelia cells (Bravo et al., 2007) thus causing cessation of feeding and death of insects (Raymond et al., 2010). Mixtures of toxins produced are strain-specific, dried spores and toxins of *Bacillus thuringiensis* (Bt) are sold in powder form as biopesticides. Other modes of action include antibiosis, competitive exclusion, antibiotics and toxin production (Ratna Kumari et al., 2014; Raymond et al., 2010) (Table 8.1).

8.3.1.2 Fungi

Fungal species specific to attacking and killing insect are known as entomopathogenic fungi, most entomopathogenic fungi belong to the class entomophthorales and hyphomycetes of zygomycota and deuteromycota respectively. The utilization of species in the genera *Beauveria, Metarhizium,* and *Isaria* as biopesticides is on the increase based on production; Amongst these species *Beauveria bassiana* is the most vastly used (Chandler, 2017).

TABLE 8.1

Bacteria Species and Their Susceptible Insect

S No.	Bacterial Species		Susceptible Pest	References
1	Bacillus thuringiensis	Bacillus thuringiensis ssptolworthi	Fall armyworm Spodoptera frugiperda	Fontana Capalbo et al. (2001)
		Bacillus thuringiensis ssp; Serovar israelensis	Aedes aegypti	Gemma Armengol et al. (2006), Ritchie et al. (2010), and Williams et al. (2014)
		Bacillus thuringiensis ssp Berliner	Aphid, Aphis gossypii Glover	Wu and Guo (2003)
2	Bacillus subtilis		Pome fruit Fire blight, Erwinia amylovora; Mosquito, Anopheles stephens	Geetha and Manonmani (2008) and Fan et al. (2017)
3	Pseudomonas		Black cut worm, Agrotis ipsilon; Wax moth, Galleria mellonella; Cotton aphid, Aphis gossypii; Cotton leaf hopper, Amrasca devastans; Tea red spider mite, Oligonychus coffeae	Awad (2012), Mahar et al. (2005), and Manjula et al. (2017)

Entomopathogenic fungi produce spores that lyse insect cuticles through the production of an array of extracellular enzymes such as chymoelastase, chitinase and protease responsible for breaking down cuticles of insects, multiplying and invading insect tissues (Ratna Kumari et al., 2014). Unlike bacteria and viruses which must be ingested, entomopathogenic fungi penetrate the insect's cuticle directly to multiply in the hemocoel (Wang et al., 2016).

The mycopesticide *Metarhizium anisopliae* has a wide host range and is found in moist soil, it has successfully been used to control the larvae and adult biting midges, the vector of *Culicoides* in livestock (Ansari et al., 2011).

Secondary metabolite and biofumigant of some fungi species are also portrayed to be insecticidal such as chloramphenicol derivatives isolated from *Acremonium vitellinum* a marine alga-derived fungi; these derivatives showed insecticidal activity against cotton bollworm, *Helicoverpa armigera* (Chen et al., 2018). Extracts of *Paecilomyces lilacinus*, *Penicillium griseofulvum*, *Beauveria bassiana*, *Metarhizium anisopliae* and *Talaromyces pinophilus* showed insecticidal activities against aphids within 72 hours; including an insecticidal azaphilone compound called chlamyphilone which was isolated from *Pochonia chlamydosporia* (Lacatena et al., 2019). Other fungi metabolite with insecticidal activities includes griseofulvin, beauvericin and leucinostatins. Spores and mycotoxin have also displayed antagonistic action against pest as observed in *M. anisopliae* which had a high mortality rate on insect population of *C. pavonana* (Melanie et al., 2018) (Table 8.2).

8.3.1.3 Virus

Baculovirus is the most represented biopesticide amongst viruses and is host specific. Two genera: Nucleopolyhedrosis viruses (NPVs) and granuloviruses (GVs) are pesticidal and the most used. These viruses have an innate ability to control the outbreaks of the larvae of lepidoptera, mosquitoes and sawfly (Williams et al., 2017). They are highly pathogenic but must be ingested by insect larvae to initiate infection; however, infection from insect to insect could also be initiated through mating and laying of eggs (Pathak et al., 2017; Senthil-Nathan, 2014). Bacuolovirus are able to produce infectious inclusion bodies which invade nucleus and other insect tissues, causing death and liquefying the insect cardavers (Szewczyk et al., 2006; Dara, 2017).

Wild strain Baculovirus are being bioengineered to increase the rate of efficiency (Froyd, 1997); some recombinant Baculovirus are bioengineered to express insect hormones, thus causing hormonal imbalance in insect and insect selective toxin (Inceoglu et al., 2001). Recombinant NPV

TABLE 8.2
Fungal Species and Their Susceptible Insect

S No	Fungi Species	Susceptible Insect	References
1	*Beauveria bassiana*	Pea leaf miner, *Liriomyza huidobrensis*	Robles-Acosta et al. (2019)
		Citrus rust mite, *Phyllocoptruta oleivora*	Wamiti et al. (2018)
		Sand fly, *Phlebotomuspapatasi*	Kaaya and Munyinyi (1995)
		Caterpillar, *Pericallia ricini*	George et al. (2013)
		Moth, *Thaumatotibia leucotreta*	De La Rosa et al. (2002)
		Mexican fruit fly, *Anastrepha ludens*	Zayed et al. (2013)
		Mosquito, *Anopheles stephensi*	Sahayaraj and Borgio (2012)
		Cat flee, *Ctenocephalides felis*	Jensen et al. (2019)
		Aphids, *Aphis fabae*	Pittarate et al. (2018)
		Ambrosia beetle, *Xylosandrus germanus*	Mondaca et al. (2020)
		Striped Rice Stem Borer, *Chilo suppressalis*	Batta (2018)
		Stored-grain pest, *Sitophilus granaries*	Tuncer et al. (2019)
2	*Paecilomyces lilacinus*	Mexican fruit fly, *Anastrepha ludens*; diamondback moth, *Plutella xylostella*; Oriental leafworm moth, *Spodopteral itura*; greenhouse whitefly, *Trialeurodes vaporariorum*; glasshouse red spider mite, *Tetranychu surticae*; the cotton aphid, *Aphis gossypii*; western flower thrips, *Frankliniella occidentalis*	Toledo-Hernández et al. (2019), Nguyen et al. (2017), and Fiedler and Sosnowska (2007)
3	*Metarrhizium anisopliae*	Coconut rhinoceros beetle, *Oryctes rhinoceros*; rice citrus rust mite, *Phyllocoptrutao oleivora*; Pea leaf miner, *Liriomyza huidobrensis*; ambrosia beetle, *Xylosandrus germanus*; brown plant hopper, *Nilaparvata lugens*; False codling moth, *Thaumatotibia leucotreta*; Striped rice stemborer, *Chilo suppressalis*; Stored-grain pest, *Sitophilus granaries*	Robles-Acosta et al. (2019), Migiro et al. (2010), Tuncer et al. (2019), Shoaib and Pandurang (2014)
4	*Lecanicelium lecanii*	The whitefly, *Bemisia tabaci*	Hanan et al. (2019)
		Green peach aphid, *Myzus persicae*	Nazir et al. (2018)
		Thrips mealy bug, *Maconellicoccus hirsutus*	Dixit et al. (2016)
		Pine bast scale, *Matsucoccus matsumurae*	Liu et al. (2014)
5	*Verticillium lecanii*	Sucking pest of okro, *Thrips tabaci*; silver leaf whitefly, *Bemisia tabaci*; green peach aphid, *Myzus persicae*	Khating et al. (2014), Sandhu et al. (2012), Vestergaard et al. (1995), Banafsheh et al. (2004), Javed et al., (2019), and Abdel-Raheem and Ahmed Al-K (2017)
6	*Zoophthora Radicans*	Diamondback moth, *Plutella xylostella*	Yeo et al. (2001)

reduced the feeding and increased insecticidal activity of *Heliothis virescens* (Rajendra et al., 2006) (Table 8.3).

8.3.1.4 Nematode

Entomopathogenic nematodes are mainly from the families of Steinernematidae and Heterorhabditidae, these nematodes are host-specific, pathogenic and symbiots of entomopathogenic bacterial *Xenorhabdus* and *Photorhabdus*. They act by releasing symbiot bacteria which are able to cause septicemia. Entomopathogenic nematodes are infectious only at their juvenile stages which are able to survive in soil for month's lying in wait for a susceptible host. Pathogenicity for pest such as weevils, gnats, white grubs have been recorded (Koul, 2011; Abbas et al., 2001) with results obtained within 24–48 hours (Table 8.4).

TABLE 8.3
Virus Species and Their Susceptible Insect

S/No.	Virus Species	Susceptible Insect	References
1	*Cydia pomonella* granulovirus	Codling moth, *C. pomonella*	Lacey et al. (2004) and Motsoeneng et al. (2019)
2	Nucleopolyhedrovirus (NeabNPV)	Balsam fir sawfly, *Neodiprion abieti*	Lucarotti et al. (2007)
3	*P. xylostella* Granulovirus	Diamondback moth, *Plutella xylostella*	Dezianian (2010) and Parnell et al. (2002)
4	Helicoverpaarmigera Nucleopolyhedrosis virus (HaNPV)	Diamondback moth, *Plutella xylostella*	Magholi et al. (2014)

TABLE 8.4
Nematode Species and Their Susceptible Insect

S No.	Active Ingredient	Susceptible Insect	References
1	*Heterorhabditis bacteriophora*	Tomato leafminer *Tuta absoluta*; leopard moth borer larvae	Gözel and Kasap (2015)
		Black vine weevil, *Otiorhynchus sulcatus*	Susurluk and Ehlers (2008)
		Japanese beetle, *Popillia japonica*	Morris and Grewal (2011)
		Citrus root weevil	Bullock et al. (1999)
2	*Steinernema carpocapsae*	*Leopard moth borer Zeuzera pyrina; Tomato leaf miner, Tuta absoluta; Japanese beetle, Popillia japonica*	Morris and Grewal (2011), Kamali et al. (2017), Ndereyimana et al. (2019), Henderson et al. (1995), Mutegi et al. (2017), and Salari et al. (2014)
3	*Phasmarhabditis hermaphrodita*	Portuguese slug, *Arion lusitanicus*	Grimm (2002) and Speiser et al. (2001)
		Grey garden slug, *Deroceras reticulatum*	Morley and Morritt (2006)
4	*S. feltiae*	Tomato leaf miner, *Tuta absoluta*; Cotton bollworm, *Helicoverpa armigera*; European corn borer, *Ostrinia nubilalis*; Fall armyworm, *Spodoptera frugiperda*; western corn rootworm, *Diabrotica virgifera*; The seedcorn maggot, *Delia platura*	Gozel and Gozel (2016), Ebrahimi et al. (2018), Riga et al. (2001)

8.3.1.5 Protozoa

Protozoa are known to cause some debilitating human diseases and are host to some pathogens; they, however, control pests by decreasing the number of progeny produced by infected insects, causing difficulty in molting and reducing feeding. Protozoa from the general *Nosema* and *Vairimorpha* are prospective biocontrol agents against lepidopteran and orthopteran insects of which hoppers are the most susceptible (Bjørnson and Oi, 2014). They do not instantly kill the insect and may take a few days for results to be appreciated, it is best to integrate protozoa with other pest management methods (Table 8.5).

8.4 CONCLUSION

The ability of beneficial microbes has been ascribed to their mode of action such as antibiosis through the production of antibiotics, production of lytic enzyme, competitive exclusion, antagonism and hyperparasitism in which the pathogen is directly attacked and killed by the microbial

TABLE 8.5

Protozoa Species and Their Susceptible Insect

S No.	Active Ingredient	Susceptible Insect	References
1	*Nosema pyrausta*	Grasshoppers, *Heteracris littoralis*; European corn borer, *Ostrinia nubilalis*	Khaled et al. (2005), Lewis et al. (1983), and Zimmermann et al. (2016)
2	*Vairimorpha necatrix*	Tomato moth, *Lacanobia oleracea*; Black Cutworm, *Agrotis ipsilon;* Corn earworm, *Helicoverpa zea;* European corn borer, *Ostrinia nubilalis*	Down et al. (2004), Grundler et al. (1987), Lewis et al. (1983), and Patel and Habib (1988)
3	*Malpighamoebamellificae*	Honeybee, *Apis mellifera*	Örösi-Pál (1963)
4	*Malameba locustae*	Grasshoppers	Henry (1968)

pesticides. Microbial pesticides have been seen to act using one or more of these mechanisms in controlling and destroying pests.

There are high potentials for microbial pesticides due to the problems associated with the use of synthetic chemicals which are not sustainable. The advantage of multiple modes of action, environmental safe nature and the host specificity of microbial pesticides gives it an edge over synthetic pesticides, integration with other IPM is sure to give comparable results to synthetic pesticides. The perception of microbial pesticide is tied to their quality and effectiveness and both needs to be continually researched. Some rural farmers are still in the dark concerning microbial pesticide and as such they still heavily depend on synthetic pesticide, collaboration of multinational companies with extension workers to bring new findings to such farmers will change the perception of microbial pesticides and increase their acceptability.

REFERENCES

Abbas, M. S., Saleh, M. M., & Akil1, A. M. (2001). Laboratory and field evaluation of the pathogenicity of entomopathogenic nematodes to the red palm weevil, Rhynchophorus ferrugineus (Oliv.) (Col.: Curculionidae). *Anzeiger fur Schdlingskunde*, 74(6), 167–168. doi: 10.1046/j.1439-0280.2001.01025.x.

Abdel-Raheem, M. & Ahmed Al-K, L. (2017). Virulence of three Entomopathogenic fungi against Whitefly, Bemisia tabaci (Gennadius) (Hemiptera: Aleyrodidae) in tomato crop. *Journal of Entomology*, 14(4), 155–159. doi: 10.3923/je.2017.155.159.

Aksoy, H. M., Tuncer, C., Saruhan, İ., Erper, İ., Öztürk, M., & Akca, İ. (2018). Isolation and characterization of bacillus megaterium isolates from dead pentatomids and their insecticidal activity to Palomena prasina nymphs. *Akademik Ziraat Dergisi*, 21–25. doi: 10.29278/azd.440586.

Ansari, M. A., Pope, E. C., Carpenter, S., Scholte, E., & Butt, T. M. (2011). Entomopathogenic fungus as a biological control for an important vector of livestock disease: The Culicoides biting midge. *PLoS One*, 6(1), e16108. doi: 10.1371/journal.pone.0016108.

Armengol, G., Hernandez, J., Velez, J. G., & Orduz, S. (2006). Long-lasting effects of a bacillus thuringiensis Serovar israelensis experimental tablet formulation for aedes aegypti (Diptera: Culicidae) control. *Journal of Economic Entomology*, 99(5), 1590–1595. doi: 10.1093/jee/99.5.1590.

Awad, H. H. (2012). Effect of bacillus thuringiensis and farnesol on haemocytes response and Lysozymal activity of the Black cut worm Agrotis ipsilon larvae. *Asian Journal of Biological Sciences*, 5(3), 157–170. doi: 10.3923/ajbs.2012.157.170.

Banafsheh, A. L., Hassan, A., & Ahmad, A. (2004). Preliminary evaluation of the effectiveness of a *verticillium lecanii* isolate in the control of *thripstabaci* (thysanoptera: thripidae. *Communications in Agricultural and Applied Biological Sciences*, 69, 201–204.

Batta, Y. A. (2018). Efficacy of two species of entomopathogenic fungi against the stored-grain pest, Sitophilus granarius L. (Curculionidae: Coleoptera), via oral ingestion. *Egyptian Journal of Biological Pest Control*, 28(1). doi: 10.1186/s41938-018-0048-x.

Bjørnson, S., & Oi, D. (2014). Microsporidia biological control agents and pathogens of beneficial insects. *Microsporidia*, 635–670. doi: 10.1002/9781118395264.ch25.

Bravo, A., Gill, S. S., & Soberón, M. (2007). Mode of action of bacillus thuringiensis cry and Cyt toxins and their potential for insect control. *Toxicon*, 49(4), 423–435. doi: 10.1016/j.toxicon.2006.11.022.

Bullock, R. C., Pelosi, R. R., & Killer, E. E. (1999). Management of citrus root weevils (Coleoptera: Curculionidae) on Florida citrus with soil-applied Entomopathogenic nematodes (Nematoda: Rhabditida). *The Florida Entomologist*, 82(1), 1. doi: 10.2307/3495831.

Chandler, D. (2017). Basic and applied research on Entomopathogenic fungi. *Microbial Control of Insect and Mite Pests*, 69–89. doi: 10.1016/b978-0-12-803527-6.00005-6.

Chen, D., Zhang, P., Liu, T., Wang, X., Li, Z., Li, W., & Wang, F. (2018). Insecticidal activities of chloramphenicol derivatives isolated from a marine alga-derived endophytic fungus, Acremonium vitellinum, against the cotton Bollworm, Helicoverpa armigera (Hubner) (Lepidoptera: Noctuidae). *Molecules*, 23(-11), 2995. doi: 10.3390/molecules23112995.

Dara, S. K. (2017). Entomopathogenic microorganisms: modes of action and role in IPM. https://ucanr.edu/blogs/blogcore/postdetail.cfm?postnum=24119. Accessed 20 December 2019.

De La Rosa, W., Lopez, F. L., & Liedo, P. (2002). Beauveria bassiana as a pathogen of the Mexican fruit fly (Diptera: Tephritidae) under laboratory conditions. *Journal of Economic Entomology*, 95(1), 36–43. doi: 10.1603/0022-0493-95.1.36.

Dezianian, A. (2010). Morphological characteristics of P. xylostella Granulovirus and effects on its larval host diamondback moth Plutella xylostella L. (Lepidoptera, Plutellidae). *American Journal of Agricultural and Biological Sciences*, 5(1), 43–49. doi: 10.3844/ajabssp.2010.43.49.

Dixit, S., Kabre, G., & Patil, V. (2016). Evaluation of entomopathogenic fungi against the Mealy bug on Custard Apple. *International Journal of Plant Protection*, 9(2), 510–513. doi: 10.15740/has/ijpp/9.2/510-513.

Down, R. E., Bell, H. A., Kirkbride-Smith, A. E., & Edwards, J. P. (2004). The pathogenicity of Vairimorpha necatrix (Microspora: Microsporidia) against the tomato moth, Lacanobia oleracea (Lepidoptera: Noctuidae) and its potential use for the control of lepidopteran glasshouse pests. *Pest Management Science*, 60(8), 755–764. doi: 10.1002/ps.872.

Dufossé, L. (2018). Microbial pigments from bacteria, yeasts, fungi, and Microalgae for the food and feed industries. *Natural and Artificial Flavoring Agents and Food Dyes*, 113–132. doi: 10.1016/b978-0-12-811518-3.00004-1.

Dunham, W., & Trimmer, M. (2018). Biological products around the world. BPIA.org. Accessed 31 December 2019.

Ebrahimi, L., Shiri, M., & Dunphy, G. B. (2018). Effect of entomopathogenic nematode, Steinernema feltiae, on survival and plasma phenoloxidase activity of Helicoverpa armigera (HB) (Lepidoptera: Noctuidae) in laboratory conditions. *Egyptian Journal of Biological Pest Control*, 28(1). doi: 10.1186/s41938-017-0016-x.

EPA. (2016). EPA: Regulating biopesticides. www.epa.gov. Accessed 22 December 2019.

Fan, H., Ru, J., Zhang, Y., Wang, Q., & Li, Y. (2017). Fengycin produced by bacillus subtilis 9407 plays a major role in the biocontrol of Apple ring rot disease. *Microbiological Research*, 199, 89–97. doi: 10.1016/j.micres.2017.03.004.

FAO/WHO. (2001). Health and nutritional properties of probiotics in food including powder milk with live lactic acid bacteria. http://www.fao.org. Accessed 22 December 2019.

Fiedler, Ż., & Sosnowska, D. (2007). Nematophagous fungus Paecilomyces lilacinus (Thom) Samson is also a biological agent for control of greenhouse insects and mite pests. *BioControl*, 52(4), 547–558. doi: 10.1007/s10526-006-9052-2.

Fontana Capalbo, D. M., Valicente, F. H., De Oliveira Moraes, I., & Pelizer, L. H. (2001). Solid-state fermentation of bacillus thuringiensis tolworthi to control fall armyworm in maize. *Electronic Journal of Biotechnology*, 4(2). doi: 10.2225/vol4-issue2-fulltext-5.

Froyd, J. D. (1997). Can synthetic pesticides be replaced with biologically-based alternatives? An industry perspective. *Journal of Industrial Microbiology and Biotechnology*, 19(3), 192–195. doi: 10.1038/sj.jim.2900421.

Geetha, I., & Manonmani, A. (2008). Mosquito pupicidal toxin production by bacillus subtilis subsp. subtilis. *Biological Control*, 44(2), 242–247. doi: 10.1016/j.biocontrol.2007.10.007.

George, J., Jenkins, N. E., Blanford, S., Thomas, M. B., & Baker, T. C. (2013). Malaria mosquitoes attracted by fatal fungus. *PLoS One*, 8(5), e62632. doi: 10.1371/journal.pone.0062632.

Gözel, Ç., & Kasap, İ. (2015). Efficacy of entomopathogenic nematodes against the tomato leafminer, Tuta absoluta (Meyrick) (Lepidoptera: Gelechiidae) in tomato field. *Turkish Journal of Entomology*, 39(3). doi: 10.16970/ted.84972.

Gozel, U., & Gozel, C. (2016). *Integrated Pest Management (IPM): Environmentally Sound Pest Management.* doi: 10.5772/63894.

Grimm, B. (2002). Effect of the nematode phasmarhabditis hermaphrodita on youngstages of the pest slug Arion lusitanicus. *Journal of Molluscan Studies*, 68(1), 25–28. doi: 10.1093/mollus/68.1.25.

Grundler, J. A., Hostetter, D. L., & Keaster, A. J. (1987). Laboratory evaluation of Vairimorpha necatrix (Microspora: Microsporidia) as a control agent for the Black cutworm (Lepidoptera: Noctuidae). *Environmental Entomology*, 16(6), 1228–1230. doi: 10.1093/ee/16.6.1228.

Gupta, P. L., Rajput, M., Oza, T., Trivedi, U., & Sanghvi, G. (2019). Eminence of microbial products in cosmetic industry. *Natural Products and Bioprospecting*, 9(4), 267–278. doi: 10.1007/s13659-019-0215-0.

Hanan, A., Nazir, T., Basit, A., Ahmad, S., & Qiu, D. (2019). Potential of Lecanicillium lecanii (Zimm.) as a microbial control agent for green peach aphid, Myzus persicae (Sulzer) (Hemiptera: Aphididae). *Pakistan Journal of Zoology*, 52(1). doi: 10.17582/journal.pjz/2020.52.1.1.131.137.

Henderson, G., Manweiler, S. A., Lawrence, W. J., Tempelman, R. J., & Foil, L. D. (1995). The effects of Steinernema carpocapsae (Weiser) application to different life stages on adult emergence of the cat flea Ctenocephalides felis (Bouche). *Veterinary Dermatology*, 6(3), 159–163. doi: 10.1111/j.1365-3164.1995.tb00060.x.

Henry, J. (1968). Malameba locustae and its antibiotic control in grasshopper cultures. *Journal of Invertebrate Pathology*, 11(2), 224–233. doi: 10.1016/0022-2011(68)90153-5.

Hubbard, M., Hynes, R. K., Erlandson, M., & Bailey, K. L. (2014). The biochemistry behind biopesticide efficacy. *Sustainable Chemical Processes*, 2(1). doi: 10.1186/s40508-014-0018-x.

Inceoglu, A. B., Kamita, S. G., Hinton, A. C., Huang, Q., Severson, T. F., Kang, K., & Hammock, B. D. (2001). Recombinant baculoviruses for insect control. *Pest Management Science*, 57(10), 981–987. doi: 10.1002/ps.393.

Javed, K., Javed, H., Mukhtar, T., & Qiu, D. (2019). Pathogenicity of some entomopathogenic fungal strains to green peach aphid, Myzus persicae Sulzer (Homoptera: Aphididae). *Egyptian Journal of Biological Pest Control*, 29(1). doi: 10.1186/s41938-019-0183-z.

Jensen, R. E., Enkegaard, A., & Steenberg, T. (2019). Increased fecundity of aphis fabae on Vicia faba plants following seed or leaf inoculation with the entomopathogenic fungus Beauveria bassiana. *PLoS One*, 14(10), e0223616. doi: 10.1371/journal.pone.0223616.

Kaaya, G. P., & Munyinyi, D. M. (1995). Biocontrol potential of the Entomogenous fungi Beauveria bassiana and Metarhizium anisopliae for tsetse flies (Glossina spp.) at developmental sites. *Journal of Invertebrate Pathology*, 66(3), 237–241. doi: 10.1006/jipa.1995.1095.

Kamali, S., Karimi, J., & Koppenhöfer, A. M. (2017). New insight into the management of the tomato leaf miner, Tuta absoluta (Lepidoptera: Gelechiidae) with Entomopathogenic nematodes. *Journal of Economic Entomology*, 111(1), 112–119. doi: 10.1093/jee/tox332.

Khaled, M., Abdel Rahman, E., & Abouziid, M. (2005). Effect of *Nosema Locustae* (Microsporida: Nosematidae) on Food Consumption and Egg Production of the Grasshopper *Heteracris littoralis* Rambur (Orthoptera: Acrididae). *Bulletin of the Entomological Society of Egypt*, 82, 71.

Khating, S. S., Kabre, G. B., & Shinde, S. R. (2014). Bioefficacy of certain insecticides and Verticillium lecanii against sucking pests of okra. *Bioinfolet*, 14, 2.

Lacatena, F., Marra, R., Mazzei, P., Piccolo, A., Digilio, M., Giorgini, M., Woo, S., Cavallo, P., Lorito, M., & Vinale, F. (2019). Chlamyphilone, a novel Pochonia chlamydosporia metabolite with Insecticidal activity. *Molecules*, 24(4), 750. doi: 10.3390/molecules24040750.

Lacey, L. A., Arthurs, S. P., Thomson, D., Fritts, Jr., R., & Granatstein, D. (2004). Codling moth granulovirus and insect-specific nematodes for control of codling moth in the Pacific Northwest. *Tilth Producers Quarterly*, 13, 10–12.

Lewis, L. C., Cossentine, J. E., & Gunnarson, R. D. (1983). Impact of two microsporidia, Nosema pyrausta and Vairimorpha necatrix, in Nosema pyrausta infected European corn borer (Ostrinia nubilalis) larvae. *Canadian Journal of Zoology*, 61(4), 915–921. doi: 10.1139/z83-120.

Linares, D. M., Ross, P., & Stanton, C. (2015). Beneficial microbes: The pharmacy in the gut. *Bioengineered*, 7(1), 11–20. doi: 10.1080/21655979.2015.1126015.

Liu, W., Xie, Y., Dong, J., Xue, J., Zhang, Y., Lu, Y., & Wu, J. (2014). Pathogenicity of three Entomopathogenic fungi to Matsucoccus matsumurae. *PLoS One*, 9(7), e103350. doi: 10.1371/journal.pone.0103350.

Lucarotti, C. J., Morin, B., Graham, R. I., & Lapointe, R. (2007). Production, application, and field performance of Abietiv™, the balsam fir sawfly nucleopolyhedrovirus. *Virologica Sinica*, 22(2), 163–172. doi: 10.1007/s12250-007-0018-z.

Magholi, Z., Abbasipour, H., & Marzban, R. (2014). Effects of Helicoverpa armigera Nucleopolyhedrosis virus (HaNPV) on the larvae of the diamondback moth, Plutella xylostella (L.) (Lepidoptera: Plutellidae). *Plant Protection Science*, 50(4), 184–189. doi: 10.17221/78/2013-pps.

Mahar, A., Darban, D., Gowen, S., Hague, N., Jan, N., Munir, M., & Mahar, A. (2005). Use of Entomopathogenic bacterium pseudomonas putida (Enterobacteriaceae) and its secretion against greater wax moth, Galleria mellonella pupae. *Journal of Entomology*, 2(1), 77–85. doi: 10.3923/je.2005.77.85.

Manjula, T., Kannan, G., & Sivasubramanian, P. (2017). Field efficacy of Pseudomonas fluorescens against the cotton aphid, aphis gossypii Glover (Hemiptera: Aphididae) in bt and non bt cotton. *Agriculture Update*, 12(4), 720–728. doi: 10.15740/has/au/12.4/720-728.

Marrone, P. G. (2019). Pesticidal natural products: Status and future potential. *Pest Management Science*. doi: 10.1002/ps.5433.

Melanie, Miranti, M., Kasmara, H., Hazar, S., & Martina, A. (2018). Insecticidal activities of crude extact of Metarhizium anisopliae and conida suspension against Crocidolomia pavonana fabricius. *IOP Conference Series: Earth and Environmental Science*, 166, 012017. doi: 10.1088/1755-1315/166/1/012017.

Migiro, L. N., Maniania, N. K., Chabi-Olaye, A., & Vandenberg, J. (2010). Pathogenicity of Entomopathogenic Fungi Metarhizium anisopliaeand Beauveria bassiana (Hypocreales: Clavicipitaceae) isolates to the adult pea Leafminer (Diptera: Agromyzidae) and prospects of an Autoinoculation device for infection in the Field. *Environmental Entomology*, 39(2), 468–475. doi: 10.1603/en09359.

Mondaca, L. L., Da-Costa, N., Protasov, A., Ben-Yehuda, S., Peisahovich, A., Mendel, Z., & Ment, D. (2020). Activity of Metarhizium brunneum and Beauveria bassiana against early developmental stages of the false codling moth Thaumatotibia leucotreta. *Journal of Invertebrate Pathology*, 170, 107312. doi: 10.1016/j.jip.2019.107312.

Morley, N., & Morritt, D. (2006). The effects of the slug biological control agent, Phasmarhabditis hermaphrodita (Nematoda), on non-target aquatic molluscs. *Journal of Invertebrate Pathology*, 92(2), 112–114. doi: 10.1016/j.jip.2006.04.001.

Morris, E. E., & Grewal, P. S. (2011). Susceptibility of the adult Japanese beetle, *Popillia japonica* to entomopathogenic nematodes. *Journal of Nematology*, 43(3–4), 196–200.

Mosa, W. F., Sas Paszt, L., Frąc, M., & Trzciński, P. (2016). Microbial products and Biofertilizers in improving growth and productivity of Apple: A review. *Polish Journal of Microbiology*, 65(3), 243–251. doi: 10.5604/17331331.1215599.

Motsoeneng, B., Jukes, M. D., Knox, C. M., Hill, M. P., & Moore, S. D. (2019). Genome analysis of a novel South African Cydia pomonella granulovirus (CpGV-SA) with resistance-breaking potential. *Viruses*, 11(7), 658. doi: 10.3390/v11070658.

Mukhtar, B., Malik, M. F., Shah, S. H., Azzam, A., Slahuddin, N., Liaqat, I. (2017). Lysine supplementation in fish feed. *International Journal of Applied Biology and Forensics* 1(2), 26–31.

Mutegi, D. M., Kilalo, D., Kimenju, J. W., & Waturu, C. (2017). Pathogenicity of selected native entomopathogenic nematodes against tomato leaf miner (*Tuta absoluta*) in Kenya. *World Journal of Agricultural Research*, 5(4), 233–239. doi: 10.12691/wjar-5-4-5.

Nazir, T., Basit, A., Hanan, A., Majeed, M., & Qiu, D. (2018). In vitro pathogenicity of some Entomopathogenic fungal strains against green peach aphid Myzus persicae (Homoptera: Aphididae). *Agronomy*, 9(1), 7. doi: 10.3390/agronomy9010007.

Ndereyimana, A., Nyalala, S., Murerwa, P., & Gaidashova, S. (2019). Potential of entomopathogenic nematode isolates from Rwanda to control the tomato leaf miner, Tuta absoluta (Meyrick) (Lepidoptera: Gelechiidae). *Egyptian Journal of Biological Pest Control*, 29(1). doi: 10.1186/s41938-019-0163-3.

Neel, J. S., & Onkar, C. S. (2017). Role of probiotics in diabetes: A review of their rationale and efficacy. *EMJ Diabet*, 5(1), 104–110.

Nguyen, H. C., Tran, T. V., Nguyen, Q. L., Nguyen, N. N., Nguyen, M. K., Nguyen, N. T., Su, C., & Lin, K. (2017). Newly isolated Paecilomyces lilacinus and Paecilomyces javanicus as novel biocontrol agents for Plutella xylostella and Spodoptera litura. *Notulae Botanicae Horti Agrobotanici Cluj-Napoca*, 45(1), 280–286. doi: 10.15835/nbha45110726.

Opender Koul, O. K. (2011). Microbial biopesticides: Opportunities and challenges. *CAB Reviews: Perspectives in Agriculture, Veterinary Science, Nutrition and Natural Resources*, 6(056). doi: 10.1079/pavsnnr 20116056.

Örösi-Pál, Z. (1963). A moeba disease in the Queen honeybee. *Journal of Apicultural Research*, 2(2), 109–111. doi: 10.1080/00218839.1963.11100069.

Panizzon, J. P., Pilz Júnior, H. L., Knaak, N., Ramos, R. C., Ziegler, D. R., & Fiuza, L. M. (2015). Microbial diversity: Relevance and relationship between environmental conservation and human health. *Brazilian Archives of Biology and Technology*, 58(1), 137–145. doi: 10.1590/s1516-8913201502821.

Parnell, M., Grzywacz, D., Jones, K., Brown, M., Oduor, G., & Ong'aro, J. (2002). The strain variation and virulence of granulovirus of diamondback moth (Plutella xylostella Linnaeus, lep., Yponomeutidae) isolated in Kenya. *Journal of Invertebrate Pathology*, 79(3), 192–196. doi: 10.1016/s0022-2011(02)00001-0.

Patel, P. N., & Habib, M. E. (1988). Protozoosis caused by Vairimorpha necatrix (Microsporia, Nosematidae) in larvae of Spodoptera frugiperda (Lepidoptera, noctuidae). *Revista Brasileira de Zoologia*, 5(4), 593–598. doi: 10.1590/s0101-81751988000400007.

Pathak, D. V., Yadav, R., & Kumar, M. (2017). Microbial pesticides: Development, prospects and popularization in India. *Plant-Microbe Interactions in Agro-Ecological Perspectives*, 455–471. doi: 10.1007/978-981-10-6593-4_18.

Pittarate, S., Thungrabeab, M., Mekchay, S., & Krutmuang, P. (2018). Virulence of aerial conidia of Beauveria bassiana produced under LED light to Ctenocephalides felis (Cat flea). *Journal of Pathogens*, 1–4. doi: 10.1155/2018/1806830.

Rajendra, W., Hackett, K. J., Buckley, E., & Hammock, B. D. (2006). Functional expression of lepidopteran-selective neurotoxin in baculovirus: Potential for effective pest management. *Biochimica et Biophysica Acta (BBA): General Subjects*, 1760(2), 158–163. doi: 10.1016/j.bbagen.2005.11.008.

Ratna Kumari, B., Vijayabharathi, R., Srinivas, V., & Gopalkrishnan (2014). Microbes as interesting source of novel insecticides: A review. *African Journal of Biotechnology*, 13(26), 2582–2592. doi: 10.5897/ajb2013.13003.

Raymond, B., Johnston, P. R., Nielsen-LeRoux, C., Lereclus, D., & Crickmore, N. (2010). Bacillus thuringiensis: An impotent pathogen? *Trends in Microbiology*, 18(5), 189–194. doi: 10.1016/j.tim.2010.02.006.

Riga, E., Whistlecraft, J., & Potter, J. (2001). Potential of controlling insect pests of corn using entomopathogenic nematodes. *Canadian Journal of Plant Science*, 81(4), 783–787. doi: 10.4141/p00-116.

Ritchie, S. A., Rapley, L. P., & Benjamin, S. (2010). Bacillus thuringiensis Var. israelensis (Bti) provides residual control of aedes aegypti in small containers. *The American Journal of Tropical Medicine and Hygiene*, 82(6), 1053–1059. doi: 10.4269/ajtmh.2010.09-0603.

Robles-Acosta, I. N., Chacón-Hernández, J. C., Torres-Acosta, R. I., Landeros-Flores, J., Vanoye-Eligio, V., & Arredondo-Valdés, R. (2019). Entomopathogenic fungi as biological control agents of Phyllocoptruta oleivora (Prostigmata: Eriophyidae) under greenhouse conditions. *Florida Entomologist*, 102(2), 303. doi: 10.1653/024.102.0203.

Sahayaraj, K., & Borgio, J. F. (2012). Screening of some Mycoinsecticides for the managing hairy caterpillar, Pericallia ricini fab. (Lepidoptera: Arctiidae) in castor. *Journal of Entomology*, 9(2), 89–97. doi: 10.3923/je.2012.89.97.

Salari, E., Karimi, J., Sadeghi-Nameghi, H., & Hosseini, M. (2014). Efficacy of two entomopathogenic nematodes Heterorhabditis bacteriophora and Steinernema carpocapsae for control of the Leopard moth borer Zeuzera pyrina (Lepidoptera: Cossidae) larvae under laboratory conditions. *Biocontrol Science and Technology*, 25(3), 260–275. doi: 10.1080/09583157.2014.971710.

Sandhu, S. S., Sharma, A. K., Beniwal, V., Goel, G., Batra, P., Kumar, A., Jaglan, S., Sharma, A. K., & Malhotra, S. (2012). Myco-biocontrol of insect pests: Factors involved, mechanism, and regulation. *Journal of Pathogens*, 1–10. doi: 10.1155/2012/126819.

Sen, T., Barrow, C. J., & Deshmukh, S. K. (2019). Microbial pigments in the food industry: Challenges and the way forward. *Frontiers in Nutrition*, 6. doi: 10.3389/fnut.2019.00007.

Senthil-Nathan, S. (2014). A review of Biopesticides and their mode of action against insect pests. *Environmental Sustainability*, 49–63. doi: 10.1007/978-81-322-2056-5_3.

Shoaib, H. S., & Pandurang, M. (2014). Effect of entomopathogenic fungi against brown plant hopper, Nilaparvata lugens (Stal.) (Hemiptera: Delphacidae) infesting rice. *IJSR*, 6(14), 1–3.

Singh, D. P., Singh, H. B., & Prabha, R. (2016). *Microbial Inoculants in Sustainable Agricultural Productivity: Vol. 1: Research Perspectives*. Springer: Berlin, Germany.

Speiser, B., Zaller, J., & Neudecker, A. (2001). Size-specific susceptibility of the pest slugs *Deroceras reticulatum* and *Arion lusitanicus* to the nematode biocontrol agent *Phasmarhabditis hermaphrodita*. *BioControl*, 46, 311–320. doi: 10.1023/A:1011469730322.

Susurluk, A., & Ehlers, R. (2008). Sustainable control of Black vine weevil larvae, Otiorhynchus sulcatus (Coleoptera: Curculionidae) with Heterorhabditis bacteriophorain strawberry. *Biocontrol Science and Technology*, 18(6), 627–632. doi: 10.1080/09583150802090026.

Szewczyk, B., Hoyos-Carvajal, L., Paluszek, M., Skrzecz, I., & Lobo de Souza, M. (2006). Baculoviruses: Re-emerging biopesticides. *Biotechnology Advances*, 24(2), 143–160. doi: 10.1016/j.biotechadv.2005.09.001.

Toledo-Hernández, R. A., Toledo, J., Valle-Mora, J., Holguín-Meléndez, F., Liedo, P., & Huerta-Palacios, G. (2019). Pathogenicity and virulence of *Purpureocillium lilacinum* (Hypocreales: Ophiocordycipitaceae) on Mexican fruit fly adults. *Florida Entomologist*, 102(2), 309. doi: 10.1653/024.102.0204.

Toledo-Hernández, R. A., Toledo, J., Valle-Mora, J., Holguín-Meléndez, F., Liedo, P., & Huerta-Palacios, G. (2019). Pathogenicity and virulence of Purpureocillium lilacinum (Hypocreales: Ophiocordycipitaceae) on Mexican fruit fly adults. *Florida Entomologist*, 102(2), 309. doi: 10.1653/024.102.0204.

Tsiamis, G., Karpouzas, D., Cherif, A., & Mavrommatis, K. (2014). Microbial diversity for biotechnology. *BioMed Research International*, 1–3. doi: 10.1155/2014/845972.

Tuncer, C., Kushiyev, R., Erper, I., Ozdemir, I. O., & Saruhan, I. (2019). Efficacy of native isolates of Metarhizium anisopliae and Beauveria bassiana against the invasive ambrosia beetle, Xylosandrus germanus Blandford (Coleoptera: Curculionidae: Scolytinae). *Egyptian Journal of Biological Pest Control*, 29(1). doi: 10.1186/s41938-019-0132-x.

Usta, C. (2013). Microorganisms in biological pest control: A review (Bacterial toxin application and effect of environmental factors). *Current Progress in Biological Research*. doi: 10.5772/55786.

Vestergaard, S., Gillespie, A. T., Butt, T. M., Schreiter, G., & Eilenberg, J. (1995). Pathogenicity of the Hyphomycete fungi verticillium lecanii and Metarhizium anisopliae to the western flower thrips, Frankliniella occidentalis. *Biocontrol Science and Technology*, 5(2), 185–192. doi: 10.1080/09583159550039909.

Vitorino, L., & Bessa, L. (2018). Microbial diversity: The gap between the estimated and the known. *Diversity*, 10(2), 46. doi: 10.3390/d10020046.

Wamiti, L. G., Khamis, F. M., Abd-alla, A. M., Ombura, F. L., Akutse, K. S., Subramanian, S., Odiwuor, S. O., Ochieng, S. J., Ekesi, S., & Maniania, N. K. (2018). Metarhizium anisopliae infection reduces trypanosoma congolense reproduction in glossina fuscipes fuscipes and its ability to acquire or transmit the parasite. *BMC Microbiology*, 18(S1). doi: 10.1186/s12866-018-1277-6.

Wang, J., St. Leger, R., & Wang, C. (2016). Advances in genomics of Entomopathogenic fungi. *Genetics and Molecular Biology of Entomopathogenic Fungi*, 67–105. doi: 10.1016/bs.adgen.2016.01.002.

Wilkins, T., & Sequoia, J. (2017). Probiotics for gastrointestinal conditions: A summary of the evidence. *AFP*, 96, 3.

Williams, G. M., Faraji, A., Unlu, I., Healy, S. P., Farooq, M., Gaugler, R., Hamilton, G., & Fonseca, D. M. (2014). Area-wide ground applications of bacillus thuringiensis Var. israelensis for the control of aedes albopictus in residential neighborhoods: From optimization to operation. *PLoS One*, 9(10), e110035. doi: 10.1371/journal.pone.0110035.

Williams, T., Virto, C., Murillo, R., & Caballero, P. (2017). Covert infection of insects by Baculoviruses. *Frontiers in Microbiology*, 8. doi: 10.3389/fmicb.2017.01337.

Wu, K., & Guo, Y. (2003). Influences of Bacillus thuringiensis Berliner cotton planting on population dynamics of the cotton Aphid, Aphis gossypii Glover, in northern China. *Environmental Entomology*, 32(2), 312–318. doi: 10.1603/0046-225x-32.2.312.

Yangilar, F., & Yildiz, P. O. (2016). Microbial pigments and the important for food industry. *Erzincan Üniversitesi Fen Bilimleri Enstitüsü Dergisi*, 9(2). doi: 10.18185/eufbed.55880.

Yeo, H., Pell, J., Walter, M., Boyd-Wilson, K., Snelling, C., & Suckling, D. (2001). Susceptibility of diamondback moth (*Plutella xylostella* (L)) larvae to the entomopathogenic fungus Zoophthora radicans (Brefeld) Batko. *New Zealand Plant Protection*, 54, 47–50. doi: 10.30843/nzpp.2001.54.3738.

Zayed, A., Soliman, M. M., & El-Shazly, M. M. (2013). Infectivity of Metarhizium anisopliae (Hypocreales: Clavicipitaceae) to Phlebotomus papatasi (Diptera: Psychodidae) under laboratory conditions. *Journal of Medical Entomology*, 50(4), 796–803. doi: 10.1603/me12244.

Zimmermann, G., Huger, A. M., Langenbruch, G. A., & Kleespies, R. G. (2016). Pathogens of the European corn borer, Ostrinia nubilalis, with special regard to the microsporidium Nosema pyrausta. *Journal of Pest Science*, 89(2), 329–346. doi: 10.1007/s10340-016-0749-4.

9 Genetically Modified Orange
From Farm to Food the Undiscovered Medicinal and Food Benefits

Charles Oluwaseun Adetunji
Edo State University Uzairue

Muhammad Akram
Government College University Faisalabad

Olugbenga Samuel Michael
Bowen University

Benjamin Ewa Ubi
Ebonyi State University

Chioma Bertha Ehis-Eriakha
Edo State University Uzairue

Arish Sohail
Government College University Faisalabad

Juliana Bunmi Adetunji
Osun State University

Hina Anwar
Government College University Faisalabad

Oluwaseyi Paul Olaniyan
Osun State University

Khurram Shahzad
Government College University Faisalabad

Mehwsih Iqbal
Dow University of Health Sciences

Wadzani Palnam Dauda
Federal University Gashua

Neera Bhalla Sarin
Jawaharlal Nehru University

DOI: 10.1201/9781003268468-9

CONTENTS

9.1   INTRODUCTION

Citrus sinensis generally called orange is a component of Rutaceae family and a chief source of vitamins, particularly vitamin C, adequate amount of calcium, niacin, thiamine, folacin, potassium and magnesium (Angew, 2007). Economically, oranges are significant fruit crops, with an approximated 60 million metric tonnes made worldwide as at 2005 for an overall worth of $9 billion. From this total, partially came fromthe United States of America and Brazil (Goudeau et al., 2008; Bernardi et al., 2010). The worldwide citrus area of agricultural land according to Food and Agriculture Organization statistics in 2009 was 9 million hectares with production set at 122.3 million tons, categorizing sweet oranges first amid all the crops of fruit (Xu et al., 2013). The genetic derivation of *Citrus sinensis* is unclear, though it is considered to be originated from the interspecific hybridization of a number of earliest citrus species (Xu et al., 2013).

Citrus fruits are the most important source of essential plant chemical nutrients and for long have been rated for their healthy nourishing and antioxidant characteristics. Scientifically it is verified that oranges being loaded with minerals and vitamins have lots of health advantages. Furthermore, it is now realized that other organically active, non-nutrient complexes established in citrus fruits such as plant chemicals antioxidants, dissolvable and in dissolvable dietary fibres are recognized to be useful in minimizing the risk for carcinoma, numerous chronic ailments like obesity, arthritis, and CHD (Crowell, 1999). A particular orange is supposed to have about one hundred and seventy plant nutrients and more than 60 flavonoids with anti-inflammatory, anti-cancer, antioxidant and blood clot-reducing properties. All these characteristics assist to support health comprehensively (Cha et al., 2001).

Fruits and vegetables play an important role in the prevention of chronic diseases. Dietary fibre is the common component of foods consisting of polysaccharides such as gums, cellulose, pectin, hemicellulose and lignin. These dietary fibres play significant roles in the disease prevention and treatment of chronic diseases such as gastrointestinal diseases, diabetes, cardiovascular diseases, cancer, and obesity and also encourage physiological functions (Adetunji et al., 2020a,b; Adetunji and Anani 2020; Adetunji and Varma, 2020; Olaniyan and Adetunji, 2021; Inobeme et al., 2021; Anani et al., 2021; Adetunji et al., 2021a,b,c,d,e,f,g,h; Jeevanandam et al., 2021; Adetunji et al., 2022a,b). The daily recommended dietary fibre intake is 25–30 g, helpful in overcoming dietary deficiency and has been linked to many physiological effects. Natural products have been playing a vigorous role in health for many years. Among the natural sources, plants have possessed chemical substances, function as drugs as compared to synthetic drugs. The herbs and plants being rich sources of bioactive phytonutrients are helpful for human health. Due to the presence of such bioactive compounds plants act as antioxidants and medicines. Among the bioactive compounds, Flavonoids are important naturally occurring polyphenolic compounds. Flavonoids are available in abundant quantity in fruits, vegetables and plant-based beverages. Flavonoids are accepted as disease-preventing and health-promoting supplements. Flavonoids pharmacologically exhibit free radical scavenging and antioxidant activities (Selmi et al., 2017).

Gene modification is the transferal technique of genes to organisms for the purpose of getting favourite characteristics biotechnologically. This transferal of genes can be done to plants or animals and known as genetically engineered food (GE) or genetically modified organisms (GMO). Genetically modified organisms are described as follows by World Health Organization: Organisms such as animals, plants or microorganisms in which the deoxyribonucleic acid has been modified in a means that does not take place normally by breeding and or usual recombination (WHO, 2016). In the same way, the Food and Agriculture Organization of the UN and the Commission of Europe describe genetically modified organisms as a product "not arise naturally by breeding and/or expected recombination" (FAO, 2016). Among the agricultural GM products producers, the United States is on top producing almost 43% of the world's genetically modified crops of different varieties. American farmers participate on harvesting herbs-resistant sugar beets, alfalfa and virus-resistant oranges and papaya on millions of acres (Hallman et al., 2013). The genes transferal technique is adopted for improving fruits quality, better fruit yield, shelf life, increasing nutritional value and making resistance to abiotic and biotic stresses. The antibiotic-resistant petunias and tobacco were the first GMO plants produced by self-governing researchers groups in 1983. The GMO tobacco, first time commercialized by China in 1990 (Bawa and Anilakumar, 2013). The Food and Drug Administration (FDA) approved the first GMO tomato species having feature of delayed ripening. Similarly a lot of transgenic crops have approved from FDA counting Canola with desired oil composition and herbs resistant soyabeans and cotton (Zhang et al., 2015).

Genetically modified foods that are accessible in the market consist of eggplants, carrots, potatoes, strawberries, and a lot more are in pipeline (Bawa and Anilakumar, 2013). Now, the pace of raise in yield of crop is not more than 1.7% while the yearly amplification in yield should have to be 2.4% to link up the demands of population expansion, advanced nutritional standards and decreasing capability of producing crops. This is an intimidating task, which seems only attainable by implication and optimization of genetics of crops combined with quantitative upgrading in management of the farming system.

The genus Citrus possesses several undesirable characteristics including salt and cold sensitivity (Garcia-Agustin and Primo-Millo, 1995; Van Le et al., 1999); they are also susceptible to diseases caused by fungi, bacteria and viruses, such as Citrus exocortis viroid (CEV), Citrus infectious variegation virus (CIVV), Citrus cachexia viroid (CCaV) and Citrus tristeza closterovirus (CTV) (Van Le et al., 1999; Greño et al., 1988). Citrus exocortis viroid. It has been highlighted that Classical genetic selection, micrografting, grafting, gene transfer methodology can enhance the cultivation of a particular species. Hence, in vitro manipulation actions could result into quick, genetic transformation and bud regeneration for effective micropropagation are essential for citrus bud regeneration for efficient micropropagation, Also, several practical advantages have been documented from in vitro culture techniques in citrus cultivation (Van Le et al., 1999; Grosser et al., 1988; Ghorbel et al., 1998) while several recent advancements have been employed in the gene transfer methodology through the classical regeneration methods to various genus and this has led to inducement of specific alteration within a short period when compared to classical genetic selection techniques (Van Le et al., 1999; Kaneyoshi et al., 1994; Gutiérrez et al., 1997).

Moreover, examples of limitation encountered in the conventional techniques during the process of citrus improvement include auto incompatibility, nucellar polyembryony, long juvenile period, and high heterozygosity (Ghorbel et al., 1999; Boscariol et al., 2003). The introduction of biotechnological techniques has played a critical role in the eradication of these highlights challenges via several techniques such as genetic transformation (Boscariol et al., 2003; Peña and Navarro, 1999; Costa et al., 2002) somatic hybridization and citrus breeding programs (Boscariol et al., 2003; Grosser and Gmitter, 1990; Mendes et al., 2001). These techniques have been affirmed by several scientist in numerous countries (Gutiérrez et al., 1997; Boscariol et al., 2003; Cervera et al., 2000; Januzzi et al., 2002).

Globally, there has been a main plunge in citrus enhancement as competition from international citrus markets because of several factors such as abiotic and biotic stress conditions and pressure

due to the high susceptibility of citrus to pests and diseases (Grosser et al., 2000; Dutt and Grosser, 2010). Numerous techniques utilized for the enhancement of citrus include conventional breeding and genetic transformation (Peña et al., 2007; Dutt et al., 2010). Moreover, the application of genetically modified of citrus has been recognized as an efficient techniques because it enhance easy introduction of a particular desire traits of interest to another elite cultivar by conventional breeding. Citrus cultivars differs when responding to genetic transformation and in vitro organogenesis. Therefore, there is a need for the optimization of cultivar-specific through invitro methodology and genetically modified techniques (Peña et al., 2007; Dutt et al., 2010).

The application of Agrobacterium-mediated transformation has been highlighted as one of the most common techniques utilized for genetic alteration of citrus utilizing epicotyl explants as target cells for integration of the T-DNA (Dutt et al., 2010; Dutt and Grosser, 2009) but unfortunately, it could not be used for alteration of any seedless cultivar. Also, some cultivars in the mandarin group remain strong to the process of transformation when this methodology is applied (Dutt et al., 2010; Khawale et al., 2006).

Therefore, this chapter intends to discuss the recent advances in the medical and nutritional benefits of genetically modified oranges.

9.2   MUTATION AND CONVENTIONAL TECHNIQUES FOR THE BREEDING OF ORANGE VARIETIES

Several methods have been discovered to be very effective in the enhancement of various types of citrus which has led to the development of new varieties. During conventional citrus breeding researcher normally blend the DNA of two different species one from the father and the other from the mother which eventually results in hybrid embryos by the process referred to as meiosis. The embryo normally generates hybrid seeds which will carry some interesting features from different orange cultivar. Moreover, some interesting varieties have been produced using the process of induced mutation or through natural techniques.

The natural techniques occur extemporaneously over time. The citrus breeder sometimes selects natural mutants and fruit which has been discovered to show some level of variation from the parent will be selected, assessed, and mass produced through the process of propagation. Also, some breeders have utilized other techniques such as zapping budwood with radiation to induce mutations or somaclonal variation which has led to the development of enhanced cultivars with interesting features.

Mutation occurs whenever the DNA disintegrates and rearranges each other or a point mutation where a single nucleotide show some level of alteration which eventually leads to enhanced varieties. The process of natural mutation has led to the production of some special species which are seedless such as Tango, Kinnow LS, N40W-6-3 (Seedless Snack), and Fairchild LS.

Furthermore, the University of Florida/Institute of Food and Agricultural Sciences (UF/IFAS) has also released some sweet oranges varieties which are seedless such as OLL-8 and Valquarius. These varieties are produced as a result of somaclonal variation. Moreover, it has been highlighted that most of the citrus cultivar has an extended juvenile period which makes most of the cultivar to take a minimum of six for their seed to germinate to flower. Interestingly, the recent advances in genetic enhancement has led to prevention of biotic and abiotic conditions which normally prevent this extended juvenile phase.

9.3   RECENT ADVANCES ON GENETICALLY MODIFIED ORANGES WITH SOME SPECIFIC EXAMPLES

It has been highlighted that somatic hybridization and genetic transformation studies have been applied as sustainable biotechnological tools in citrus breeding programs in numerous countries.

Also, the application of genetic transformation techniques has been used for the transfer of particular desired traits without changing the genetic background (Marques et al., 2011). The application of Agrobacterium has been highlighted in the genetical transformation of citrus plant with explants obtained from seedlings germinated in vitro or under a greenhouse environment (Almeida et al., 2003). Another example of transformation techniques utilized in the improvement of citrus includes the application of techniques such as RNA silencing. There is a need to optimize the process of vitro regeneration before the process of transforming genes can take place. In view of the aforementioned, several authors have reported the various genetically modified process towards the enhancement of the citrus plant.

RNA interference (RNAi) has been recognized as a posttranscriptional gene-silencing process triggered by the double-stranded RNA. It has been utilized as a knockdown technology for the evaluating the role of numerous gene in organism. This techniques has been utilized in plant for the production of transgenic that could generate hairpin RNA. RNAi has several benefits which includes the genetical enhancement of plants which form an advantage of cosuppression and antisense-mediated gene silencing when compared to effectiveness and high stability (Kusaba, 2004). Soler et al. (2012) discovered that *Citrus tristeza* virus responsible for the viral diseases of citrus plant possesses three silencing suppressor proteins at intercellular level (p20 and p25) and that could suppresses the host viral defense system. It was observed that there was a complete transformation of Mexican lime when amplified with an intron-hairpin vector having full-length untranslatable versions of the genes p23, p25 and p20 derived from *Citrus tristeza* virus strain T36 which play a crucial role in the silencing of the expression of these gene in *Citrus tristeza* virus affected cells. The result obtained shows that the 3 transgenic lines available stimulates the high level of resistance to viral infection without the presence of any symptom, absence of virus aftergraft inoculation with CTV-T36 whether in transgenic scion or non-transgenic rootstock. It was observed that the buildup of transgene-derived siRNAs was mandatory but might not be adequate for the development of a high resistance to Citrus tristeza virus. Also, it was observed that the presence of a divergent *Citrus tristeza* virus strain resulted into overcoming of the resistance which play a vital function of the sequence in the mode of action. The application of RNAi showed that this one of the techniques that could ensure the transgenic resistance to *Citrus tristeza* virus and also validated that aiming concurrently by RNA interference (RNAi) the three viral silencing suppressors seem perilous for this function, though the contribution of coexisting RNAi modes of action cannot be excepted.

Orbović et al. (2011) evaluated the influence of seed age on shoot regeneration capability and the transformation of "Flame" and "Duncan" grapefruit cultivars in addition to "Hamlin" sweet orange cultivar. The authors applied the process of genetically transformation to the citrus explants as stated in a protocol developed by 93 with using *A. tumefaciens* strain EHA105 (Hood et al., 1993) which possesses a binary vector obtained from pD35s (Dutt, et al., 2010). Their study showed that there was variation in the transformability and regeneration potential of citrus juvenile explants which also changes during the fruits harvest season. This make it difficult to establish a specific methodology for the genetic transformation of citrus but maximum transformation effectiveness swill need a flexible methodology that could account for timing of seed collection and cultivar variability. Khan et al. (2012) genetically modifies the transformation of leaf segments from "Valencia" sweet orange (*C. sinensis* L. Osbeck) utilizing green fluorescence protein (gfp) as an important marker (Khan et al., 2012). This a typical example of generating transgenic plant using leaf segments as explants.

Moreover, apartfrom *A. tumefaciens,* some other researchers have reported the application of *A. rhizogenes* for the manifestation of rol genes and for effective delivery of foreign genes to susceptible plants (Christey, 2001). It has been highlighted that the hairy root possess the T-DNA segment of Ri-plasmid inside its nuclear genomes (Chilton et al., 1982). Also, the *A. rhizogenes* could facilitates the movement of T-DNA of binary vectors in trans which play a crucial role in the collection of genetically modified plants from screened hairy roots (Christey, 2001). The application of hairy root play a crucial role in the monoclonal antibody production (Wongsamuth and Doran, 1997), genetical alteration and mass production of phytochemicals (Shanks and Morgan, 1999), phytoremediation

(Nedelkoska and Doran, 2000), large scale secondary metabolite production (Choi et al., 2000). Furthermore, the utilization of *A. rhizogenes* has been acclaimed to effective for protecting binary vectors with preferred gene constructs (Christey, 2001) for the purpose of genetic transformation (Kumar et al., 2006).

Chávez-Vela et al. (2003) applied *A. rhizogenes* A4 agropine-type strain for the development and during the process of the transformation system. It was observed that A4 entails wild-type plasmid pRi A4 which play a crucial role in the binary vector pESC4 and hairy-root genotype. The author utilized a transgenic sour orange (*C. aurantium* L.) plants and 75-day-old sour orange seedlings were regenerated from *A. rhizogenes* transformed roots the result obtained indicates that 91% of all the explants generated transformed roots with an average of 3.6 roots per explant. Similarly, the transformation of *A. rhizogenes* strain A4, possessing the binary vector pESC4 with nos-npt II, cab-gus genes and the wild-type plasmid pRiA4 were employed for the production of genetically modified orange from transgenic Mexican lime (*C. aurantifolia* (Christm.) Swing) plants. Their result showed that more than 300 Mexican lime transgenic plants were derived from the genetically modified experiment performed. Furthermore, it was observed that 60 out of these cultivars were confirmed to improve in their growing condition when transferred to the soil (Pérez-Molphe-Balch and Ochoa-Alejo, 1998).

Apart from the method of indirect gene transfer, some scientists have employed the direct gene transfer method for the genetic modification of citrus plants. The process of particle bombardment was applied on the Carrizo citrange (*C. sinensis Poncirus trifoliata*) thin epicotyl sections after the optimizing the condition involved in the process of for transient gene expression for the enhancement of citrus oranges. The maximum conditions that enhance the transient GUS expression includes culture of explants in a high osmolarity medium (sorbitol) 4 hours prior, M-25 tungsten particles, 9 cm distance between DNA/particle holder and specimen. Moreover, an average of 102 blue spots per bombardment (20 explants/plate) were obtained. Their study showed the suitability of the procedure for the genetical transformation of sweet orange (*C. sinensis*) and Carrizo citrange.

The process of electroporation has been highlighted as an operative direct gene transfer system utilized for the transformation of genetical make up of citrus fruits. Hidaka and Omura (1993) have validated the application of electroporation techniques for the gene transformation in citrus plant. The authors applied the protopalat derived from embryogenic callus of "Ohta" ponkan (*C. reticulata* Blanco) and the process of electroporation with exponential decay pulses was carried out in the solution that contains the CaMV 35S promoter (pBI221) together with β-glucuronidase (GUS) chimeric gene, It was observed that an anehnaced GUS activity was recorded in the cells when performed using fluorometric assay. Moreover, it has been observed that the application of Plasmid DNA encoding the nondestructive selectable marker improved the activity of green fluorescent protein gene that was inserted using polyethylene glycol into protoplasts of "Itaborai" sweet orange derived from an embryogenic nucellar-derived suspension culture. After the process of protoplast cultivation in the submerged medium and their relocation to the solid medium, the transformed callus were recognized through manifestation of the green fluorescent protein, which separated physically from the tissue that were not transformed and they were later cultivated somatic embryogenesis induction medium. The transgenic plantlets were obtained later from in vitro rooting of shoots and from the germinating somatic embryos.

In eukaryotes, the process of methylation and histone acetylation have been identified to play a crucial role in directing various plant defense and the rate of their growth. The process of histone alteration is regulated by several histone modification gene families. It has been noted that there is no single study on genome-wide characterization of histone modification that involves genes that are somehow related to various citrus species. In view of the aforementioned, Xu et al. (2015), performed a study on histone alteration and their role in the development of citrus fruits. The result obtained from some of the recent studies established that the sequenced sweet orange genome databases possess a cumulative of 136 *CsHMs* (*Citrus sinensis* histone modification genes) that contain 16 *CsHDACs* (histone deacetylase genes), 47 *CsHMTs* (histone methyltransferase genes), 50

CsHATs (histone acetyltransferase genes), and 23 *CsHDMs* (histone demethylase genes). The identified genes were later grouped into 11 gene families. The evaluation of these 111 genes was carried out with gene structures, conserved domain compositions of proteins, chromosome locations and phylogenetic comparison respectively. Furthermore, in order to have a better understanding of these genes played in the citrus fruit development selection of 42 *CsHMs* with high mRNA abundance in fruit tissues was carried out so as to evaluate their appearance profiles at six stages of fruit growth. The result obtained showed that numbers of genes were manifested in high quantity in flesh of ripening fruit while the other citrus fruits showed enhanced expression levels as the citrus plant developed. Also, the manifestation of all 136 *CsHMs* response to the infection of blue mould (*Penicillium digitatum) which is responsible for the post-harvest losses of most citrus crops showed strong changes to their* expression levels during the fruit-pathogen infection. Their study established that the various comprehensive evaluation performed on the histone modification gene families in sweet orange affirms their role during the development of citrus fruit and the role they played by inducing a high resistance to blue mould infection.

It has been shown that sweet orange cultivars do not have resistance to Huanglongbing (HLB) which constitutes a phloem-limited bacterial disease. Therefore, there is a need to develop varieties that is more resistant to HLB. Dutt et al. (2015) 'Valencia' and 'Hamlin' which are typical examples of sweet orange that showed the expression of *Arabidopsis thaliana NPR1* gene under the influence of a phloem specific *Arabidopsis* SUC2 (*AtSUC2*) promoter and constitutive CaMV 35S promoter were performed. The result obtained indicated that the manifestation of *AtNPR1 in trees with the ordinary* phenotic expression demonstrates more resistance to HLB while Phloem-specific expression of NPR1 also showed strong resistance against invading pathogens. Furthermore, it was observed that the genetically modified trees showed a decrease in diseases severity and only a small few lines still retained diseases free after 3 years of planting in a presence of heavily infested field site. The manifestation of *NPR1* gene induced expression of numerous native genes that are entailed in the plant defense signalling pathways. The *AtNPR1* gene being plant derivative can assist as a constituent for the growth of an all plant T-DNA resultant consumer-friendly genetically modified tree.

In plant breeding programs, the application of genetic modification has been recognized sustainable biotechnological method that could increase the quantity and quality of crop yield and enhance the level of disease resistance. Also, the utilization of Targeted genome engineering is assumed to play a vital role in the development of various varieties and genome editing technologies for the production of transgenic citrus trees using transcription activator-like effector nucleases, clustered regularly interspaced short palindromic repeat (CRISPR)/Cas9/single guide RNA (sgRNA), zinc finger nucleases. Moreover, there is no report yet on the application of current genome editing for the improvement of sweet orange. In view of the aforementioned, Jia and wang (2014) utilized an innovative tool, and Xcc-facilitated agroinfiltration, for the improvement of transient protein manifestation in sweet orange leaves. The authors, later utilized Xcc-facilitated agroinfiltration for effective delivery of Cas9, together with artificial sgRNA targeting the *CsPDS* gene inside a sweet orange. The result obtained showed that there was a mutation of *CsPDS* gene at the target site of the treated orange leaves. Furthermore, it was observed that the mutation rate of 3.2%–3.9% was observed when Cas9/sgRNA system was employed. Their study is the first report on the application of Cas9/sgRNA system for targeted genome alteration of the role of gene present in citrus plant.

Genetically modified science has been identified as a sustainable solution that could also help in the mitigation of various challenges against citrus greening (Allen, 2016). These diseases affected majority of the US citrus industry without any apparent solution. Moreover, a particular bacterium has been identified as the causal agent for the following diseases such Huanglongbing or citrus greening. The disease is normally spread by Asian citrus psyllid which normally affects the whole tree and eventually has an adverse effect on the taste of the citrus by becoming bitter (UF/IFAS Citrus Extension, 2016a,b). The worst part of this situation was that most farmers do not have sufficient money to prevent the incident of this citrus greening because there is no cure presently. This

situation has affected the rate of citrus production which led to the dropping of over 100 million boxes in Florida during the 2014 season and farmers reported that almost 80% of their citrus tree have been affected by HLB (Singerman and Useche, 2016). The incident of this diseases aggravated the situation by ending almost $9 billion citrus industry in Florida (Voosen, 2014). Moreover, the scientist in this region have invented some sustainable biotechnological techniques for the revitalization of the citrus industry in this region through the introduction of trees that are resistance to citrus greening, and citrus canker (Allen, 2016). Moreover, numerous experts have suggested that the industry should expect the introduction of genetically modified citrus tree as a sustainable solution to all the highlighted problems provided that the consumers could demonstrate a stronger interest to purchase and consume orange juice derived from genetically modified orange (Voosen, 2014).

The University of Florida/Institute of Food and Agricultural Sciences (UF/IFAS) research team has discovered several improvement on citrus. They explore the application of genetically modified technology to produce new tools that could led to the breeding of conventional plant.

These Biotechnology techniques used in the breeding of new transgenic orange varieties has been affirmed to be sustainable because they could mitigates against several diseases that normally affect the mass cultivation of several orange species. This could be linked to their capability to produce an immunity against any infection. Moreover, known as systemic acquired resistance (SAR) has been highlighted to prevent subsequent infection by enhancing the immune system of the affected plant. It has been observed that SAR gene commonly referred to as NPR1 derived from *Arabidopsis*, a mustard plant has been confirmed to demonstrate high biological control effectiveness against huanglongbing. This was achieved by insertion of NPR1 into the citrus plant using *Agrobacterium* as a vector which enhances the immunity of the plant by preventing any external infection. This is also facilitated by the presence of a promoter which is a switch that regulates the manifestation of any particular gene at any particular location.

Some of their discovery on transgenic sweet orange trees includes: Regulation of a phloem-targeted promoter or constitutive promoter, improvement of the citrus tree immunity to foreign diseases, pests and infection most especially a high tolerance to HLB. They have achieved huge success by this gene which allows the persistence of HLB resistance for two and a half years when an experiment was performed in a greenhouse (Citrus Industry, 2015).

Moreover, they have been able to generate transgenic rootstocks which are HLB-tolerant cultivars. Much effort has also been put in place for the addition of this gene into rootstocks in commercial scions which will serve as additional protection against pathogens and other infections thereby strengthening the immune system of citrus plant. In addition, these scientists have obtained several genes from plants such as soybean, grapes, beans, tomato, pepper, and tobacco for their capability to produce resistance against HLB. They have been able to produce transgenic citrus plants either rootstocks or scion cultivars that are fabricated through a process referred to as "gene stacking." In this process, they have identified several genes of inters that are involved in several mechanisms of action which are incorporated into the citrus plant. This technology was established because they realized that the cultivation of citrus is a long-term investment and there is a probability that pathogenic microorganisms, pathogens and HLB disintegrate the resistance of a single gene. Therefore, the addition of several genes could enhance the mechanism of action against the invading pathogen or pests which eventually led to stability in the long-term resistance.

Moreover, they have discovered edible and eco-friendly citrus produced through a special protoplast where its cell wall was removed through the application of digestive enzymes. This process enables the application of genetic engineering for the incorporation of edible plant-derived.

Transgenes in the absence of either viral or bacterial DNA sequences. They inserted a plant-derived visual reporter gene which has been selected based on special interest unswervingly into an isolated citrus protoplast with the assistance of electroporation or polyethylene glycol.

Additionally, they have been able to isolate another reporter gene referred to as Ruby obtained from blood orange which results in the fabrication of purple anthocyanin. They utilized a citrus-derived, embryo-specific promoter, which enables them to generate a DNA transformation construct that

permits anthocyanin to be communicated in the preliminary stages of plant regeneration (purple embryos) but which is consequently prevented in the mature or developing plant. Their finding showed that by exploring the ruby gene, they can obtain phenotypically normal transgenic citrus plants entailing specific and interesting DNA of special interest most especially from edible plants which will improve consumer acceptability of these citrus fruits.

Furthermore, they have been able to use genetically modified technology to minimize juvenility which is a common setback in most citrus plants. This was carried out through the identification of a key gene referred to as FT that generates a protein that initiates the floral process.

This particular protein of interest is normally available at minimal levels in juvenile plants and the quantity increases as the plant attains maturity. They have been able to isolate floral induction gene either with CFT protein or Citrus FT derived from Clementine with the help of genetically modified technology added to Carrizo rootstock genome. Interestingly, they engineered the gene to turn on at a very high level most of the time rather than switching on whenever the plant matures. This resulted in an improvement in the level of FT protein level and rapid flowering in 3-month-old Carrizo plantlets. This will also minimize the time involved in the evaluation of the new created germplasm (Citrus Industry, 2015).

Ruth et al. (2017) carried out a research study that focus on the innovative features that were presented by Rogers to validate if the application of genetically modified science could bring about a sustainable solution to the problem of citrus greening that affected most parts of the US resident. The study was carried out by collating necessary information from 1,051 respondents across various parts of the US. The various respondent was later distributed into various geographical regions while demographic variation among the respondent in various region was observed. Conversely, the information was collated from respondents from several regions who had neutral knowledge about genetically modified science, observability, compatibility, complexity, and trialability. The result obtained from the western region attests that there are benefits to the application of utilizing genetically modified science while the Midwest attests they would prefer to consume genetically modified citrus when compared to the northern region. Their study showed that extension agents could play a crucial role in the education of most people in these areas that are affected by citrus pests and diseases about the benefits of genetically modified citrus in the future.

Singerman and Useche (2017) carried out a study that specializes in the growers' survey-based evaluation of the impact of citrus greening on the rate of citrus processes in Florida as well as growers' evaluation regarding the level of potential concerns, information and their general believe pertaining their acceptability of genetically modified technology as a means of managing the diseases. Moreover, the authors also evaluate growers' preferences on certain characteristics of a Huanglongbing-resistant tree. It was discovered that the most significant heterogeneity in producers' apprehensions and trait favourites could be found in the some are such as safety, gene origin and environment.

9.4   ANTICANCER ACTIVITY OF ORANGE

Cancer is a group of diseases that affect human health leading to death. According to 2012 study, there were about 14 million cases of cancer and 8 million deaths due to cancer. Cancer is a disease of uncontrolled growth of cells as a result of proliferation, so the only treatment is to stop proliferation. Nowadays and in past days chemotherapy is used to prevent and reduce the rate of growth of cancer which is very expensive. The main cause of the increasing cancer rate is diet. We use less fruits and vegetables in diet. It is studied that most of cancer preventive chemoprotective agents are naturally present in fruits and vegetables. So, peoples with increased use of fruits and vegetables have low risk of cancer. Many chemoprotective constituents like flavones, vitamins and limonene are present in orange juice and orange peel.

The main cause of cancer in the human body is oxidative stress thus antioxidants can play important role in protection from cancer. Flavones present in orange juice are highly antioxidant.

TABLE 9.1

The Chemoprotective Constituents of Available in Various Orange Varieties and Their Modes of Action

S/N	Chemoprotective Constituents	Causative Enzyme	Cancer Sites	References
1	Hesperidin	Azoxymethane COX-2, IL-6	Colon, bladder, esophagus	Tanaka et al. (1997)
2	Limonene	DMBA Benzo pyrene	Buccal cavity, skin Forestomach, lungs	Miller et al. (1992)
3	Auraptene	Azoxymethane	Colon	Miyagi et al. (2000)
4	Perillyl alcohol	P13K/Akt	Brain tumors	Jayaprakasha et al. (2013)
5	Blood orange EO		Colon	Murthy et al. (2012)
6	Alpha pinene		Lungs	Zhang et al. (2015)
7	Citral	ALDH1A3	Stem cells	Thomas et al. (2016)

Carotenes like beta carotenes and lutein that are actually colouring agents also have antioxidant protection along with other benefits. Folate is needed for DNA synthesis and its mild deficiency leads to altered DNA and decrease the expression of tumour suppressor proteins. It is also reported that orange juice has some phytochemicals that can inhibit the initiation stage of colon cancer. As flavones are glycosides so they are not digested by small intestine, they go to colon where their break down occur by hydrolysis, so these are highly active against colon carcinogens. For different chemoprotective constituents and their action see Table 9.1.

Citrus flavonoids has been highlighted to inhibit cancer through the following process which includes apoptosis, selective cytotoxicity and antiproliferative actions (Elangovan et al., 1994). Moreover, Citrus flavonoids have been highlighted to possess antimutagenic effect which prevents DNA from destruction by their potential to enthrall ultraviolet light (Stapleton and Walbot, 1994). They have the capability to prevent and neutralize free radicals that stimulates the process of mutation whenever they are formed near DNA. This has also been affirmed in mice body exposed to X-ray (Shimoi et al., 1994). Moreover, citrus flavonoids could prevent the DNA by interrelating unswervingly with the tumoral agents when induced by chromosomal aberrations by bleomycin (Heo et al., 1994). Bracke et al. (1989) reported the modes of action of citrus flavonoids against cell proliferation and tumoral growth in cardiac, rat malignant cells and rat malignant cells. Oranges have been highlighted to possess several biological constituets such as amino acids, folic acid, iron, amino acids, chlorine, beta-carotene, manganese, pectin, zinc, pectin, sodium, potassium, manganese, folic acid, calcium, iodine. Cha et al. (2001) stated that a single oranges poseses over 60 flavonoids and 170 phytonutrients that poseeses anti-inflammatory, antioxidant properties, blood clot inhibiting.

9.5 ANTIOXIDANT PROPERTIES

Orange is virtually referred to as a natural source of bioactive components involved in promoting a good health system. These bioactive components are mostly antioxidants whose capacity needs to be increased or boosted. However, enhancing particular antioxidants concentration in the fruits of orange is attainable with metabolic engineering which will help strengthen the fruits to build a sustainable health system (Pons et al., 2014). Pons and colleagues (2014) performed an experiment on improving the beta carotene level in the fruits of orange via the obstruction of beta carotene hydroxylase gene (Csb-CHX) with the assistance of RNA intervention. The authors also concurrently express in excess the rate-limiting gene (Flowering locust T) responsible for the flowering transition in the engineered pulp sweet orange to produce fruits within a limited

time frame. On the other hand, obstructing the beta carotene hydroxylase gene gave a golden phenotype juvenile orange with a substantial increase in the beta carotene level to about 36-fold (Pons et al., 2014). The level at which the beta carotene was able to prevent the oranges from being exposed to oxidative stress was also determined through experimental studies using animal models by *Caenorhabditis elegans*. It was observed from Pons et al. (2014) studies that the engineered golden phenotype orange was able to increase the antioxidant capacity by 20% greater than that seen in the isogenic control orange. It was then concluded that genetic engineering has the potential to elevate the nutritional capacity of oranges by improving the antioxidant capacity to fight oxidative stress.

Polyamine oxidase (PAO) is the fundamental enzyme responsible for the catabolism of polyamine for the production of H_2O_2. Wang and Liu (2016) reveal the explanation behind the efficacy of $CsPAO_4$ gene in the catabolism of polyamine and its ability to suppress the growth of plant expose to salt stress. It was observed that $CsPAO_4$ uses spermidine and spermine present in plasma membrane of the citrus as their substrate for catabolism at the terminus. Furthermore, the genetic engineered citrus displayed an overexpression in the $CsPAO_4$ gene which is evident in increasing the activity of polyamine oxidase (PAO) thereby elevating the production of H_2O_2 with parallel reduction in spermidine and spermine level (Wang and Liu, 2016). The authors also observed that under a stress condition, the seeds of engineered citrus had a better growth than the wild type. The H_2O_2 was higher in the transgenic citrus seed with inhibition of it vegetative germination and root elongation leading to apoptosis. Application of catalase from external source was able to scavenge the H_2O_2 thereby leading to partial recovery of the vegetative growth and root elongation. Hence, spermine was able to suppress the growth of the engineered citrus using the $CsPAO_4$ gene to induce oxidative stress and then lower the production of polyamine oxidase (Wang and Liu, 2016).

9.6   ANTI-CANKER

Fu et al. (2011) reported on the Ectopic expression of MdSPDS1 gene in *Citrus sinensis* ability to suppress canker vulnerability, hydrogen peroxide (H_2O_2) production and transcriptional modification. It is known that polyamines are substance with wide physiological processes. Fu and colleagues reported on the introduction of an apple spermidine synthase gene (MdSPDS1) into *Citrus sinensis* through an Agrobacterium-mediated transformation of embryogenic calluses. It was observed that the two transgenic lines (TG_4 and TG_9) were less susceptible to *Xanthomonas axonopodis pv. citri* (Xac), like the wild type (WT) to citrus canker (Fu et al., 2011). The authors also showed that there was higher free spermine (Spm) and polyamine oxidase (PAO) activity when there was Xac attack on TG_9 cell line more significantly in comparison with WT where there was an apparent overexpression and H_2O_2 accumulation. Hence, prior treatment of TG9 cell line of leaves with guazatine acetate a substance that suppress the activity of PAO and the accumulation H_2O_2 thereby revealing the plant to more disease symptoms than the controls when exposed to Xac. Furthermore, the TG_9 lines in the transgenic citrus was upregulated with the help of pathogenesis-related protein and jasmonic acid synthesis on exposure to Xac than what was observed in the WT (Fu et al., 2011). Moreover, mRNA levels of most of the defense-related genes involved in the synthesis of pathogenesis-related protein and jasmonic acid were upregulated in TG9 than in the WT regardless of Xac infection. The authors conclude that MdSPDS1 gene was over-expressed in the transgenic citrus to fight against the development of cancer.

9.7   CONCLUSION AND FUTURE RECOMMENDATION

This study provided detailed information about the nutritional and medical benefits of genetically modified orange. Also, during this study special attention was laid on some nutritional benefits of biological and pharmacological components of oranges such as antihyperlipidemic anti-obesity, antioxidant, anti-diabetic, antiulcer and many more. Moreover, some recent and novel Biotechnological

techniques which are based on genetically modified science were also highlighted in details. Moreover, there is a need to strengthen policy that will enhance the introduction of these technologies to the farmers most especially those that dwell in developing countries about the advantages, profits and yield involved in genetically modified crops. Moreover, more transgenic orange tree that has been tested to entail a high resistance to diseases and pests should be distributed to the farmers. This chapter has established that the application of genetic manipulation methods could lead to drastic improvement of plants. The application of plant transformation has given enough room for genetic modification of one or more traits while maintaining the uniqueness of the original citrus cultivar. Also, there is a need to strengthen the development of citrus plant that could resist the adverse effect of pests and diseases, climate changes, and abiotic stress using genetically modified techniques.

REFERENCES

Adetunji C, Anani O. (2020). Bio-fertilizer from Trichoderma: boom for agriculture production and management of soil- and root-borne plant pathogens. In: *Innovations in Food Technology: Current Perspectives and Future Goals*, Mishra P, Mishra RR, Adetunji CO (Ed.), pp. 245–256. https://doi.org/10.1007/978-981-15-6121-4

Adetunji CO, Varma A. (2020). Biotechnological application of *Trichoderma*: A Powerful fungal isolate with diverse potentials for the attainment of food safety, management of pest and diseases, healthy planet, and sustainable agriculture. In: *Trichoderma: Agricultural Applications and Beyond. Soil Biology* Manoharachary C, Singh HB, Varma A. (Eds.), vol 61. Cham: Springer. https://doi.org/10.1007/978-3-030-54758-5_12

Adetunji CO, Roli OI, Adetunji JB. (2020a). Exopolysaccharides derived from beneficial microorganisms: antimicrobial, food, and health benefits. In: *Innovations in Food Technology*, Mishra P, Mishra RR, Adetunji CO (Eds.), Singapore: Springer. https://doi.org/10.1007/978-981-15-6121-4_10

Adetunji CO, Akram M, Imtiaz A, Bertha EC, Sohail A, Olaniyan OP, Zahid R, Adetunji JB, Enoyoze GE, Sarin NB. (2020b). Genetically modified cassava; the last hope that could help to feed the world: recent advances. In: *Genetically Modified Crops - Current Status, Prospects and Challenges*, Kishor P, Kavi B, Rajam MV, Pullaiah T (Eds.). https://www.springer.com/gp/book/9789811559310.

Adetunji CO, Inobeme A, Jeevanandam J, Yerima MB, Thangadurai D, Islam S, Oyawoye OM, Oloke JK, Olaniyan OT. (2021a). Isolation, screening, and characterization of biosurfactant-producing microorganism that can biodegrade heavily polluted soil using molecular techniques. In: *Green Sustainable Process for Chemical and Environmental Engineering and Science. Biosurfactants for the Bioremediation of Polluted Environments*, Inamuddin, Adetunji CO (Eds.), pp. 53–68. London: Elsevier. https://doi.org/10.1016/B978-0-12-822696-4.00016-4

Adetunji CO, Jeevanandam J, Anani OA, Inobeme A, Thangadurai D, Islam S, Olaniyan OT. (2021b). Strain improvement methodology and genetic engineering that could lead to an increase in the production of biosurfactants. In: *Green Sustainable Process for Chemical and Environmental Engineering and Science*, Inamuddin, Adetunji CO, Asiri AM (Eds.), pp. 299–315. London: Elsevier.

Adetunji CO, Jeevanandam J, Inobeme A, Olaniyan OT, Anani OA, Thangadurai D, Islam S. 2021c. Application of biosurfactant for the production of adjuvant and their synergetic effects when combined with different agro-pesticides. In: *Green Sustainable Process for Chemical and Environmental Engineering and Science*, Inamuddin, Adetunji CO, Asiri AM (Eds.), pp. 255–277. London: Elsevier.

Adetunji CO, Olaniyan OT, Anani OA, Inobeme A, Ukhurebor KE, Bodunrinde RE, Adetunji JB, Singh KRB, Nayak V, Palnam WD, Singh RP. 2021d. Bionanomaterials for green bionanotechnology. In: *Bionanomaterials: Fundamentals and Biomedical Applications*. IOP Publishing. https://doi.org/10.1088/978-0-7503-3767-0ch10

Adetunji CO, Inobeme A, Olaniyan OT, Olisaka FN, Bodunrinde RE, Ahamed MI. (2021e). Microbial Desalination. In: *Sustainable Materials and Systems for Water Desalination. Advances in Science, Technology & Innovation (IEREK Interdisciplinary Series for Sustainable Development)*, Inamuddin KA (Eds.), Cham: Springer. https://doi.org/10.1007/978-3-030-72873-1_13

Adetunji JB, Adetunji CO, Olaniyan OT. (2021f). African Walnuts: A Natural Depository of Nutritional and Bioactive Compounds Essential for Food and Nutritional Security in Africa. In: *Food Security and Safety*, Babalola OO (Eds.), Cham: Springer. https://doi.org/10.1007/978-3-030-50672-8_19

Adetunji CO, Olaniyan OT, Anani OA, Olisaka FN, Inobeme A, Bodunrinde RE, Adetunji JB, Singh KRB, Palnam WD, Singh RP. (2021g). Current scenario of nanomaterials in the environmental, agricultural, and biomedical fields. Nanomaterials in Bionanotechnology. In: *Nanomaterials in Bionanotechnology: Fundamentals and Applications*, 1st edn., Chapter 6. CRC Press. https://doi.org/10.1201/9781003139744-6

Adetunji CO, Kremer RJ, Makanjuola R, Sarin NB. (2021h). Application of molecular biotechnology to manage biotic stress affecting crop enhancement and sustainable agriculture. *Advances in Agronomy*. 168, 39–81. https://www.sciencedirect.com/science/article/pii/S0065211321000304?dgcid=author

Adegbola P, Aderibigbe A, Hammed W, Omotayo, T. (2017). Antioxidant and anti-inflammatory medicinal plants have potential role in the treatment of cardiovascular disease: A review. *American Journal of Cardiovascular Drugs*. 7(2): 19–32.

Adetunji CO, Olaniyan OT, Adetunji JB, Osemwegie OO, Ubi BE. 2022a. African mushrooms as functional foods and nutraceuticals. In: *Fermentation and Algal Biotechnologies for the Food, Beverage and Other Bioproduct Industries*. 1st edn., p. 19. First Published 2022. CRC Press. https://doi.org/10.1201/9781003178378-12.

Adetunji CO, Ukhurebor KE, Olaniyan OT, Ubi BE, Oloke JK, Dauda WP, Hefft DI. 2022b. Recent Advances in Molecular Techniques for the Enhancement of Crop Production. In: *Agricultural Biotechnology, Biodiversity and Bioresources Conservation and Utilization*, 1st Edn, p. 20. First Published 2022. CRC Press. https://doi.org/10.1201/9781003178880-12.

Allen, G. (2016). After a sour decade, Florida citrus may be near a comeback. *NPR*. Retrieved from http://www.npr.org/sections/thesalt/2016/12/04/503183540/after-a-sourdecade-floridacitrus-may-be-near-acomeback?

Almeida WAB, Mourao Filho FAA, Mendes BMJ, Pavan A, Rodriquez APM (2003). Agrobacterium-mediated transformation of Citrus sinensis and Citrus limonia epicotyl segments. *Scientia Agricola*. 60(1): 23–29.

Anani OA, Jeevanandam J, Adetunji CO, Inobeme A, Oloke JK, Yerima MB, Thangadurai D, Islam S, Oyawoye OM, Olaniyan OT. 2021. Application of biosurfactant as a noninvasive stimulant to enhance the degradation activities of indigenous hydrocarbon degraders in the soil. In: *Green Sustainable Process for Chemical and Environmental Engineering and Science. Biosurfactants for the Bioremediation of Polluted Environments*, pp. 69–87. https://doi.org/10.1016/B978-0-12-822696-4.00019-X

Angew ON. (2007). Functional foods. *Trends in Food Science and Technology*. 30: 19–21.

Anoosh E, Mojtaba E, Fatemeh S. (2010). Study the effect of juice of two variety of pomegranate on decreasing plasma LDL cholesterol. *Procedia Social and Behavioral Sciences*. 2: 620–623.

Arora, M, Kaur P. (2007). Phytochemical screening of orange peel and pulp. *International Journal of Research in Engineering and Technology*. eISSN: 2319–1163.

Avoseh O, Oyedeji O, Rungqu P, Nkeh-Chungag B, Oyedeji A. (2015) *Cymbopogon* species; ethnopharmacology, phytochemistry and the pharmacological importance. *Molecules*. 20: 7438–7453.

Azab A, Nassar A, Azab AN. (2016). Anti-inflammatory activity of natural products. *Molecules*. 21(10): 1321. doi: 10.3390/molecules21101321.

Bawa AS, Anilakumar KR. (2013). Genetically modified foods: Safety risks and public concerns-a review. *Journal of Food Science and Technology*. 50(6): 1035–1046.

Bennink MR, Om, AS. (1998). Inhibition of colon cancer (CC) by soy phytochemicals but not by soy protein (abstr). *Federation of American Societies for Experimental Biology Journal*. 12: A655.

Bernardi J, Licciardello C, Russo MP, Chiusano ML, Carletti G, Recupero GR, Marocco A. (2010). Use of a custom array to study differentially expressed genes during blood orange (*Citrus sinensis L. Osbeck*) ripening. *Journal of Plant Physiology*. 167: 301–310.

Bevan MW, Chilton MD. (1982). Multiple transcripts of T-DNA detected in nopaline crown gall tumors. *Journal of Molecularand Applied Genetics*. 1(6): 539–546.

Block G. (1991). Vitamin C and cancer prevention: The epidemiologic evidence. *American Journal of Clinical Nutrition*. 53: 270s–282s.

Block G, Patterson B, Subar A. (1992). Fruit, vegetables, and cancer prevention: A review of the epidemiological evidence. *Nutrition and Cancer* 18: 1–29.

Boscariol RL, Almeida WAB, Derbyshire MTVC, Mourão Filho FAA, Mendes BMJ. (2003). The use of the PMI/mannose selection system to recover transgenic sweet orange plants (Citrus sinensis L. Osbeck). *Plant Cell Reports*. 22(2): 122–128.

Bracke ME, Vyncke B, Van Larebeke NA, Bruyneel EA, De Bruyne GK, De Pestel GH, et al. (1989). The flavonoid tangeretin inhibits invasion of MO_4 mouse cells into embryonic chick heart *in vitro*. *Clinical and Experimental Metastasis*. 7: 283–300.

Busing A, Drotleff AM, Ternes W. (2012). Identification of α-tocotrienolquinone epoxides and development of an efficient molecular distillation procedure for quantitation of α-tocotrienoloxidation products in food matrices by high-performance liquid chromatography with diode array and fluorescence detection. *Journal of Agricultural and Food Chemistry*. 60: 8302–8313.

Centers for Disease Control and Prevention (CDC). (2001). Mortality from coronary heart disease and acute myocardial infarction: United States, 1998. *Morbidity and Mortality Weekly Report*. 50: 90–93.

Cervera M, Ortega C, Navarro A, Navarro L, Peña L. (2000). Generation of transgenic citrus plants with the tolerance-to-salinity gene HAL2 from yeast. *Journal of Horticultural Science and Biotechnology*. 75(1): 26–30.

Cha JY, Cho YS, Kim I, Anno T, Rahman SM, Yanagita T. (2001). Effect of hesperetin, a citrus flavonoid on the liver triacylglycerol content and phosphatidate phosphohydrolase activity in orotic acid-fed rats. *Plant Foods for Human Nutrition*. 56: 349–358.

Chalova VI, Crandall PG, Ricke SC. (2010). Microbial inhibitory and radical scavenging activities of cold-pressed terpeneless Valencia orange (Citrus sinensis) oil in different dispersing agents. *Journal of the Science of Food and Agriculture*. 90(5): 870–876.

Chaudhary SC, Siddiqui MS, Athar M, Alam MS. (2012). D-Limonene modulates inflammation, oxidative stress and Ras-ERK pathway to inhibit murine skin tumorigenesis. *Human & Experimental Toxicology*. 31: 798–811.

Chaudhury A, Duvoor, C, Reddy Dendi, VS, Kraleti S, ChadaA, Ravilla R, Marco NSMontales MT, Kuriakose K, Sasapu A, Beebe A, Patil N, Musham CK, Lohani, GP, Mirza W. (2017). Clinical review of antidiabetic-drugs: Implications for type 2 diabetes mellitus management. *Frontier Endocrinology (Lausanne)*. 8: 1–12.

Chávez-Vela NA, Chávez-Ortiz LI, Balch EP-M. (2003). Genetic transformation of sour orange using Agrobacterium rhizogenes. *Agrocencia*. 37(6): 629–639.

Chen H, Ward MH, Graubard BI. (2002). Dietary patterns and adenocarcinoma of the esophagus and distal stomach. *American Journal of Clinical Nutrition*. 75: 137–144.

Chen TC, Fonseca CO, Schönthal AH. (2015). Preclinical development and clinical use of perillyl alcohol for chemoprevention and cancer therapy. *American Journal of Cancer Research*. 5: 1580–1593.

Chiba H, Uehara M, Wu J, Wang X, Masuyama R, Suzuki K, et al. (2003). Hesperidin, a citrus flavonoid, inhibits bone loss and decreases serum and hepatic lipids in ovariectomized mice. *Journal of Nutrition*. 133: 1892–1897.

Chilton MD, Tepfer DA, Petit A, David C, Casse-Delbart F, Tempé J. (1982). Agrobacterium rhizogenes inserts T-DNA into the genomes of the host plant root cells. *Nature*. 295(5848): 432–434.

Choi JS, Yokozawa T, Oura H. (1991). Antihyperlipidemic effect of flavonoids from *Prunus davidiana*. *Journal of Natural Products*. 54: 218–224.

Choi SM, Son SH, Yun SR, Kwon OW, Seon JH, and Paek KY. (2000). Pilot-scale culture of adventitious roots of ginseng in a bioreactor system. *Plant Cell, Tissue and Organ Culture*. 62(3): 187–193.

Christey MC (2001). Use of Ri-mediated transformation for production of transgenic plants. *In Vitro Cellular and Developmental Biology—Plant*. 37(6): 687–700.

Codoner-franch P, Valls-belles V. (2010). Citrus as functional foods. *Current Topics in Nutraceutical Research*. 8(4): 173–184.

Costa MGC, Otoni WC, Moore GA (2002). An evaluation of factors affecting the efficiency of Agrobacterium-mediated transformation of Citrus paradisi (Macf.) and production of transgenic plants containing carotenoid biosynthetic genes. *Plant Cell Reports*. 21(4): 365–373.

Crowell PL. (1999). Prevention and therapy of cancer by dietary monoterpenes. *The Journal of Nutrition*. 129(3): 775S–778S.

Da Silva EJA, Oliveiraand AS, Lapa AJ. (1994). Pharmacological evaluation of the anti-inflammatory activity of a citrus bioflavonoid, hesperidin, and the isoflavonoids, duartin and claussequinone, in rats and mice. *Journal of Pharmacy and Pharmacology*. 46: 118–122.

Diodati JG, Dakak N, Gilligan DM, Quyyumi AA. (1998). Effect of atherosclerosis on endothelium-dependent inhibition of platelet activation in humans. *Circulation*. 98(1): 17–24.

Doll R, Peto R. (1981). Causes of cancer: Quantitative estimates of avoidable risks of cancer in the US today. *JNCI*. 66: 1193–1308.

Donth MY. (2014). Targeting inflammation in the treatment of type 2 diabetes: Time to start. *Nature Reviews Drug Discovery*. 13: 465–476.

Dutt, M, Grosser, JW (2009) Evaluation of parameters affecting Agrobacterium-mediated transformation of citrus. *Plant Cell, Tissue and Organ Culture*. 98(3): 331–340.

Dutt M, Grosser, JW. (2010) An embryogenic suspension cell culture system for Agrobacterium-mediated transformation of citrus. *Plant Cell Reports.* 29(11): 1251–1260.

Dutt M, Lee DH, Grosser JW (2010). Bifunctional selection-reporter systems for genetic transformation of citrus: Mannose- and kanamycin-based systems. *In Vitro Cellular and Developmental Biology—Plant.* 46(6): 467–476.

Dutt M, Barthe G, Irey M, Grosser J. (2015).Transgenic Citrus Expressing an *Arabidopsis* NPR1 Gene Exhibit Enhanced Resistance against Huanglongbing (HLB; Citrus Greening). *PLoS One.* 2015; 10(9): e0137134. doi: 10.1371/journal.pone.0137134.

Elangovan V, Sekar N, Govindasamy S. (1994). Chemopreventive potential of dietary bioflavonoids against 20-methylcholanthreneinduced tumorigenesis. *Cancer Letters.* 87, 107–113.

Food and Agriculture Organization of the United Nations. (2016). http://wwwfaoorg/docrep/005/y2772e/ y2772e04htm.

Fraley RT. (1983). Liposome-mediated delivery of tobacco mosaic virus RNA into petunia protoplast: Improved conditions for liposome-protoplast incubations. *Plant Molecular Biology.* 2(1): 5–14.

Fu X-Z, Chen C-W, Wang Y, Liu J-H, Moriguchi T. (2011) Ectopic expression of MdSPDS1 in sweet orange (Citrus sinensis Osbeck) reduces canker susceptibility: involvement of H_2O_2 production and transcriptional alteration.

Gao K, Henning SM, Niu Y, Youssefian AA, Seeram NP, Xu A, et al. (2006). The citrus flavonoid naringenin stimulates DNA repair in prostate cancer cells. *Journal of Nutritional Biochemistry.* 17: 89–95.

Garcia-Agustin P, Primo-Millo E. (1995). Selection of a NaCl-tolerant citrus plant. *Plant Cell Reports.* 14(5): 314–318.

Ghorbel R, Juárez J, Navarro L, Peña L (1999). Green fluorescent protein as a screenable marker to increase the efficiency of generating transgenic woody fruit plants. *Theoretical and Applied Genetics.* 99(1–2): 350–358.

Ghorbel R, Navarro L, Duran-Vila N. (1998). Morphogenesis and regeneration of whole plants of grapefruit (Citrus paradisi), sour orange (C. aurantium) and alemow (C. macrophylla). *Journal of Horticultural Science and Biotechnology.* 73(3): 323–327.

Goudeau D, Uratsu SL, Inoue K, daSilva FG, Leslie A, Cook D, Reagan L, Dandekar AM. (2008). Tuning the orchestra: Selective gene regulation and orange fruit quality. *Plant Science.* 174: 310–320.

Greño V, Navarro L, Duran-Vila N. (1988). Influence of virus and virus-like agents on the development of citrus buds cultured in vitro. *Plant Cell, Tissue and Organ Culture.* 15(2): 113–124.

Grosser JW, Gmitter FG. (1990). Protoplast fusion and citrus improvement. *Plant Breeding Reviews.* 8: 339–374.

Grosser JW, Gmitter Jr FG, Chandler JL. (1988). Intergeneric somatic hybrid plants of Citrus sinensis cv. Hamlin and Poncirus trifoliata cv. Flying Dragon. *Plant Cell Reports.* 7(1): 5–8.

Grosser JW, Ollitrault P, Olivares-Fuster O. (2000). Somatic hybridization in citrus: An effective tool to facilitate variety improvement. *In Vitro Cellular and Developmental Biology—Plant.* 36(6): 434–449.

Gutiérrez EMA, Luth D, Moore GA. (1997). Factors affecting Agrobacterium-mediated transformation in Citrus and production of sour orange (Citrus aurantium L.) plants expressing the coat protein gene of Citrus tristeza virus. *Plant Cell Reports.* 16(11): 745–753.

Havsteen B. (1983). Flavonoids, a class of natural products of high pharmacological potency. *Biochemical Pharmacology.* 32: 1141–1148.

Hallman WK, Cuite CL, Morin XK. (2013). Public Perceptions of Labeling Genetically Modified Foods. 1–42. http://humeco.rutgers.edu/documents_PDF/news/GMlabelingperceptions.pdf

Heo HY, Lee SJ, Kwon CH, Kin SW, Sohn DH, Au WW. (1994). Anticlastogenic effects of galangin against bleomycin induced chromosomal aberrations in mouse spleen lymphocytes. *Mutation Research: Fundamental and Molecular Mechanisms of Mutagenesis.* 311: 225–229.

Herrera-Estrella L, Block MD, Messens E, Hernalsteens JP, Mon- tagu MV, Schell J. (1983). Chimeric genes as dominant selectable markers in plant cells. *EMBO Journal.* 2(6): 987–995.

Hertog MG, Hollman PCH, Katan MB, Kromhout D. (1993). Dietary antioxidant flavonoids and risk of coronary heart disease. *Lancet.* 342: 1007–1011.

Hidaka T, Omura M. (1993). Transformation of Citrus protoplasts by electroporation. *Journal of the Japanese Society Horticulture Science.* 62(2): 371–376.

Higashimoto M, Yamato H, Kinouchi T, Ohnishi Y. (1998). Inhibitory effects of citrus fruits on themutagenicity of 1-methyl-1,2,3,4-tetrahydro-beta-carboline-3-carboxylic acid treated with nitrite in the presence of ethanol. *Mutation Research.* 415: 219.

Hood EE, Gelvin SB, Melchers LS, Hoekema A. (1993) New Agrobacterium helper plasmids for gene transfer to plants. *Transgenic Research.* 2(4): 208–218.

Inobeme A, Jeevanandam J, Adetunji CO, Anani OA, Thangadurai D, Islam S, Oyawoye OM, Oloke JK, Yerima MB, Olaniyan OT. (2021). Ecorestoration of soil treated with biosurfactant during greenhouse and field trials. In: Green Sustainable Process for Chemical and Environmental Engineering and Science. *Biosurfactants for the Bioremediation of Polluted Environments*, pp. 89–105. https://doi.org/10.1016/B978-0-12-822696-4.00010-3

Jacobs LR, Lupton, JR. (1986). Relationship between colonic luminal pH, cell proliferation, and colon carcinogenesis in 1,2-dimethylhydrazinetreated rats fed high-fiber diets. *Cancer Reserch*. 46: 1727–1734.

Januzzi Mendes BM, Luciana Boscariol R, Mourão Filho FDAA, Bastos de Almeida WA (2002). Agrobacterium-mediated genetic transformation of "Hamlin" sweet orange. *Pesquisa Agropecuaria Brasileira*. 37(7): 955–961.

Jayaprakasha GK, Murthy KNC, Uckoo RM, Patil BS. (2013). Chemical composition of volatile oil from *Citrus limettioides* and their inhibition of colon cancer cell proliferation. *Industrial Crops and Products*. 45: 200–207.

Jeevanandam J, Adetunji CO, Selvam JD, Anani OA, Inobeme A, Islam S, Thangadurai D, Olaniyan OT. (2021). High industrial beneficial microorganisms for effective production of a high quantity of biosurfactant. In: *Green Sustainable Process for Chemical and Environmental Engineering and Science*, 2021, pp. 279–297. London: Elsevier.

Jia H, Wang N. (2014). Targeted genome editing of sweet orange using Cas9/sgRNA. *PLoS One*. 9(4): e93806. doi: 10.1371/journal.pone.0093806.

Jia SS, Xi GP, Zhang M, Chen YB, Lei B, Dong XS, Yang YM. (2013). Induction of apoptosis by D-limonene is mediated by inactivation of Akt in LS174T human colon cancer cells. *Oncology Reports*. 29: 349–354.

Kanakavalli K, Thillaivanan S, Parthiban P. (2014). Hypolipidemic activity of NathaichooriChooranam (NC) (Siddha drug) on high fat diet induced hyperlipidemic rats. *International Journal of Pharmaceutical Sciences and Research*. 2: 104–109.

Kaneyoshi J, Kobayashi S, Nakamura Y, Shigemoto N, Doi Y. (1994). A simple and efficient gene transfer system of trifoliate orange (Poncirus trifoliata Raf.). *Plant Cell Reports*. 13(10): 541–545.

Kanungo SK, Panda DS, Swain SR, Barik BB, Tripathi DK. (2007). Comparative evaluation of hypolipidemic activity of some marketed herbal formulations in triton induced hyperlipidemic rats. *Pharmacology Online*. 3: 211–221.

Khan, EU, Fu XZ, Liu J-H. (2012). Agrobacterium-mediated genetic transformation and regeneration of transgenic plants using leaf segments as explants in Valencia sweet orange. *Plant Cell, Tissue and Organ Culture*. 109(2): 383–390.

Khawale RN, Singh SK, Garg G, Baranwal VK, Ajirlo SA. (2006). Agrobacterium-mediated genetic transformation of Nagpur mandarin (Citrus reticulata Blanco). *Current Science*. 91(12): 1700–1705.

Khodabakhsh P, Schafaroodi H, Asgarpanah J. (2015). Analgesic and anti-inflammatory activities of Citrus aurantium L. blossoms essential oil (neroli): Involvement of the nitric oxide/cyclic–guanosine monophosphate pathway. *Journal of Natural Medicines*. 69(3): 324–331. doi: 10.1007/s11418-015-0896.

Kim DH, Jung EA, Sohng IS, Han JA, Kim TH, Han MJ. (1998). Intestinal bacterial metabolism of flavonoids and its relation to some biological activities. *Archives of Pharmacal Research*. 21: 17–23.

Kumar V, Sharma A, Prasad BCN, Gururaj HB, Ravishankar GA (2006). Agrobacterium rhizogenes mediated genetic transformation resulting in hairy root formation is enhanced by ultrasonication and acetosyringone treatment. *Electronic Journal of Biotechnology*. 9(4): 1–9.

Kummer R, Fachini-Queiroz FC, Estevao-Silva CF, Grepan R, Silva EL, Bersani- Amado CA, Cuman KN. (2013). Evaluation of anti-inflammatory activity of Citrus latifolia Tanaka essential oi and Limonene in experimental mouse models. *Evidence-Based Complementary and Alternative Medicine*. 859083: 1–8.

Kurowska EM, Spence JD, Jordan J, Wetmore S, Freeman DJ, Piché LA, et al. (2000). HDL-cholesterol-raising effect of orange juice in subjects with hypercholesterolemia. *American Journal of Clinical Nutrition*. 72: 1095–1100.

Kusaba M (2004). RNA interference in crop plants. *Current Opinion in Biotechnology*. 15(2): 139–143.

Le Marchand L, Murphy SP, Hankin JH, Wilkens LR, Kolonel LN. (2000). Intake of flavonoids and lung cancer. *Journal of National Cancer Institute*. 92: 154–160.

Liu Y, Heying E, Tanumihardjo SA. (2012). History, global distribution, and nutritional importance of citrus fruits. *Comprehensive Reviews in Food Science and Food Safety*. 11: 530–545.

Manthey JA, Guthrie N, Grohmann K. (2001). Biological properties of citrus flavonoids pertaining to cancer and inflammation. *Current Medicinal Chemistry*. 8: 135–153.

Marques NT, Nolasco GB, Leitão JP (2011). Factors affecting in vitro adventitious shoot formation on internode explants of Citrus aurantium L. cv. Brazilian. *Scientia Horticulturae*. 129: 176–182.

Mauro M, Catanzaro I, Naselli F, Sciandrello G, Caradonna F. (2013). Abnormal mitotic spindle assembly and cytokinesis induced by D-limonene in cultured mammalian cells. *Mutagenesis.* 28: 631–635.

McArdle MA, Finucane OM, Connaughton RM, McMorrow AM, Roche HM. (2013). Mechanisms of obesity-induced inflammation and insulinresistance: Insights into the emerging role of nutritional strategies. *Frontier Endocrinology (Lausanne).* 4: 1–23.

Mendes BMJ, Mourão Filho FDAA, Farias PCDM, Benedito VA. (2001). Citrus somatic hybridization with potential for improved blight and CTV resistance. *In Vitro Cellular and Developmental Biology: Plant,* 37(4): 490–495.

Miller EG, Gonzales-Sanders AP, et al. (1992). Inhibition of hamster buccal pouch carcinogenesis by limonin 17-β-D-glucopyranoside.

Miyagi Y, Om A-S, et al. (2000). Inhibition of azoxymethane-induced colon cancer by orange juice. *Nutrition and Cancer.* 36(2): 224–229.

Murthy KNC, Jayaprakasha GK, Patil BS. (2012). D-Limonene rich volatile oil from blood oranges inhibits angiogenesis, metastasis and cell death in human colon cancer cells. *Life Sciences.* 91: 429–439.

Nedelkoska TV, Doran PM. (2000). Hyperaccumulation of cadmium by hairy roots of Thlapsi caerulescens. *Biotechnology and Bioengineering.* 67(5): 607–615.

Nielsen LB, Stender S, Kjeldsen K. (1993). Effect of lovastatin on cholesterol absorption in cholesterol-fed rabbits. *Pharmacology Toxicology.* 72: 148–151.

Nowlin, SY, Hammer MJ, Melkus GD. (2012). Diet, inflammation, andglycemic control in type 2 diabetes: An integrative review of the literature. *Journal of Nutrition and Metabolism.* 2012: 542698.

Orbović V, Dutt M, Grosser JW. (2011). Seasonal effects of seed age on regeneration potential and transformation success rate in three citrus cultivars. *Scientia Horticulturae.* 127(3): 262–266.

Organization WH. (2000). *Obesity: Preventing and Managing the Global Epidemic.* Geneva, Switzerland: World Health Organization.

Olaniyan OT, Adetunji CO. (2021). Biochemical Role of Beneficial Microorganisms: An Overview on Recent Development in Environmental and Agro-Science. In: *Microbial Rejuvenation of Polluted Environment. Microorganisms for Sustainability,* Adetunji CO, Panpatte DG, Jhala YK (Eds.), vol 27. Singapore: Springer. DOI:10.1007/978-981-15-7459-7_2.

Osarumwense PO. (2017). Anti-inflammatory activity of methanoilc and ethanolic extracts of *Citrus sinensis* peel (L) Osbeck on Carrageenan induced Paw Oedema in Wistar rats. *Journal of Applied Science Environment and Management.* 21(6): 1223–1225.

Ozaki Y, Ayano S, Inaba N, Miyake M, Berhow MA, et al. (1995). Limonoid glucosides in fruit, juice and processing by-products of satsuma mandarin (Citrus unshiumarcov.). *Journal of Food Science.* 60: 186–189.

Pallavi M, Ramesh CK, Krishna V, Parveen S, Swamy NL. (2017). Quantitative phytochemical analysis and antioxidant activities of some Citrus fruits of South India. *Asian Journal of Pharmaceutical and Clinical Research.* 10: 198–205.

Pallavi M, Ramesh CK, Krishna V, Parveen S. (2018). Peels of citrus fruits: A potential source of anti-inflammatory and anti-nociceptive agents. *Pharmacognosy Journal.* 10(6): s172–s178.

Pan Z, Liu Q, Yun Z, Guan R, Zeng W, Xu Q, et al. (2009). Comparative proteomics of a lycopene-accumulating mutant reveals the important role of oxidative stress on carotenogenesis in sweet orange (*Citrus sinensis* [L.] osbeck). *Proteomics.* 9(24): 5455–5470.

Pan MH, Li S, Lai CS, Miyauchi Y, Suzawa M, Ho CT. (2012). Inhibition of citrusflavonoids on 12-O-tetradecanoylphorbol-acetate-induced skin inflammationand tumorigenesis in mice. *Food Science and Human Wellness.* 1: 65–73.

Park HJ, Jung UJ, Cho SJ, Jung HK, Shim S, Choi MS. (2013). Citrus unshiu peelextract ameliorates hyperglycemia and hepatic steatosis by alteringinflammation and hepatic glucose- and lipid-regulating enzymes in db/dbmice. *Journal of Nutritional Biochemistry.* 24: 419–427.

Patel DK, Patel KA, Patel UK, Thounaoja MC, Jadeja RN, Ansarullah, et al. (2009). Assessment of lipid lowering effect of *Sidarhomboidea.* Roxb methanolic extract in experimentally induced hyperlipidemia. *Journal of Young Pharmacists.* 1: 233–238.

Peluso MR. (2006). Flavonoids attenuate cardiovascular disease, inhibit phosphodiesterase, and modulate lipid homeostasis in adipose tissue and liver. *Experimental Biology and Medicine (Maywood).* 231: 1287–1299.

Peña L, Navarro L. (1999). Transgenic citrus. In: *Transgenic Trees,* Bajaj YPS (Ed.), pp. 39–54. Berlin, Germany: Springer.

Peña L, Cervera M, Ghorbel R, et al. (2007). Genetic transformation. In: *Citrus Genetics, Breeding, and Biotechnology,* Khan IA (Ed.), pp. 329–344. Wallingford, UK: CABI International.

Pérez-Molphe-Balch E, Ochoa-Alejo N. (1998). Regeneration of transgenic plants of Mexican lime from Agrobacterium rhizogenes-transformed tissues. *Plant Cell Reports*. 17(8): 591–596.

Pons E, Alquezar B, Rodrıguez A, Martorell P, Genoves S, Ramon D, Rodrigo MJ, Zacarıas L, Pena L. (2014) Metabolic engineering of b-carotene in orange fruit increases its *in vivo* antioxidant properties. *Plant Biotechnology Journal*. 12: 17–27.

Ray DK, Mueller ND, West PC, Foley JA. (2013). Yield trends are insufficient to double global crop production by 2050. *PLoS One*. 8(6): e66428.

Reece EA, Homko CJ. (1998). Diabetes mellitus in pregnancy. What are the best treatment options? *Drug Safety*. 18: 209–220.

Ruth TK, Lamm AJ, Rumble JN, Ellis JD. (2017). Conversing about citrus greening: Extension's role in educating about genetic modification science as a solution. *Journal of Agricultural Education*. 58(4): 34–49.

Sakata K, Hirose Y, Qiao Z, Tanaka T, Mori H. (2003). Inhibition of inducible isoforms of cyclooxygenase and nitric oxide synthase by flavonoid hesperidin in mouse macrophage cell line. *Cancer Letters*. 199: 139–145.

Selmi S, Rtibi K, Grami D, Grami D, Sebai H, Marzouki L. (2017). Protective effects of orange (*Citrus sinensis L.*) peel aqueous extract and hesperidin on oxidative stress and peptic ulcer induced by alcohol in rat. *Lipids in Health and Disease* 16: 152. https://doi.org/10.1186/s12944-017-0546-y

Shanks JV, Morgan J. (1999). Plant "hairy root" culture. *Current Opinion in Biotechnology*. 10(2): 151–155.

Shaw JE, Sicree RA, Zimmet PZ. (2010). Global estimates of the prevalence of diabetes for 2010 and 2030. *Diabetes Research and Clinical Practice*. 87: 4–14.

Shimoi K, Masuda S, Furogori M, Esaki S, Kinae N. (1994). Radioprotective affect of antioxidative flavonoids in c-ray irradiated mice. *Carcinogenesis*. 15: 2669–2672.

Singerman A, Useche P. (2016). Impact of citrus greening on citrus operations in Florida (FE983). Retrieved from UF/IFAS website: http://edis.ifas.ufl.edu/fe983.

Singerman A, Useche P. (2017). Florida citrus growers' first impressions on genetically modified trees. *AgBioForum*. 20(1): 67–83.

Soler N, Plomer M, Fagoaga C, et al., (2012). Transformation of Mexican lime with an intron-hairpin construct expressing untranslatable versions of the genes coding for the three silencing suppressors of Citrus tristeza virus confers complete resistance to the virus. *Plant Biotechnology Journal*. 10: 597–608.

Sood S, Bansal S, Arunachalam M, Gill NS, Manoj B. (2009). Therapeutic potential of citrus medica L. peel extract in carrageenan induced inflammatory pain in rat. *Research Journal of Medicinal Plants*. 3: 123–133.

Stapleton AE, Walbot V. (1994). Flavonoids can protect maize DNA from the induction of ultraviolet radiation damage. *Plant Physiology*. 105: 881–889.

Tanaka T, Makita H, Kawabata K, Mori H, Kakumoto M, et al. (1997). Chemoprevention of azoxymethane-induced rat colon carcinogenesis by the naturally occurring flavonoids, diosmin and hesperidin. *Carcinogenesis*. 18: 957–965.

Thomas I, Gregg B. (2017). Metformin; a review of its history and future: From lilacto longevity, Pediatr. *Diabetes* 18: 10–16.

Thomas ML, De Antueno R., et al. (2016). Citral reduces breast tumor growth by inhibiting the cancer stem cell marker $ALDH_1A_3$. *Molecular Oncology*. 10(9): 1485–1496.

Tsai SH, Lin-Shiau SY, Lin JK. (1999). Suppression of nitric oxide synthase and the down-regulation of the activation of NFkappaB in macrophages by resveratrol. *British Journal of Pharmacology*. 126: 673–680.

UF/IFAS Citrus Extension. (2016a). Citrus greening (Huanglongbing). Retrieved from http://www.crec.ifas.ufl.edu/extension/greening/index.shtml.

UF/IFAS Citrus Extension. (2016b). Symptoms. Retrieved from http://www.crec.ifas.ufl.edu/extension/greening/symptoms.shtml.

Van Le B, Thanh Ha N, Anh Hong LT, Trân Thanh Vân K. (1999). High frequency shoot regeneration from trifoliate orange (Poncirus trifoliata L. Raf.) using the thin cell layer method. *Comptes Rendus de l'Academie des Sciences*. 322(12): 1105–1111.

Verlecar XN, Jena KB, Chainy GB. (2007). Biochemical markers of oxidative stress in *Pernaviridis* exposed to mercury and temperature. *Chemico-Biological Interactions*. 167: 219.

Voosen, P. (2014). Can genetic engineering save the Florida orange? *National Geographic*. Retrieved from http://news.nationalgeographic.com/news/2014/09/140914-florida-orange-citrus-greening-gmo-environment-science/.

Wahyuni FS, Ali DAI, Lajis NH. (2017). Dachriyanus anti-inflammatory activity of isolated compounds from the stem bark of *Garcinia cowa* Roxb. *Pharmacognosy Journal*. 9(1): 55–57.

Wald NJ, Law MR. (1995). Serum cholesterol and ischaemic heart disease. *Atherosclerosis*. 118: S1–S5.

Wang W, Liu J-H. (2016).*CsPAO₄* of *Citrus sinensis* functions in polyamine terminal catabolism and inhibits plant growth under salt stress. *Scientific Reports*. 6(31384): 1–15.

Wongsamuth R, Doran PM. (1997). Hairy root as an experimental system for production of antibodies. In: *Hairy Roots, Culture and Applications*, Doran MP (Ed.), pp. 89–97. Amsterdam, The Netherlands: Harwood.

World Health Organization. (2016). http://wwwwhoint/foodsafety/areaswork/food-technology/faq-genetically-modified-food/en/.

Xu Q, Chen LL, Ruan X, Chen D, Zhu A, Chen C, Bertrand, D. (2013). The draft genome of sweet orange (*Citrus sinensis*). *Nature Genetics*. 45(1): 59–68.

Xu J, Xu H, Liu Y, Wang X, Xu Q, Deng X. (2015). Genome-wide identification of sweet orange (*Citrus sinensis*) histone modification gene families and their expression analysis during the fruit development and fruit-blue mold infection process. *Frontiers in Plant Science*. 6: 607. doi: 10.3389/fpls.2015.00607.

Zeiher AM, Drexler H, Saurbier B, Just H. (1993). Endothelium-mediated coronary blood flow modulation in humans. Effects of age, atherosclerosis, hypercholesterolemia, and hypertension. *Journal of Clinical Investigation*. 92: 652–662.

Zhang Z., Guo S., Liu X., Gao, X. (2015). Synergistic antitumor effect of α-pinene and β-pinene with paclitaxel against non-small-cell lung carcinoma (NSCLC). *Drug Research*. 65: 214–218.

10 Advancing Aquaculture with Artificial Intelligence

Nivedita Patel, Shireen Patel, and Pranav Parekh
Nirma University

Manan Shah
Pandit Deendayal Petroleum University

CONTENTS

10.1 Introduction 190
10.2 Smart Fish Farm Systems 190
10.3 Artificial Intelligence for the Development of Aquaculture Systems 192
    10.3.1 Technologies Employed in Support of Artificial Intelligence 192
        10.3.1.1 Data Acquisition for Aquaculture Processing....................192
        10.3.1.2 Computerized Models 192
        10.3.1.3 Decision Systems 193
10.4 Artificial Intelligence Concepts Commonly Used in Aquaculture Systems.....194
    10.4.1 Expert Systems (ES) or Knowledge Based Systems (KBS)................194
    10.4.2 Neural Networks...195
10.5 Processes Prevalent in Aquaculture Systems...196
    10.5.1 Management of Fish Feed 196
10.6 Disease Detection in Fish...198
10.7 Water Quality Monitoring...199
10.8 Swarm Intelligence in Fishing 200
    10.8.1 Need for Optimization in Aquaculture..200
    10.8.2 Swarm Intelligence..201
    10.8.3 Particle Swarm Optimization (PSO) 201
    10.8.4 Applications of PSO in Aquaculture 203
    10.8.5 Swarm Robotics...205
10.9 Challenges and Future Scope...207
Declaration..207
    Authors Contribution 207
    Acknowledgments..207
    Availability of Data and Material...208
    Competing Interests 208
    Funding 208
    Consent for Publication..208
    Ethics Approval and Consent to Participate 208
References..208

DOI: 10.1201/9781003268468-10

189

10.1   INTRODUCTION

As posited by the NOAA, aquaculture is the cultivational method responsible for achieving feats such as increased food production, habitat restoration, and replete populations of threatened and endangered species. This cultivational method does not strictly fit its norm since it involves the cultivation of freshwater and seawater populations under controlled conditions. Such populations include fish, mollusks, crustaceans, algae, aquatic plants, and other underwater organisms. Fish supplies have burgeoned as a result of aquaculture (Naylor et al., 2000). This however has been considered by them as a mixed blessing that depends greatly on aquatic populations that are being impacted. The conservation of aquatic biodiversity is equally important (Diana, 2009) as the increasing use of aquaculture is evident (Bostock et al., 2010). Aquaculture as an industrial sector has been recognized, along with its global status and trends. It seems conclusive that aquaculture is an essentially growing field that will require technical feedback for its prime functioning.

The benefits of aquaculture are vast as it involves the breeding of specific plant and animal species under controlled environments which can lead to the development of healthier habitats, provision of an additional food source and commercial products, and engendering species that are close to extinction. Productions through aquaculture systems are a major source of income for many countries as well since they contribute to the economy of the country through exports (Li et al., 1997). The sector of food fish production has benefited greatly from aquaculture with a consistent increase in the average growth rate. Moreover, aquaculture provides vast economic and environmental benefits by reducing the sea trade deficit, increasing jobs, and reducing fishing pressure and pollution.

In the modern era technology has become an essential tool that is used to increase the productivity and efficiency of almost every sector and aquaculture is no exception as well (Patel et al., 2020; Shah et al., 2020; Ahir et al., 2020; Kundalia et al., 2020). Artificial Intelligence is one such field that is collaborated and used in aquaculture to further its progress (Gupta et al., 2020; Naik et al., 2020; Parekh et al., 2020; Patel et al., 2020). AI is the field of study that is essential for its techniques in solving problems and developing tools that are commonly associated with or done using human intelligence (El-Gayar, 1997; Kakkad et al., 2020; Pandya et al., 2020; Sukhadia et al., 2020; Jani et al., 2020). The following sections illustrate and elucidate the AI technologies that are used in improving aquaculture systems, and discuss the other information technology systems that are used in collaboration with AI. They also elucidate specific aquacultural processes that are improved using AI and discuss the importance and procedures of optimization in an aquaculture system. Also, swarm intelligence, an application of AI, has also been discussed along with its incorporation in aquaculture.

10.2   SMART FISH FARM SYSTEMS

Precision First Farming may be defined in a cyclical operational process realized in four phases: Observe, Interpret, Decide, Act (Føre et al., 2018; Talaviya et al., 2020; Pathan et al., 2020). In the first phase, observations are made in order to represent the assessments of the general population in the fish farms. Common tools utilized in the process are submerged cameras which are further processed with the help of computer vision technologies. Computer vision is capable of an array of things such as counting, quality inspection (Misimi et al., 2008), gender classification (Zion et al., 2008) and monitoring of welfare and behavior (Papadakis et al., 2012; Zion, 2012). In combination with other AI technologies it is further capable of assessing many things like sea lice infestations, feed pellet quantities (Skøien et al., 2014), as well as behavioral changes in terms of both swimming behavior (Pinkiewicz et al., 2011) or feeding behavior (Føre et al., 2011).

In the second phase, the observations are interpreted. Conventionally, this has been done by knowledgeable farmers based on their experience. However, modern research aims to automate this process with the help of artificial intelligence technologies that involve mathematical modeling. After this, decisions are made by humans as this is a highly complex task and finally, in the fourth

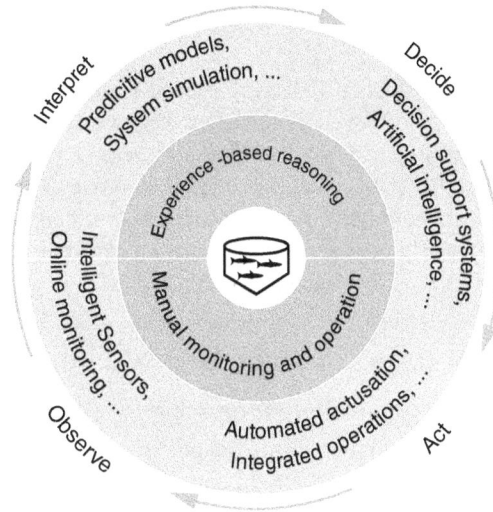

FIGURE 10.1 Representation of influence of the four stage PFF model (outer cycle) on traditional fish farming models (inner cycle) (Føre et al., 2018).

stage, these decisions are acted upon to incite the desired biological response from the fish farms (Figure 10.1).

The addition of this knowledge to aquaculture is primarily what has led to the commercialization and industrialization of this sector. Currently, by the addition of artificial intelligence technologies in this particular sector, there are anticipated benefits expected to change aquaculture process control systems. Increased process efficiency, reduced water loss, reduced energy wastage, reduced labor costs, reduced disease, and better understanding of the process are some of these anticipated benefits (Mustafa, 2016). Intelligent fish farm systems mainly help farmers with three major problems among the many challenges that they face.

Firstly, by monitoring the water quality constantly, control of water quality can be monitored in order to lower fish mortality and simultaneously increase the feed ratio. Feed management is important because it has been recorded that 70% of costs in aquaculture are attributed to feed and hence, conservation and management of feed is bound to have economic benefits and increase profitability. It has been found that the high mortality rate of fish farms is attributed to poor water quality a majority of the time and hence, this is quite important. Besides this, disease management is another aspect of smart aquaculture that focuses on immediate disease detection and rapid reporting to make informed decisions about how to control it or combat it.

For optimization, it is necessary to get an insight into collected datasets and statistics, so that we can turn mined knowledge into better procedures and systems today (Costa and Rihtar, 2016). Existing data as well as information technology can be used for data management and implementation of computerized models for decision support (El-Gayar, 1997). The automation, standardization and intellectualization of aquaculture as well as animal husbandry improves the level of production management when it is successfully implemented. Aquaculturists have started to recognize the potential of controlling the environmental conditions and system inputs (e.g. water, oxygen, temperature, feed rate and stocking density) in order to be able to get the required or anticipated growth rates of cultured species and process outputs.

In order to be able to observe the successful implementation of artificial intelligence in aquaculture, the major factors that are to be considered in the use and design of a process control software that is based on artificial intelligence are the functionality or intuitiveness and the compatibility (Lee, 2000).

10.3 ARTIFICIAL INTELLIGENCE FOR THE DEVELOPMENT OF AQUACULTURE SYSTEMS

Planning for aquaculture development as well as improved management for aquaculture facilities has further emphasized the need to develop and adapt advanced Information Technology methodologies such as Artificial Intelligence, IoT, etc. The advances in such IT fields have had a huge impact on the field of aquaculture. (El-Gayar, 1997). Huge improvements in the efficiency and sustainability of global aquaculture have been obviated through Artificial Intelligence and its underlying associated systems. The underlying associated systems include data management, computerized models, decision support systems, and image recognition.

10.3.1 TECHNOLOGIES EMPLOYED IN SUPPORT OF ARTIFICIAL INTELLIGENCE

10.3.1.1 Data Acquisition for Aquaculture Processing

The first step for the application of Artificial Intelligence in an aquacultural outset is the acquisition of data and its management . Due to the high damage incurred as a result of an inappropriate decision and the variable nature of the production process, the timely availability of data is of extreme importance. Apart from the timely availability, it is also necessary that the data is not only appropriate but also accurate. A menu-driven, pond-oriented, microcomputer program called FISHY 2.0 allowed fish farmers to accumulate and analyze data regarding fish production operations. This database technology was used alongside a spreadsheet called LOTUS@123 which enabled data logging alongside a data management system (El-Gayar, 1997). The current methods of data acquisition and data monitoring comprise IoT concepts such as Wireless sensor networks, fog computing, and cloud computing for the acquisition and monitoring of data in a coherent and contiguous manner.

A Recirculation Aquaculture System (RAS) is an essential technology that brings about high-density fish farming in an environmentally friendly manner. Data acquisition and data management through fog computing technology are necessary for the implementation of RAS (Al-Hussaini et al., 2018). The usage of temporary storage of data is made possible through such a technology which also enables the quick access of data when required to process. A traceability system for RAS is implemented through Wireless Sensor Networks (WSN) due to its comparable data accuracy as well as easy configuration (Qi et al., 2011). A hardware platform is developed that automatically monitors and transmits data to the software for processing or the fog for temporary storage.

10.3.1.2 Computerized Models

The processes following the acquisition and management of data are associated with how that particular data is processed. The requirement of computer modeling is to process such data in an efficient manner and to make intelligent predictions through the data. They provide a more precise and quantitative method in aquacultural research and also facilitate the evaluation of interactions of the aquacultural systems under observation with other externalities. Computerized modeling helps put together laboratory, field and theoretical studies to identify the areas where data is lacking, inconsistent or sparse (El-Gayar, 1997).

The computerized models are further classified as economic, biological, bio-economic and biophysical. A bio-economic model comprises a fusion of physical, biological and fusion elements to assist decision-making systems and producers in identifying operation management and optimal production system designs. A bio-economic model includes various different biophysical components that include economic strategies for its optimization (Brown, 2000). An integrated bioeconomic model consists of decision-making modeling and biological process modeling. Decision-making modeling, as the name suggests, is the optimization of a decision maker to make smart decisions. On the other hand, biological process modeling must capture the dynamic nature of biological processes involved and must allow for dynamism in biological processes and human decisions.

The need for bioeconomic models (BEM) for fisheries is primarily due to the presence of external factors which directly affect the biological side which further affects the economic profitability of the fisheries. BEM for fisheries is developed to capture long term, such as investments as well as short-term behavior such as population dynamics. Biological components for such a model include the factor of availability of stock assessments which is the segregation of fish stock and data based on the variety of species available in the fisheries. The economic components of such a model primarily include fleet and effort dynamics, price dynamics and cost dynamics. Despite the functioning of the individual components, the integration of the different components is essential to the success for the functioning of the fisheries. The integration should be such that there is an efficient dynamic interplay between the human and biological parts of the system (Prellezo et al., 2012).

10.3.1.3 Decision Systems

Besides efficient modeling, we require a system that facilitates better decision-making while encountering complex decision problems. A decision support system (DSS) is a system that enables decision makers in making informed decisions as such systems can combine technical and domain knowledge of the project under study, in a single package (Mathisen et al., 2016). A DSS system can contribute a significant improvement to the already developed computerized models, and it helps the decision makers in using the collected data and computerized models in an effective manner. To further elaborate on this point, an aquacultural development decision system (ADDSS) consists of three main components which are the dialog component, the database component, and the modeling component (El-Gayar and Leung, 2000). The dialog component provides a graphical user interface so that the user of the system can interact with the two other underlying components. The two other underlying components are the database component which provides a database management system (DBMS) to store the relevant data, and a modeling component which is a collection of the relevant models that are essential for the planning of the aqua cultural system.

The aqua farm project is one such implementation that provides simulation along with the dialog, database and modeling components. To further elaborate on this point, it is stated that an aqua farm project enables the simulation of the chemical, physical and biological unit processes and the simulation of faculty and fish culture management. The aqua farm is a modular programming architecture based on the fundamentals of Object Oriented Programming (OOP) that is suited for the development of a complex, aqua cultural model (Ernst et al., 2000). The five major components for an aqua farm system, as opposed to the DDS system, now include the simulation managers, domain experts, GUI and data managers, facility managers and facility components.

Decision support systems are used for cage aquaculture which involves the implementation of a fishery system and growing of fishes within a net cage that also allows the free flowing of water. Two additional steps involved in cage aquaculture is the selection of the best site and calculating a sustainable holding density for that particular site (Halide et al., 2009). Site selection and fisheries management is efficiently carried out by various Multi Criteria Decision Making (MCDM) techniques. The goal programming (GP) technique is one of the oldest MCDM techniques and it can be linear, nonlinear, fuzzy or fractional in nature. Another division of the MCDM techniques consists of the Multi Criteria Decision Analysis (MCDA) which is further subdivided into Multi Attribute Utility Theory (MAUT) and the Analytic Hierarchy Process (AHP). Despite their widespread applications in other fields, GP and MAUT have not thus far been used extensively in aquaculture systems. However, on the other hand AHP is used essentially in cage aquaculture for fisheries management (Mardel et al., 2004) and site selection (Vafaie et al., 2015). The AHP analysis tool assigns suitability through a 'good', 'bad' or 'medium' score thereby enabling site selection necessary for cage aquaculture.

Holding capacity, on the other hand, refers to the maximum value of fish biomass permissible for an aquaculture cage. It gives an idea of the maximum value of fish production that can be sustained for a particular period of time, given that certain constraints are kept constant (Halide et al., 2009). The solution to such a problem is provided by the Modelling Ongrowing Monitoring (MOM) model (Stigebrandt et al., 2004).

A DSS system can be implemented using fuzzy logic and a fuzzy inference system (FIS) as well. Fuzzy logic is implemented over the process information obtained from an operator rather than a predefined mathematical model (Rana and Rani, 2015). Due to this a fuzzy logic system often delivers faster solutions as compared to a DSS system or a computerized model. One such objective of fuzzy logic in aquaculture is the design of an algorithm required for making feeding decisions during the feeding process of the fishes in a fishery (Zhou et al., 2018). The decision making was primarily performed by an adaptive network based fuzzy inference system (ANFIS) which works on the basis of fuzzy control rules to achieve on demand feeding. The performance for such a model is evaluated by model parameters such as feed conversion rate, specific growth rate, weight gain rate and water quality parameters.

10.4 ARTIFICIAL INTELLIGENCE CONCEPTS COMMONLY USED IN AQUACULTURE SYSTEMS

AI is a field of intersection of expert systems and knowledge based systems (KBS), Logic and linguistics, computer vision and deep learning and machine learning algorithms. The following fields have a significant importance in the development of aquaculture systems. The primary objective of such a system is to implement a management model and a Decision support system (DSS) that is both inferential, smart and intuitive in nature. The other benefits of such a system is the reduced energy and water losses, increased process efficiency and reduced labor costs (Lee, 2000).

10.4.1 EXPERT SYSTEMS (ES) OR KNOWLEDGE BASED SYSTEMS (KBS)

In China, aquaculture plays an important role in generating income for farmers, however the rapid growth of aquaculture has led to the occurrence of infectious diseases that pose a great challenge to system owners. The problems and the different techniques to provide solutions regarding a fish disease diagnosis program using expert systems is described in Zeldis and Prescott (2000). On the other hand, a web based application of an expert system, called Fish Expert is used to treat fish diseases by mimicking human fish disease expertise (Li et al., 2002). The domain information involved for implementation of such a system involves the inspection of the pond and the fish, and the examination of the water quality to determine if the problem is a water quality problem or a parasite problem. The fish expert system design includes the use of a database for the storage of information, a knowledge base that consists of the modelled rules for the fish diagnosis disease and an inference process that is used to match the facts against the patterns. The knowledge base and inference engine plays an essential role in making the expert system intelligent while the database is concerned primarily with data storage and data management.

Expert systems that are used in aquaculture can be case based ES, rule based ES, neural networks based ES and fuzzy logic ES. Fish Vet is another effective fish disease diagnosis software program that can be implemented mainly using rule based expert systems or case based expert systems (Zeldis and Prescott, 2000). A rule based expert system consists of a database of rules relating to the problem domain in question. In Fish Vet, the primary objective of using a rule based expression is to cut down the problem space to a diminished size through which all the species specific diseases are thrown out that do not belong to the species in question. A case based ES, on the other hand, consists of a large number of previously solved cases of fish diseases. When a new case is presented to the user by the system, the database of a case based ES is searched to locate similar cases and present them to the user. A notion of similarity between the actual problems from the case base are used to deduce a solution to the problem at hand.

Apart from the databases that are essential to an expert system, an inference engine is necessary as well. An inference engine for an expert system is analogous to that of a human brain. Disease diagnosis in eels that is carried out by an expert system called SEDPA comprises an inference

engine that implements fuzzy logic using a fuzzy logic controller (Gutiérrez-Estrada et al., 2005). Fuzzy logic consists of fuzzy sets that are divided into geometric partitions and any value of the dataset can be described as a function of its membership in the different sets (Zeldis and Prescott, 2000). The fuzzy logic controller consists of an input fuzzy set and an output fuzzy set whose relationship in data is captured and stored in a fuzzy associative memory (FAM). A FAM is created by a human expert and three such memories are used by three such domain experts. The eel disease data is given as an input to a fuzzy logic controller and an output is drawn from it through the fuzzy associative memory. Apart from this, the other principal components of the SEDPA inference engine are the augmented transition network (ATN), which is a syntactic analysis methodology to access the data stored in the domain databases and a system of uncertainty transmission that is used to manage uncertainty using the Dempster Shafer theory.

An expert system based over a neural network refers to an artificially intelligent system that works with a neural network at its core. An intelligent aqua cultural growth model is implemented over a neural network taking into account the various quality parameters (Deng et al., 2010). A feed forward neural network is used to achieve the expression of the knowledge base of the growth model. Taking into the different parameters such as salinity, dissolved oxygen, temperature, pH, ammonia nitrogen and time. The output of such a model is the body length of the aquaculture organisms. Hence, with the help of expert systems and neural networks, the system has predicted the forecast of the growing length of the organisms.

10.4.2 Neural Networks

Neural networks are one of the most widely used Artificial Intelligence techniques that are used in aquaculture. A neural network consists of a sequence of layers where each layer consists of a PE which corresponds to the response variables of every layer. A transfer function is a mathematical equation associated with the PE of every layer. Usage of a neural network involves training of the data set which is a vector in the data matrix (Suryanarayana et al., 2008). A neural network having such a structure is used for fish stock forecast, determination of life cycle and age of a fish and fish identification. Neural networks are acclaimed for great success in fish stock forecast which refers to the automatic counting of fishes of a particular species by a three layer feed forward neural network (FFNN). The abundance of fish depends on the parameters such as distribution, stages of growth, recruitment, dynamics of eggs, death due to predator, closure of fishery, biomass availability, unorganized fishing and natural deaths. Fish identification refers to the detection of a fish of a particular species or formulation of clusters based on the different types of fishes present in the pond. Another important use of a neural network is in modeling such as the eco path modeling of a river system and the fish yield map modeling .

Temperature is a principal factor which is very important for shrimp growth since it provides a superior environment for growth and reproduction. The control of pond water temperature can be achieved through neural networks and mathematical models. Regulation of temperature is essential since the most essential parameter of a solar thermal system is the storage tank temperature. A mathematical model based on energy balance is developed to simulate the thermal behavior of an open pond upon which the ANN is applied. The NN controller is fitted to regulate the input temperature with respect to the value of the temperature required, which is read by a thermostat (Atia et al., 2011). Temperature control can also be performed through a fuzzy logic controller in which an input is converted into a fuzzy set which is subsequently processed by the inference controller that derives its knowledge from the rule stated in the rule based database (Atia et al., 2011).

Another important use case of neural networks in aquaculture is that of fish stock prediction. The modeling of a neural network and the prediction of fish yield pertaining to a data set of 59 African lakes is developed from a combination of six variables, which are depth, latitude, altitude of presence of lake, catchment area over maximum area, conductivity and fishing effort . The neural network algorithm used is the backpropagation algorithm in stochastic approaches which gives us the

predictive power required for fish stock prediction. The back propagation algorithm was also used in the prediction of three commercially important marine fishes on the basis of five different variables which are the latitude and longitude, biomass upon abundance ratio, sampling month and depth of the water column (). A three layered perceptron model is used where the input layer comprises five nodes each representing a single node for the variable stated above and output layer is a binary output, 1 or 0 where 1 represents that the fish is of the desired species and 0 represents otherwise. Another technology used for fish stock prediction is discriminate analysis (DA). DA uses classification functions to classify linearly modelled groups which thereby helps us in discriminating the groups and classifying them accordingly (Baldridge and Palmer, 2009). However, it is observed in many distinct comparative studies that neural networks provide a more accurate classification as compared to DA.

Another modelling procedure that is carried out for intelligent aquaculture systems involves the use of geographical information systems (GIS). A GIS model is developed to provide modelling to the data pertaining to a particular geography, and which varies from geography to geography. A neural network can be used alongside with a GIS for predictive mapping. For example, in the mapping and predictive modelling of fish and decapod assemblages, ANN is used for predictive modelling while a GIS is used for mapping of occurrences of the species present in the river water (Joy and Death, 2004). Remote sensing and GIS are used along with neural networks for shrimp culture. GIS integrated with remote sensing plays an important role in providing information on water quality, productivity, land or water cover, coastal infrastructure and tidal influence. In the following case, through remote sensing and imagery, we can obtain the generated data which is modelled using a GIS to provide suitable information which is finally used by a neural network in performing predictive modelling (Table 10.1).

10.5 PROCESSES PREVALENT IN AQUACULTURE SYSTEMS

10.5.1 MANAGEMENT OF FISH FEED

The majority of economic cost related to aquaculture comes from fish feed. Therefore, the importance of proper management of fish feed is great. This includes monitoring of feeding activity, estimation of feed, removal of excess feed and reduction in the wastage of feed. The monitoring of fish tanks can be done via different techniques. An underwater camera system was used for the monitoring of coral reef fish in real time (Jan et al., 2007), a digital stereo-video camera system for three-dimensional monitoring of Pacific bluefin tuna (Torisawa et al., 2011) and a baited camera called BotCam for the monitoring of bottomfish species (Merritt et al., 2011). Wireless Sensor Networks can also be used to detect changes in dissolved oxygen levels and monitor fish activities which ultimately determine fish feed. The most popular techniques used today are telemetry and computer vision based monitoring.

Fish telemetry is particularly useful considering it allows remote measurements through the use of a battery driven electronic tag that contains sensors (according to the required parameter) along with a processor and or storage unit used for logging (Øyvind Aas-Hansen et al., 2010). The incorporation of these sensors into radio and acoustic transmitters has provided novel insight into the environmental associations as well as their different behaviours and reaction to stressors (Cooke et al., 2013).

Telemetry is further of two types: acoustic telemetry and radio telemetry. Radio telemetry consists of detecting the fish based off of a radio tag that is attached to the fish. The radio tag is capable of identifying the frequency and the pulse rate and the frequency (Moore et al., 1997). Acoustic telemetry tags monitor parameters and convert electrical waveforms to frequency modulated acoustic signals that are transmitted through the water. The receiving system receives it after demodulating it (Dewar et al., 1999).

Estimation of fish feed is crucial to increase in efficiency. The intake of fish is dependent on their individual activities as well as factors such as temperature, dissolved oxygen etc. Due to this

TABLE 10.1

An Overview of Different Types of Intelligent Aquaculture Systems

Sr. No.	Objective	AI Technology Used	Other Technologies Used	Domain under Study	References
1.	A fish feeding system	A fuzzy logic controller used alongside a rule-base expert system	Sensors, site technology	Tilapia production	Soto-Zarazúa et al. (2010)
2.	Fish farming	Case based reasoning expert systems and machine learning modules.	Sensors, physical modelling, site technology	A DSS that increases export of farmed fish in Norway	Tidemann et al. (2012)
3.	Modelling and prediction of algal blooms	The neural network shell EXPLORER	Site technology and data	Algal blooms and phytoplankton in freshwater systems.	Recknagel et al. (2001)
4.	Water quality prediction	Genetic algorithm optimization as a search technique, support vector regression for predictive modelling	Wireless network system for data acquirement	Water quality which is essential for intensive breeding of river crab.	Liu et al. (2011)
5.	Aquaculture site selection	Fuzzy logic implemented over a MCDM model, Analytical Hierarchy Process (AHP)	GIS for obtaining geographical information about different sites	Suitable site selections along the coast for carrying out shrimp culture	Vafaie et al. (2015)
6.	Fish stock prediction and allowable catch evaluation	An intelligent expert system called SimerFish	A software application for data processing, user interface and additional databases	Recommendations on the management of biological resources for the pelagic fish stock present in the Caspian and Azov seas	Sazonova et al. (1999)
7.	Monitoring of an aquaculture system	A fuzzy controller and and its inference system	PID controller, sensor networks which follow the Zigbee protocol, a temperature sensor and other site technology	The aquaculture system under study that requires monitoring	Chang et al. (2013)

complexity of the problem, it is not possible to solve feed intake problems by mathematical models. Therefore, models such as back propagation neural networks have to be implemented (Chen et al., 2020). In a proposed MEA-BPNN model, data acquisition is the first stage.

It was observed that the fish either swim freely for continuous feeding, return to their original position after feeding, eat food that simply falls in front of them, or do not respond to bait. If fish do not respond it is possible that there is excess feed in the tank and this leads to wastage of feed. A BPNN with the structure 4–10–1 was used with the four input factors being water temperature, dissolved oxygen, the average weight, and the number. Ten hidden nodes were present in the model and the output was the feed intake. The model was further optimized by the mind evolution algorithm (MEA) that is used for convergence and alienation. The experimental results showed that the Spearman correlation coefficient between actual values and the predicted values was up to 0.994, which is a successful and a strong correlation.

Further, remote feeding could be done which largely reduces the labor cost in aquaculture. A trigger button is implemented through a web application that controls the feed. By using such an

application on a smartphone, it is easier to start and stop feeding. Moreover, adapting feeding technologies can also be used. These adaptive feeders continue to feed the fish until the feed is no longer dropped based on the Fish Learning Index (FLI) value. The FLI value is based on two factors: Time Feeder Inder (TFI) and the smart feeder index (SFI) and it acts as a threshold value for controlling the feed. It is determined by fish activities and the time period since last feed was conducted (AlZubi et al., 2016).

So, the management of fish feeding can be potentially optimized and has been done such as the study of the intelligent feeding of Rohu (Saikia et al., 2013) or the development of an intelligent system used for deployment in eel culturing (Chang et al., 2013).

10.6   DISEASE DETECTION IN FISH

Among production limiting challenges, disease takes the lion share by causing multibillion-dollar loss annually (Assefa and Abunna, 2018). Fish disease is another substantial source of monetary loss to aquaculturists. Disease outbreaks and subsequent death leads to investment loss, increase in cost due to treatment, and decreased growth. It also leads to food safety issues (Zhang et al., 2012). Fish disease problems tend to be less significant in nature backgrounds due to the fact that predators prey on sick animals easily and also because these fish are less crowded in natural surroundings. However, under no natural conditions, the bacteria and parasites can cause substantial problems and hence, disease detection is important to the process of aquaculture.

Fish Detection is normally a process that requires expertise. It involves pond inspection, general fish inspection, water quality inspection, and ultimately recommendation of treatments and prevention mechanisms. Fish disease management and control requires assistance in the form of smart systems as the use of AI technology has the ability to do what can be done with human expertise.

The fish disease management system presented (Li and Gatlin, 2006; Zhang et al., 2012) consists of different modules:

1. The water eutrophication model that is used to asses/rank the trophic status of pond water
2. Factor forecasting model which gives more insight into the complex interactions between the water factors and dynamically predicts their short term variation.
3. Fish disease risk early warning models to estimate the degree of fish disease risk and issue a warning signal.
4. EW/Response model: This module consists of decision making and action implementation. The thresholds to trigger actions are defined according to the early warning levels.

An example system, Fish Expert (Li et al., 2002), was developed using a combination of Internet technologies as well as SQL programming. A server side database kept records of all the information that was used for diagnosis of a fish disease. There were separate databases like symptoms database, fish medicine database, fish disease database, etc. for each purpose. The users were expected to query this system and it would match the input facts with patterns and make suggestions based on the rules that were applicable.

Therefore, Expert Systems contain an inference engine, an explicative subsystem (provides information on the disease), a proposal engine, and a learning subsystem (incorporates new learned information into the system) besides the user interface (Gutiérrez-Estrada et al., 2005). SEDPA, another example of a similar system used in eel rearing, has an inference engine which is the brain of the system that has been implemented using a fuzzy logic controller. It also uses the ATN syntactic analysis method in order to assure that the information that is introduced to the expert system is coherent.

Similarly, expert systems are implemented using fuzzy neural networks and artificial neural networks (Deng et al., 2013b). In the latter, the diagnosis system uses the old cases of fish disease to train the neural network and further disease diagnosis is achieved through these neural networks. Therefore, while lack of human expertise increases loss in aquaculture, these expert systems can

TABLE 10.2

Disease Detection in Fish Using AI Techniques

No.	Technology for Inference System	Fish Specimen	Accuracy	References
1.	Artificial Neural Networks	Grass Carp	90%	Deng et al. (2013a)
2.	Fuzzy Logic	Discus Fish	90.32%	Hanafia et al. (2015)
3.	Nearest Neighbour Search	N/A	93.57%	Wang et al. (2009)
4.	Case Based Reasoning	N/A	N/A	Sun and Li (2016)
5.	Fuzzy Logic	Shrimp	N/A	Rao et al. (2019)

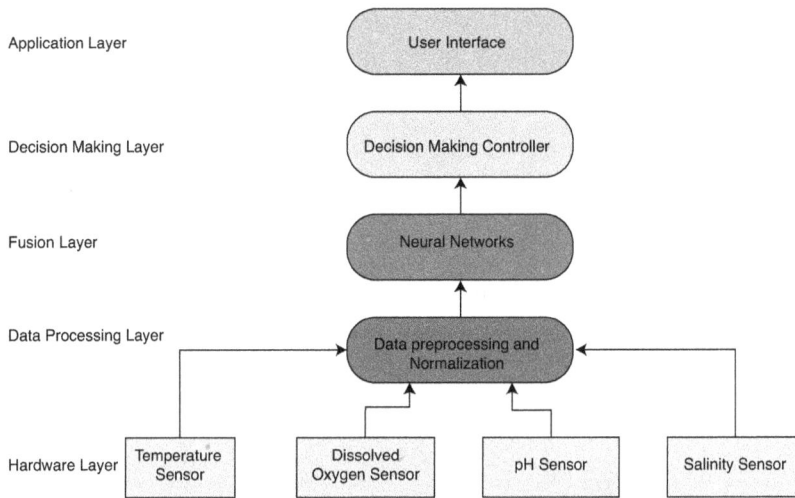

FIGURE 10.2   Layers of water quality assessment model.

help with the process of accurate diagnosis and timely treatments. It can also reduce labor costs (Table 10.2).

10.7   WATER QUALITY MONITORING

Water Temperature, Dissolved Oxygen, pH, Ammonia and nitrates concentration, salinity, and alkalinity are some of the important parameters that help determine the quality of water (Simbeye and Yang, 2014). This is because parameters like water temperature have the capability to directly affect feed utilization, animal health, growth rates, and quality of fish stock. Furthermore, once water quality deteriorates and organisms are forced to survive in poor environments, they are easily susceptible to disease and consequently death (Khotimah, 2015). Therefore water quality prediction plays an important role in modern intensive aquaculture management.

Water Quality monitoring has significantly been done using WSN (Zhang et al., 2011) largely as it allows real time monitoring and enables remote monitoring in some cases (Wang et al., 2009). WSN were also used for monitoring water turbidity (Rocher et al., 2017). However, by the introduction of AI technologies, it is possible to form systems that can be used for the processing of data and decision making so that different results of water quality triggers different levels of alarm and suggestions.

Therefore, this type of monitoring system will consist of four parts: hardware layer, data processing layer, decision making layer, and application layer (Figure 10.2).

An environment monitoring system and control is implemented with artificial neural networks (Gustilo and Dadios, 2011). Sensor readings are taken from five different sensors that measure critical parameters- salinity, temperature, pH, dissolved oxygen, and volume. These values are accepted by the ANN which then gives the state of each input parameter along with the overall water quality index (wqi). It also provides methods of correction in the case the parameters are not in optimal conditions.

The input parameters depend on the system selection. In (Simões et al., 2008) turbidity, total phosphorus and dissolved oxygen concentrations were normalized on a scale from 0 to 100 and translated into statements of water quality (excellent, good, regular, fair and poor). The parameters such as dissolved oxygen (DO), acidity (pH), and temperature were used in another model (Salim et al., 2016).

The quality assessment system can also be implemented using a fuzzy inference system (Hernández et al., 2012). The Fuzzy inference systems (FIS) provided a non-linear relationship between input variables and Water Quality Index.

The input set for an FIS would be as follows:

$$z^1 = \left[\mu_{temp}^i, \mu_{sal}^j, \mu_{DO}^k, \mu_{pH}^i \right]$$

The output rules function is as follows:

$$\mu_R = \min \left\{ \mu_{temp}^i, \mu_{sal}^j, \mu_{DO}^k, \mu_{pH}^i \right\}$$

The rule sets that are defined for classification were used and based on this the water quality status were classified as poor, regular, good, and excellent. Similarly, the FIS used in the water monitoring system for eels gives an output in the form of a survival rate of the eels (Wahjuni et al., 2016).

10.8 SWARM INTELLIGENCE IN FISHING

10.8.1 NEED FOR OPTIMIZATION IN AQUACULTURE

Optimization is extremely important in almost all fields ranging from technology to industrial applications. Optimization means improving productivity, utilization, efficiency, and strength of a system. The aim is to always maximize or minimize the factors that affect a particular system. The need for optimization in aquaculture has increased over the years due to problems faced by the conventional system majorly in the decision-making process. The mathematical models can be used to improve the process of feeding, harvesting, and seed selection (Cobo et al., 2015). Generally, there is one objective function that has to be optimized, few decision variables, and some constraints that limit the values that could be assigned to decision variables. But in real-world applications, optimization of multiple objectives at the same time might be required. This could lead to different objectives conflicting with each other. Sometimes, it is also difficult to get an optimal solution for a single objective function. Thus, metaheuristic algorithms have been gaining attention since the last few years. Balancing intensification and diversification, the two characteristics of metaheuristic algorithms could help achieve a global optimal solution to a problem (Yang, 2011). Intensification is to search in a local region and try to find a solution from that region, whereas diversification means to search on a global scale to generate diverse solutions.

Metaheuristic algorithms are classified on the basis of a number of criteria (Blum and Roli, 2003). Metaheuristics can be classified as nature-inspired or non-nature inspired depending on the origins of the algorithms. Ant algorithms and genetic algorithms are examples of nature-inspired algorithms. Another classification is based on the number of solutions that are used at the same time. It could be a single solution or an entire population. Thus, categorizing into trajectory-based,

and population-based algorithms respectively. Particle Swarm Optimization (PSO) algorithm is an example of population-based metaheuristics.

Swarm Intelligence (SI) algorithms are the subset of population-based algorithms (Cobo et al. 2015). Due to the capability of metaheuristics to work with complex problems, its acceptance in the management of aquaculture has been increasing. The following sections will introduce swarm intelligence, SI technique called PSO, and how the management of aquaculture farms has improved with swarm intelligence and swarm robotics.

10.8.2 Swarm Intelligence

A swarm is a group of similar agents who interact among themselves locally and, with their environment, in order to emerge a global behavior (Ahmed and Glasgow, 2012). Swarm Intelligence, an emerging and a new branch of Artificial Intelligence (AI), is a paradigm that is used to model the collective behavior of a number of social insects such as ants, wasps, bees, and also from other animal groups like schools of fish and flocks of birds. Swarm intelligence was first introduced by Beni and Wang (1993) for cellular robotic systems. They introduced a set of algorithms where the agents use nearest neighbor interaction to self-organize.

Swarms perform actions to solve complex behavior and show intelligence in forming self-organizational and decentralized systems (Cobo et al., 2015). These groups of agents show intelligent behavior by interacting and communicating with their environment locally even though they have limited capabilities on their own. The nature of interactions that occur among the agents has also been classified into two categories (Ahmed and Glasgow 2012). Swarms might communicate among themselves either directly or indirectly. Direct interaction might happen through audio or visual contact. An example for this is a honey bees' waggle dance. An indirect interaction occurs when a single agent changes or modifies the environment followed by other agents responding to the new surroundings. Such an interaction is seen when the ants leave pheromone trails while searching for food.

Swarm-based algorithms are population-based and nature-inspired algorithms that have proved to be cost-efficient, robust and fast to be able to solve complex problems. These algorithms are used for situations that do not have global information about an environment. In such cases, groups of agents are made to interact with each other and the environment at a local scale, whose behavior is then used to solve an objective at a global level. The two most popular and widely used algorithms based on the behavior of natural swarms are Ant Colony Optimization (ACO) and Particle Swarm Optimization (PSO). ACO is based on the behavior of ants, where artificial ants generate solutions to problems and communicate via methods similar to the ones used by real ants (Dorigo et al., 2006). PSO was introduced by Kennedy and Eberhart (1995) to understand the concept of optimization of continuous non-linear functions with the help of the behavior showcased by the flocks of birds.

Artificial Bee Colony, Particle Swarm Optimization, Ant Colony Optimization, Cat Swarm Optimization and Artificial Immune System are a few of examples of swarm intelligence models. As this paper discusses aquaculture, we will mainly focus on particle swarm intelligence as it has been inspired by schools of fish along with flocks of birds. Hence, the following section talks about PSO in detail and how the conventional PSO must be modified in order to be able to work in a dynamic environment.

10.8.3 Particle Swarm Optimization (PSO)

Particle Swarm Optimization is a technique based on the intelligent behavior of flocks of birds or fish schools, providing optimal solutions to real-life problems. Kennedy and Eberhart were the first to introduce PSO in 1995. PSO algorithm is connected to two methodologies, artificial intelligence and evolutionary computation (Kennedy and Eberhart, 1995). PSO has been used to evolve the

structure of neural networks, in reactive power plants, learning to play games and tracking dynamic systems.

Particle swarm optimization is a population-based algorithm that generates solutions that can be represented as a set of points in the n-dimensional solution space (Ahmed and Glasgow, 2012). The behavior of an entire population of particles is observed and analyzed to get the best solution. Each particle has a velocity vector that represents its speed in each dimension. Here, each particle is considered as an individual solution whose memory and knowledge is constantly improved by updating its current memory to the historically best solution. Here, historically the best solution refers to the best position a particle attains which is better than any previously attained location. A fitness function is evaluated for any given problem in order to get the best optimal solution (Cobo et al., 2015; Ahmed and Glasgow, 2012).

The process starts by choosing a random swarm of agents in the search space. All the candidates chosen are set in motion throughout the solution space looking for an optimum influenced by global and local factors. Hence, all particles have their respective velocity vector $V_i(t)$ and position vector $X_i(t)$, where i stands for a particular particle (Xin-She Yang, 2011; Blackwell et al., 2008; Cobo et al., 2015). $X_i(0)$ and $V_i(0)$ are initial vectors that are randomly selected from the search space. Slowly these particles evolve in terms of their memory of the best solution, and their knowledge of their neighbor's or the global best. At each iteration, all moving particles evaluate their current position with respect to the fitness function and then compare it to that of the other individuals and their own previous best positions. Therefore, at the end of each iteration, the index of the swarm's global optima is updated if there is a new position better than the previous index value of the swarm's global best. This procedure continues until there are no longer improvements in the solutions, or if the function was to be evaluated only for a specific number of iterations.

The new velocity vector can be evaluated using the formula

$$V_i(t+1) = V_i(t) + \alpha\varepsilon[g^* - X_i(t)] + \beta\varepsilon_2[X_i^* - X_i(t)]$$

Where:
g^* stands for the value of the current global best
X_i^* stands for a particle's own best-known position
V_i represents the velocity vector
X_i represents the location vector
· t represents the time
i stands for a specific particle
α and β are the acceleration factors or also known as the learning parameters, whose value is generally taken as 2. These parameters control the importance of a particle's own experience versus the social experience of the entire swarm.

ε_1 and ε_2 are two random vectors whose values are in the range [0,1] and are helpful in the random search strategy of the PSO algorithm (Figure 10.3).

Then, the new position can be modified as

$$X_i(t+1) = X_i(t) + V_i(t+1)$$

This is the standard PSO algorithm that has undergone a number of improvements, one of which is using the inertia function, $\theta(t)$. So, $V_i(t)$ is replaced by $\theta(t)V_i(t)$. θ can have any value between 0 and 1 (Shi and Eberhart, 1998).

The standard PSO needs to be modified in order to work in dynamic environments. Outdated memory and diversity loss are the two problems that need to be addressed (Blackwell et al., 2008). If a problem changes then the algorithm should be able to update the memory accordingly before searching for the optimal solution. Hence, a lot of enhanced versions of PSO have been proposed.

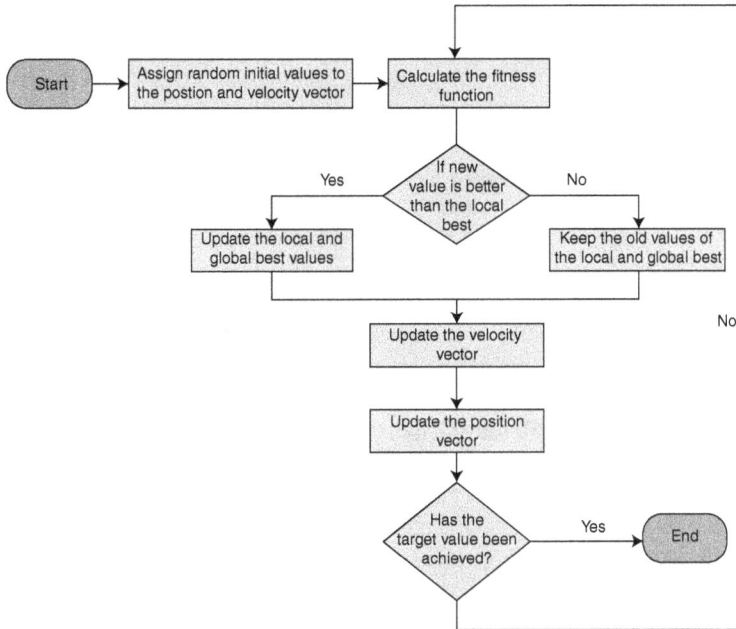

FIGURE 10.3 Flowchart for the working of PSO algorithm.

Multi-swarm PSO (MPSO), and speciation-based (SPSO) are the two algorithms that work in dynamism. In MPSO, a swarm is divided into multiple sub-swarms that are allocated to different areas of a specific region to maintain diversity among the swarm. While SPSO already has a specific number of particles that are then dynamically allotted to swarms. PSO has also been advanced based on the behavior of fish schooling (FSPSO) (Yan et al., 2015). The main objective of using FSPSO is to achieve an improved global best position and eliminate weak particles. Area-oriented particle swarm optimization (AOPSO) has been proposed to improve the learning strategy of the original PSO (Liu et al., 2019). MBPSO is an algorithm that has been inspired by the migration behavior of animals that is used to improve the convergence speed and avoid the standard PSO from being trapped in local optimum (Jiang et al., 2019). Enhanced partial search optimization (EPS-PSO) is another modification that has been proposed to improve the efficiency and convergence of PSO (Fan and Jen, 2019).

10.8.4 APPLICATIONS OF PSO IN AQUACULTURE

In recent years, aquaculture has become a significant industry for sustaining the demand for fish. Hence, efficient management of fish farms has become one of the most crucial factors in order to maintain the economy of the fish industry. There are a number of internal factors such as fingerling weights or feeding rates, and external factors like social behavior of individuals or water temperature, that have an impact on aquaculture. Complex interactions of biological, economical, technical, and environmental aspects have generated a need for a model that could make efficient decisions regarding feeding rates, diet composition, and harvesting time (Cobo et al., 2015, 2019). As the management of fish farming is complex, using classical optimization methods would not be feasible due to their limitations. Many real-world problems are multimodal and highly non-linear, where the task of finding an optimal solution is not easy and too complex. Thus, using classical optimization methods could have a risk of being trapped in local optima or might face problems in handling the constraints. Such optimization problems would require a method that would give accurate results in

a non-linear, stochastic, and complex environment. Hence, a more powerful optimization method is PSO which has become one of the most used algorithms in management of the aquaculture farms.

PSO can be used for production planning and for making the process of decision-making more efficient. A model had been proposed for optimal planning in the production of sea breams in floating cages (Cobo et al., 2015). Biological submodel and economic submodel of farming in sea cages have been interrelated in order to consider the implications resulting from any changes that might occur in the market or the farm. Swarm intelligence techniques have been applied in order to determine the harvesting strategy for maximizing the gross margin and eliminating the operational risks faced by aquaculture managers and enterprises. Gross margin calculated in a particular time period is taken as the objective function that has to be optimized. For this, the cost of feed and the cost of stocking is considered. The results obtained after implementing this model show that the optimization technique applied generates a near-to-optimal solution. Similarly, PSO has been applied for determining optimal production techniques for the production of gilthead seabream in floating cages in two different places, the Mediterranean and the Canary Islands (Llorente and Luna, 2014). But, the algorithm was used only for optimizing a single production unit. Difficulty increased with introducing more cages. Moreover, industrial dynamics or the geographical location of aquaculture farms hasn't been analyzed, opening up the path for future research.

A model-driven Decision Support System (DSS) for fish farming in floating cages has been also implemented that uses the PSO algorithm to understand the operational risks as well as to support strategic decisions such as production capacity, feed, harvest weight, and site selection (Cobo et al., 2019). DSS developed is applied to the sea bream production facility in Spain which is able to determine the cultivation strategies that would maximize the present profits of the farmer. However, it is seen that the present model design has a few limitations that need to be addressed in future research. The future work might apply the DSS to other species like catfish and salmon or could integrate DSS with ERP (Enterprise Resource Planning) software. Such works will allow the uncertainty of the aquaculture process to reduce more and more.

Prediction and detection of algal blooms using PSO were employed in ToloHarbour, Hong Kong (Chau, 2005). Precise prediction could help farmers take precautionary measures quickly in enough time. Artificial Neural Networks (ANNs) could be used as they prove to be cost-effective solutions. ANNs, especially back-propagation (BP) algorithms have been used widely in water projects. However, their slow convergence and easy entrapment of local minimum have generated a need for a better method. Hence, PSO has been used to train multi-layer perceptrons to determine and forecast the algal bloom dynamics with different lead times and input variables. As per the results, it was found that the performance of PSO-based perceptron was better than that of BP-based perceptron with respect to speed and accuracy.

In a lot of places, it has been identified that fish farming faces negative effects due to poor management techniques. Due to this, the risk of spreading fish diseases within and between different farms might increase. Hence, a model that integrates both ABMs (Agent-Based Models) and PSO algorithms for farms in Norwegian fjords has been proposed (Alaliyat et al., 2019). ABMs are used to analyze the fish disease dynamics and PSO is used to attain the optimal value of fish farm locations and fish density. The results demonstrated that PSO was able to converge at a rapid pace in only 18 iterations and was able to get the optimal value of fish density three times larger than the original value while the risk of infection stays at the same level. The future work includes this method to be applied to more complex scenarios and more fish farms.

Dissolved oxygen is an important parameter of water quality in aquaculture. It affects the environment of water as well as the growth of all water animals. Hence, predicting the concentration of dissolved oxygen in a pond will help let the farmer know if the oxygen content is lacking. Fuzzy neural networks along with the PSO algorithm have been applied and compared with the BP algorithm (Deng et al., 2006). The model is implemented for a pond at Dalian. About 300 groups of data have been collected over a period of 3 years from which 260 groups were selected as simulating data. The 240 groups of data have been set for training the network model and 20

others for testing. The results showed that the neural network model trained by PSO is more stable, noise insensitive, more precise and faster than that trained by BP. Determining dissolved oxygen content is helpful for river crabs as increased amounts of dissolved oxygen reduce their mortality rates, reduce river crab stress, lead to high reproduction and increase overall quantity and quality. Thus, predicting the oxygen content in water will let the managers take precautionary measures beforehand. A model that integrates least squares support vector regression (LSSVR) and improved particle swarm optimization (IPSO) algorithm has been proposed (Liu et al., 2013a,b). Its results were compared with that of standard support vector regression (SVR) and BP neural network. It was found that the former method had more accuracy and better performance than that of the latter. Another model proposed is the one that combines wavelet analysis (WA) with LSSVR and improved Cauchy particle swarm optimization (CPSO) for predicting dissolved oxygen in crab culture ponds (Liu et al., 2014). This hybrid model was able to increase the overall performance for both linear and non-linear patterns. Cauchy distribution is broader and hence has a better searching ability than the conventional PSO algorithm. Using this model proved to be more adequate than other traditional methods applied for the crab culture dissolved oxygen prediction. BPNN optimized by PSO and coupled with the Kriging interpolation method has also been applied to the crab pond in Gaocheng town in China (Chen et al., 2016). WA has also been used for de-noising. Kriging method is used for three-dimensional predictions, thus achieving more accurate results than the traditional one-dimensional prediction model. However, even this model has its own limitations that need to be solved in the future. Time and breeding regions' limitation needs to be addressed along with factors such as carbon and nitrogen that have not been considered.

Despite all the limitations and challenges faced by various models discussed above, the PSO algorithm has been used widely by researchers to improve the current aquaculture scenario and work on more improved models.

10.8.5 Swarm Robotics

Swarm robotics is an application of swarm intelligence that has taken inspiration from the behavior of social animals to form robotic systems or physically embodied agents that interact among themselves and the environment to generate their collective behavior (Şahin et al., 2008; Brambilla et al., 2013). The three main properties seen in the natural swarms that should be present in the swarm robots are robustness, scalability, and flexibility. Robustness is the ability to cope with any malfunctions or loss of individuals, scalability is the ability to operate with different group sizes, and flexibility is the ability to perform well in different environments and with tasks of different natures. As interactions among the individuals and between individuals and the environment are probabilistic and not straightforward, analysis and modeling of the swarms are required. Swarm robotic systems have been modeled at three different levels: sensor-based modeling, microscopic modeling, and macroscopic modeling (Sahin et al., 2008). In sensor-based modeling, interactions of robot-robot, individual robots, and between robot environments are sensed and actuated. Such kind of modeling is used when the experiments are to be held in simulation and its output has to be in accordance with that of the physical robots. Microscopic modeling is similar to sensor-based modeling, but the modeling is carried out at an individual level. Here, the states of an individual can evolve with time. However, macroscopic modeling is done at the swarm level and these models need to be solved only once to get the required steady state.

A microscopic as well as macroscopic model has been proposed for jellyfish bloom detection using UAVs (Unmanned Aerial Vehicles) (Aznar et al., 2017). It has been assumed that there is GPS positioning in the simulator and the UAVs have local communication mechanisms among nearby agents. MASON simulator has been used for the simulation. It is found that the macroscopic model is able to evaluate the global swarm behavior effectively. However, the proposed model is to be implemented on the real platform in the future in addition to increasing the communication range of the agents.

The biggest problem faced by underwater swarm robotics is the absence of wide bandwidth and reliable communication systems that are used to control their behavior. Mostly, SONAR is used by underwater communications which is a very slow method (Waduge and Joordens, 2017) which leads to time delays and minimal information. Hence, either a control method that could work with out-of-date or minimal information must be proposed, or the need for communication can be eliminated. The control method proposed is based on modern communication systems where each system knows what other systems are up to (Joordens and Jamshidi, 2009). The strategy is to eliminate the need for a central command robot as all the robots will have the same knowledge which will let them take further actions on their own. Another approach is to use wireless battery charging and wireless programming that will cancel the need for an underwater communication system and keep the bio-inspired fish robot waterproof at all times (Waduge and Joordens, 2017). A fish eye sensor is embedded in the robot that tells the robot where the object is. This helps the robot be self-sufficient. A fish eye consists of four IR (Infrared) sensors that help the robot roughly detect the positions of other robots in the swarm and help them make decisions from the gathered data (Elege et al., 2017). The position of the object detected is determined in terms of whether the object is in the front, behind, lower, or higher. Such sensor-based fish eye is useful for fish robots that have low processing power as the eye is a cost-effective and efficient feature. Further, 3D printing can be used for the designing and manufacturing of a simple robotic fish along with the integration of IR sensors (Kiebert and Joordens, 2016) (Figure 10.4).

Area Extended Particle Swarm Optimization (AEPSO), an enhanced version of PSO, has been applied to swarm robotics as AEPSO provides better communication capability, dynamic velocity adjustment, and dynamic neighborhood topology over PSO (Atyabi and Powers, 2010). A simulated environment consisting of 50 survivors, 5 robots, and 50 obstacles with random positioning initially is developed. The robots are required to locate the survivors. After, the number of survivors becomes 15 in a new scenario called population density. Then, there is a dynamic environment where the survivors are randomly moving without any knowledge of the robots' whereabouts. The experimental results show that AEPSO performed better than the basic PSO in various scenarios and also in a more complicated environment that reflected dynamism.

Integration of Virtual Reality (VR) in swarm robotics has also been explored for situations where the testing environments are difficult to set up (Joordens and Jamshidi, 2018). A robotic fish was developed for fish schooling. But, the problem was that the robot cannot be used with real fish unless it behaves like one. Also, a fish robot won't be able to behave like a real fish if it does not know what it is supposed to monitor. Hence, the use of VR can help develop a digital twin of the robotic fish developed, reducing the difficulty and the development time. Doing so will also help us decide what systems can be created physically. However, VR needs to be applied to different fish sizes and types in the future.

FIGURE 10.4 Fish robot body (Waduge and Joordens, 2017).

As swarm robotics is still an emerging technology, facing and solving the issues and challenges faced by the technology right now would be beneficial in the later stage. Like any other environment, swarm robotics also requires certain security services to be met in order to be safe and secure. These services include confidentiality, integrity, authentication, and availability (Higgins et al., 2009). However, swarm robotics like other systems too face challenges in integrating these services. Few challenges faced by the fish robots are the same as those faced by other related technologies like multi-robot systems, MANETs, and software agents (Higgins et al., 2008). Still, there are other security problems that are faced by only the swarm robotic environment. Resource constraints, control, communication, intrusion detection, and swarm mobility are some of these issues. With simple robots comes the need to limit resources that could jeopardize the security of a system. Some of the swarm robots used are autonomous which means that there is no need for a control mechanism to look after them. This might cause a problem when these autonomous systems might go out of control. Any communication, whether implicit or explicit, has a risk of being tampered with by the attacker. Hence, the sensors used in fish robots need to be thoroughly investigated. Also, sometimes a rogue agent might intrude on the swarm location that could be malicious for the entire swarm behavior. This kind of intrusion also needs to be addressed in the future. Providing security in a mobile environment is difficult, as putting any bounds on the movement of the swarm agents could give rise to more security issues. The problems discussed here along with a few others need to be solved in the future as they might have an impact on the practical implications of swarm robotics.

10.9   CHALLENGES AND FUTURE SCOPE

Artificial Intelligence has the potential to revolutionize the fishing industry as it could help reduce the wasted feeds, help make data-driven decisions for optimizing feeding schedules and help in detecting any spread of disease among the aquaculture animals. However, few challenges AI faces to be incorporated fully by the aqua farms. Firstly, farmers are too concerned about their data being used and are quite aware of the data security issues which make them not fully trust or accept the technology. Secondly, proper training and knowledge need to be provided to all farmers to expose them to the new technology and make them comfortable with the idea of using it. Thirdly, even with all data and resources sometimes people wouldn't be able to use them efficiently and effectively, which would make all the data useless. Thus, all these challenges need to be tackled for advancing the use of AI in aquaculture.

Despite all challenges, AI has a huge scope in the aquaculture industry as the demand for fish is increasing rapidly. Hence, the need to manage and improve the production of fish is necessary and has to be adopted by more companies in the future. Drones, robots, sensors, augmented reality, virtual reality and blockchain are some of the technologies that are being researched and trying to be incorporated with the current technologies being used in the farms.

DECLARATION

AUTHORS CONTRIBUTION

All the authors make a substantial contribution to this manuscript. NP, SP, PP, and MS participated in drafting the book chapter. NP, SP, PP, and MS wrote the main chapter. All the authors discussed the results and implications of the book chapter at all stages.

ACKNOWLEDGMENTS

The authors are grateful to the Department of Computer engineering, Nirma University and Department of Chemical Engineering, School of Technology, Pandit Deendayal Petroleum University for the permission to publish this research.

AVAILABILITY OF DATA AND MATERIAL

All relevant data and material are presented in the main paper.

COMPETING INTERESTS

The authors declare that they have no competing interests.

FUNDING

Not Applicable

CONSENT FOR PUBLICATION

Not applicable.

ETHICS APPROVAL AND CONSENT TO PARTICIPATE

Not applicable.

REFERENCES

A digital stereo-video camera system for three-dimensional monitoring of free-swimming Pacific bluefin tuna, Thunnusorientalis, cultured in a net cage.

Aas-Hansen, Ø., Evensen, T., Tennøy, T., Bjørnsen, J., Sæther, B-S., Karlsson-Drangsholt, A., Sivertsen, A., Koren, C., Alfredsen, J. A., Toften, H., Bégout, M. L. & Damsgård, B. (2010). Recent developments in smart-tag fish telemetry for monitoring fish welfare in aquaculture.

Ahir, K., Govani, K., Gajera, R., & Shah, M. (2020). Application on virtual reality for enhanced education learning, military training and sports. *Augmented Human Research*, 5(1). doi: 10.1007/s41133-019-0025-2.

Ahmed, H. & Glasgow, J. (2012). *Swarm Intelligence: Concepts, Models and Applications*. doi: 10.13140/2.1.1320.2568.

Alaliyat, S., Yndestad, H., & Davidsen, P. I. (2019). Optimal fish densities and farm locations in Norwegian fjords: A framework to use a PSO algorithm to optimize an agent-based model to simulate fish disease dynamics. *Aquaculture International*, 27(3), 747–770. doi: 10.1007/s10499-019-00366-6.

Al-Hussaini, K. & Zainol, S. & Ahmad, R. B., & Daud, S. (2018). IoT monitoring and automation data acquisition for recirculating aquaculture system using fog computing. *Journal of Computer Hardware Engineering*, 1, 1–6. 10.63019/jche.v1i2.610.

AlZubi, H. S., Al-Nuaimy, W., Buckley, J., & Young, I. (2016). An intelligent behavior-based fish feeding system. *2016 13th International Multi-Conference on Systems, Signals & Devices (SSD)*. doi: 10.1109/ssd.2016.7473754.

Assefa, A., & Abunna, F. (2018). Maintenance of fish health in aquaculture: Review of epidemiological approaches for prevention and control of infectious disease of fish. *Veterinary Medicine International*, 2018, 1–10. doi: 10.1155/2018/5432497.

Atia, D. M., Fahmy, F. H., Ahmed, N. M., & Dorrah, H. T. (2011). Artificial intelligence techniques based on aquaculture solar thermal water heating system control. *Renewable Energy and Power Quality Journal*, 1027–1034. doi: 10.24084/repqj09.536.

Atyabi, A., & Powers, D. M. (2010). The use of area extended particle swarm optimization (AEPSO) in swarm robotics. *2010 11th International Conference on Control Automation Robotics & Vision*. doi: 10.1109/icarcv.2010.5707854.

Aznar, F., Pujol, M., & Rizo, R. (2017). A swarm behaviour for jellyfish bloom detection. *Ocean Engineering*, 134, 24–34. doi: 10.1016/j.oceaneng.2017.02.009.

Baldridge, J., & Palmer, A. (2009). How well does active learningactually work? *Proceedings of the 2009 Conference on Empirical Methods in Natural Language Processing Volume 1- EMNLP'09*. doi: 10.3115/1699510.1699549.

Beni, G. & Wang, J. (1993) Swarm intelligence in cellular robotic systems. In: Dario, P., Sandini, G., & Aebischer, P. (Eds.) *Robots and Biological Systems: Towards a New Bionics?* NATO ASI Series (Series F: Computer and Systems Sciences), vol. 102. Springer: Berlin, Heidelberg. doi: 10.1007/978-3-642-58069-7_38.

Blackwell, T., Branke, J., & Li, X. (2008). Particle swarms for dynamic optimization problems. *Natural Computing Series*, 193–217. doi: 10.1007/978-3-540-74089-6_6.

Blum, C., & Roli, A. (2003). Metaheuristics in combinatorial optimization. *ACM Computing Surveys*, 35(3), 268–308. doi: 10.1145/937503.937505.

Bostock, J., McAndrew, B., Richards, R., Jauncey, K., Telfer, T., Lorenzen, K., Little, D., Ross, L., Handisyde, N., Gatward, I., & Corner, R. (2010). Aquaculture: Global status and trends. *Philosophical Transactions of the Royal Society B: Biological Sciences*, 365(1554), 2897–2912. doi: 10.1098/rstb.2010.0170.

Brambilla, M., Ferrante, E., Birattari, M., & Dorigo, M. (2013). Swarm robotics: A review from the swarm engineering perspective. *Swarm Intelligence*, 7(1), 1–41. doi: 10.1007/s11721-012-0075-2.

Brown, D. R. (2000). A review of bio-economic models. *Cornell African Food Security and Natural Resource Management (CAFSNRM) Program*, 102.

Chang, B., Zhang, X., et al. (2013). Aquaculture monitoring system based on fuzzy-PID algorithm and intelligent sensor networks. *2013 Cross Strait Quad-Regional Radio Science and Wireless Technology Conference.* doi: 10.1109/csqrwc.2013.6657435.

Chau, K. W. (2005). Algal bloom prediction with particle swarm optimization algorithm. *Computational Intelligence and Security*, 645–650. doi: 10.1007/11596448_95.

Chen, L., Yang, X., Sun, C., Wang, Y., Xu, D., & Zhou, C. (2020). Feed intake prediction model for group fish using the MEA-BP neural network in intensive aquaculture. *Information Processing in Agriculture*, 7(2), 261–271. doi: 10.1016/j.inpa.2019.09.001.

Chen, Y., Xu, J., Yu, H., Zhen, Z., & Li, D. (2016). Three-dimensional short-term prediction model of dissolved oxygen content based on PSO-BPANN algorithm coupled with Kriging interpolation. *Mathematical Problems in Engineering*, 2016, 1–10. doi: 10.1155/2016/6564202.

Cobo, A., Llorente, I., & Luna, L. (2015). Swarm intelligence in optimal management of aquaculture farms. *Handbook of Operations Research in Agriculture and the Agri-Food Industry*, 221–239. doi: 10.1007/978-1-4939-2483-7_10.

Cobo, Á., Llorente, I., Luna, L., & Luna, M. (2019). A decision support system for fish farming using particle swarm optimization. *Computers and Electronics in Agriculture, 161*, 121–130. doi: 10.1016/j.compag.2018.03.036.

Cooke, S. J., Midwood, J. D., Thiem, J. D., Klimley, P., Lucas, M. C., Thorstad, E. B., Eiler, J., Holbrook, C., & Ebner, B. C. (2013). Tracking animals in freshwater with electronic tags: Past, present and future. *Animal Biotelemetry, 1*(1), 5. doi: 10.1186/2050-3385-1-5.

Costa, J., & Rihtar, M. (2016). Data analytics in aquaculture. SIKDD 2016.

Deng, C., Wei, X. J., & Guo, L. X. (2006). Application of neural network based on PSO algorithm in prediction model for dissolved oxygen in fishpond. *2006 6th World Congress on Intelligent Control and Automation.* doi: 10.1109/wcica.2006.1713821.

Deng, C., Gao, Y., Gu, J., Miao, X., & Li, S. (2010). Research on the growth model of aquaculture organisms based on neural network expert system. *2010 Sixth International Conference on Natural Computation.* doi: 10.1109/icnc.2010.5584492.

Deng, J., Kang, B., Tao, L., Rong, H., & Zhang, X. (2013a). Effects of dietary cholesterol on antioxidant capacity, non-specific immune response, and resistance to Aeromonas hydrophila in rainbow trout (Oncorhynchus mykiss) fed soybean meal-based diets. *Fish & Shellfish Immunology, 34*(1), 324–331. doi: 10.1016/j.fsi.2012.11.008.

Deng, C., Wang, W., Gu, J., Cao, X., & Ye, C. (2013b). Research of fish disease diagnosis expert system based on artificial neural networks. *Proceedings of 2013 IEEE International Conference on Service Operations and Logistics, and Informatics.* doi: 10.1109/soli.2013.6611483.

Dewar, H., Deffenbaugh, M., Thurmond, G., Lashkari, K., & Block, B. (1999). Development of an acoustic telemetry tag for monitoring electromyograms in free-swimming fish. *Journal of Experimental Biology, 202*(19), 2693–2699. doi: 10.1242/jeb.202.19.2693.

Diana, J. S. (2009). Aquaculture production and biodiversity conservation. *BioScience, 59*(1), 27–38. doi: 10.1525/bio.2009.59.1.7.

Dorigo, M., Birattari, M., & Stutzle, T. (2006). Ant colony optimization. *IEEE Computational Intelligence Magazine, 1*(4), 28–39. doi: 10.1109/ci-m.2006.248054.

Elege, N., Solapurkar, S., & Joordens, M. (2017). Eye sensor for swarm robotic fish. *2017 12th System of Systems Engineering Conference (SoSE).* doi: 10.1109/sysose.2017.7994956.

El-Gayar, O. F. (1997). The use of information technology in aquaculture management. *Aquaculture Economics & Management, 1*(1–2), 109–128. doi: 10.1080/13657309709380207.

El-Gayar, O. F., & Leung, P. (2000). ADDSS: A tool for regional aquaculture development. *Aquacultural Engineering, 23*(1–3), 181–202. doi: 10.1016/s0144-8609(00)00043-1.

Ernst, D. H., Bolte, J. P., & Nath, S. S. (2000). AquaFarm: Simulation and decision support for aquaculture facility design and management planning. *Aquacultural Engineering, 23*(1–3), 121–179. doi: 10.1016/s0144-8609(00)00045-5.

Fan, S. S., & Jen, C. (2019). An enhanced partial search to particle swarm optimization for unconstrained optimization. *Mathematics, 7*(4), 357. doi: 10.3390/math7040357.

Føre, M., Alfredsen, J. A., & Gronningsater, A. (2011). Development of two telemetry-based systems for monitoring the feeding behaviour of Atlantic salmon (Salmo salar L.) in aquaculture sea-cages. *Computers and Electronics in Agriculture, 76*(2), 240–251. doi: 10.1016/j.compag.2011.02.003.

Føre, M., Frank, K., Norton, T., Svendsen, E., Alfredsen, J. A., Dempster, T., Eguiraun, H., Watson, W., Stahl, A., Sunde, L. M., Schellewald, C., Skøien, K. R., Alver, M. O., & Berckmans, D. (2018). Precision fish farming: A new framework to improve production in aquaculture. *Biosystems Engineering, 173*, 176–193. doi: 10.1016/j.biosystemseng.2017.10.014.

Gupta, A., Dengre, V., Kheruwala, H. A., & Shah, M. (2020). Comprehensive review of text-mining applications in finance. *Financial Innovation, 6*(39). doi: 10.1186/s40854-020-00205-1.

Gustilo, R. C., & Dadios, E. (2011). Optimal control of prawn aquaculture water quality index using artificial neural networks. *2011 IEEE 5th International Conference on Cybernetics and Intelligent Systems (CIS)*, 266–271. doi: 10.1109/iccis.2011.6070339.

Gutiérrez-Estrada, J., Sanz, E. D., López-Luque, R., & Pulido-Calvo, I. (2005). SEDPA, an expert system for disease diagnosis in eel rearing systems. *Aquacultural Engineering, 33*(2), 110–125. doi: 10.1016/j.aquaeng.2004.12.003.

Halide, H., Stigebrandt, A., Rehbein, M., & McKinnon, A. (2009). Developing a decision support system for sustainable cage aquaculture. *Environmental Modelling & Software, 24*(6), 694–702. doi: 10.1016/j.envsoft.2008.10.013.

Hanafiah, N., Sugiarto, K., Ardy, Y., Prathama, R., & Suhartono, D. (2015). Expert system for diagnosis of discus fish disease using fuzzy logic approach. *2015 IEEE International Conference on Computer and Communications (ICCC)*. doi: 10.1109/compcomm.2015.7387540.

Higgins, F., Allan T., & Keith M. M. (2008). Security challenges for Swarm robotics. Technical Report. University of London, Royal Holloway, Department of Mathematics.

Higgins, F., Allan, T., & Keith, M. M. (2009). Threats to the swarm: Security considerations for swarm robotics. *International Journal on Advances in Security, 2*(2&3), 288–297.

Jan, R.-Q., Shao, Y.-T., Fan, Y., Tu, Y.-Y., Tsai, H.-S., & Shao, K.-T. (2007). An underwater camera system for Real-Time coral reef fish monitoring. *The Raffles Bulletin of Zoology, 14*. 273–279.

Jani, K., Chaudhuri, M., Patel, H., & Shah, M. (2020). Machine learning in films: An approach towards automation in film censoring. *Journal of Data, Information and Management, 2*(1), 55–64. doi: 10.1007/s42488-019-00016-9.

Jiang, S., Jiang, J., Zheng, C., Liang, Y., & Tan, L. (2019). An improved PSO algorithm with migration behavior and asynchronous varying acceleration coefficient. *Intelligent Computing Methodologies*, 651–659. doi: 10.1007/978-3-030-26766-7_59.

Joordens, M. A., & Jamshidi, M. (2009). Underwater swarm robotics consensus control. *2009 IEEE International Conference on Systems, Man and Cybernetics*. doi: 10.1109/icsmc.2009.5346165.

Joordens, M., & Jamshidi, M. (2018). On the development of robot fish swarms in virtual reality with digital twins. *2018 13th Annual Conference on System of Systems Engineering (SoSE)*. doi: 10.1109/sysose.2018.8428748.

Joy, M. K., & Death, R. G. (2004). Predictive modelling and spatial mapping of freshwater fish and decapod assemblages using GIS and neural networks. *Freshwater Biology, 49*(8), 1036–1052. doi: 10.1111/j.1365-2427.2004.01248.x.

Kakkad, V., Patel, M., & Shah, M. (2020). Biometric authentication and image encryption for image security in cloud framework. *Multiscale and Multidisciplinary Modeling, Experiments and Design, 2*(4), 233–248. doi: 10.1007/s41939-019-00049-y.

Kennedy, J., & Eberhart, R. (1995). Particle swarm optimization. *Proceedings of ICNN'95- International Conference on Neural Networks*, vol. 4. doi: 10.1109/icnn.1995.488968.

Khotimah, W. N. (2015). Aquaculture water quality prediction using smooth SVM. *IPTEK Journal of Proceedings Series 1*, 342–345.

Kiebert, L., & Joordens, M. (2016). Autonomous robotic fish for a swarm environment. *2016 11th System of Systems Engineering Conference (SoSE)*. doi: 10.1109/sysose.2016.7542928.

Kundalia, K., Patel, Y., & Shah, M. (2020). Multi-label movie genre detection from a movie poster using knowledge transfer learning. *Augmented Human Research, 5*(1). doi: 10.1007/s41133-019-0029-y.

Lee, P. G. (2000). Process control and artificial intelligence software for aquaculture. *Aquacultural Engineering, 23*(1–3), 13–36. doi: 10.1016/s0144-8609(00)00044-3.

Li, P., & Gatlin, D. M. (2006). Nucleotide nutrition in fish: Current knowledge and future applications. *Aquaculture, 251*(2–4), 141–152. doi: 10.1016/j.aquaculture.2005.01.009.

Li, H. L., et al. (1997). Constructions having improved penetration resistance. U.S. Patent No. 5,591,933. 7 January 1997.

Li, D., Fu, Z., & Duan, Y. (2002). Fish-expert: A web-based expert system for fish disease diagnosis. *Expert Systems with Applications, 23*(3), 311–320. doi: 10.1016/s0957-4174(02)00050-7.

Liu, S., Tai, H., Ding, Q., Li, D., Xu, L., & Wei, Y. (2013a). A hybrid approach of support vector regression with genetic algorithm optimization for aquaculture water quality prediction. *Mathematical and Computer Modelling, 58*(3–4), 458–465. doi: 10.1016/j.mcm.2011.11.021.

Liu, S., Xu, L., Li, D., Li, Q., Jiang, Y., Tai, H., & Zeng, L. (2013b). Prediction of dissolved oxygen content in river crab culture based on least squares support vector regression optimized by improved particle swarm optimization. *Computers and Electronics in Agriculture, 95*, 82–91. doi: 10.1016/j.compag.2013.03.009.

Liu, S., Xu, L., Jiang, Y., Li, D., Chen, Y., & Li, Z. (2014). A hybrid WA–CPSO-LSSVR model for dissolved oxygen content prediction in crab culture. *Engineering Applications of Artificial Intelligence, 29*, 114–124. doi: 10.1016/j.engappai.2013.09.019.

Liu, T., Chen, J., Rong, Y., Zheng, Y., & Tan, L. (2019). An improved PSO algorithm with an area-oriented learning strategy. *Intelligent Computing Methodologies*, 640–650. doi: 10.1007/978-3-030-26766-7_58.

Llorente, I., & Luna, L. (2014). Economic optimisation in seabream (Sparus aurata) aquaculture production using a particle swarm optimisation algorithm. *Aquaculture International, 22*(6), 1837–1849. doi: 10.1007/s10499-014-9786-2.

Maravelias, C., Haralabous, J., & Papaconstantinou, C. (2003). Predicting demersal fish species distributions in the Mediterranean Sea using artificial neural networks. *Marine Ecology Progress Series, 255*, 249–258. doi: 10.3354/meps255249.

Mardel, S., Pascoe, S., & Herrero, I. (2004). Management objective importance in fisheries: An evaluation using the analytic hierarchy process (AHP). *Environmental Management, 33*(1), 1–11. doi: 10.1007/s00267-003-3070-y.

Mathisen, B. M., Haro, P., Hanssen, B., Björk, S., & Walderhaug, S. (2016). Decision support systems in fisheries and aquaculture: A systematic review. arXiv preprint arXiv:1611.08374.

Merritt, D., Donovan, M., Kelley, C., Waterhouse, L., Parke, M., Wong, K., & Drazen, J. (2011). BotCam: A baited camera system for nonextractive monitoring of bottomfish species. *Fishery Bulletin. 109*. 56–67.

Misimi, E., Erikson, U., & Skavhaug, A. (2008). Quality grading of Atlantic salmon (Salmo salar) by computer vision. *Journal of Food Science, 73*(5), E211–E217. doi: 10.1111/j.1750-3841.2008.00779.x.

Moore, A., Potter, E. C. E., & Elson, J. (1997). Fish Tracking Technology Development.

Mustafa, F. H. (2016). A review of smart fish farming systems. *Journal of Aquaculture Engineering and Fisheries Research*, 193–200. doi: 10.3153/jaefr16021.

Naik, B., Mehta, A., & Shah, M. (2020). Denouements of machine learning and multimodal diagnostic classification of Alzheimer's disease. *Visual Computing for Industry, Biomedicine, and Art, 3*(1). doi: 10.1186/s42492-020-00062-w.

Naylor, R., Goldburg, R., Primavera, J., Kautsky, N., Beveridge, M., Clay, J., Folke, C., Lubchenco, J., Mooney, H., & Troell, M. (2000). Effect of aquaculture on World Fish supplies. *Nature, 405*. 1017–1024. doi: 10.1038/35016500.

Pandya, R., Nadiadwala, S., Shah, R., & Shah, M. (2020). Buildout of methodology for meticulous diagnosis of K-complex in EEG for aiding the detection of Alzheimer's by artificial intelligence. *Augmented Human Research, 5*(1). doi: 10.1007/s41133-019-0021-6.

Papadakis, V. M., Papadakis, I. E., Lamprianidou, F., Glaropoulos, A., & Kentouri, M. (2012). A computer-vision system and methodology for the analysis of fish behavior. *Aquacultural Engineering, 46*, 53–59. doi: 10.1016/j.aquaeng.2011.11.002.

Parekh, P., Patel, S., Patel, N., & Shah, M. (2020). Systematic review and meta-analysis of augmented reality in medicine, retail, and games. *Visual Computing for Industry, Biomedicine, and Art, 3*(1). doi: 10.1186/s42492-020-00057-7.

Patel, H., Prajapati, D., Mahida, D., & Shah, M. (2020). Transforming petroleum downstream sector through big data: A holistic review. *Journal of Petroleum Exploration and Production Technology, 10*(6), 2601–2611. doi: 10.1007/s13202-020-00889-2.

Pathan, M., Patel, N., Yagnik, H., & Shah, M. (2020). Artificial cognition for applications in smart agriculture: A comprehensive review. *Artificial Intelligence in Agriculture, 4*, 81–95. doi: 10.1016/j.aiia.2020.06.001.

Pinkiewicz, T., Purser, G., & Williams, R. (2011). A computer vision system to analyse the swimming behaviour of farmed fish in commercial aquaculture facilities: A case study using cage-held Atlantic salmon. *Aquacultural Engineering, 45*(1), 20–27. doi: 10.1016/j.aquaeng.2011.05.002.

Prellezo, R., Accadia, P., Andersen, J. L., Andersen, B. S., Buisman, E., Little, A., Nielsen, J. R., Poos, J. J., Powell, J., & Röckmann, C. (2012). A review of EU bio-economic models for fisheries: The value of a diversity of models. *Marine Policy, 36*(2), 423–431. doi: 10.1016/j.marpol.2011.08.003.

Qi, L., Zhang, J., Xu, M., Fu, Z., Chen, W., & Zhang, X. (2011). Developing WSN-based traceability system for recirculation aquaculture. *Mathematical and Computer Modelling, 53*(11–12), 2162–2172. doi: 10.1016/j.mcm.2010.08.023.

Rana, D., & Rani, S. (2015). Fuzzy logic based control system for fresh water aquaculture: A MATLAB based simulation approach. *Serbian Journal of Electrical Engineering, 12*(2), 171–182. doi: 10.2298/sjee1502171r.

Rao, M., Pankyamma, V., & Debbarma, J. (2019). Aquaculture Asia Rao et al. 2017. XXI, 24–26.

Recknagel, F. (2001). Applications of machine learning to ecological modelling. *Ecological Modelling, 146*(1–3), 303–310. doi: 10.1016/s0304-3800(01)00316-7.

Rocher, J., Taha, M., Parra, L., & Lloret, J. (2017). Design and deployment of a WSN for water turbidity monitoring in fish farms. *2017 10th IFIP Wireless and Mobile Networking Conference (WMNC)*. doi: 10.1109/wmnc.2017.8248854.

Şahin, E., Girgin, S., Bayindir, L., & Turgut, A. E. (2008). Swarm robotics. *Natural Computing Series*, 87–100. doi: 10.1007/978-3-540-74089-6_3.

Saikia, S. K., Majumder, S., Nandi, S., & Saha, S. K. (2013). Feeding ecology of the freshwater fish rohuLabeo rohita (Hamilton 1822): A case of intelligent feeding in the periphyton-based environment. *Zoology and Ecology, 23*(4), 266–274. doi: 10.1080/21658005.2013.859849.

Salim, T. I., Haiyunnisa, T., & Alam, H. S. (2016). Design and implementation of water quality monitoring for eel fish aquaculture. *2016 International Symposium on Electronics and Smart Devices (ISESD)*, 208–213. doi: 10.1109/isesd.2016.7886720.

Sazonova, L., Osipov, G., & Godovnikov, M. (1999). Intelligent system for fish stock prediction and allowable catch evaluation. *Environmental Modelling & Software, 14*(5), 391–399. doi: 10.1016/s1364-8152(98)00100-5.

Shah, D., Dixit, R., Shah, A., Shah, P., & Shah, M. (2020). A comprehensive analysis regarding several breakthroughs based on computer intelligence targeting various syndromes. *Augmented Human Research, 5*(1). doi: 10.1007/s41133-020-00033-z.

Shi, Y., & Eberhart, R. (1998). A modified particle swarm optimizer. *1998 IEEE International Conference on Evolutionary Computation Proceedings. IEEE World Congress on Computational Intelligence (Cat. No.98TH8360)*. doi: 10.1109/icec.1998.699146.

Simbeye, D. S., & Yang, S. F. (2014). Water quality monitoring and control for aquaculture based on wireless sensor networks. *Journal of Networks, 9*(4). doi: 10.4304/jnw.9.4.840-849.

Simões, F. D., Moreira, A. B., Bisinoti, M. C., Gimenez, S. M., & Yabe, M. J. (2008). Water quality index as a simple indicator of aquaculture effects on aquatic bodies. *Ecological Indicators, 8*(5), 476–484. doi: 10.1016/j.ecolind.2007.05.002.

Skøien, K. R., Alver, M. O., & Alfredsen, J. A. (2014). A computer vision approach for detection and quantification of feed particles in marine fish farms. *2014 IEEE International Conference on Image Processing (ICIP)*, 1648–1652. doi: 10.1109/icip.2014.7025330.

Soto-Zarazúa, G. M., Rico-García, E., Ocampo, R., Guevara-González, R. G., & Herrera-Ruiz, G. (2010). Fuzzy-logic-based feeder system for intensive tilapia production (Oreochromis niloticus). *Aquaculture International, 18*(3), 379–391. doi: 10.1007/s10499-009-9251-9.

Stigebrandt, A., Aure, J., Ervik, A., & Hansen, P. K. (2004). Regulating the local environmental impact of intensive marine fish farming. *Aquaculture, 234*(1–4), 239–261. doi: 10.1016/j.aquaculture.2003.11.029.

Sukhadia, A., Upadhyay, K., Gundeti, M., Shah, S., & Shah, M. (2020). Optimization of smart traffic governance system using artificial intelligence. *Augmented Human Research, 5*(1). doi: 10.1007/s41133-020-00035-x.

Sun, M., & Li, D. (2016). Aquatic animal disease diagnosis system based on Android. *Computer and Computing Technologies in Agriculture, IX*, 115–124. doi: 10.1007/978-3-319-48357-3_12.

Suryanarayana, I., Braibanti, A., Sambasiva Rao, R., Ramam, V. A., Sudarsan, D., & Nageswara Rao, G. (2008). Neural networks in fisheries research. *Fisheries Research, 92*(2–3), 115–139. doi: 10.1016/j.fishres.2008.01.012.

Talaviya, T., Shah, D., Patel, N., Yagnik, H., & Shah, M. (2020). Implementation of artificial intelligence in agriculture for optimisation of irrigation and application of pesticides and herbicides. *Artificial Intelligence in Agriculture, 4*, 58–73. doi: 10.1016/j.aiia.2020.04.002.

Tidemann, A., Bjørnson, F. O., & Aamodt, A. (2012). Operational support in fish farming through case-based reasoning. *Advanced Research in Applied Artificial Intelligence*, 104–113. doi: 10.1007/978-3-642-31087-4_12.

Torisawa, S., Kadota, M., Komeyama, K., Suzuki, K., & Takagi, T. (2011). A digital stereo-video camera system for three-dimensional monitoring of free-swimming Pacific bluefin tuna, Thunnus orientalis, cultured in a net cage. *Aquatic Living Resources*, *24*(2), 107–112. doi: 10.1051/alr/2011133.

Vafaie, F., Hadipour, A., & Hadipour, V. (2015). Gis-based fuzzy multi-criteria decision making model for coastal aquaculture site selection. *Environmental Engineering and Management Journal*, *14*(10), 2415–2425. doi: 10.30638/eemj.2015.258.

Waduge, T. J., & Joordens, M. (2017). Fish robotic research platform for swarms. *2017 25th International Conference on Systems Engineering (ICSEng)*. doi: 10.1109/icseng.2017.22.

Wahjuni, S., Maarik, A., & Budiardi, T. (2016). The fuzzy inference system for intelligent water quality monitoring system to optimize eel fish farming. *2016 International Symposium on Electronics and Smart Devices (ISESD)*. doi: 10.1109/isesd.2016.7886712.

Wang, Z., Wang, Q., & Hao, X. (2009). The design of the remote water quality monitoring system based on WSN. *2009 5th International Conference on Wireless Communications, Networking and Mobile Computing*. doi: 10.1109/wicom.2009.5303974.

Yan, X., He, F., Chen, Y., & Yuan, Z. (2015). An efficient improved particle swarm optimization based on prey behavior of fish schooling. *Journal of Advanced Mechanical Design, Systems, and Manufacturing*, *9*(4). doi: 10.1299/jamdsm.2015jamdsm0048.

Yang, X. (2011). Metaheuristic optimization. *Scholarpedia*, *6*(8), 11472. doi: 10.4249/scholarpedia.11472.

Zeldis, D., & Prescott, S. (2000). Fish disease diagnosis program: Problems and some solutions. *Aquacultural Engineering*, *23*(1–3), 3–11. doi: 10.1016/s0144-8609(00)00047-9.

Zhang, B., Song, X., Zhang, Y., Han, D., Tang, C., Yu, Y., & Ma, Y. (2012). Hydrochemical characteristics and water quality assessment of surface water and groundwater in Songnen plain, Northeast China. *Water Research*, *46*(8), 2737–2748. doi: 10.1016/j.watres.2012.02.033.

Zhou, C., Lin, K., Xu, D., Chen, L., Guo, Q., Sun, C., & Yang, X. (2018). Near infrared computer vision and neuro-fuzzy model-based feeding decision system for fish in aquaculture. *Computers and Electronics in Agriculture*, *146*, 114–124. doi: 10.1016/j.compag.2018.02.006.

Zion, B. (2012). The use of computer vision technologies in aquaculture: A review. *Computers and Electronics in Agriculture*, *88*, 125–132. doi: 10.1016/j.compag.2012.07.010.

Zion, B., Alchanatis, V., Ostrovsky, V., Barki, A., & Karplus, I. (2008). Classification of guppies' (Poecilia reticulata) gender by computer vision. *Aquacultural Engineering*, *38*(2), 97–104. doi: 10.1016/j.aquaeng.2008.01.002.

11 A Computational Approach for Prediction and Modelling of Agricultural Crop Using Artificial Intelligence

Het K. Patel
Vellore Institute of Technology

Manan Shah
Pandit Deendayal Petroleum University

CONTENTS

DOI: 10.1201/9781003268468-11

11.1   INTRODUCTION

In the 19th century, the idea of revolution from machine to intelligent machines was deployed to eradicate human labour by means of computational intelligence (Shah et al., 2020; Patel et al., 2020b; Ahir et al., 2020). In artificial Intelligence based agriculture, there is a quick accommodation in various farming techniques. The advanced intelligent farming solution, contends the farmer to do more efficient work in less time, intensifying the quality, and securing a quick GTM (go-to-market strategy) strategy for crops (Dharmaraj and Vijayanand, 2018). There are several ways for predicting and modelling crops based on various intelligent techniques. In the last twenty years, fuzzy sets have procured a demanding research interest, and its increasing applicability to help the farmer, make the correct decision for their cultivated crops (Kakkad et al., 2019; Pandya et al., 2020; Sukhadia et al., 2020). The most useful capability of fuzzy sets is to represent and manage uncertainty, and also ensure that incomplete information is valued and provide a solution for crucial agricultural issues. The complex process of yield production is solved by the methods of Fuzzy cognitive maps (FCM) (Jani et al., 2020; Parekh et al., 2020). FCM is a mixture of cognitive map theories and fuzzy logic, which is useful for representing an expert's knowledge. This developed FCM model comprises nodes mapped with directed edges. For example, cotton is considered the best outcome of the FCM model, hence the organic matter, pH, Ca, K, Na, and cotton yield will act as the nodes and the weighted relationship between the cotton yield and soil properties will be illustrated by the directed edges (Papageorgiou et al., 2011).

In agriculture, the fuzzy expert system provides advanced expert knowledge to farmers, and hence they can take the appropriate right decision. To improve the expert system in agriculture, the crop protection expert system plays a vital role. The arising uncertainty during crop plantation, crop disease management, crop production marketing, etc. is constituted from the reality that many decision-enhanced activities are imprecise and vague. To eliminate the vagueness, fuzzy logic plays a vital role and allows the expert system to efficiently minimize the uncertainty (Figure 11.1).

The prescience knowledge of problems of the crop yield will enhance the subsequent profit and intensifies the crop yield. The most effective tool for the prediction and modelling of crop yields is Artificial Neural Network (ANN). The ANN uses the feedforward and back propagation algorithm to train the model that predicts the crop based on certain parameters like rainfall, soil moisture, temperature, pH, nitrogen, potassium, phosphate, etc. To plant a specific crop, ANN can be useful to find all the parameters quantity for that specific crop. The accurate prediction of the crop is done with the

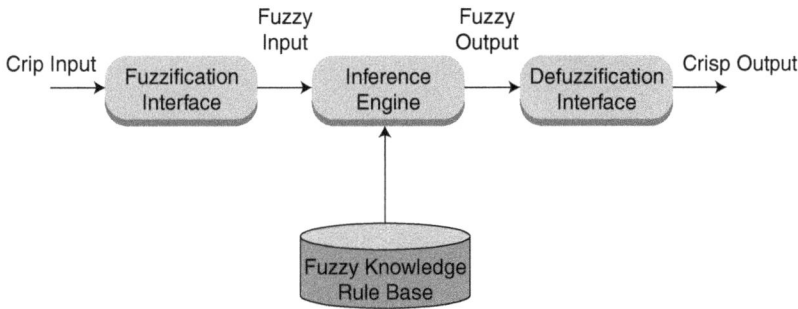

FIGURE 11.1 Fuzzy expert system components.

help of a clustering algorithm called Self Organizing Map (SOM) followed by the multilayer neural network to find the best combinations of the crop for the specific soil (Ogwueleka, 2016).

To increase crop production now and in the future, constant attempts are needed to sustain soil fertility. The genetic algorithm, an NP-hard problem, influences the agricultural crop yield, maintaining soil fertility (Olakulehin and Elijah, 2014) (Figure 11.2).

The crop-planning issues can be solved using different approaches. The first approach is the ε-constrained method, in which any one objective function is selected and optimized while all other objective functions, by applying upper bounds are transformed to some constraints. The second approach is the Weighted Sum Method (WSM), in which all the objective functions are interlinked with the weighting coefficient and finally it tries to minimize/maximize the weighted sum.

Annual Crop Planning (ACP) is an optimization problem in which the various crops which are to be grown are allocated for the appropriate agricultural land. There are many swarm intelligence techniques that provide the solution to these ACP problems. The main metaheuristics of swarm intelligence are (1) Cuckoo Search (CS); (2) Glow worm swarm optimization (GSO); (3) Firefly Algorithm (FA).

The crop prediction is more accurate when Artificial Bee Colony (ABC), a hybridization technique along with a fuzzy clustering algorithm is implemented. Initially, fuzzy clustering is done and then with the use of an artificial bee colony algorithm the best suitable seasonal crop is selected. The efficiency of this method is higher than the clustering algorithms like k-means, k-median, k-mode, etc.

11.2 FUZZY SET IN AGRICULTURE

Modelling, Predicting and management of agricultural crops are composite conceptual process, where there are large number of input variables which are taken into consideration and interrelated for decision making and system analysis (Patel et al., 2020a; Kundalia et al., 2020). Almost every process in agricultural sector include ambiguity, uncertainty, and incomplete information. Fuzzy sets came into existence to solve this problem of uncertainty, also assures that all the incomplete information is considered which is crucial to take decisions in agricultural sectors (Pathan et al., 2020; Talaviya et al., 2020).

This present chapter aims to a refinement of review studies by applying Fuzzy sets (FS), Fuzzy logic (FL), Fuzzy expert system (FES), Fuzzy Inference System (FIS), and Fuzzy cognitive mapping (FCM) to the exploration of modelling, predicting and management of the agricultural crops (Papageorgiou et al., 2011).

11.2.1 FUZZY SETS IN AGRICULTURAL SUSTAINABILITY

The idea of sustainability has procured expanding attention from economists who are continuously working in areas related to agricultural crops and farming land. It is asserted that fuzzy sets

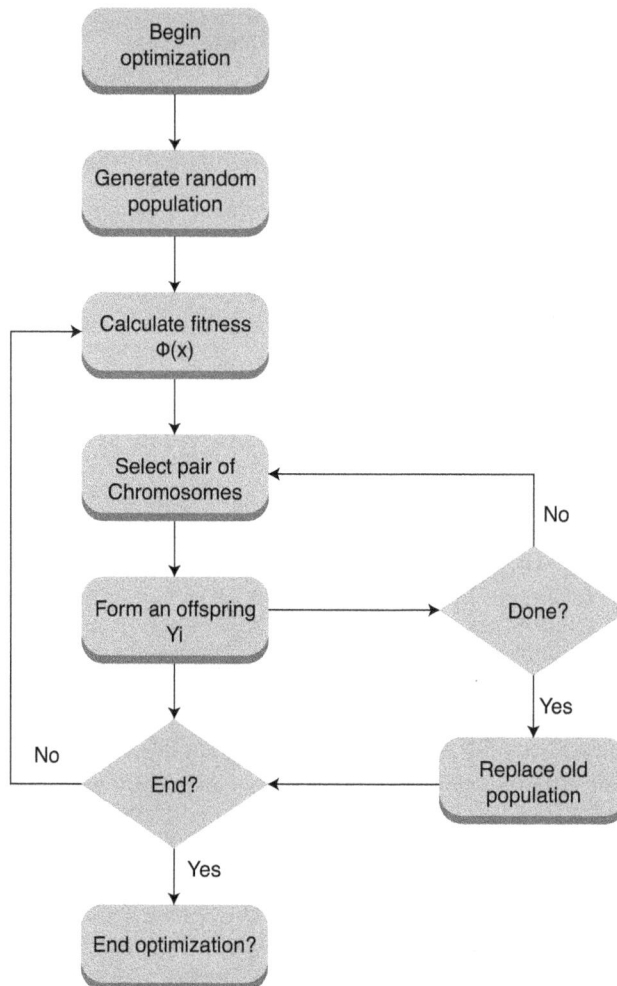

FIGURE 11.2 Genetic algorithm flowchart.

methods provide important leads to overcome three difficulties in sustainable modelling of agricultural crops: (1) by use of the linguistic variables the problem of nonfigurative units could be handled fruitfully; (2) with the use of fuzzy sets the vague or the imprecise information can still be taken into consideration; (3) relationship among the dimensions of sustainability can be included to the model by fuzzy rule base.

11.2.1.1 Sustainability in Agriculture

There has been considerable engrossment in merging the concepts of agricultural sustainability with the agricultural systems (Barrow, 1990). There are three sustainability dimensions: (1) environmental dimension; (2) social dimension; (3) economic dimension. In order to enhance agricultural sustainability, along with the farmers, the system analyst also must subsume the information from all the three sustainability dimensions into their effective decision. The models of agricultural sustainability are in huge demand due to the multidimensional nature of taking decisions, which are assimilated into both private and public decision-making systems.

In the public sphere, to stabilize the contradictory outcome goals of the different parts of society, decision models which are capable to merge all three dimensions of sustainability can only aid in selecting the agricultural land. In the private sector, with the help of decision support systems, the

best sustainable and efficient farming systems from all the available farming options are selected by the farmer.

11.2.1.2 Modelling Sustainability with Fuzzy Sets

There are in general three problems that add difficulty to facilitating actual decision support: (1) different type of scales is used for estimating the information on the dissimilar sustainability dimensions; (2) vague or imprecise information is available to pilot the decision maker; (3) because of the interrelation between variables which are incorporated in the sustainability models, the performance of the model can't be affected by the single course of action. The fuzzy sets and their associated fuzzy methods have ability to rectify the each of three problems (Gandhi et al., 2020; Panchiwala and Shah, 2020; Parekh et al., 2020). The application of fuzzy sets assisting in the modelling of the agricultural sustainability to surmount three challenges is described in the following section.

11.2.1.2.1 Nonfigurative Units

As the decision variables are measured using different scales, the issue of nonfigurative units evokes in many applications where there is a need for decision analysis by means of multiple criteria. Also, all the decision variables measured on variety of scales must be facilitated in the model of sustainability. In the social dimension, decision criteria are measured in the range of 0–20, depicting a qualitative assessment. All the decision criteria, such as cost, profit, etc are measured on a monetary scale in the case of economic dimension. In case of environmental dimensions such as soil loss due to erosion, the agricultural chemicals used in water supply, etc, are measured in parts per billion and tons per acre. Therefore, both quantitative variables (i.e. both physical and monetary units) and qualitative variables should be combined to achieve a high-performance model of sustainability.

The most common approach for modelling multiple criteria decisions would be to normalize all the variables to a similar scale (Yoon and Hwang, 1995). After rescaling, the normalized measurement is used to rank alternatives. Also, the ranking of alternatives depends highly on the selection of normalization technique.

Another approach would be to represent all the decision variables as linguistic variables . In the model for every linguistic variable, the base variable is determined using the normalized scale. Many base variables are measured on a ratio scale (i.e. in physical and monetary units), while if the base variable uses an interval scale, then more qualitative decision variables are modelled. This approach has a major advantage because there isn't any need to apply any random transformation. The measured base variables are in fact most appropriate for the specific problem.

11.2.1.2.2 Imprecise or Vague Information

With the increasing complexity in the models of agricultural sustainability, all the imprecise variables must be included in the decision variables to achieve more precision. To measure the sustainability variables, two fuzzy uncertainties are present: informational fuzziness and intrinsic fuzziness. With the help of fuzzy sets, both imprecise and important information can be fused with the models of agricultural sustainability.

There is a difference between fuzzy uncertainty and stochastic uncertainty. Fuzzy uncertainty represents the interpretation and measurement of events, whereas stochastic uncertainty represents the lack of preceding information for any events, such as rainfall prediction or price prediction at the time of harvest. In fuzzy uncertainty, the result of an event is prior known, while there is ambiguity in characterization of an event.

There are some decision variables that are fundamental fuzzy as they are suitable to the concept of agricultural sustainability. For example, "life quality", "landscape attractiveness", and "farmers dignity" are concepts in which it is understood in qualitative terms rather than quantitative base variables.

Even though fundamental fuzzy linguistic variables are measured qualitatively, they can also be measured with the help of an artificial quantitative base variable. For example, the health of any

person can be evaluated on a scale between 0 and 25. Fuzzy sets represent several labels such as "very serious", "serious", "normal", "good", "fit", for linguistic variables. These variables can be constructed in which every base variable has the membership degree in the fuzzy sets.

The informational fuzziness and uncertainty evoke in sustainability models of agriculture. This uncertainty arises when the decision variables are quantitative instead of qualitative, nonetheless understanding the resulting measurement is a subject of concern. For example, if there is an erosion rate of nine tons per hector per year can also have a non-zero membership degree which categorize into both a normal and high rate of soil loss. In the sustainability model, the decision variables have both the types of intrinsic fuzziness and informational fuzziness. With the help of the linguistic variables and fuzzy sets, the decision variables can be fused with the quantitative models of sustainability.

In addition, a fuzzy rule base describes how to model the interrelationships among the linguistic variables and uses those linguistic variables just to measure agricultural sustainability.

11.2.1.2.3   Interrelationships in Sustainability

The decision variables mapped with three dimensions of sustainability can be linked when a single decision can affect the dimensions of sustainability and the decision variables. In crop production, the choice of farming method has an adverse effect on all three dimensions. It affects the economic dimension by investments and profits, social dimensions by the lifestyle and confidence of the farmer, and the environmental dimension by using agricultural chemicals, and fossil fuel.

Fuzzy rule base pictures a brief idea of the interrelationships among the variables of the sustainability models. With the help of fuzzy rule base, inference engine can be proved useful for natural language processing (Zimmerman, 1987). It is possible to rank the farming system using fuzzy rule base.

11.3   CREATING FUZZY LOGIC MODELS

In 2012, Bosma proposed an approach for the development of fuzzy logic models (FLMs), which sets out tactical decisions where motives and human motorist are the most primal variables. Also, ten steps modelling approach is proposed in which fuzzy rules and membership functions are determined. This method is also used in the hybrid agriculture system. The fuzzy sets (FS) maintain a sequential structure of five subsets:

- Primary production factors
- Product opportunities
- Product options
- Reference frame of farmers'
- the final output layers

This transparent structure allows stakeholders' participation to strengthen the ability of the modeller to empathize with the experts' reasoning.

11.3.1   FLMs and Image Processing Techniques

For the management and planning of irrigation, the utmost significant parameter is the land usage information. Estimating the crop water needs for irrigation, image classification of satellite data proves to be a very efficient technique to create crop area thematic maps (Murmu and Biswas, 2015). Various fuzzy classifiers were examined and concluded that fuzzy logic (FL) uses general rules resulting in less time-consuming rather than any conventional methods. Fuzzy c-means classifier can prove to be a good tool for generating models which are used for multivariate studies by

recognizing appropriate parameters and using the interpretation of satellite data. A fuzzy inference System (FIS) is merged with an image processing method to advance a decision support system to monitor and measure the qualitative grading of pulverized rice. The percentage of broken kernels (PBK), and the degree of milling (DOM) are the quality indices and are divided into five subclasses. Now both the satellite images and the two quality indices were used by the classifier to generate one output variable (Quality). All the indices, images and output variables are in the form of a triangle membership function. Fuzzy classification of remotely perceived data is used in order to estimate parameters of potato crops (Pandey et al., 2013). A FIS is executed to measure the parameters of the potato crop such as plant height, leaf index, soil moisture, and biomass.

11.3.2 Fuzzy Expert System in Agriculture

Expert system in agriculture creates the mainstay for precision agriculture. In order to achieve higher quality and quantity of crop, expert knowledge is required to make an effective choice during sowing, irrigation management, storage, pest management, land preparation, fertilizer management, etc. Fuzzy expert system guides the farmer to take the right decision in terms of crop modelling. Expert system is significantly used for crop protection and pest management. For crop protection and pest/fertilizer management, knowledge of weeds, plant pathology, entomology, nematology, nutritional disorders, and several techniques is required. Uncertainty is very much challenged from the time of sowing, to pest management, or to nutritional disorders, etc. As there are many decision activities in the agricultural sector considered often imprecise or else based on some perception, uncertainty is compounded. This vagueness, imprecision and lack of knowledge can be handled by fuzzy logic. When dealing with ambiguous data or insufficient knowledge, fuzzy logic allows the expert systems to perform optimally. In the Fuzzy expert system, fuzzy logic is used rather than Boolean logic and is focused in the direction of numerical processing.

Lotfi A. Zadeh proposed the theory of fuzzy sets and fuzzy logic (Zadeh, 1965). Soft computing is an effective approach to overcome imprecision, sub-optimality, and thus provides efficient solutions. Fuzzy logic itself is a sub-part of soft computing (Chen and Chen, 1994). Fundamentally in fuzzy sets, all the elements must have a degree of membership in between 0 and 1. Whereas in classical sets, all the elements must have membership degree as either 0 or 1. If the value is zero, then it is outside the set, and if a value is one then it is fully inside the set. The fuzzy logic, rules which are used for reasoning and the membership degrees are all used by the Fuzzy expert System to handle the uncertain information.

As shown in Figure 11.1, fuzzy experts system comprises of fuzzification, followed by inference engine where knowledge base is provided, and finally defuzzification. It uses fuzzy logic for reasoning the data in the inference system. As the inference engine has set of programs, the knowledge base provides submissive data structures. The knowledge base is created by Knowledge engineer, who gathers knowledge from the entire domain expert and the knowledge is converted into production rules which provides knowledge base.

Fuzzification is the process in which the real values are converted into linguistic values having membership function vacillating from 0 to 1. Fuzzy reasoning and fuzzy "IF-THEN" rules are considered as the two pillars of fuzzy expert system and is the most effective approach for modelling of the agricultural crop. With the help of "IF-THEN" rules, the linguistic variables can make a conclusion with respect to all the combinations of conditions. For a fuzzy logic system, the rule base is formed by these collections of if-then rules, and appropriate conclusion is fetched with the relevant inference procedure. Thus, the result obtained in form of one fuzzy subset is assigned to every output variable. Now, for every output variable, all the fuzzy subsets are merged into a single fuzzy subset. Defuzzification is the process in which the set fuzzy output is converted to a crisp output which describes the primal fuzzy set. The inference engine has the tendency to take both the fuzzy or crisp inputs, and output is provided in the form of fuzzy sets .

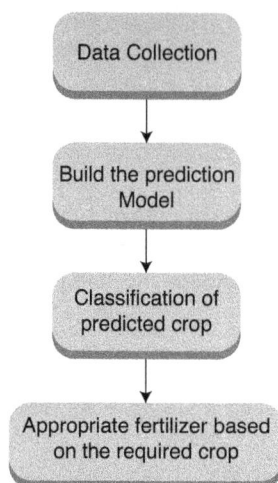

FIGURE 11.3   Model design flow chart.

11.3.3   Neural Network Approach in Agricultural Sector

The most powerful approach for prediction and modelling of crop yield is Artificial Neural Network (ANN). The ANN design is based on the learning function exactly the same as the human brain, and hence it can recognize and predict patterns. The ANN functioning is the same as the human brain; hence it receives the signals, sums up, and is redirected back when the threshold is achieved (Parekh et al., 2020; Shah et al., 2020; Patel et al., 2020a). Farming itself is the mainstay that contributes more than 50% to the economy. The ANN model can be useful for the farmers to take correct decisions by predicting the crop yield based on various parameters including climatographic factors. The local weather conditions and climatographic factors play a vital role in the quality of crop yield.

As illiteracy is the main reason among farmers for lower quality crop, the prediction of crop yield on particular soil and the actual fertilizer needed for that specific crop is estimated by the ANN model. Feed-forward back propagation network is the most common type of ANN, in which the crop performance factors and the climatographic factors are chosen to be the input parameters. Based on the input parameters the model is trained using feed-forward neural network and to minimize error, back propagation algorithm is applied. Self-organizing map is a clustering technique used to categorize the different types of soil based on their geology. To develop this model, the following four steps are required: (1) Data collection; (2) Build/Develop the prediction model; (3) Classification of predicted crop; (4) Appropriate fertilizer based on the required crop (Figure 11.3).

11.4   AGRICULTURE CROP PREDICTION USING ARTIFICIAL NEURAL NETWORK

The use of systems trained by ANN models is increasing rapidly. Agricultural crop prediction is an important aspect in order to get the better quality crop in lesser time. The crop modelling and prediction system are built using the ANN model, which gives better efficiency than regression models. In an Artificial neural network, the model is trained with the help of feed-forward neural network and back-propagation algorithm. The model's architecture and design are demonstrated in the following part.

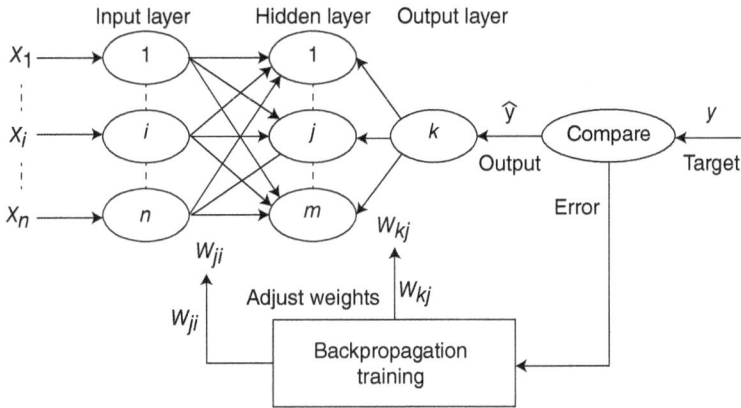

FIGURE 11.4   Feed-forward neural network using back-propagation algorithm.

11.4.1   Feed-Forward Neural Network Using Back Propagation Algorithm

This Feed-forward network architecture using a back propagation algorithm is the most common type of ANN which is used for different tasks. Feed-forward ensures that how this network will process by recalling the patterns and the back propagation algorithm, a supervised training technique, ensures to minimize the error by updating the weight. The difference in weights (i.e. error) is calculated and all the weights need to be adjusted in the backward direction using back-propagation algorithm i.e. from output/target layer to hidden layers, and from hidden layers to the input layer. Initially, random weights are taken based on the input parameters and a bias and learning rate is selected. Training of this network begins with random weights and the error is minimized by adjusting the weights while considering the goal (Figure 11.4).

The gradient-descent method is used for back propagation to secure the mean squared error. The error function is defined as:

$$y_i = \sum_{j=1}^{n} \omega_{ij} x_j \tag{11.1}$$

$$E = (\hat{y} - y)^2 \tag{11.2}$$

Here, E is the square error, $\hat{y}$ is the desired output, where y is the definite output of neuron, n denotes the total number of inputs to the network, w is the weight for every corresponding input parameter.

A node output (Oi) is calculated by applying the activation/transfer function to the weighted values. There are many activation functions like Step function, sign function, sigmoidal function, etc. Commonly, the Sigmoidal function is used as an activation function. The sigmoidal function is represented as

$$\phi(x) = \frac{1}{1 + e^{-ax}} \tag{11.3}$$

Using the chain rule, to update the weight based upon the error, the partial derivative is taken with respect to every weight of the input parameter.

$$\frac{\partial E}{\partial w_i} = \frac{\partial E}{\partial y} \frac{\partial y}{\partial net} \frac{\partial net}{\partial \omega_i} \tag{11.4}$$

TABLE 11.1
Database for the Prediction System

Crop	pH	N	P	K	Depth	Temp	Rainfall
Sugarcane	6.5–7.5	175	100	100	60	20–50	750–1,200
Cotton	7–8.5	100	50	50	30	27–33	700–1,200
Bajra	7–8.5	40	20	25	15	28–32	400–750
Jowar	6.0–8.5	80	40	40	50–20	25–30	800–1,000
Corn	7.5–8.5	100	25	0	20–50	13–30	500–600
Soybeans	6.5–7.5	30	75	15	15–20	25–33	700–1,000
Wheat	5.5–8.5	100	50	50	50–20	22–25	1,000–1,500
Groundnut	6–7.5	25	50	30	20	24–27	500–1250
Rice	6–8.5	100	50	50	15–20	16–22	25–180

The weights of all the input parameters are updated and using back propagation the network is again trained until the error tends to zero (Dahikar and Rode, 2014).

11.4.2 MODEL DESIGN WITH GUI

There are four processing steps to develop these agricultural crops predicting models.

11.4.2.1 Data Collection

For different types of crops, the standard data which is suitable based on the different environmental and climatographic conditions is given in Table 11.1.

11.4.2.2 Build/Develop the Prediction Model

There are three stages to develop this analysing and prediction model. They are as follows:

Step 1: For the finest possible optimum configuration, the various prototypes of ANN are merged together. Many different algorithms such as Delta-bar-delta, Quickprop, Silva and Almeida's Algorithm, etc. are used to train the model. Using Silva and Almeida's algorithm, the model improves the back-propagation algorithm by updating the learning rates of each weight w_i for input parameter x_i. Using the Delta-bar-delta algorithm, on studying the sign of gradient, the learning rates are controlled.

Step 2: The number of hidden layers is carefully observed and decided using the trial and error method.

Step 3: Similarly, certain parameters including bias and the initial weight are initialized randomly (Figure 11.5).

11.5 CLASSIFICATION OF PREDICTED CROP

After training the model using ANN, the prediction of the crop is done. The input parameters are passed to the system to predict which type of crop is more suitable based on the favourable climatographical and environmental factors. With the help of GUI, the prediction of crop from the input parameters is clearly explained in Figure 11.6. The model has been trained with the input parameters. The input parameters have to be entered to test the sample. By clicking on the test button, the predicted crop is found out, which is displayed in Figure 11.6.

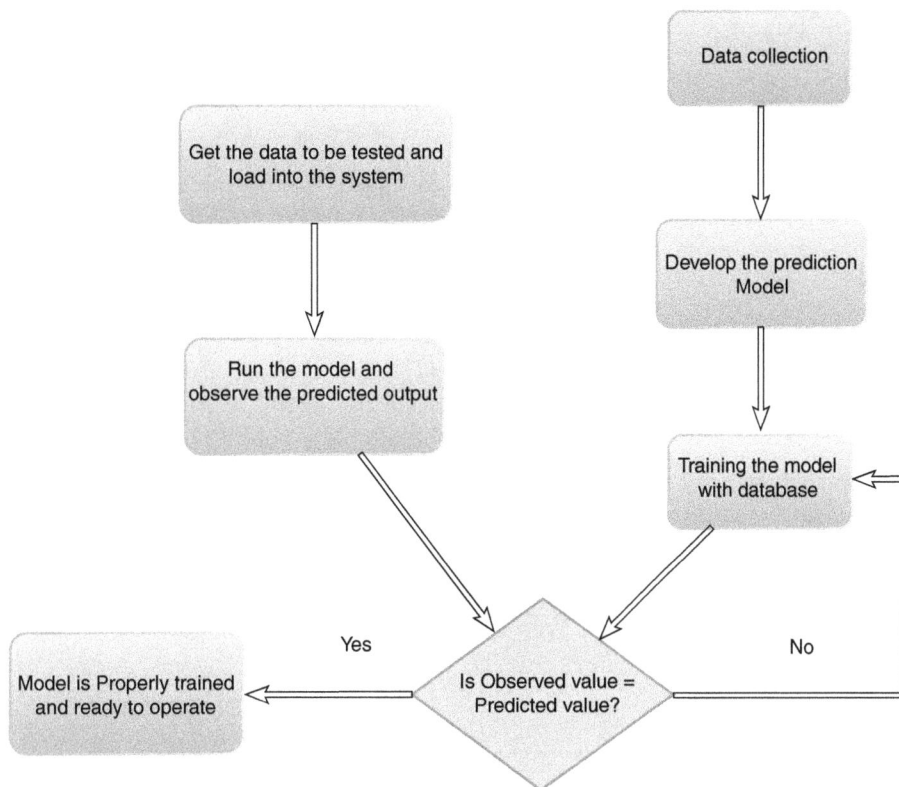

FIGURE 11.5 Flowchart of the model.

TABLE 11.2
Appropriate Fertilizer Based on the Deficit

Sr. No.	Deficit	Advised Fertilizer
1.	Phosphorus	Ammonium Hydrogen Phosphate, Calcium Hydrogen Phosphate or Superphosphate, Ammonium Phosphate.
2.	Potassium	Potassium Nitrate (), Potassium Chloride, Potassium Sulphate.
3.	Nitrogen	Sodium Nitrate, Urea, Ammonium Sulphate.

11.6 APPROPRIATE FERTILIZER BASED ON THE REQUIRED CROP

There are chances of the situation when the farmers don't want to cultivate the best-predicted crop, instead the farmer wants to cultivate another crop. The deficit of fertilizer for the specific crop is also predicted from the model. Farmers will try to eradicate the deficit of that fertilizer by adding the suggested organic and composite fertilizer (Table 11.2).

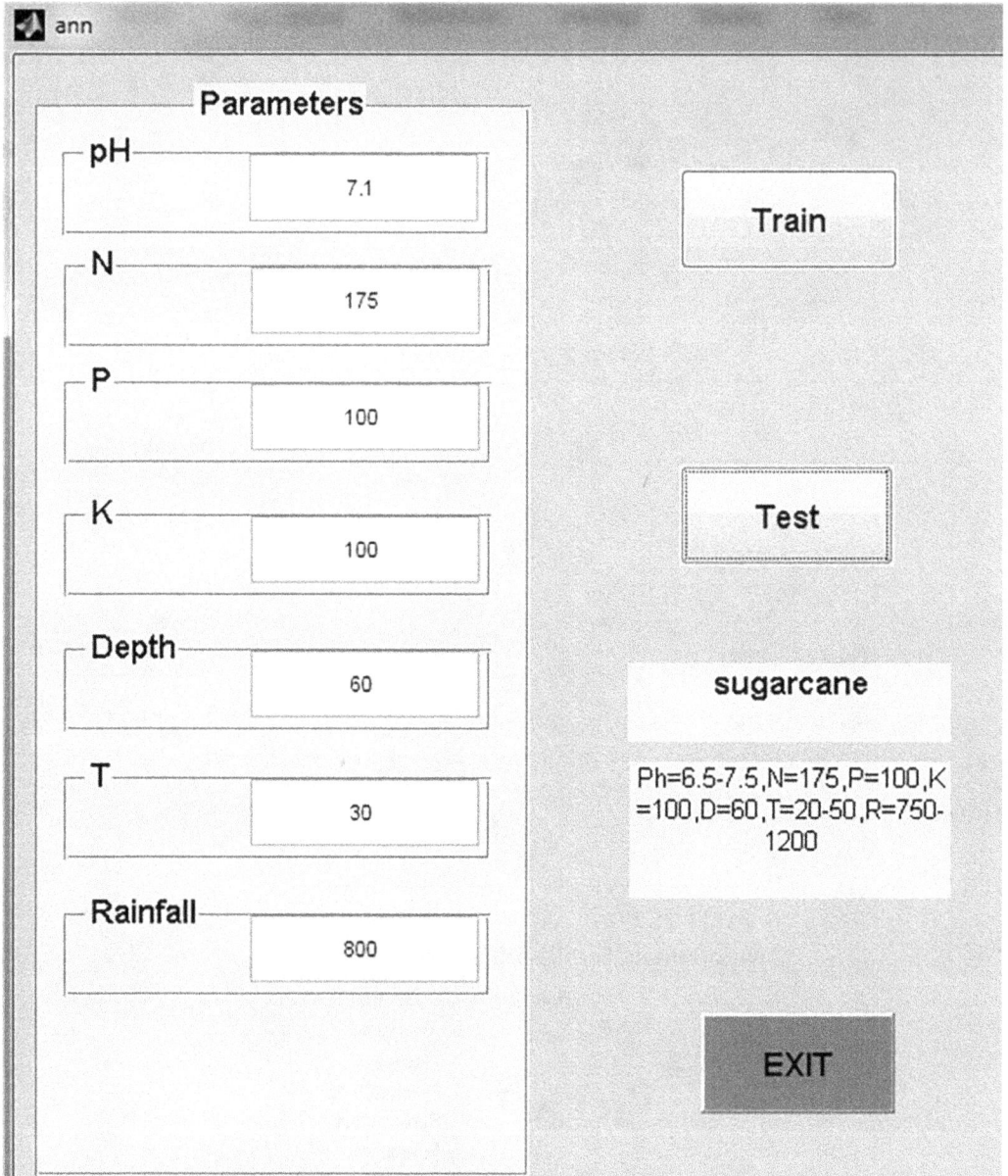

FIGURE 11.6 Prediction model using GUI

Here phosphorous, potassium, and nitrogen are the three most significant minerals used for crop cultivation (Giritharan and Koteeshwari, 2016). With the help of GUI, the appropriate fertilizer needed to grow the sugarcane is elaborated in Figure 11.7.

11.6.1 AGRICULTURAL CROP GROWTH AND SOIL FERTILITY USING SELF-ORGANIZING MAPS AND MULTILAYER FEED-FORWARD NEURAL NETWORK USING BACK PROPAGATION ALGORITHM

In agricultural engineering, the application of artificial neural network is enormously expanding. The biggest challenge is to handle the large database. To model, analyze, and cluster the different

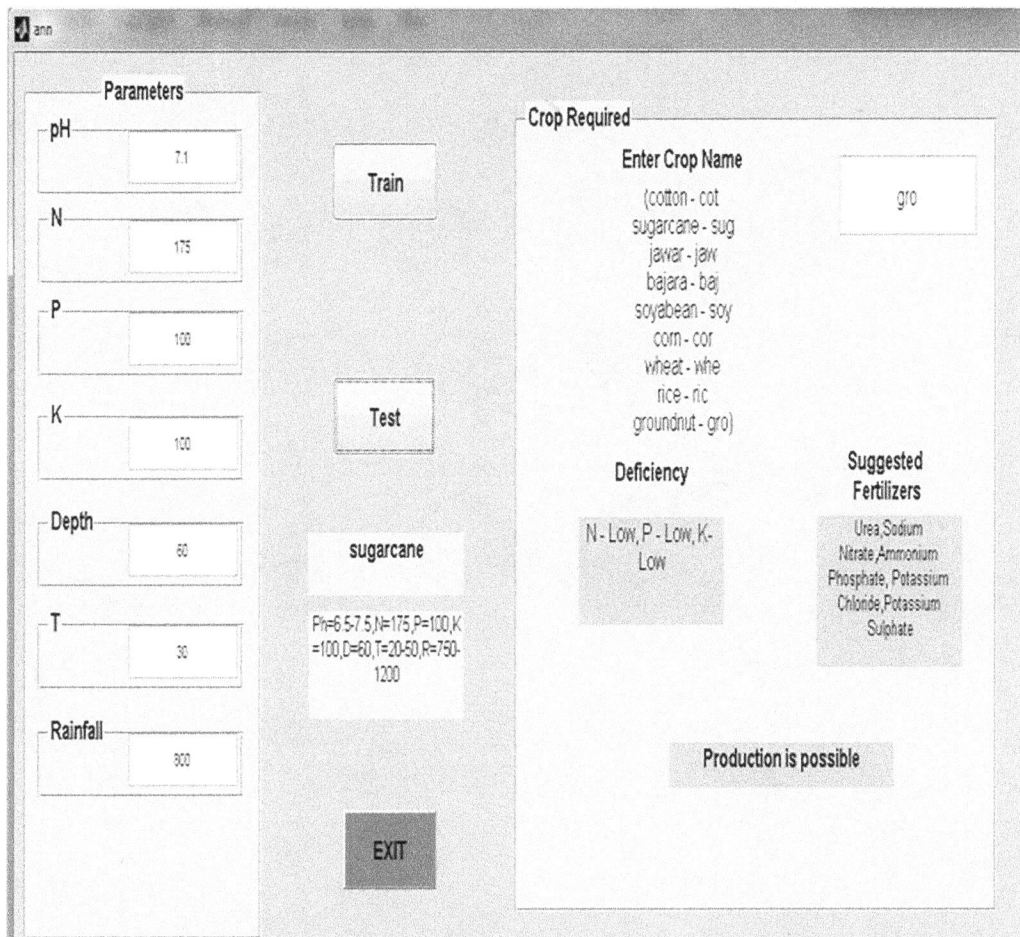

FIGURE 11.7 Appropriate fertilizer needed for specific sugarcane production using GUI.

enormous database, Self-Organizing map (SOM) can play a vital role. Of various ANN models and algorithms, SOM is the most popular clustering and modelling algorithm which is based on unsupervised learning. The inherent features of the data can also be identified using SOM, and hence it is also known as Self-Organizing Feature Map. A low-dimensional representation of the training sample is produced by SOM, called a map (Figure 11.8).

A topology sustaining mapping is delivered by SOM, which converts higher dimensional data to low dimensional representation or map.

Figure 11.9 gives a brief idea of how the architecture of the prediction model works using SOM Clustering and Multi-layer Perceptron Neural Network (MLPNN). The working of SOM comprises five major steps:

1. Initialization: weight vectors are initialized with random values.
2. Sapling: from input space, select an input training sample.
3. Matching: neuron whose weight vector is closest to the input vector is considered as winning neuron and is known as Best Matching Unit (BMU).
4. Updating: weight is updated by $\Delta W_{ji} = \eta(t)T_{j, I(x)}(t)(x_i - w_{ij})$.
5. Continuation: repeat step 2 until there is no change in the feature map.

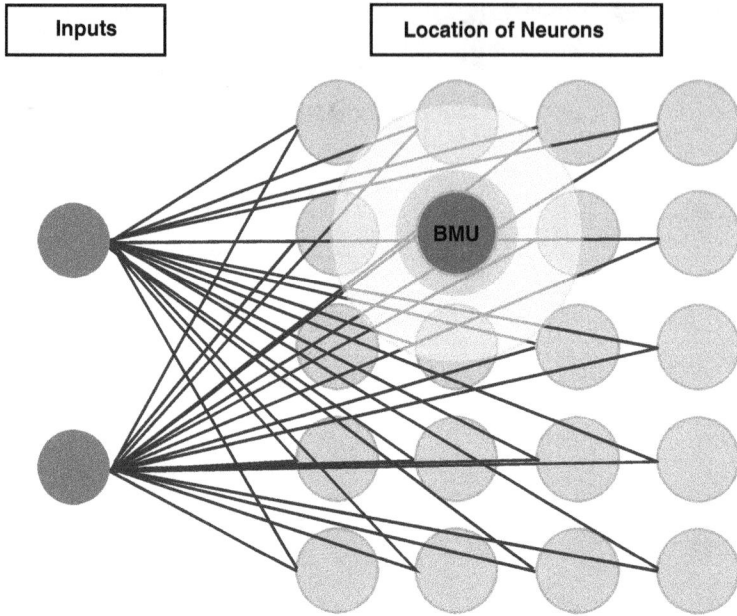

FIGURE 11.8 Self-organizing map.

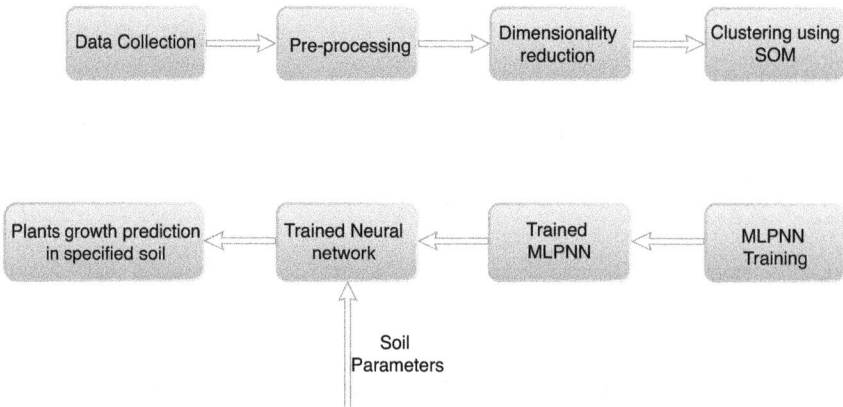

FIGURE 11.9 Prediction model using SOM and MLPNN

Using Best Matching Unit (BMU), the distance between each weight vector and sample vector is measured, by the means of commonly used Euclidean distance. If a node is closer to BMU, then it learns more by altering its weight and if the neighbour is far away from BMU, it learns less (Mokarram et al., 2018).

11.6.1.1   Clustering Using Self-Organizing Map on IRIS Dataset

The simple SOM clustering is applied and the following things are inferred from it.

A colour with higher intensity is applied, if lower average distance and a colour with lesser intensity is applied, if higher average distance (i.e. there is a difference in surrounding weight). The higher or lesser colour intensity of species is shown in Figure 11.10. The part with lesser intensity represents a cluster, while the part with higher intensity represents segregation of clusters.

||||After clustering, the clusters of the species are formed which is illustrated in Figure 11.11.

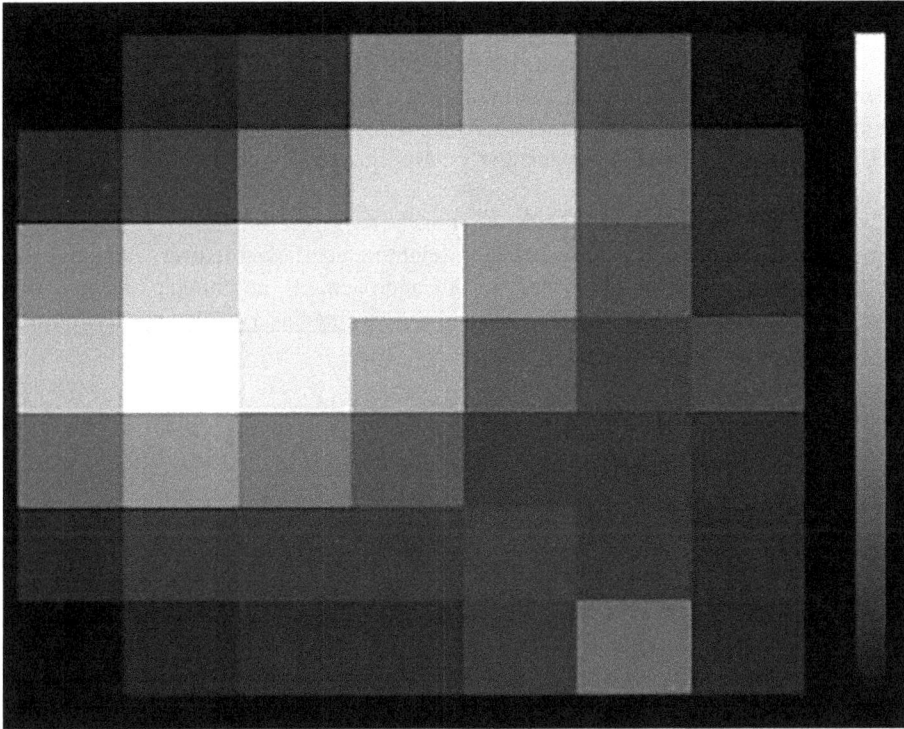

FIGURE 11.10 Cluster division based on density

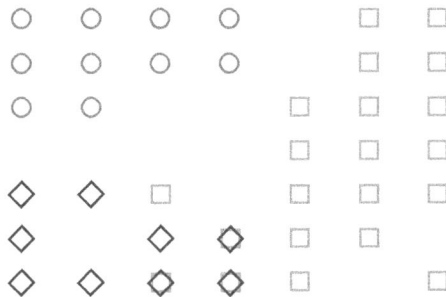

FIGURE 11.11 Three cluster predominant in three different zones (Round → Iris-setosa, Rectangle and star → Iris-virginica, Square→ Iris-versicolour).

11.6.1.2 Limitations of Self-Organizing Maps

The SOM when applied on the categorical data doesn't operates softly, and performs even worst in case of the mixed-data (i.e. numerical, categorical, etc). The efficiency of the model is decreased if the data is evolving slowly.

11.7 EVOLUTIONARY COMPUTING IN AGRICULTURE

The Evolutionary Algorithms are used to solve the NP-Hard problems which cannot be solved in the polynomial time. The early similarities among the natural selection procedure and a learning process which led to the evolution of so called "Evolutionary algorithms". The main goal of evolutionary computing is to simulate the evolutionary algorithms and its paradigm into a computer.

11.7.1   Multi-Objective Optimization Methods

In order to elucidate multi-objective solutions, two most widely used multi-objective optimization technique is illustrated. First conventional method is weighted sum method (WSM) and second conventional method is ε-constrained method. Along this, nondominated sorting genetic algorithm (NSGAII), a very popular evolutionary method, is also illustrated.

11.7.1.1   Weighted Sum Method (WSM)

The motive behind this method is to associate a weighting coefficient to every objective function, and maximize/minimize the weighted sum. In this approach, all the multiple objective functions are converted into a single objective functions, and are executed as a single objective optimization problem. The function can be given by

$$F(x) = \sum_{i=1}^{k} w_i f_i(x) \tag{11.5}$$

Where, $f_i(x) = i^{\text{th}}$ objective function

k = number of objectives, $w_i \geq 0$ and $\sum_{i=1}^{k} w_i = 1$. The weights of all the objective function denotes the effectiveness of the functions. Due to change in the weighting coefficient, the solutions will vary according. Using different weights, a trade-off set of solution is generated by WSM.

11.7.1.2   ε-Constraint Method

In ε-Constraint method, by setting an upper bound, all the objective functions are transformed into different constraints except the one which is selected for optimization. The problem can be expressed as:

Minimize : $f_l(x)$

Subject to : $f_k(x) \leq \varepsilon_k$, for all $k = 1, \ldots, n, i \neq l$

$x \in S$

where $l \in \{1, \ldots, n\}$

Here, by this method every time a single objective model is solved. In this method, the upper limit of ε_k is adjusted based on the primal value of a single objective function k. The Pareto optimal set is generated altering the level of ε_k.

11.7.1.3   Non-Dominated Sorting Genetic Algorithm (NSGAII)

In 1994, a Non-dominated sorting genetic algorithm (NSGAII) was proposed in measure for multi-objective optimization. The two single objective optimization technique GA and NSGA-II differs in allocating of fitness for every individual. After many improvements, NSGAII is considered as the latest form. In NSGAII, the non-domination level over the population size of P, is used to measure the fitness of every individual. To generate a child population of size P, operators like recombination, selection and mutation are selected. And hence, the population of P parents and P child is sorted based on the non-domination level. Until the size surpasses the total population size P, by augmenting the solutions successively the new population set of parents is formed. If the total solutions at a given non-domination level surpasses the number which can be adjusted in the new population set of parents, the use of crowded comparison operator is used.

11.7.2   Evolutionary Strategies

Ingo Rechenberg found that the optimization problems of hydrodynamics can be solved using mathematical programming techniques (Rao, 1996). In 1964, with the development of optimization

algorithms, "evolution strategy" was introduced (Fogel, 1998). Initially, the evolution strategy consists of mutation of single parent to yield an offspring and it was known as $(1+1)$-ES. Further, the best from parent and its offspring was picked up to be a parent for the next iteration.

A new individual is yielded using Eq. (11.6):

$$\bar{x}^{t+1} = \bar{x}^t + N\,(0,\bar{\sigma}) \tag{11.6}$$

Where, t=current iteration (or generation),

$N(0,\bar{\sigma})$=vector of independent Gaussian numbers having median value as zero and standard deviation given as $\bar{\sigma}$.

In an evolution strategy, an individual has the set of decision variables. No modification or encoding is applied to the decision variables. Therefore, if there are real numbers in decision variable, then all the real numbers are combined together to form a single vector for every individual.

The example of a $(1+1)$-ES is given below:

Example 11.1

Suppose we want to maximize:

$$f(y_1,\,y_2) = 100\left(y_1^2 - y_2\right)^2 + \left(1 - y_1\right)^2$$

Give that: $-2.048 \le y_1,\,y_2 \le 2.048$

Let us assume that our population consists of arbitrary generated individual, given below

$$\bar{y}^t,\,\bar{\sigma} = (-1.0,1.0),\,(1.0,1.0)$$

Let us now suppose that the mutations generated are the following:

$$y_1^{t+1} = y_1^t + N(0,1.0) = -1.0 + 0.61 = -0.39$$

$$y_2^{t+1} = y_2^t + N(0,1.0) = 1.0 + 0.57 = 1.57$$

Now, on comparing the parent with its offspring:

Parent: $\text{Parent}: f(x_t) = f(-1.0,1.0) = 4.0$

Child: $\text{Child}: f(x_{t+1}) = f(-0.39,1.57) = 201.416$

Since: $201.416 > 4.0$ the offspring will replace its parent in the following generation and offspring will act as parent in the next generation.

In 1973, Rechenberg gave a rule to regulate standard deviation (SD) in such a way that evolution strategy could be converged to primal globally. The rule is named as "1/5 success rule", and can be expressed as following:

$$\sigma(t) = \begin{cases} \sigma(t-n)/c & \text{if } p_s > 1/5 \\ \sigma(t-n)*c & \text{if } p_s < 1/5 \\ \sigma(t-n) & \text{if } p_s = 1/5 \end{cases} \tag{11.7}$$

Where, n=no. of decision variables,

t=current generation,

p_s=relative frequency mutations in which the parent is replaced by its offspring

c=0.817 (constant given by Schwefel (1981)).

The value of $\sigma(t)$ is accustomed for every mutation.

After the concept of population set was introduced, many modified evolution strategies were developed over many years. The $(\mu+\lambda)$ – ES and the (μ, λ)–ES are the latest modified approach of the evolution strategy. Here μ denotes number of parents are mutated to generate λ number of offspring. Initially, from the both parents and offspring, μ number of finest individuals are selected. Furthermore, the next finest individuals are only selected from the offspring generated.

Along with decision variables of the given problem, parameters (like standard deviation) of algorithm have also advanced in the modern evolution strategies. It is known as "Self-adaptation". The mutation in the parents is done using Eqs. (11.8) and (11.9).

$$\sigma'(i) = \sigma(i) \times \exp \tag{11.8}$$

$$x'(i) = x(i) + N\big(0, \sigma'(i)\big) \tag{11.9}$$

Where, τ and τ' gives the proportionality constants which are expressed in terms of n.

The evolution strategies are used in Biochemistry, Routing and networking, Optics, Engineering design, etc.

11.7.3 Evolutionary Programming

Lawrence J. Fogel introduced in the 1960s an approach called "evolutionary programming", in which intelligence is seen as an adaptive behaviour (Rechenberg, 1973). In order to match genetic operators, the social relationships between offspring and their parents are highlighted by evolutionary programming (same as genetic algorithm).

The algorithm of evolutionary is almost identical to evolution strategy. Alike evolution strategy, the population set is selected and mutated in order to generate offspring. In evolutionary programming, several genres of mutation operators are there and no recombination, therefore only the same species can interbreed. In contrast to evolution strategy, in this method parents can generate maximum of only one offspring. In evolutionary programming, whether the parent is eligible or not eligible is decided based on the probabilistic manner, rather than a deterministic manner (like in evolution strategy).

Finally, no encoding is used in this case (similarly to the evolution strategy) and emphasis is placed on the selection of the most appropriate representation of the decision variables.

In the below example, it is clearly explained how evolutionary programming works. Consider finite automaton shown in Figure 11.12. The transition table for the given finite automaton is given below:

Current State	A	A	B	B	C	C
Input Symbol	0	1	1	0	0	1
Next State	B	C	B	C	C	A
Output Symbol	a	B	c	b	a	b

Considering the type of problem at hand, several mutation operators are possible. For example: change an output symbol, change a transition, add a state, delete a state and change the initial state. The goal is to make this automaton able to recognize a certain set of inputs (i.e., a certain regular expression) without making a single mistake.

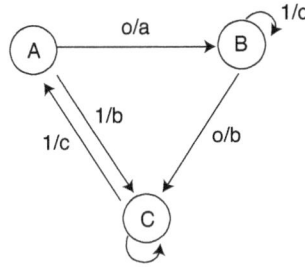

FIGURE 11.12 Finite automata having three states. The symbols which are present on the right side of "/" are the output symbols, and which are on the left of "/" are input symbols. A is the initial state.

The evolutionary programming approach is used in Generalization, Traveling salesperson problem (TSP), Games, Forecasting, Route planning, Automatic control, Pattern recognition, Artificial Neural networks training (ANN) (Fogel, 1998).

11.7.4 GENETIC ALGORITHM

In the early 1960s, the Genetic algorithm (GA) (initially known as "genetic reproductive plans") was discovered by John H. Holland. Solving the machine learning problems was the main motivation of discover genetic algorithm. GA is enthused by the process of natural evolution (Holland, 1975). By modelling all the evolutionary process such as crossover, mutation, selection, a chromosome population is evolved from one generation to another. For discrete optimization problems, the chromosomes are coded in binary, whereas for continuous optimization problems, real-value encoding is done .

In GA initially, an arbitrarily generated population set of solutions/chromosomes is selected. For every solution, a fitness value is measured which depicts their individual asset. The pair of solutions at every generation is selected stochastically from the population set. Then offspring solutions are produced by these selected pairs of solutions using crossover and mutation. Now, the new population set is formed by these offspring solutions, which depicts the population set for the next generation. Until the appropriate fitness value is not found, this process of producing new generations will be continued, and once found the process is stopped.

The different efficient techniques of selection are random selection and roulette wheel selection. In random selection, the fitness values are not considered, while in roulette wheel selection, fitness values are considered. After the pair of solutions are selected, then by crossover process, offspring solutions are generated which are recombination of their parent solutions. There are three techniques to perform recombination: (1) uniform crossover; (2) arithmetic crossover; and (3) n-point crossover. The genes from the offspring solutions have a likelihood of enduring mutations, so the risk of premature convergence is highly reduced.

Firstly, from the total samples, a population set of n solutions/chromosomes is generated (i.e. $chrom_1$, $chrom_2$, ..., $chrom_n$). The gene in every $chrom_i$ is characterized by an arbitrarily created number vacillating in range 0 to 1. For each $chrom_i$, fitness value is premeditated, and the solution having the best fitness value is stored in the variable *bestIndividual*.

Algorithm 11.1

1. Generate an initial random population of *n* individuals = *population* (for $i = 1, ..., n$)
2. Initialize another population of size *n*, i.e. *newPopulation*
3. Evaluate the fitness of each individual $population_i$, i.e. *population.fitness_i*
4. Determine the best individual from *population* using *population.fitness_i* = *bestIndividual*

5. Set crossover rate = *cRate*
6. Set mutation rate = *mRate*
7. for *i* till *maxNoOfGenerations* do
   7.1. *count* = 0
   7.2. while *count<n* do
      7.2.1. Select parents
      7.2.2. Perform crossover using *cRate*
      7.2.3. Perform mutation using *mRate*
      7.2.4. Add offspring to *newPopulation*
      7.2.5. *count = count*+ 2
   7.3. end while
   7.4. *population = newPopulation*
   7.5. *bestIndiv* = find_Best_Individual(*population*)
   7.6. if *bestIndiv.fitness* better then *bestIndividual.fitness* then
      7.6.1. *bestIndividual = bestIndiv*
   7.7. end if
8. end for
9. return *bestIndividual*

For every iteration, using roulette wheel selection, two parent solutions are selected from the old population set, until the novel population set is generated. Let us assume any arbitrary value *arb*. The crossover rate (*cRate*) and mutation rate (mRate) is predetermined and if *arb* is less than *cRate*, then crossover is performed for every index of the adjacent genes of the parent solutions. If *arb* is less than *cRate*, then to generate the offspring solutions adjacent genes are exchanged. If *arb* is less than *mRate*, then mutation is carried out on every gene and a new arbitrary number is assigned. After completion of mutation, the offspring are added to population set.

From the new population set, again the best solution is fetched and compared with the *bestIndividual*. If any improved solution is found then the *bestIndividual* is updated with the improved solution. Finally, the *bestIndividual* is returned when the stopping criterion is met.

11.8   SWARM INTELLIGENCE

An Annual Crop Planning (ACP) is influenced by various factors such as inadequate land for different competing crop, seasonal factors, production factors, better quantity and quality of crop in quick time, etc. ACP is an NP-Hard optimization problem that determines the suitable crop among different competing crops considering various soil factors and restricted land. In the agricultural sector, there is tremendous wastage of water. In order to overcome these problems, Swarm Intelligence (SI) techniques plays a vivacious role in annual crop planning.

The study of Swarm Intelligence is enthused by witnessing the biological agent's intelligent behaviour patterns of swarm within their environment. These swarms interrelate in an autonomous way to accomplish certain tasks. These led to the growth of new optimization algorithms. These SI algorithms have provided effectual results to many real-world NP-Hard problems. There are three metaheuristic algorithms of SI which are beneficial for ACP problem. The three-metaheuristic algorithm is explained in the subsection below.

11.8.1   SWARM INTELLIGENCE TECHNIQUES FOR ANNUAL CROP PLANNING

The three SI techniques used for optimizing the annual crop planning problem are: (1) Cuckoo Search (CS); (2) Firefly Algorithm (FA); (3) Glow-worm Swarm Optimization (GWO). In Section 5.1.4 the statistical comparison of CS, FA, GSO and Genetic Algorithm (GA) is illustrated on the basis of Best Fitness Values (BFV) along with 95% confidence Interval (CI).

11.8.1.1   Cuckoo Search (CS)

Cuckoo search is stimulated based on the biology of a bird species, named cuckoo (Yang, 2010). After reproduction, these birds evacuate their own eggs in the nest of some other bird species, which is considered as host. Some host birds abandon their nest and shift to the new nest, while some aggressively banish the alien eggs if the intrusion is determined. There are three instructions which are overseen by the CS algorithm.

1. At a time only one egg is laid by each bird. The laid egg is placed randomly in the nest of the host bird.
2. For next generation, the nest with the maximum fitness values is carried forward.
3. The number of nests of the host bird is fixed. The probability (p_a) is a constant set in the interval [0,1], which is represented when the intrust ion is determined by the host nest.

Levy flights are used to perform random-walk to produce a novel solution. The equation to find Levy flights of cuckoo k is given as

$$x_k(t+1) = x_k(t) + s\delta \tag{11.10}$$

Here δ is the measurement of movement direction and is measured from a normal distribution having mean and standard deviation as 0 and 1. In the above equation, the step size is denoted by s. It is very tricky to determine s. If the s value is too small, then the term $x_k(t)$ will be almost equivalent or very near to $x_k(t+1)$ and hence it will be more significant. If the s value is very large, then the term $x_k(t+1)$ will be far away from $x_k(t)$.

The most effective algorithm to compute the s value is Mantegna's Algorithm. According to Mantegna's algorithm, s will be computed as

$$s = \frac{u}{|v|^{1/\beta}} \tag{11.11}$$

From normal distribution, u and v are pinched and $0 < \beta \le 2$.

Algorithm 11.2

1. Generate an initial random solution of n host birdnests$=nest$ (for $i=1, 2, ..., n$)
2. Evaluate the fitness of $nest_i$ i.e. $(nest_i)$.
3. Find the best fitness ($bestFitness$) and best nest ($bestNest$) from $nest$
4. $bestFitnessOverall = bestFitness$
5. $bestNestOverall = bestNest$
6. while $t < noOfIterations$ do
    6.1. Generate $newNest$, using $nest$ and $bestNest$ in performing levy flights
    6.2. Get $bestNest$ by performing these steps
        6.2.1. if $f(newNest_i) > f(nest_i)$ then
            $f(nest_i) = f(newNest_i)$
            $nest_i = newNest_i$
        6.2.2. end if
            Evaluate $(nest_i)$ to determine $bestFitness$ and $bestNest$
    6.3. $t = t + n$
    6.4. Generate $newNest$, using $nest$ and p_a. Here a fraction of the worst solutions is replaced with new solutions for each $nesti$
    6.5. Determine $bestNest$ again using step 6.2.
    6.6. $t = t + n$

6.7. if *bestFitness > bestFitnessOverall* then
          6.7.1. *bestFitnessOverall = bestFitness*
          6.7.2. *bestNestOverall = bestNest*
  6.8. end if
7. end while
8. return *bestNestOverall*

Initially, from the total sample, n host bird nests are generated randomly (i.e. $nest_0$, $nest_1$, ..., $nest_n$). The egg in each $nest_i$ is characterized by an arbitrarily created number vacillating in range 0 to 1. For each $nest_i$, fitness value is premeditated. Two terms *bestNest* and *bestFitness* are determined by the best fitness value of the population and its corresponding nest.

For every iteration, using *nest* and *bestNest* new population set of nests (*newNest*) is produced. Every $newNest_i$ is determined from the $nest_i$ and *bestNest* using equation 8. From *newNest*, the solution of best nest is compared with the *bestNest* and if solution of the best nest of *newNest* is more efficient than the *bestNest* than *bestNest* value is updated. After implementing intrusion, for every egg of the *newNest*, if p_a < random, a random new value for egg is generated and the best nest of *newNest* is again compared with the *bestNest* to check if any more efficient solution is there or not. If the stopping criteria is matched, the *bestNest* will be the best solution according to the CS algorithm.

11.8.1.2 Firefly Algorithm (FA)

Fireflies has ability to secrete light. Firefly Algorithm (FA) is stimulated based on the attractiveness property among the fireflies (Yang, 2010). There are two instructions which are overseen by FA algorithm.

1. The fireflies are attracted towards the firefly whose ability to secrete light is brightest.
2. The brightness of fireflies moving randomly fluctuates based on the distance. (assumed that brightness is usually reduced with distance).

Two parameters required to implement the FA algorithm are as follows:
  1. Attractiveness: This property of a firefly is measured from Eq. (11.12).

$$\beta(r) = \beta_0 \exp^{-\gamma r^2} \tag{11.12}$$

    where r = distance between two fireflies
    β_0 = initial attractiveness
    γ = coefficient of absorption.

2. Movement: The movement property depends on the attractiveness of the firefly. The movement between the less attractive firefly i and the more attractive firefly j is represented in Eq. (11.13).

$$x_i = x_i + \beta_0 \exp^{-\gamma r_{ij}^2} \left(x_j - x_i \right) + \alpha \left(\text{rand} - \frac{1}{2} \right) \tag{11.13}$$

Here, the first term x_i represents the existing position of firefly, the second term represents the attractiveness between the two fireflies, and the third term is the arbitrary modification in the firefly's movement having scaling factor as α. The cartesian distance r_{ij} is calculated by

$$r_{ij} = \sqrt{\sum_{k=1}^{d} \left(x_{ik} - x_{jk} \right)^2} \tag{11.14}$$

Algorithm 11.3

Initialize α, β_0, and *no of Iterations*
1. Initialize *n* fireflies =*firefly Location* (for $i=1 \ldots n$)
2. The light intensity of *firefly Location* $s_{i=}$*firefly Fitnes* s_i
3. for *l* till *no of Iterations* do
    3.1. for *i* till *n* do
        3.1.1. *firefly Fitnes* $s_{i=}$Evaluate(*firefly Location* s_i)
    3.2. end for
    3.3. Sort *firefly Locations* and *firefly Fitness* according to *firefly Fitness*
    3.4. *best Firefly Fitness* = *firefly Fitness* 0
    3.5. *best Firefly Location* = *firefly Location* s_0
    3.6. Move fireflies to new locations by performing these steps
        3.6.1. for *i* till *n* do
            3.6.1.1. for *j* till *n* do
                3.6.1.1.1. if *firefly Fitnes* s_i< *firefly Fitnes* s_j then
                    3.6.1.1.1.1. Calculate r_{ij}
                    3.6.1.1.1.2. Calculate (r)
                    3.6.1.1.1.3. Update *fireflyLocations* $_i$
                3.6.1.1.2. end if
            3.6.1.2. end for
        3.6.2. end for
4. end for
5. return *bestFireflyLocation*

Initially, from the total sample, the population of n fireflies are generated randomly (i.e. firefly$_0$, firefly$_1$, …, firefly$_n$). In each firefly, *fireflyLocation*$_{ik}$ ($\forall k=1,\ldots, p$), is visualized by an arbitrary produced number vacillating in range 0 to 1. The fitness value, or light intensity, is premeditated as *fireflyfitness*$_i$.

In every iteration, the fireflies are arranged in order from highest fitness to lowest fitness. For every firefly i the fitness value is equated with all the other fireflies *j* ($j=1, 2, …, n$) of the population, and if the *fireflyfitness*$_i$ is less than fireflyfitness$_j$ of the population than the firefly *i* moves towards firefly *j* using Eq. (11.11). After the stopping criteria is met, the solution generated will be the sorted order of *fireflylocations*.

11.8.1.3 Glow-Worm Swarm Optimization (GSO)

The Glow-worm has the ability to attract other glow-worms by secreting luminescent property knows as luciferin. Glow-worms are more attractive when they secret more luciferin. Glow-worm has a tendency to move in a direction where there is brighter glow-worm within their range.

At time *t*, the luciferin level is $l_i(t)$ for glow-worm *i*, position is $x_i(t)$ having range for vision as $r_i(t)$. This entire variable changes when there is a movement in glow-worms. The luciferin is updated and is measured by the Eq. (11.15).

$$l_i(t+1) = (1-\rho)l_i(t) + \gamma I(x_i(t)) \tag{11.15}$$

Here ρ ($0 < \rho < 1$) is called decay constant and γ is called enhancement constant. At time *t*, $I(x_i(t))$ is known as estimation of objective function.

The neighbours $N_i(t)$ should be governed in order to alter the glow-worm's position. If the luciferin level $l_i(t)$ of glow-worm *i* is less than luciferin level of glow-worm *j* and if glow-worm *j* is within glow-worm i's range than glow-worm *j* is considered a s the neighbour of glow-worm *i*. Based on the roulette wheel selection, from the neighbours of *i* ($N_i(t)$), a glow-worm *j* is selected. From the movement Eq. (11.16), the Glow-worm *i* is moved towards Glow-worm *j*.

$$x_i(t+1) = x_i(t) + st * \left\{ \frac{\left(x_j(t) - x_i(t)\right)}{x_j(t) - x_i(t)} \right\} \tag{11.16}$$

Here, st is called the constant step size.

Finally, the vision range also gets modified using Eq. (11.17).

$$r_i(t+1) = \min\left\{ r_s, \max\left[0, r_i(t) + \beta\left(N_d - |N_i(t)|\right) \right] \right\} \tag{11.17}$$

Here, N_d, β, and r_s are constant values. The maximum vision range for the glow-worm i is denoted by r_s. The rate of change in neighbourhood range is given by β. The maximum neighbour i can have is given by N_d.

Initially, from the total sample, the population of n glow-worm is generated randomly (i.e. glow-worm$_0$, glowworm$_1$, ..., glowworm$_n$). Every element that each glow-worm comprises, $glowworm_{ik}$ ($\forall k = 1, ..., p$), is visualized by an arbitrary produced number vacillating in range 0 to 1. The best fitness value is premeditated and stored as $bestfitness_i$ and its corresponding solution is stored in the $bestLocation$ For every Iteration, the ith glow-worm having the best fitness value from the new population set is compared with the $bestfitness$. If the fitness value for glow-worm i is more efficient than $bestfitness$ then the $bestfitness$ and the $bestLocation$ values are updated and the process is repeated until the stopping criteria are met.

Algorithm 11.4

1. Generate a population of n glow-worms $= glowworm$ (for $i = 1, ..., n$)
2. Initialize the best fitness overall $= bestFitness$
3. Initialize the best location overall $= bestLocation$
4. while t till $noOfIterations$ do
    4.1. for i till n do
        4.1.1. Update luciferin of $glowworm_i$
    4.2. end for
    4.3. for i till n do
        4.3.1. Find $N(t)$
        4.3.2. for each $glowworm_j \in N(t)$ do
            4.3.2.1. Find probability:
                $pij(t) = lj(t) - li(t) \Sigma lk(t) - li(t) k \in Ni(t)$
        4.3.3. end for
        4.3.4. Select $glowworm_j$ using roulette wheel selection with $p_i(t)$
        4.3.5. Update $glowworm_i$ location
        4.3.6. Update vision range
    4.4. end for
    4.5. for i till n do
        4.5.1. if $glowworm_i.fitness > bestFitness$ then
            4.5.1.1. $bestFitness = glowworm_i.fitness$
            4.5.1.2. $bestLocation = glowworm_i.Location$
        4.5.2. end if
    4.6. end for
    4.7. $t = t + 1$
5. end while
6. return $bestLocation$

11.8.1.4 Case Study: Comparison of CS, FA, GSO and Genetic Algorithm (GA)

All the non-heuristic parameters used for the execution of the algorithms are given in Tables 11.3 and 11.4. In Table 11.4, the upper and lower bound settings are given for different plot types. Table 11.4 also contains cost of irrigated water, fraction values of land coverage, operational costs for each crop, and lower bound and upper bound settings. In order to evaluate the efficiency of the solution over larger solution space, there are huge differences between upper bounds and lower bounds.

For all meta-heuristic algorithms, the initial parameters for this study are set as shown in Table 11.5.

To fairly compare the meta-heuristic algorithms, the population size in every algorithm is set to 20. The max_no of Generations for GA algorithm and no of iterations for GSO, FA, and CS algorithm warranted the execution for 100,000 objective functions evaluations. Using the arbitrarily generated population set, every algorithm was executed for 100 times.

It is concluded from Table 11.6 that the execution time for GSO algorithm is fastest (As shown in Figures 11.13 and 11.14). The lesser execution time in GSO is because the glow-worm is allowed to have a maximum threshold number of neighbours. Therefore, for every glow-worm the vision range decreases, as the iterations are augmented. Due to this the glow-worms are more separated when searching for local neighbourhood in the solution space. Finally, the number of glow-worms who were in the search of neighbours decreased because of the separation, and therefore the execution time in boosted.

TABLE 11.3
Upper and Lower Bounds for Each Plot Type

Plot Types	Bounds (ha)	
	Ub_Pk	Lb_Pk
Single-crop	1,700	10
Double-crop	1,740	50

TABLE 11.4
Non-Heuristic Specific Parameters for the Execution of the Algorithms

Crops	Ubk_{ij}	Lbk_{ij}	Fk_{ij}	C_IRk_{ij}	Ok_{ij}
Lucerne (y)	1,700	10	1	877.26	6259.52
Tomato (s)	1,740	10	1	685.11	71478.00
Pumpkin (s)	1,740	10	1	451.66	10408.80
Maize (s)	1,740	10	1	613.90	3924.09
Groundnut (s)	1,740	10	1	502.08	5025.24
Sunflower (s)	1,740	10	1	292.13	3701.61
Barley (w)	1,740	12.5	1	413.68	4124.88
Onion (w)	1,740	12.5	1	221.00	23739.30
Potato (w)	1,740	12.5	1	186.10	22758.12
Cabbage (w)	1,740	12.5	1	172.94	23720.00

$Fk_{ij} \in [0,1]$. C_IRk_{ij} represents the cost of irrigated water in terms of per crop per hectare (ZAR ha^{-1}). Ok_{ij}, operational cost is third of the producer price over a ton of yield (ZAR ha^{-1}).

TABLE 11.5

Initial Parameters of Meta-Heuristic Algorithms (All the data collected from 21,22,23)

CS	FA
$n=20$	$n=20$
no of iterations $=100,000$	no of iterations $=5,000$
$p_a=0.25$	$\alpha = 0.25$
	$\beta_0 = 0.2$
	$\gamma = 1$
GSO	**GA**
$n=20$	$n=20$
no of iterations $=5,000$	max no of Generations $=5,000$
$l_0=1, r_0=1.2$	cRate $=0.8$
$rs=1.5, \rho=0.4$	mRate $=0.05$ $(1/n)$
$\gamma=0.6, \beta=0.08$	
$st=0.3, Nd=10$	

TABLE 11.6

Statistics of the Average Execution Times (AVG) in milliseconds (ms), and the 95% Confidence Interval (95% CI) Values of Each Heuristic Algorithm

Statistics	Algorithms			
	FA	CS	GA	GSO
AVG (ms)	3,455	884	915	751
At 95% CI	AVG ± 6	AVG ± 2	AVG ± 3	AVG ± 3

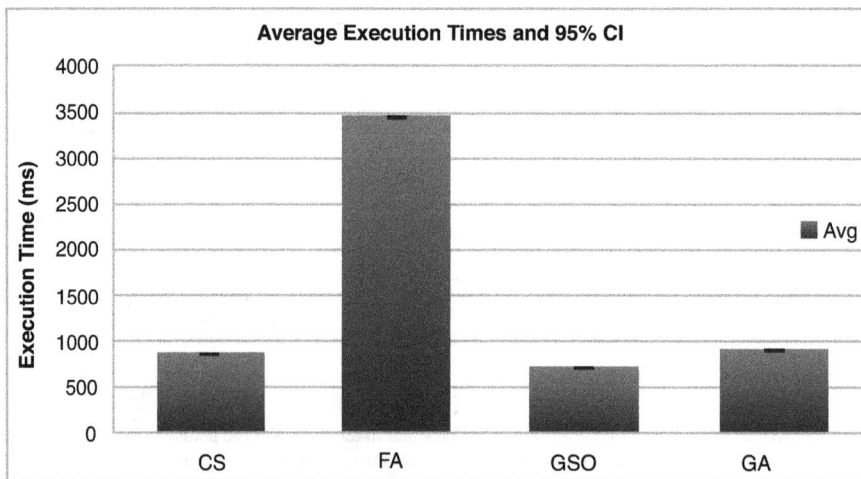

FIGURE 11.13   The average execution times, in milliseconds (ms), and the 95% CI values of the algorithms.

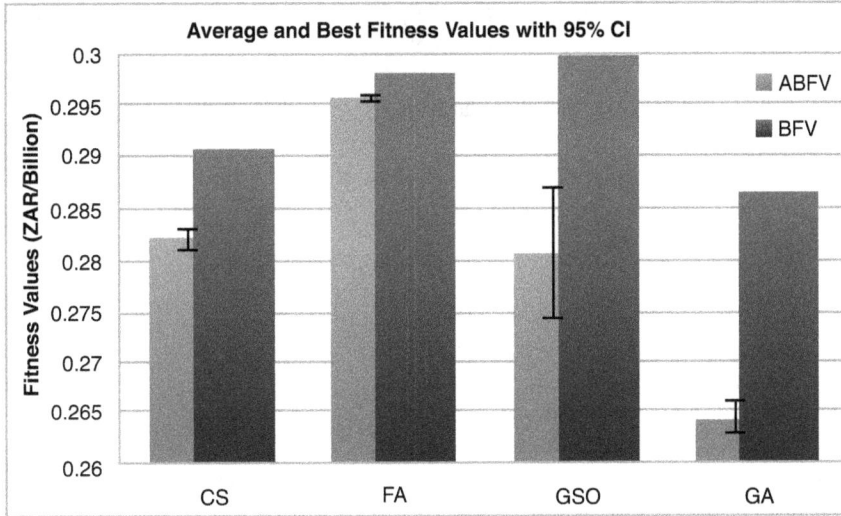

FIGURE 11.14 A comparison of best and average fitness values, along with the 95% CI estimates.

11.8.2 CROP CLASSIFICATION USING ARTIFICIAL BEE COLONY OPTIMIZATION

A novel approach of extracting the finest set of features by computing Gray Level Co-occurrence Matrix (GLCM) and Artificial Bee Colony (ABC) algorithm as a distance classifier . For classifying the crop, several diverse distance metrics are trained. As ABC does not have any binary version, the individual depiction is binarized to select the finest set of features which are used in classifying the crop. The Euclidean or Manhattan distance metrics are used to differentiate the selected features.

The ABC is heuristic approach which is based on the behaviour of bees foraging. This algorithm comprises the population set nb (bees), $i=1,..., nb$, $x_i \in R^n$, which denotes different positions of the source food of every bees (Karaboga, 2005). In this algorithm three genres of bees plays vital role in order to get convergence nearest to the best optimal solution: *scout bees, onlooker bees, and employed bees.*

Using Eq. (11.18), the *employed bees* look for the novel neighbourfood source which is nearest to their hive and compare with the previous one. The bee remembers the best food source from the two.

$$v_i^j = x_i^j + \emptyset_i^j \left(x_i^j - x_k^j \right) \tag{11.18}$$

Where, $k \in \{1, 2,..., nb\}$ and $j \in \{1, 2,..., n\}$ are arbitrarily selected indexes where $k \neq i$. $\emptyset_i^j$ is selected as random between $[-a, a]$.

Further using a fitness function, the value or quality of every source of food is calculated based on the quantity of information. Now, employed bees share this information with the onlooker bees by returning to the dancing area in their respective hives.

By observing the dance of the employed bees, the onlooker bees identify the information such as size of the food source, where it is located, and whether that is quality information or not. By probabilistic value, the onlooker bees select the food source based on the information shown by the employed bee. Using Eq. (11.19), the probabilistic value is calculated as

$$p_i = \frac{\text{fit}_i}{\sum_{k=1}^{nb} \text{fit}_i} \tag{11.19}$$

Where, fit_i=fitness value of solution i

nb=is the number of food sources that are equal to the number of *employed bees.*

Finally, by using Eq. (11.20), the *scout bees* help by generating arbitrarily new solutions when there is no option to improve food source anymore.

$$x_i^j = x_{min}^j + \text{rand}(0,1)\left(x_{max}^j - x_{min}^j\right) \tag{11.20}$$

By considering all the three types of bees, the Pseudocode of the ABC Algorithm is shown below.

Algorithm 11.5

1   Initialize the population of solutions $x_i \forall i$, $i = 1,\dots, nb$.
2   Evaluate the population $x_i \forall i$, $i = 1,\dots, nb$.
3   for *cycle* = 1 to maximum cycle number *MCN* do
4   Produce and evaluate new solutions v_i using Eq. (11.1).
5   Apply the greedy selection process.
6   Calculate the probability values p_i for the solutions x_i by Eq. (11.2).
7   Produce and evaluate new solutions v_i for the solutions x_i selected depending on p_i.
8   Apply the greedy selection process.
9   Replace the abandoned solutions with a new one x_i by using Eq. (11.3).
10 Memorize the best solution achieved so far.
11 *cycle* = *cycle* + 1
12 end for

In divergence, where satellite images are found with Synthetic Aperture Radar devices, or multi-spectral, the below approach is beneficial when executing with satellite images, which are attained from Google Earth. These satellite images contain data only from the noticeable part of electromagnetic spectrum. The image is segmented into different classes of crops represented by several polygons which are shown in Figure 11.15. Every polygon refers to a unique class, which makes the feature extraction part much easier. From each colour channel that uniquely defines the patterns in the feature set, a set of eight features is selected by the GLCM method.

Using ABC algorithm, the finest set of features from GLCM is selected. If the input patterns are given by $X = \{x_1, \dots, x_p\}$ where $x_i \in R^n$ and $i = 1 \dots, p$, and set of desired classes $d = \{d_1, \dots, d_p\}$

FIGURE 11.15   Segmentation of satellite image into various different polygon.

where $d \in N$, identify the set of features $G \in \{0, 1\}^n$ considering that $\min(F(X|G, d))$ is minimized (Garro et al., 2016).

The solution to above-stated problem is denoted in terms of set of features and an array $A \in R^n$ which contains the information of source food's position. Every value of array A_k where $k = 1,\ldots,$ nb is binarized as per Eq. (11.21) by means of a threshold level t, so that the finest set of features is selected which is defined as $G^k = T_t(A^k)$, $k = 1,\ldots, n$; values whose component is set to 0, indicates, that feature will not be selected to make up the set of features and vice-versa.

$$T_{th}(x) = \begin{cases} 0, & x < th \\ 1, & x \geq th \end{cases} \tag{11.21}$$

By using Eq. (11.22), the ability of all individuals is calculated by means of the classification error (CER) function, which determines how many samples of the crop are incorrectly classified.

$$F(X \mid G, d) = \frac{\sum_{i=1}^{p} \left(\left| \arg\min_{k=1}^{K} \left(D\left(x_i \mid G, c^k\right)\right) - d_i \right| \right)}{tsc} \tag{11.22}$$

Where, tsc = total number of samples of a crop to be classified,

D = distance measure,

K = number of classes,

c = center of each class.

DECLARATION

AUTHORS CONTRIBUTION

All the authors make a substantial contribution to this manuscript. HP and MS participated in drafting the manuscript. HP and MS wrote the main manuscript. All the authors discussed the results and implications of the manuscript at all stages.

ACKNOWLEDGEMENTS

The authors are grateful to the Department of Computer Engineering, Vellore Institute of Technology and Department of Chemical Engineering School of Technology, Pandit Deendayal Petroleum University for the permission to publish this research.

AVAILABILITY OF DATA AND MATERIAL

All relevant data and material are presented in the main paper.

COMPETING INTERESTS

The authors declare that they have no competing interests.

FUNDING

Not Applicable

CONSENT FOR PUBLICATION

Not applicable.

Ethics Approval and Consent to Participate

Not applicable.

REFERENCES

Ahir, K., Govani, K., Gajera, R., & Shah, M. (2020). Application on virtual reality for enhanced education learning, military training and sports. *Augmented Human Research*, 5(1). doi: 10.1007/s41133-019-0025-2.

Barrow, C. J. (1990). Sustainable agricultural systems. In C. A. Edwards, R. Lal, P. Madden, R. H. Miller & G. House (Eds.), *Soil and Water Conservation Society*. Ankeny, LA. ISBN: 0 935734 21 X, US$ 40.00 (hardback), xvi + 696 pp. *Land Degradation & Development*, 2, 328–328. doi: 10.1002/ldr.3400020410.

Bosma, R., Van den Berg, J., Kaymak, U., Udo, H., & Verreth, J. (2012). A generic methodology for developing fuzzy decision models. *Expert Systems with Applications*, 39(1), 1200–1210. doi: 10.1016/j. eswa.2011.07.126.

Chen, C. L., & Chen, W. C. (1994). Fuzzy controller design by using neural network techniques. *IEEE Transactions on Fuzzy Systems*, 2(3), 235–244. doi: 10.1109/91.298452.

Dahikar, S. S., Rode, S. V., & Deshmukh, P. (2015). An artificial neural network approach for agricultural crop yield prediction based on various parameters. *International Journal of Advanced Research in Electronics and Communication Engineering (IJARECE)*, 4(1), 94–98.

Dharmaraj, V., & Vijayanand, C. (2018). Artificial intelligence (AI) in agriculture. *International Journal of Current Microbiology and Applied Sciences*, 7(12), 2122–2128. doi: 10.20546/ijcmas.2018.712.241.

Fogel, D. B. (1998). *Evolutionary Computation. The Fossil Record. Selected Readings on the History of Evolutionary Algorithms*. The Institute of Electrical and Electronic Engineers: New York.

Gandhi, M., Kamdar, J., & Shah, M. (2020). Preprocessing of non-symmetrical images for edge detection. *Augmented Human Research*, 5(1). doi: 10.1007/s41133-019-0030-5.

Garro, B. A., Rodríguez, K., & Vázquez, R. A. (2016). Classification of DNA microarrays using artificial neural networks and ABC algorithm. *Applied Soft Computing*, 38, 548–560. doi: 10.1016/j.asoc.2015.10.002.

Holland, J. H. (1975). *Adaptation in Natural and Artificial Systems*. University of Michigan Press: Ann Arbor, MI.

Jani, K., Chaudhuri, M., Patel, H., & Shah, M. (2020). Machine learning in films: An approach towards automation in film censoring. *Journal of Data, Information and Management*, 2(1), 55–64. doi: 10.1007/s42488-019-00016-9.

Kakkad, V., Patel, M., & Shah, M. (2019). Biometric authentication and image encryption for image security in cloud framework. *Multiscale and Multidisciplinary Modeling, Experiments and Design*, 2(4), 233–248. doi: 10.1007/s41939-019-00049-y.

Karaboga, D. (2005). An idea based on honey bee swarm for numerical optimization. Technical report, Computer Engineering Department, Engineering Faculty, Erciyes University.

Kundalia, K., Patel, Y., & Shah, M. (2020). Multi-label movie genre detection from a movie poster using knowledge transfer learning. *Augmented Human Research*, 5(1). doi: 10.1007/s41133-019-0029-y.

Mokarram, M., Najafi-Ghiri, M., & Zarei, A. (2018). Using self-organizing maps for determination of soil fertility (case study: Shiraz plain). *Soil and Water Research*, 13(1), 11–17. doi: 10.17221/139/2016-swr.

Murmu, S., & Biswas, S. (2015). Application of fuzzy logic and neural network in crop classification: A review. *Aquatic Procedia*, 4, 1203–1210. doi: 10.1016/j.aqpro.2015.02.153.

Ogwueleka, F. N. (2016). Crop growth prediction using self-organizing map and multilayer feed-forward neural network. *American-Eurasian Journal of Sustainable Agriculture*, 5(2), 168–176.

Olakulehin, O. J., & Elijah, O. O. (2014). A genetic algorithm approach to maximize crop yields and sustain soil fertility. *Net Journal of Agricultural Science*, 2(3), 94–103.

Panchiwala, S., & Shah, M. (2020). A comprehensive study on critical security issues and challenges of the IoT world. *Journal of Data, Information and Management*, 2(4), 257–278. doi: 10.1007/s42488-020-00030-2.

Pandey, A., Prasad, R., Singh, V., Jha, S., & Shukla, K. (2013). Crop parameters estimation by fuzzy inference system using X-band scatterometer data. *Advances in Space Research*, 51(5), 905–911. doi: 10.1016/j. asr.2012.10.018.

Pandya, R., Nadiadwala, S., Shah, R., & Shah, M. (2020). Buildout of methodology for meticulous diagnosis of K-complex in EEG for aiding the detection of Alzheimer's by artificial intelligence. *Augmented Human Research*, 5(1). doi: 10.1007/s41133-019-0021-6.

Papageorgiou, E., Markinos, A., & Gemtos, T. (2011). Fuzzy cognitive map based approach for predicting yield in cotton crop production as a basis for decision support system in precision agriculture application. *Applied Soft Computing*, 11(4), 3643–3657. doi: 10.1016/j.asoc.2011.01.036.

Parekh, V., Shah, D., & Shah, M. (2019). Fatigue detection using artificial intelligence framework. *Augmented Human Research*, 5(1). doi: 10.1007/s41133-019-0023-4.

Patel, H., Prajapati, D., Mahida, D., & Shah, M. (2020a). Transforming petroleum downstream sector through big data: A holistic review. *Journal of Petroleum Exploration and Production Technology*, 10(6), 2601–2611. doi: 10.1007/s13202-020-00889-2.

Patel, D., Shah, D. & Shah, M (2020b). The intertwine of brain and body: A quantitative analysis on how big data influences the system of sports. *Annals of Data Science*, 7, 1–16. doi: 10.1007/s40745-019-00239-y.

Pathan, M., Patel, N., Yagnik, H., & Shah, M. (2020). Artificial cognition for applications in smart agriculture: A comprehensive review. *Artificial Intelligence in Agriculture*, 4, 81–95. doi: 10.1016/j.aiia.2020.06.001

Rao, S. S. (1996). *Engineering Optimization. Theory and Practice*, 3rd edition. John Wiley & Sons, Inc: Hoboken, NJ.

Rechenberg, I (1973). *Evolutions strategies: Optimierung technischer Systeme nach Prinzipien der biologischen Evolution*. Frommann–Holzboog: Stuttgart, Germany.

Schwefel, H. P. (1981). *Numerical Optimization of Computer Models*. John Wiley & Sons, Inc: Hoboken, NJ.

Shah, D., Dixit, R., Shah, A., Shah, P., & Shah, M. (2020). A comprehensive analysis regarding several breakthroughs based on computer intelligence targeting various syndromes. *Augmented Human Research*, 5(1). doi: 10.1007/s41133-020-00033-z.

Sukhadia, A., Upadhyay, K., Gundeti, M., Shah, S., & Shah, M. (2020). Optimization of smart traffic governance system using artificial intelligence. *Augmented Human Research*, 5(1). doi: 10.1007/s41133-020-00035-x.

Talaviya, T., Shah, D., Patel, N., Yagnik, H., & Shah, M. (2020). Implementation of artificial intelligence in agriculture for optimisation of irrigation and application of pesticides and herbicides. *Artificial Intelligence in Agriculture*, 4, 58–73. doi: 10.1016/j.aiia.2020.04.002.

Yang, X. S. 2010. *Nature-Inspired Metaheuristic Algorithms*, 2nd edition. Luniver Press: United Kingdom.

Yoon, K. P., & Hwang, C. (1995). *Multiple Attribute Decision Making*. SAGE Publications, Inc. doi: 10.4135/9781412985161.

Zadeh, L. (1965). Fuzzy sets. *Information and Control*, 8(3), 338–353. doi: 10.1016/s0019-9958(65)90241-x.

12 Artificial Intelligence in Crop Monitoring

Dineesha Soni, Priya Patel, and Manan Shah
Pandit Deendayal Petroleum University

CONTENTS

12.1 INTRODUCTION

Majority of the occupying area approximately more than 40% in India is the agricultural sector. Agriculture is considered to be the most important part of India playing a vital role in increasing India's economy and reserving around 52% of total number of jobs available in India. Hence, due to increased agricultural farming, according to previous data, India was among the top two farm producers in the whole world. The contribution of the agricultural sector is very prominent for the livelihood, economical growth and upliftment of rural India. Although it is primarily dependent on monsoons, it has a major contribution to the capital formation of India. Furthermore in today's era, although industries play a vital role in economic growth, agricultural sector becomes an important part in providing raw materials to industries. Apart from India, it has an importance in sharing national income, providing a source of employment, supplying raw materials, and earning of foreign exchange in the international market (Kekane, 2013). Also, India is rapidly progressing against undernutrition. Improving diets, maternal health, care practices, and income generations are the

major outcomes due to agricultural practice (Kadiyala et al., 2014). Thus, upgradation should be undertaken and modern technologies must be implemented for further increase in agriculture products and economic development of nation.

For the increased development and upgradation in the agricultural sector, one of the methods harming less to the environment and fulfilling our demands is Artificial Intelligence (AI). The following chapter discusses in detail the role, importance, and various methods pertaining to AI used in the field of Crop Monitoring in the agricultural sector.

Artificial Intelligence (AI), also known as Machine Intelligence, performs a major role in decision making, speech recognition, visual perception, and translation between languages. It is considered as theory and development of a computer system that performs various tasks normally requiring human intelligence. Problem solving, Reasoning, Planning, Learning, Knowledge, and ability to manipulate and move objects are the primary aspects of this swiftly advancing technology. This advancing technology gives a great opportunity and can have significant impacts on human lives which helps them to grow socially, economically, and technically. In this modern world, this technology has delivered accurate, precise, and substantial outcome so that we can assure its capability and actions.

But the most important factor affecting the overall growth of the product is its monitoring. Hence modern techniques must be implemented resulting in higher crop production and thus indirectly resulting in economic growth of the country using Artificial Intelligence.

The persons who first proposed the idea for the overall management of crops using AI were Mckinion and Lemmon. In the year 1985 auordance with the paper "Expert systems for Agriculture" (McKinion and Lemmon, 1985; Banerjee et al., 2018). One must understand the main aim and importance of crop monitoring before going to the various methods implemented for it. Crop Monitoring mainly discusses determining the main tool for early detection of any kind of disease on crop and eventually improving the quality of the crop (López-Granados et al., 2016). AI proposes various techniques for the automated assessment of health of the crops. Its development at various stages is an important aspect of agricultural management and for the farmers. This can be clearly described from the given Figure 12.1.

Over the last many decades, numerous techniques have been implemented and invented as a purpose to help farmers to increase the production of crops along with the major important factor i.e. maintaining the health of crop by improving its quality, protecting them from various diseases, by removing weeds and unwanted wasted, by precise detection and analysis of micron parts of the crops like leaves. Thus the overall monitoring of crop is maintained by the latest techniques of AI which is shown in Figure 12.2.

Discussing the different techniques, our chapter includes major and important three various sectors and its techniques which highly involves in monitoring of crop: these three techniques can be termed as follows.

  i. Field Mapping
 ii. Precision Agriculture
iii. Remote Sensing Image Analysis

Let's have a brief look at each technique and then further details with its pictorial presentations and with tables are discussed later.

Under the topic of field mapping, the image of different crops are captured under white light and UVA light to check crop conditions with the help of methods like drone and copters systems during cultivation, estimation of area where the crop requires fertilizer, water and pesticides can be made (Dharmaraj and Vijayanand, 2018). Some other major methods which have been discussed in this chapter are aerial photography, crop scouting in the context of image analysis under field mapping.

Precision agriculture, developed during the mid-1970s to early 1980s focuses on soil sampling and soil surveying considerably resulting in a better awareness of soil and crop condition variability

FIGURE 12.1   Describing the pictorial view of crop monitoring using AI technique.

FIGURE 12.2   Automation technique in crop monitoring.

within fields. Also Geographic Information Systems (GIS), microcomputers, Global Positioning Systems (GPS) played a vital role in the upliftment of farm machinery with computerized controllers and sensors (Robert, 2002). Furthermore, UAV models and SSM techniques for PA are also discussed.

Remote Sensing Image Analysis uses radiant energies for the extraction of various information on the ground as well as on an environmental basis. Apart from this various geospatial techniques, aerial photography, various sensors including ANN are used along with their DataAnalysis and Algorithms for the monitoring of soil and its mapping and various detection of other measures eventually helping in monitoring of crop (Zhang and Kovacs, 2012). Let's have a look at each technique in detail.

12.2   FIELD MAPPING

Field mapping is helpful to monitor changes of vegetation, to observe erosion patterns, to detect crop disease and stress and to do many more things so that one can increase efficiency and can improve crop yield in less time by focusing on the field. Field mapping can be done by various techniques. Here, for example, aerial photography and UAV (Unmanned Autonomous Vehicle) methods are discussed.

12.2.1   AERIAL PHOTOGRAPHY

Aerial photography is a technique that is easily used by crop scientists and farmers. Using aerial photography they can understand and study the factors affecting crop performance. For monitoring momentary changes of vegetation and observing patterns of erosion small-format aerial photography is a suitable tool. Other methods like hot-air blimp and Kite Aerial Photography (KAP) are also adopted (Aber et al., 2019). Another application of aerial photography is to identify crop disease and stress. For that Black-and-white aerial photography, visible coloraerial photography, Black-and-white infrared aerial photography or color infrared aerial photography is used to identify crop disease as well as to interpret crop changes. Multi-temporal photography also helps to study the epidemiology of disease and gives existing data on crop performance through which one can monitor the effects of methods and changes in management practice (Steven and Clark, 2013).

12.2.2   UAV METHOD

UAV (Unmanned Autonomous Vehicle) method is practiced in many research fields because of its ability to capture high-resolution aerial images. In the domain of agriculture, UAV can be used for several purposes. For example, multispectral image analysis system based on UAV, with semantic computing can be used to detect unhealthy agricultural area and to check health condition of crop too. To study the health condition of plant, the environmental indices like normalized difference water index (NDWI), normalized difference vegetation index (NDVI) and soil adjust vegetation index (SAVI) can be calculated from multispectral image using pixel values (red, blue, green and IR pixel). Using these three indices, Multi-spectrum semantic space is created and health condition of area and crop is interpreted (Wijitdechakul et al., 2016). Following Figure 12.3 shows one of the UAV models.

12.3   PRECISION AGRICULTURE (PA)

In PA, methods like GPS, GIS and SSM were developed which was information technology to collect data from various sources and to make decision accordingly. The application of PA in crop monitoring can be described as improved crop quality, improved sustainability, food safety, environmental protection, etc. highly précised and developed sensors are invented for GPS receiver and

FIGURE 12.3   UAV large scale surveying drone model used in monitoring of crop.

correction signal subscription and for GIS software for data storage, processing, analysis and making prescription.

12.3.1   DRONE SYSTEMS

Drone system is considered one of the best tools in PA which fetches data from drone image and provide alerts in real time which gives momentum in precision farming. Here sensors sense the image with the help of image sensing through UVA and white light and generate various levels of readiness for fruits and crops (Dharmaraj and Vijayanand, 2018). These help in identifying various abnormalities present in crop and gives optimal data for further solutions. The problems which can be dealt with are precise capture of the individual leaf of crop, the no. of crops requiring the amount of water, various growth stages, stress level of plants, spatial information of farm land through multispectral images, etc. using the main fundamental as AI. Further, the N_2 level of the soil can also be powered by various drone systems. In the case of highly infected plants with the help of infrared by scanning crops in both RGB (Red Green Blue), multispectral images can be generated. Also, the specific spotted point of the infected part can be viewed and its optimal solution can be generated through drones. This can be shown in Figures 12.4a, b & 12.5.

12.3.2   USING GPS

These systems also include GPS & GIS which are the major participants in precision farming: with the help of highly developed sensors, they acquire spatial data and make all the processing and its utilization for further identification and solution. These techniques improve crop quality and improve sustainability at lower risk along with crop safety and environmental protection (Robert, 2002). Although GPS requires sensors that are costly but are too precise, have a wide range, various input methods, and functions at right, faster application, etc.

(a)   (b)

FIGURE 12.4   (a and b) Pictorial view of crop monitoring using various agro drones.

FIGURE 12.5   Example of drones using various cameras and microcontrollers.

12.3.3   UAV MODELS

One of the most useful and upcoming techniques which is already accepted by developed countries is UAV (Unmanned Aerial Vehicle). At first, these techniques were used in civilian and military applications (Van Blyenburgh, 1999). Then after technical analysis, this technique was implemented and applied to various agricultural operations, especially like crop monitoring by Yamaha (Giles and Billing, 2015). The UAVs are aircraft equipped with various developed sensors and cameras use for the purposes like crop height estimations, pesticide spraying, soil and field analysis, etc. (Mogili and Deepak, 2018; Anthony et al., 2014; Huang et al., 2009; Primicerio et al., 2012). Using UAVs for pesticide spraying numerous health-related problems can be solved which were faced by the farmers working manually. As hardware, it is an autonomous mini rotorcraft vehicle containing software for flight controlling, aerodynamic modeling, etc. These also include automated sprinkling system with multi spectral camera controlled by a radio channel and without a human pilot, this aircraft will eventually change all the traditional farming techniques (Figure 12.6).

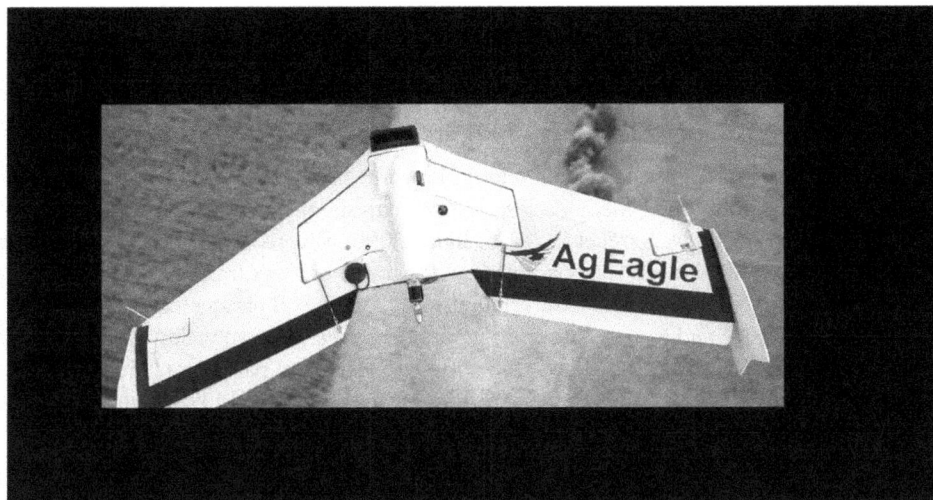

FIGURE 12.6   UAV model for precise farming.

12.3.4   Site-Specific Management

SSM, also known as precision agriculture is mostly used at a spatial scale smaller than that of the whole field. SSM is a commercial technology using GPS for crop management. This includes managing crops by collecting data using information system, analyzing them, determining their various benefits and side effects and then act accordingly (Plant, 2001). This concept is also termed as 'farming by soil types'. Its major application is in the section of fertilizer purpose for the crops like corn, soybean, wheat, etc. further applications are sampling, fertilizing, harvesting, tillage, planting, etc. nowadays it is also applicable to the variety of plants like sugar beet, cotton, rice, peanut, potato, sugarcane, palm oil, banana, etc. (Robert, 2002). These methods mentioned above are useful in PA. Its various advantages are profitability, sustainability, efficiency, etc.

12.4   REMOTE SENSING

In the agriculture sector, remote sensing has been applied for a very long time. Airplanes, balloons and helicopters, satellites, and sensors like optical and near IR sensors are adopted by remote sensing platforms. Various information such as biomass, disease, water stress, Leaf Area Index (LAI) can be derived from images captured using these sensors, and this information is helpful in yield forecasting, crop management, environmental protection, etc. (Zhang and Kovacs, 2012). Let's have a look at some of the applications of remote sensing.

12.4.1   Airborne Technique

From both digital and video cameras, Airborne is a new remote sensing technique. To identify spatial plant growth variability, airborne digital imagery can be a helpful data source. Let's take a Grain Sorghum as an example. During the 1998 growing season, around five times color-infrared (CIR) images were captured from that grain sorghum field. Based on digital images taken at previous stages of grain sorghum, patterns of plant growth that may be observed later can be revealed. The main benefit is that the plant growth variation can be observed during the growing season and

to correct the problems, proper actions can be taken accordingly. Airborne digital imagery provides helpful data for both during-season and after-season management of crop (Yang et al., 2000).

12.4.2 ANN MODEL FOR CROP YIELD RESPONDING TO SOIL PARAMETERS

Crop yield estimation using ANN (Artificial Neural Network) has been proved self-adaptive and more powerful than the traditional methods like linear analysis and simple non-linear analysis (Jiang et al., 2004; Simpson, 1994; Baret et al., 1995; Jiang, 2000). By preparing a Back Propagation (BP) neural network, an ANN model responding to soil parameters was trained to estimate crop yield. The designed BP-ANN model shows the relationship between soil parameters and crop yield. Soil moisture, P, N, K, SOM and N_{total} are taken as input nodes, and the crop yield will be only one node present in the output layer. Result shows that with increasing soil parameters crop yield also increases and for different soil parameters the sensitivity level of crop yield keeps changing (Liu et al., 2005).

Some of the various methods which can be helpful in the field of crop monitoring can be depicted and described with their brief methodologies and their disadvantages as follows:

Sr No.	System / Meyhod / Tool	Technology	Description	Limitation	Reference
1.	Fuzzy Logic	SRC (DSS)	Elaborate decision-making in soil risk characterization.	Requires huge data.	López et al. (2008)
2.	ANN	SEA prediction	Study carried out to forecast the activity of enzymes.	Priority of classification is more over performance improvement of sand.	Tajik et al. (2012)
3.	ANN	DEM	From soil attributes, it predicts high-resolution soil textures by calculating hydrological parameters.	Could not generate precise prediction due to poor generation of over-fitted ANN.	Zhao et al. (2009)
4.	ANN	Remote sensing image analysis.	Analysis of remotely sensed data using algorithms and important architectures.	Only applicable to limited number of inputs.	Chang and Islam (2000)
5.	Precision Agriculture	VRT	For precise sensing of various crops.	Requires accurate mapping of crop growth, infectious diseases, etc	Zhang and Kovacs (2012)
6.	Precision Agriculture	LARS (Low Altitude Remote Sensing System)	Acquires Earth Surface images at low altitude using UAS systems.	Disadvantages like its maintenance and installation	Zhang and Kovacs (2012)
7.	Precision Agriculture	WSN System using IEEE technology	Provides motion detection, agricultural Data monitoring over long distances using camera sensors.	Energy consumption of sensors.	Garcia-Sanchez et al. (2011)

12.5   CONCLUSION

This book chapter discusses and focuses mainly on the application of artificial intelligence and its various methods and functions which can be applied in the field of agriculture sector for upgrading farming and eventually increasing farm products. By overcoming the traditional methods and upgrading its

efficiency up to 90% AI has reflected its major results in every sector of Agriculture. Main focus of this chapter is on Crop Monitoring. Using AI, not only the this sector but also the majority occupied sector specially in India can be greatly developed and improved. Some of the other advantages like decrease in environmental antagonistic impacts has also reduced in a great amount. Furthermore gradually farm products increases the economic development of country using Robots and hence decreasing hand efforts of farmers and achieving tasks in lesser period of time.Many Agro-based businesses have increased in a big amount all over the areas in few past years by implementing these technologies of AI. Giving up marking accurate and correct results, and its potentiality, AI has drastically strengthened economically as well as eco- friendly outcomes. Although not a wider area has been discussed in the paper, but some of the greatly demanded tasks can be easily reviewed in this paper.

DECLARATION

AUTHORS CONTRIBUTION

All the authors make a substantial contribution to this manuscript. DS, PP and MS participated in drafting the manuscript. DS, PP and MS wrote the main manuscript. All the authors discussed the results and implication on the manuscript at all stages.

ACKNOWLEDGEMENTS

The authors are grateful to Department of Information and Communication Technology, Department of Computer Engineering and Department of Chemical Engineering School of Technology, Pandit Deendayal Petroleum University for the permission to publish this research.

AVAILABILITY OF DATA AND MATERIAL

All relevant data and material are presented in the main paper.

COMPETING INTERESTS

The authors declare that they have no competing interests.

FUNDING

Not Applicable

CONSENT FOR PUBLICATION

Not applicable.

ETHICS APPROVAL AND CONSENT TO PARTICIPATE

Not applicable.

REFERENCES

Aber, J. S., Marzolff, I., Ries, J. B., & Aber, S. E. (2019). Introduction to small-format aerial photography. *Small-FormatAerialPhotographyandUASImagery*, 1–10. https://doi.org/10.1016/b978-0-12-812942-5.00001-x.
Anthony, D., Elbaum, S., Lorenz, A., & Detweiler, C. (2014). On crop height estimation with UAVs. *IEEE/RSJ International Conference on Intelligent Robots and Systems (IROS 2014)*, Chicago, Illinois, USA, pp. 4805–4812.

Banerjee, G., Sarkar, U., Das, S., & Ghosh, I. (2018). Artificial intelligence in agriculture: A literature survey. *International Journal of Scientific Research in Computer Science Applications and Management Studies*, *7*(3), 1–6.

Baret, F., Clevers, J., & Steven, M. (1995). The robustness of canopy gap fraction estimates from red and near-infrared reflectances: A comparison of approaches. *Remote Sensing of Environment*, *54*(2), 141–151. https://doi.org/10.1016/0034-4257(95)00136-o.

Chang, D., & Islam, S. (2000). Estimation of soil physical properties using remote sensing and artificial neural network. *Remote Sensing of Environment*, *74*(3), 534–544. https://doi.org/10.1016/s0034-4257(00)00144-9.

Dharmaraj, V., & Vijayanand, C. (2018). Artificial intelligence (AI) in agriculture. *International Journal of Current Microbiology and Applied Sciences*, *7*(12), 2122–2128. https://doi.org/10.20546/ijcmas.2018.712.241.

Garcia-Sanchez, A., Garcia-Sanchez, F., & Garcia-Haro, J. (2011). Wireless sensor network deployment for integrating video-surveillance and data-monitoring in precision agriculture over distributed crops. *Computers and Electronics in Agriculture*, *75*(2), 288–303. https://doi.org/10.1016/j.compag.2010.12.005.

Giles, D. K., & Billing, R. C. (2015). Deployment and performance of a UAV for crop spraying. *Chemical Engineering Transactions*, *44*, 307–322.

Huang, Y., Hoffmann, W. C., Lan, Y., Wu, W., & Fritz, B. K. (2009). Development of a spray system for an unmanned aerial vehicle platform. *Applied Engineering in Agriculture*, *25*(6), 803–809. https://doi.org/10.13031/2013.29229.

Jiang, D. (2000). Study on crop yield forecasting model using remote sensed information supported by artificial neural network [Unpublished doctoral dissertation]. Institute of Geographic Science and Natural Resources Research, Chinese Academy of Sciences.

Jiang, D., Yang, X., Clinton, N., & Wang, N. (2004). An artificial neural network model for estimating crop yields using remotely sensed information. *International Journal of Remote Sensing*, *25*(9), 1723–1732. https://doi.org/10.1080/0143116031000150068.

Kadiyala, S., Harris, J., Headey, D., Yosef, S., & Gillespie, S. (2014). Agriculture and nutrition in India: Mapping evidence to pathways. *Annals of the New York Academy of Sciences*, *1331*(1), 43–56. https://doi.org/10.1111/nyas.12477.

Kekane, M. A. (2013). Indian agriculture- status, importance and role in Indian economy. *International Journal of Agriculture and Food Science Technology*, *4*(4), 343–346.

Liu, G., Yang, X., & Li, M. (2005). An artificial neural network model for crop yield responding to soil parameters. *Advances in Neural Networks* – ISNN 2005, 1017–1021. https://doi.org/10.1007/11427469_161.

López, E. M., García, M., Schuhmacher, M., & Domingo, J. L. (2008). A fuzzy expert system for soil characterization. *Environment International*, *34*(7), 950–958. https://doi.org/10.1016/j.envint.2008.02.005.

López-Granados, F., Torres-Sánchez, J., De Castro, A., Serrano-Pérez, A., Mesas-Carrascosa, F., & Peña, J. (2016). Object-based early monitoring of a grass weed in a grass crop using high resolution UAV imagery. *Agronomy for Sustainable Development*, *36*(4). https://doi.org/10.1007/s13593-016-0405-7.

McKinion, J., & Lemmon, H. (1985). Expert systems for agriculture. *Computers and Electronics in Agriculture*, *1*(1), 31–40. https://doi.org/10.1016/0168-1699(85)90004-3.

Mogili, U. R., & Deepak, B. B. (2018). Review on application of drone systems in precision agriculture. *Procedia Computer Science*, *133*, 502–509. https://doi.org/10.1016/j.procs.2018.07.063.

Plant, R. E. (2001). Site-specific management: The application of information technology to crop production. *Computers and Electronics in Agriculture*, *30*(1–3), 9–29. https://doi.org/10.1016/s0168-1699(00)00152-6.

Primicerio, J., Di Gennaro, S. F., Fiorillo, E., Genesio, L., Lugato, E., Matese, A., & Vaccari, F. P. (2012). A flexible unmanned aerial vehicle for precision agriculture. *Precision Agriculture*, *13*(4), 517–523. https://doi.org/10.1007/s11119-012-9257-6.

Robert, P. C. (2002). Precision agriculture: A challenge for crop nutrition management. *Progress in Plant Nutrition: Plenary Lectures of the XIV International Plant Nutrition Colloquium*, 143–149. https://doi.org/10.1007/978-94-017-2789-1_11.

Simpson, G. (1994). Crop yield prediction using a CMAC neural network. *Proceedings of the Society of Photo-Optical Instrumentation Engineers*, *2315*, 160–171.

Steven, M.D., & Clark, J.A. (2013). *Applications of Remote Sensing in Agriculture*, Elsevier B.V., USA.

Tajik, S., Ayoubi, S., & Nourbakhsh, F. (2012). Prediction of soil enzymes activity by digital terrain analysis: Comparing artificial neural network and multiple linear regression models. *Environmental Engineering Science*, *29*(8), 798–806. https://doi.org/10.1089/ees.2011.0313.

Van Blyenburgh, P. (1999). UAVs: An overview. *Air & Space Europe*, *1*(5–6), 43–47. https://doi.org/10.1016/s1290-0958(00)88869-3.

Wijitdechakul, J., Sasaki, S., Kiyoki, Y., & Koopipat, C. (2016). UAV-based multispectral image analysis system with semantic computing for agricultural health conditions monitoring and real-time management. *2016 International Electronics Symposium (IES)*, Denpasar, Indonesia, pp. 459–464.

Yang, C. H., Everitt, J. H., Bradford, J. M., & Escobar, D. E. (2000). Mapping grain sorghum growth and yield variations using airborne multispectral digital imagery. *Transactions of the ASAE, 43*(6), 1927–1938. https://doi.org/10.13031/2013.3098.

Zhang, C., & Kovacs, J. M. (2012). The application of small unmanned aerial systems for precision agriculture: A review. *Precision Agriculture, 13*(6), 693–712. https://doi.org/10.1007/s11119-012-9274-5.

Zhao, Z., Chow, T. L., Rees, H. W., Yang, Q., Xing, Z., & Meng, F. (2009). Predict soil texture distributions using an artificial neural network model. *Computers and Electronics in Agriculture, 65*(1), 36–48. https://doi.org/10.1016/j.compag.2008.07.008.

13 Application of Microbial Enzyme in Food Biotechnology

Oseni Kadiri
Edo State University Uzairue

Temitope Omolayo Fasuan
ObafemiAwolowo University

Charles Oluwaseun Adetunji
Edo State University Uzairue

Deepak Panpatte
College of Agriculture, Dongarshelki Tanda
Vasantrao Naik Marathwada Agricultural University

Olugbenga Samuel Michael
Bowen University

Daniel Ingo Hefft
Reaseheath College

Juliana Bunmi Adetunji
Osun State University

CONTENTS

13.1 INTRODUCTION

Microorganisms have been applied in numerous fermentation processes most especially from the ancient time which are utilize for the production of numerous food items (Soccol et al., 2005). It has been identified that microbial enzymes play a crucial role in several industries that based their product on food. This might be linked to the fact that animal and plant enzymes are more stable. They could be fabricated through the application of fermented methods which is more economical, high consistency and require minimal time while the process involved in the modification and the optimization of parameters are very simple (Gurung et al., 2013).

Several enzymes are utilized in numerous industrial sector, for the example the application of amylolytic enzymes in detergent, textile, food, paper industries (Pandey et al., 2000). Moreover,

DOI: 10.1201/9781003268468-13

some other application include their usage in the production of maltose syrups, glucose syrups, high fructose corn syrups, and crystalline glucose. Also, some other enzymes such as xylanases, proteases and lipases have a wider utilization in the food sector (De Souza and Magalhães, 2010).

Enzymes could be referred to as green" biological catalysts have been identified to possess the capacity to influence the way that most of our food are being prepared. Moreover, that application of enzymes have been highlighted in different feed and food industries which entail baking, oils and fats, dairy, dietary supplements, brewing, vegetable processing, meat, juice and beverages (Robinson, 2015 and Schäfer, 2007). The utilization of beneficial microorganism and enzymes have been recognized as a tradition techniques that play a crucial role most especially in the processing of food such as cheese, bread baking, wine and beer respectively (Fernandes 2010).

The application of Biotechnology has been shown to play a crucial role in the enhancement of processing of raw materials most especially their effective transformation to food products which could eventually led to a high level of nutritional value in the fermented food product (Underkofler et al., 1958). Fermentation has been identified as a microbial biotechnology through which natural renewable substrates are transformed into value-added products which entails polymers, enzymes, alcohols, and organic acids respectively.

Moreover, some examples of fermentation products are proton sinks such as lactic acid and ethanol but the NADH is reprocessed to NAD+. This enable the cell to continue generating energy through the process of glycolysis by substrate-level phosphorylation. Therefore, it has been identified that microorganism portends the capability to produce numerous end and by product that could uphold energy balance. The application of genetical engineering could led to the enhancement of some beneficial strains and thereby improving the quantity and quality relevant industrial products which produced through food processing techniques (Campbell-Platt, 1994).

Moreover, it has been discovered that the development of end or by-products depend on different types of environmental conditions and the types of microbial strain utilized. Therefore, there is a need to select a better and enhanced microbial strains that could led to production of optimized condition during the process of fermentation. The application of recombinant DNA technology and mutation could led to the production of desired product in term of quality and quantity (Demirci et al., 2014).

Specific example of some condition that could influence the development of fermented product and development of microorganisms include the composition of the fermentation medium, temperature, dissolved oxygen, and pH. Therefore, there is a need to performed optimization of these crucial parameters so as to establish the best fermentation condition that could led to increase in food processing. Hence, it has become paramount to evaluate the metabolic pathways. Adequate information about the biochemical alteration in most fermented foods could led to the generation of better strains that could led to the production desired products. Some other factors that could affect productivity includes fermentation modes, the types of media, and types of strains. Also, in order to achieve a higher productivity this might depend on the following types of fermentation which includes continuous, Fed-batch, batch, could be led to the production of a better food products. It has been observed that the application of continuous and batch types of fermentation could led to prevention of the substrate limitation most especially during the process of fermentation.

Furthermore, the application of cell immobilization, could led to improvement in the enhancement of biomass concentration most especially in the bioreactor, thereby facilitating the concentration of the enzyme or the biocatalyst available in the bioreactor. There is need to extract, purify the microbial product after fermentation but the process involve in the purification of most and product have been established to be very expensive. Moreover, microbial end-products could be derived from the biomass, intracellular and extracellular products. The application of homogenization, filtration, and extraction of product whether solid and liquid are typical illustration of recovery techniques (Demirci et al., 2014).

Enzymes have been identified as a products derived from living microorganism with high application in the industry. The application and relevance of enzyme have been documented for numerous year which might be linked to their high catalytic activities. The activities of enzymes depends

on several factors such as substrate, inhibitors, temperature, pH. These is a need to optimize these parameters in order to have an effective food products. Enzymes could be recovered through the application of enzyme immobilization in order to reduce the cost effectiveness. Enzyme could be derived from mammalian tissues, microorganism and plants respectively. Moreover, the enzymes obtained from microorganism is more preferred when compared to other enzymes derived from other sources. This might be linked to their high level of specificity and availability. It has been stated that over 500 commercial products are obtained through the application of enzymes. They are utilized in the production of detergent, paper, foods, pharmaceuticals, cosmetics, and leather (Johannes and Zhao 2006). Moreover, typical example of enzyme include rennet which is a milk clotting enzymes, lipase, proteases, pectic, amylases (Demirci et al., 2014).

13.2   MICROBIAL ENZYMES FOR FOOD APPLICATION

Microorganism and enzymes application in the processing of food is a well-known approach which has been adequately documented over the century. Beer brewing, bread baking, cheese and wine production are processes which have been known to employ enzymes and microorganism for ages. The use of enzymes in biotechnology is improving the way by which raw materials can be converted to diverse food products with improved nutritional value, functionality, flavour and texture. Some common enzymes sources includes production from microorganisms like yeasts, bacteria, fungi, and actinomycetes. Animal and plant are other sources for enzyme production.

Specifically, some globally use enzymes viz pectinase, lipase, α-amylase, protease, and glucoamylase use in food-processing industries. In the production light quality beer, glucoamylase changes the residual dextrins into the corn syrup form which are fermentable sugars. In a related fashion, starch is converted to dextrins by α-amylase to produce corn syrup for various applications. In both instance, the food product sweetness is enhanced after the enzyme activities. Lipases are flavour enhancer and also reduce cheese ripening time, and produce better quality fat products. In the production of fruit juice, pectinase is use in the extraction, clarification, and filtration process. Chymosin enzyme degrades kappa-caseins during the process of milk curdling. Proteases enzymes of fungal and bacterial are employ in the production of texturized proteins, meat extenders, fish meal, and meat extracts. In the instance of whey and milk products, lactase breaks down the lactose to produce polyactide. Glucose oxidases inhibit the Maillard reactions process through its conversion of glucose into gluconic acid. In wine making, maturation time is minimize with the help of acetolactate decarboxylase which converts acetolactate to acetoin. In grain processing, cellulose aid in cellulose to glucose conversion in cell walls thereby facilitating better cellular product extraction and increase nutrient release. Enzyme application is improving the way food manufacturing industry operates, while lowering production costs and improving production processes

The application of lipases, catalase, proteases, esterases, and lactase in dairy technology are well established. Proteinases has found application in acceleration of cheese ripening in recent times, along with it use as modifying agent of functional properties and in dietetic products preparation. Milk coagulation, using rennet was one of its earliest applications in cheese manufacturing. Proteinase accelerates the hydrolysis of protein, enhances cheese ripening and enhances cheese ripening, flavour and texture. Transglutaminase is another enzyme which is used to improve the functional properties of dairy product. It aids in peptide bonds cross-linking. Transglutaminase have also been used texture improvers in the meat and catering industry.

Typical source of the Lactase (β-galactosidase) enzyme are the K. lactis, A. niger and A. oryzae. Lactase promotes milk lactose hydrolysis to monosaccharaides such as glucose, and galactose. This allows the production of milk-based products which is suitable for consumption by consumer suffering from lactose intolerance. Lactose intolerance also results in milk sweetness thereby minimizing the need for sugar addition.

In wine making, combine action of β-glycosidases and exoglycosidases enables molecules responsible for aroma compounds to be released through the liberation of β-glucosides and sugars.

These enzymes invariably cleave the intersugar linkage of glycosides. The protease enzyme tenderizes meat by retaining myofibrillar proteins while decreasing the sum total of connective tissue. Carbohydrases, lipases, and protease are enzymes used in fish protein hydrolysates preparation. Some authors have reported the use of trans-glutaminases in meat processing (Ashie and Lanier, 2000; Kieliszek and Misiewicz, 2014). Trans-glutaminases transformed low quality collagen to highly nutritive products, when the appropriate amino acids wee added. The enzymes; pullulanases, α-glucosidases, amylglucosidases, α- and β-amylases are required for the hydrolysis of starch to glucose units (Blanco et al., 2014).

The α-acetolactate decarboxylase is an important enzyme in the beer making process as it is involved in the conversion of acetolactate to acetoin. Acetolactate gives beer a butterscotch tastes while is rather tasteless. α-acetolactate decarboxylase also speeds up the beer maturation process (Dulieu et al., 2000) (Table 13.1).

13.3   MICROBIAL ENZYMES IN FOOD BIOTECHNOLOGY AND PROCESSING

Over time, microorganisms had found applications in the production of enzymes for various industrial purposes. Enzymes obtained from microbial origin had been found to be beneficial than those from plants and animals. The advantages of microbial enzymes over those from plant and animal include lower cost of production, can be produced on large-scale, short culture development, less material involved, as well as environmentally friendly (Hasan et al., 2006; Mishra et al., 2017). Some good examples of the frequently utilized bacterial-sourced food enzymes are presented in Table 13.2.

TABLE 13.1
Example of Enzymes Used in Processing

Enzyme	Enzyme Class	Role
Lactase	Hydrolase	Hydrolysis of lactose, whey and galactooligosaccharides synthesis.
Amylase	Hydrolase	Saccharification and liquefaction of starch
Invertase	Hydrolase	Hydrolysis of sucrose and sugar syrups production
Arabinose isomerase	Isomerases	Galactose isomerization to tagatose
Cellobiose	Isomerases	Lactose isomerization to lactulose
Acetolactate decarboxylase	Lyases	Fastening beer maturation
Glucose oxidase	Oxidoreductases	Improved dough strength and handling characteristics
Cyclodextrin glycosyltransferase	Transerases	Cyclodextrins production
Fructosyl transferase	Transerases	Prebiotics production and synthesis of fructooligosaccharides
Transglutaminase	Transerases	Cross-linking function. Application in dough and meat processing
Lipoxygenase	Oxidoreductases	Dough strengthening, bread whiting
Laccases	Oxidoreductases	Improve dough strength, stabilization of wine colour during winemaking, increased storage life of beer, improve vegetable oils flavour and in cork stopper preparation
Glucanase	Hydrolase	Breaks cereal cell walls, beer making
Lipase	Hydrolase	Aromatic molecules synthesis, improves lipid digestion in young animals, in situ emulsification for the conditioning of dough, cheese flavour
Phospholipase	Hydrolase	Dough conditioning
Proteases (e.g, papain)	Hydrolase	Meat tenderizer, haze formation prevention in brewing, improvement of flavour in cheese and milk, low allergenic infant food formulation, milk clothing, protein hydrolysis, enhanced digestibility and utilization
Pectinase	Hydrolase	Fruit juice clarification, viscosity reduction
Peptidase	Hydrolase	Cheese ripening

TABLE 13.2
Microbial Enzymes and Food Use

Microbial Enzyme	Use	Reference
Glucoamylase	Beak making, production of beer	Blanco et al. (2014); Raveendran et al. (2018)
lactase (β-galactosidase)	Wheat processing, an ingredient in prebiotic	Gibson and Wang (1994); Madden (1995)
Rhizomucormiehei	Precipitation of protein in milk	Madden (1995)
Endothiaparasitica	Precipitation of protein in milk	Madden (1995)
Rhizomucorpusillus	Precipitation of protein in milk	Madden (1995)
Tyrosinases	Manufacturing of mixed-melanins, protein cross linking, phenolic based bio-sensors	Fairhead and Thony-Meyer (2012)
Amylase (from *Aspergillusoryzae*)	Baking, brewing, cakes, starch syrups, fruit juices, production of ethanol	Maryam et al. (2017)
Proteases (from *Aspergillusniger*)	Tendering of meat, hydrolysis of whey, bread making	Aruna et al. (2014); Miguel and Martins-Meyer (2013)
Pectinases (from *Aspergillusniger*)	Clarification of juice, to increase the yield of juice	Pasha et al. (2013)
Glucose oxidase (from *Penicilliumnotatum*)	To remove oxygen, dough stability, flavour enhancer	Zhu et al. (2006); Maryam et al. (2017)
Catalase (from *Aspergillusniger*)	Mayonnaise, production of cheese	Sîrbu (2011); Raveendran et al. (2018)
Invertase (from *Saccharomyces cerevisiae*)	Hydrolysis of sucrose, invert sugar syrup production	Singh and Kumar (2019)
Lactase (from *Saccharomyces fragilis*)	Hydrolysis of lactose, hydrolysis of whey	Singh and Kumar (2019)
Lipase (from *Bacillus, Pseudomonas, Burkholderia*)	Production of cheese	Aravindan et al. (2007); Jooyendeh et al. (2009)
Amyloglucosidase	Conversion of starch to sugar	Al-Maqtari et al. (2019)
Phospholipase	Production of cheese, lipolysis of milk fat	Law (2009); Al-Maqtari et al.(2019)
Papain	Tendering of meat	Al-Maqtari et al. (2019)
Inulinases	Fructose syrups production	Al-Maqtari et al. (2019)
Cellulase	Production of animal feed, fruit juice clarification	Grassin and Fauquembergue (1996); Sukumaran et al. (2005)
Xylanase	Production of beer, juice clarification	Dervilly et al. (2002); Camacho and Aguilar (2003)
Laccase	Removal of polyphenol from wine, baking	Labat et al. (2000); Tanriöven and Ekşi (2005)
Peroxidase	Flavour development, enhance the nutritional quality of food	Regaldo et al. (2004)
Phytase	Dairy	Maryam et al. (2017)
Hemicellulase	Baking, mayonnaise production	Maryam et al. (2017)
Glucose isomerase	Starch	Maryam et al. (2017)
Chymosin	Mayonnaise	Maryam et al. (2017)
β-Glucosidase	Aroma enhancer in wine production	Caldini et al. (1994); Gunata et al. (1990)
Pectin esterase	Enhancement in the clarification of cider	Uhlig (1998)

13.4 CONCLUSION AND FUTURE RECOMMENDATION TO KNOWLEDGE

This chapter has provided holistic information on the application of microbial enzymes in food processing. Detailed information on the application of genetic engineering and mutation as a sustainable biotechnology techniques that could enhanced several process involved in the production of food were highlighted. Several factors could led to increase in the production of fermented foods were discussed. Specific examples of fermented foods that are produced using microbial enzymes were also highlighted. The application of these enzymes derived from microorganism should also be sources from different sources such as marine area and high temperature environment. This will led to the production of a more a stable enzyme that could withstand a very high temperature. This have several application in the production of fermented food product even at a very high temperature. The utilization of various agricultural and industrial wastes could serves as an alternative to the costly and synthetic media so as to increase the production of a more cheaper, and cost feasible enzymes with differs application of different industrial products most especially that are food based through the process of microbial fermentation. This will also increase the production of a high quality food products such as dairy, cereals, brewing, beverages, wine making, milling, cheese making respectively.

REFERENCES

Al-Maqtari, Q. A., Al-Ansi, W., Mahdi, A. A. (2019). Microbial enzymes produced by fermentation and their applications in the food industry – a review. *International Journal of Agriculture Innovations and Research*, 8(1), 62–82.

Aravindan, R., Anbumathi, P., Viruthagiri, T. (2007). Lipase applications in food industry. *Indian Journal of Biotechnology*, 6, 141–58.

Aruna, K., Shah, J., Birmole, R. (2014). Production and partial characterization of alkaline protease from Bacillus tequilensis strains CSGAB 0139 isolated from spoilt cottage cheese. *International Journal of Applied Biology and Pharmaceutical Technology*, 5, 201–21.

Blanco, C. A., Caballero, I., Barrios, R., Rojas, A. (2014). Innovations in the brewing industry: Light beer. *International Journal of Food Sciences and Nutrition*, 65(6), 655–60.

Caldini, C., Bonomi, F., Pifferi, P. G., Lanzarini, G., Galante, Y. M. (1994). Kinetic and immobilization studies on fungal glycosidases for aroma enhancement in wine. *Enzyme and Microbial Technology*, 16, 286–91.

Camacho, N. A., Aguilar, O. G. (2003). Production, purification and characterization of a low-molecular-mass xylanase from Aspergillus sp. and its application in baking. *Applied Biochemistry and Biotechnology*, 104(3), 159–72.

Campbell-Platt, G. (1994). Fermented foods-a world perspective. *Food Research International*, 27(3), 253–7.

De Souza, P. M., Magalhães, P. O. (2010). Application of microbial α-amylase in industry –A review. *Brazilian Journal of Microbiology*, 41(4), 850–61. https://doi.org/10.1590/S1517-83822010000400004.

Demirci, A., Izmirlioglu, G., Ercan, D. (2014). *Fermentation and Enzyme Technologies in Food Processing. Food Processing: Principles and Applications*, 2nd ed. New York: Wiley, pp. 107–36.

Dervilly, G., Leclercq, C., Zimmerman, D., Roue, C., Thibault, J. F., Sauliner, L. (2002). Isolation and characterization of high molar mass water-soluble arabinoxylans from barley and barley malt. *Carbohydrate Polymers*, 47(2), 143–9.

Dulieu, C., Moll, M., Boudrant, J., Poncelet, D. (2000). Improved performances and control of beer fermentation using encapsulated α-acetolactate decarboxylase and modeling. *Biotechnology progress*, 16(6), 958–65.

Fairhead, M., Thony-Meyer, L (2012). Bacteria tyrosinases: Old enzymes with new relevance to biotechnology. *New Biotechnology*, 29(2), 183–91.

Fernandes, P. (2010). Enzymes in food processing: A condensed overview on strategies for better biocatalysts. *Enzyme Research*, 2010, 1–19. https://doi.org/10.4061/2010/862537.

Gibson, G. R., Wang, X. (1994). Regularory effects of bifidobacteria on the growth of other colonic bacteria. *Journal of Applied Microbiology*, 77, 412–20.

Grassin, C., Fauquembergue, P. (1996). Fruit juices. In: Godfrey T, West S, editors. *Industrial Enzymology*. London, UK: MacMillan Press, pp. 225–264.

Gunata, Y. Z., Bayonove, C. L., Cordonnier, R. E., Arnaud, A., Galzy, P. (1990). Hydrolysis of grape monoterpenyl glycosides by *Candida molischiana*and *Candida wickerhamii*b-glucosidases. *Journal of the Science of Food and Agriculture*, 50, 499–506.

Gurung, N., Ray, S., Bose, S., Rai, V. (2013). A broader view: Microbial enzymes and their relevance in industries, medicine, and beyond. *BioMed Research International*, 2013, 1–18. https://doi.org/10.1155/2013/329121.

Hasan, F., Shah, A. A., Hameed, A. 2006. Industrial application of microbial lipase. *Enzyme and Microbial Technology*, 39(2), 235–51.

Johannes, T. W., Zhao, H. (2006). Directed evolution of enzymes and biosynthetic pathways. *Current Opinion in Microbiology*, 9(3), pp. 261–7.

Jooyendeh, H., Kaur, A., Minhas, K. S. (2009). Lipases in the dairy industry: A review. *Journal of Food Science and Technology*, 46(3), 181–9.

Kieliszek, M., Misiewicz, A. (2014). Microbial transglutaminase and its application in the food industry. A review. *Folia Microbiologica*, 59(3), 241–50.

Labat, E., Morel, M. H., Rouau, X. (2000). Effects of laccase and ferulic acid on wheat flour doughs. *Cereal Chemistry*, 77, 823–28.

Law, B. A. (2009). Enzymes in dairy product manufacture. In: Van Oort M, Whitehurst RJ, editors. *Enzymes in Food Technology*. Oxford, UK: Wiley-Blackwell.

Madden, D. (1995). *Food Biotechnology: An Introduction*. Washington, DC: International Life Sciences Institute.

Maryam, B. M., Datsugwai, M. S. S., Shehu, I. (2017). The role of biotechnology in food production and processing. *Engineering and Applied Sciences*, 2(6), 113–24

Miguel, Â. S. M., Martins-Meyer, T. S. (2013). Veríssimo da Costa Figueiredo E, Lobo BWP, Dellamora-Ortiz GM. Enzymes in bakery: Current and future trends. In: Muzzalupo I, editor. *Food Industry*. Rijeka, Croatia: InTech, pp. 287–321.

Mishra, S., Ray, R. C., Rosell, C. M., Panda, D. (2017). *Microbial Enzyme Technology in Food Applications*, Ray RC, Rossell CM, editors. Boca Raton, FL: CRC Press, pp. 1–17.

Pandey, A., Nigam, P., Soccol, C. R., Soccol, V. T., Singh, D., Mohan, R. (2000). Advances in microbial amylases. *Biotechnology and Applied Biochemistry*, 31(2), 135. https://doi.org/10.1042/ba19990073.

Pasha, K. M., Anuradha, P., Subbarao, D. (2013). Applications of pectinases in the industrial sector. *Journal of Pure and Applied Science & Technology*, 16, 89–95.

Raveendran, S., Parameswaran, B., Ummalyma, S. B., Abraham, A., Mathew, A. K., Madhavan, A., Rebello, S., Pandey, A. (2018). Applications of microbial enzymes in the food industry. *Food Technology and Biotechnology*, 56(1), 16–30.

Regaldo, C., García-Almendárez, B. E., Duarte-Vázquez, M. A. (2004). Biotechnological applications of peroxidizes. *Phytochemistry Reviews*, 3(1–2), 243–56.

Robinson, P. K. (2015). Enzymes: Principles and biotechnological applications. *Essays in Biochemistry*, 59, pp. 1–41.

Singh, P., Kumar, S. (2019). Microbial enzyme in food biotechnology. In: Mohammed Kuddus (Ed.), *Enzymes in Food Biotechnology*. San Diego, CA: Elsevier, pp. 19–28.

Sîrbu, T. (2011). The searching of active catalase producers among the microscopic fungi. *Analele Universitatii din Oradea, Fascicula Biologie*, 2, 164–7.

Soccol, C. R., Rojan, P. J., Patel, A. K., Woiciechowski, A. L., Vandenberghe, L. P. S, Pandey, A. (2005). Glucoamylase. In: Pandey A, Webb C., Soccol CR, Larroche C, editors. *Enzyme Technology*. New Delhi: Asiatech Publishers Inc, pp. 221–38.

Sukumaran, R. K., Singhania, R. R., Pandey, A. (2005). Microbial cellulases - Production, applications, and challenges. *Journal of Scientific Research*, 64, 832–44.

Tanrıöven, D., Ekşi, A. (2005). Phenolic compounds in pear juice from different cultivars. *Food Chemistry*, 93, 89–93.

Uhlig, H. (1998). *Industrial Enzymes and Their Applications*. New York: John Wiley & Sons, Inc.

Underkofler, L., Barton, R., Rennert, S. (1958). Production of microbial enzymes and their applications. *Applied Microbiology*, 6(3), p. 212.

Zhu, Z., Momeu, C., Zakhartsev, M., Schwaneberg, U. (2006). Making glucose oxidase fit for biofuel cell applications by directed protein evolution. *Biosensors and Bioelectronics*, 21, 2046–51.

14 Nanosensor Technology for Smart Intelligent Agriculture

Suresh Kaushik
Indian Agricultural Research Institute

CONTENTS

14.1 INTRODUCTION

Nanotechnology is the manipulation and use of matter with dimensions smaller than 100 nm. Recent advances in nanotechnology have earned strength in the current agriculture system. Nanotechnology can boost agricultural production by improving nutrient use efficiency with nano-formulations of fertilizers, agrochemicals for crop enhancement, detection and treatment of diseases, host-parasite interactions at the molecular level using nanosensors, plant diagnostics, post-harvest management

DOI: 10.1201/9781003268468-14

of vegetables, reclamation of salt-affected soils, and contaminants removal from soil and water. Nanotechnology has offered huge potential to reduce the impact of major stresses on crop productivity while optimizing the use of limited resources such as nutrients and water.

For a continuously growing human population, we need a huge increase in crop productivity to meet the food demand in the coming decades. The global rise in the human population has enhanced the need for qualitative and quantitative improvement in crop productivity. But abiotic and biotic stresses impair growth and yield, leading to major crop losses all over the world (Joshi et al. 2018; Suzuki et al. 2014). Both biotic and abiotic stresses lead to a massive loss in crop yield, leading to a decrease in agricultural production worldwide. Abiotic stresses such as heat (Mickelbart et al. 2015), salinity (Suzuki et al. 2014), drought (Fahad et al. 2017), flooding (de San Celedonio et al. 2018) and frost (Guillaume et al. 2018; Mickelbart et al. 2015) are the main factors for economic losses in agriculture while biotic factors such as bacterial and viral diseases, insect infestation are leading to devastating crop losses (Chakraborty and Newton 2011; Fisher et al. 2012; Scholthof et al. 2011). The loss of agricultural products can be minimized by adopting modern technology such as smartphones with nanosensors to detect crop stress at an early stage.

Precision agriculture or smart agriculture, allows farmers to maximize yield using minimal resources such as seeds, water, and fertilizers. Precision farming has been a long-desired goal to maximize output while minimizing input through monitoring environmental variables and applying targeted action (Kaushik and Djiwanti 2019). Smart agriculture has roots going back to the 1980s when global positioning system capability became accessible for civilian use along with the use of computers and remote sensing devices to measure highly localized environmental conditions to determine whether crops are growing at maximum efficiency or precisely identifying the nature and location of problems.

Early smart agriculture users adopted crop yield monitoring to produce fertilizer recommendations. But for more accurate recommendations during the 1990s using several sensing technologies and data generated, more variables are included into a crop model for fertilizer applications, watering, and even peak yield harvesting. Hence, precise management of limited resources and costly agrochemicals such as nutrients, pesticides, and water offers the opportunity to enhance crop productivity by minimizing resource losses through the use of remote sensing methods for crop monitoring (Hatfield et al. 2008; Padilla et al. 2018). Enhancing agricultural productivity will require innovative and new technological approaches such as nanosensors for managing crop stressors and resource use efficiency.

With the help of nanosensors, plant stresses such as biotic, abiotic, and nutritional deficiency are monitored using various electronic devices including smartphones and cameras. Nanosensors communicate with and actuate electronic devices for improving crop productivity by optimization and automation of water and agrochemical allocation. Thus, these nanosensors will be able to report crop health status for precise and efficient use of resources. This technology is highly beneficial for detecting the onset of biotic and abiotic stress at an early stage. Nanotechnology has distinct advantages to engineer smart plant sensors because engineered nanomaterials can be embedded in plants for monitoring signaling molecules by near-infrared cameras in real-time (Giraldo et al. 2019). To fabricate such nanobiosensors, existing tools and technologies such as microfluidics, plasmonic nanosensors, Surface-Enhanced Raman Scattering Nanosensors (SERS), fluorescence, chemiluminescence, quartz crystal microbalance, molecular imprinted polymers, advanced electrochemical measurements coupled with customized nanomaterials and nanocomposites are used. The real-time crop monitoring technology ensures high agriculturally produce with the precise use of costly agrochemicals. It also helps minimize the loss of limited resources such as water or nutrients. Therefore, the advancement in nanotechnology helps reduce the impact of biotic and abiotic stress on agricultural produce and enhance the optimal use of limited resources. In this chapter, we discuss how nanotechnology-based sensors in plants can enable communication and actuation of electronic devices for optimizing crop growth and yield in response to resource scarcity or stresses.

TABLE 14.1

Development of Nanosensor for Plant Signaling Molecules Related to Health Status

Nanotechnology-Based Sensor, Probe, Indicator, Reporter	Plant Molecular Targets
SWCNT, HyPer, roGFP-Orp1	H_2O_2
SWCNT, GCaMP3, YC3.6, R-GECO1	Ca^{2+}
SWCNT	NO
SWCNT, FLIPglu-2μVΔ13, FLIPglu-600μΔ13, BA-QD	Glucose
FLIPsuc-90μΔ1	Sucrose
SWCNT	Ethylene
Ag NPs	Methyl salicylate
Jas9-VENUS	Jasmonic acid
ABAleon2.1, ABACUS1	Abscisic acid
GFP H148D	H^+ gradient (pH)

Source: Adapted from Giraldo et al. (2019).

14.2 PLANT SIGNALING MOLECULES FOR MONITORING CROP HEALTH

Detection of analytes in the living plant is optically difficult due to photosynthetic pigments and thick tissues. Hence, nanosensors are more suitable for *in vivo* studies of cellular signaling due to the ease of embedding them into plant tissues. There are several key signaling molecules such as sugars (sucrose and glucose), calcium ions (Ca^{2+}), reactive oxygen species (ROS), molecular oxygen, nitric oxide (NO), hydrogen peroxide (H_2O_2), volatile organic compounds (VOC), adenosine triphosphate (ATP), strigolactones, dopamine, phytohormones including abscisic acid (ABA), jasmonic acid, methyl salicylate and ethylene (Table 14.1), which have been reported to be monitored by engineered nanomaterials (ENMs) and genetically encoded based nanosensors. These plant signaling molecules have the potential to report the onset of crop health changes in real-time by detecting resource deficit or plant stress. Hence, these signaling molecules and chemical traits help to diagnose specific biological or environmental stress.

Plant signaling molecules have the potential to report the onset of crop health changes in real-time by detecting resources deficit or plant stresses. Hence, these chemical traits help to diagnose specific biological or environmental stressor. Both Ca^{2+} and ROS are at the forefront of plant stress signaling (Gilroy et al. 2014). Ca^{2+} is involved in most stress signaling pathways and is evolutionarily conserved among different plant species. ROS has a dual role in plants, acting as toxic molecules at high levels and playing a signaling role in a broad range of plant stress responses at low levels (Mittler 2017; Suzuki et al. 2013). Ca^{2+} and ROS signature signals have been associated with specific plant stress responses (Kiegle et al. 2000; Mittler 2017). NO, ABA, ethylene, and methyl salicylate have been reported to be more specific indicators of plant stress types. NO, ethylene, and methyl salicylate are mainly involved in plant pathways defense responses (Delledonne et al. 1998; Lin et al. 2017; van Loon et al. 2006). ABA is an early signal of water stress (Delledonne et al. 1998; Kim et al. 2010; Lin et al. 2017; van Loon et al. 2006; Yoshida et al. 2014). The plant hormones jasmonate coordinates both biotic and abiotic stress responses including salinity and freezing tolerance, drought, and wounding responses (Howe et al. 2018). Sugars including glucose and sucrose are also important plant molecules that regulate a broad range of physiological and developmental changes (Rolland et al. 2006; Zhu et al. 2017). Other important signaling molecules that can be future targets for sensor development are isoprenes and salicylic acids. Salicylic acid is a signaling molecule involved in plant pathways defense (Klessig et al. 2000). Plant volatile organic compounds including isoprenes are associated with high light, temperature, and water stress (Singsaas and Sharkey 1998). Hence, developing and applying nanosensing approaches for real-time monitoring of these

key chemical signaling molecules in plans will improve our understanding of plant stress communication and enable crop health status monitoring in the field.

Recently, engineers and researchers from the Massachusetts Institute of Technology (MIT) have developed a way to track how plants respond to stress such as infection, injury and light damage using nanosensors made of carbontubes. These nanosensors embedded in plants' leaves, use hydrogen peroxide signals to distinguish between different types of stress and even between different species of plants such as spinach and strawberry. Plants use H_2O_2 to communicate within their leaves, sending out a distress signal that stimulates leaf cells to produce compounds that will help them repair damage or fend off predators like insects. Thus, such nanosensors can be used to study a living plant's responses, communicating the specific types of stress in real-time (Lew et al. 2020). This type of nanosensor could be used to study how plants respond to different types of stress, thus, helping the researcher to develop new strategies to improve crop yield.

It has been found that the cytosolic steady-state glucose levels depend on external supply in both roots and leaves. The cytosolic glucose levels are lower in the roots as compared to epidermal and stomatal guard cells (Deuschle et al. 2006). Fluorescence resonance energy transfer (FRET) glucose nanosensors have been used in plants to study glucose concentrations in the cytosol of roots and leaves. Glucose transporters have been identified at the vacuole and the inner envelope of plastids. The steady-state glucose levels are governed by the flux rate and compartmentalization. The flux in sugar-producing mesophyll cells is expected to be distinct from that in the epidermal and stomatal guard cells due to different cells exhibiting different responses. The FRET-based saccharide sensors have been used to monitor relative changes in the steady-state sugar levels within a subcellular compartment (Chaudhuri et al. 2008; Deuschle et al. 2006; Lager et al. 2006). Recently, the fluorescence-based sucrose sensors by using invertase-nanogold clusters embedded in the inner epidermal membranes of onions (*Allium cepa* L.). This sensor indicated a detection limit of 2 nM with a linear dynamic range of 2.25–42.5 nM for sucrose detection (Bagal-Kestwal et al. 2015).

Molecular oxygen plays a vital role in respiratory metabolism. The availability of oxygen in the atmosphere is an essential substrate for plant metabolism. Plants are not photosynthetically active at night and the roots rely on the supply of oxygen from the environment. Earlier work on oxygen sensing was performed by detecting a current flow caused by the chemical reduction of oxygen similar to Clark-type electrode sensors, but subsequently, photoluminescence quenching was used as a convenient method to detect oxygen (Demas et al. 1999). Such a type of nanosensor is now known as a probe encapsulated by biologically localized embedding (PEBBLE). Probes encapsulated in nanoparticles have a protective shell that reduces interactions, retains stability, and prevents interferences with other proteins (Lakowicz et al. 2006). Similarly, a microbead-based probe having a size ranging from 40 to 300 nm per particle was recently developed for application in an algae-based system which included merits of single fabrication, long-term storage, and excellent brightness (Schmälzlin et al. 2005).

H_2O_2 and NO are critical signaling molecules in numerous pathways in plants such as plant responses to various stressors, cell deaths, growth, and development (Neill 2002; Quan et al. 2008). It is very difficult to obtain an absolute signal so that the probe intensity can be calculated. Hence, ratiometric sensors are used, in which one of the two distinct fluorophores reports an analytes-independent reference signal and the second one sensing signal to include analyte. Kwak et al. (2017) developed the first SWNTs (single-walled nanotubes) ratiometric fluorescent sensors (Kwak et al. 2017) with advances in separating and sorting SWANTs as distinct chiral species with characteristics emission wavelengths (Flavel et al. 2014; Liu et al. 2011; Tvrdy et al. 2013). Single-chirality SWNTs are independently functionalized to recognize H_2O_2, NO, or no analyte to create optical nanosensors responses from the ratio of distinct emission peaks. Semiconducting SWNTs are excellent signal transducers for nanosensors due to their fluorescence stability, lifetime, and emission in the NIR region. There has been significant interest in developing nanosensors by noncovalent complexation of various wrappings with SWNTs such that SWNTs' corona can recognize an analyte. For NO detection, the ss(AT)15 wrapped SWNTs were developed as a florescent

detector (Kim et al. 2010), which undergoes a strong quenching at wavelengths above 1,100 nm in the presence of dissolved NO in both extracted chloroplast and leaves (Giraldo et al. 2014).

Strigolactones are plant hormones that inhibit plant shoot branching (Gomez-Roldan et al. 2008; Umehara et al. 2008). These molecules are carotenoid-derived terpenoid lactones containing a labile ether bond, which is hydrolyzed in the rhizosphere. The use of small-molecule nanosensors presents unique opportunities for the *in vivo* study of plant signaling mechanisms. Tsuchiya et al. (2015) demonstrated the use of a fluorescence turn-on probe, Yoshimu Lactone Green (YLG), to study the signaling mechanism of strigolactones (Tsuchiya et al. 2015). YLG was designed to be recognized by strigolactone receptors. The use of nanosensors allowed for the identification of strigolactone receptors and also the establishment of a small-molecules reporter system (Tsuchiya et al. 2015).

In plants, dopamine and other catecholamines play important roles in growth, development, synthetic pathways, and metabolism (Kulma and Szopa 2007). Active research on neurotransmitters is required to understand fully their role in plants. The use of Corona Phase Molecular Recognition (CoPbMoRs) sensors could significantly aid the study of neurotransmitters in plants. Kruss et al. (2014) demonstrated that introducing 100 µM of dopamine increases the fluorescence of the construct ss(GT)-15-wrapped around HiPCO SWNTs by 58%–80% through an increase in the florescence quenching yield (Kruss et al. 2014). Recently, Wong et al. (2017) reported the implantation of ss(GT)15-wrapped SWNT nanosensors into wild-type spinach plants (*Spinacia oleracea L.*) which allowed the plants to transducer information on the uptake of dopamine (100 µM) from soil (Wong et al. 2017).

Phytoestrogens are naturally developed in various plants such as fruits, and cabbage soybeans (Cederroth and Nef 2009; Cos et al. 2003). They play important role in plants' defense systems, especially fungi. They are known as dietary estrogen because they mimic the actions of estrogen hormones in the human body. Nanoparticles-conjugated FRET probes based on human estrogen receptor alpha and ligand-binding domain have been used to detect phytoestrogens (Dumbrepatil et al. 2010). The FRET probe generated fluorescence signals approximately six times greater than those of individual FRET probes in a wide dynamic concentrations range (10^{-18}–10^{-1} M) of phytoestrogens such as daidzein, genistein, and resveratrol. This sensor demonstrated a stable florescence-time profile of more than 30 hours at room temperature.

Ethylene is a small-molecule hormone that is crucial for plants, especially for fruits and their ripening process. This can be detected by chemiresistive sensors using a Cu(I) complex with SWNTs. SWNTs are chosen for their high sensitivity to the electronic environment. The Cu(I) complex is responsible for capturing the ethylene to trigger the structural change on the surface of SWNTs. The sensor demonstrated a reliable ethylene response toward different fruit types and showed a detection limit as low as 0.5 ppm (Esser et al. 2012). There are several other studies for ethylene detection using different detection materials (Chauhan et al. 2014; Krivec et al. 2015; Mirica et al. 2013).

Volatile organic compounds are also a group of signal molecules that plants emit under the attack by insects and can be detected by chemiresistive sensors. Weerakoon et al. (2011) demonstrated using a poly(3-hexylthiophene) coated chemiresistive sensors to monitor the volatile compound gamma-terpinene molecules which are emitted during the infestation process. This volatile molecule interacts with the conducting poly(3-hexylthiophene), resulting in an increase in resistance. This system indicated a detection limit down to 36 ppm which is sufficient for insect infestation detection (Weerakoon et al. 2011).

Adenosine triphosphate (ATP) is a ubiquitous molecule in all living cells. It acts as the energy currency of cells and plays an important role in signal transduction. In plants, it is synthesized in chloroplast and mitochondria. Micro RNA (miRNA) is a class of small noncoding RNAs that have been shown to regulate gene expression in the plant development process. Although miRNA has been successfully demonstrated to be used for the detection of various diseases like cancer, its implication for crop science is still limiting. Monitoring gene expression in desired tissue is important to study basic plant biology research and crop production. This is monitored generally using traditional methods such as northern hybridization, quantitative PCR, microarrays.

14.3 NANOSENSORS TECHNOLOGY

A nanobiosensor is a self-contained analytical device that incorporates a biologically active material in intimate contact with an appropriate transducer to qualitatively or quantitatively sense chemical or biochemical phenomena occurring at the sensor surface (Kumar and Arora 2020). It converts a biological recognition response into an electrical signal, which is further processed to be represented as output display. There exists immense possibility and flexibility of developing a desired method using a wide range of nanomaterials and compatible transduction mechanisms that can be favorable integrated with various biomolecules to achieve desired performance levels (Patolsky and Lieber 2005; Swanson et al. 2011; Wujcik et al. 2014). Hence, nanobiosensors have emerged as an important method or alternative strategy for disease monitoring and diagnosis in plants.

A nano biosensors may make use of desired transduction method such as quartz crystal microbalance (Zadran et al. 2012), electrochemiluminescence (Chen et al. 2013; Maxwell et al. 2002), surface plasmon resonance (Huang and Murray 2002; Xia et al. 2011), loop-mediated isothermal amplification (Chaudhuri et al. 2008; Deuschle et al. 2006), fluorescence (Chen et al. 2013; Cognet et al. 2007), lateral flow for development of plant nanobiosensors. Although a variety of biomolecules are available for desired analyte biorecognition for biosensor fabrication, few biosensing methods are currently developed for plant biosensing. To achieve on-site, *in-vivo* and online testing and point-of-care devices, newer materials of nanobiosensors are being adopted for the fabrication of nanobiosensors for various applications including plant disease monitoring in real-time to know the crop health status.

Nanomaterials have been ubiquitously demonstrated to show unique optical, electronic, physical or mechanical properties as compared to bulk characteristics. These nanomaterials may include carbon-based materials, metals, and organic or inorganic materials. Common structural types of nanomaterials may include nanotubes, dendrimers, quantum dots, nanoparticles, nanowires, and fullerens. Nanomaterials have revolutionized the performance of sensors and biosensors through multi-dimensional roles ranging from providing stable support for the immobilization of biomolecular receptors, ease of immobilization, efficient signal transduction, signal enhancer, label, sensing as sensing receptor and ability to deal as well as a catering variety of signal ranging (optical, magnetic, electrochemical, thermal). Applications of nanomaterials include clinical diagnostics in human, detection of pathogens, viruses, bacteria, fungi, environmental monitoring, food quality control, etc. These nanosensors have also successfully demonstrated several potential applications in agriculture.

Sensor is a term used for any probe, reporter, indicator, molecular sensor or sensing device. A nanosensor has typically three main modules: a transducer, a receptor and a detector with a digital output. The target molecule connects with the receptor and the biological detecting component identifies a biological molecule through a reaction. The transducer converts changes to a signal quantified by the detector. The nanosensor has the advantages such as high specificity and sensitivity, rapid response, condensed size, portability, and real-time analysis. Nanosensor technology has been improved the specificity and sensitivity, performance, and quality by employing various types of nanosensors (Table 14.2).

14.3.1 ELECTROCHEMICAL NANOSENSORS

Such nanosensors can monitor any changes in dimension, charge distribution, electric properties and shape after complex formation on the electrode. Typically, these nanosensors consist of a working electrode, counter electrode, and reference electrode. Electrochemical electrodes designated with nanomaterials have the merits of high active surface area that permits high sensitivity and a broader range of concentration detection of plant-related redox and ion species. Hence, electrochemical detection is an attractive alternative method for the detection of redox active species in plants. These types of nanosensors have advantage of simplicity, ease of miniaturization, high sensitivity, relatively low cost, and capability for direct data analysis.

TABLE 14.2

Nanosensors and Their Types

Nanosensors	Analytes	Mechanisms	Merits	Demerits
Electrochemical	ROS, glucose, VOCs, enzymes, auxin, plant thiols, heavy metals	Consists of a working electrode, counter electrode and reference electrode. Based on the electrochemical response or electrical resistance change of materials caused by reaction with analytes	Simplicity, high sensitivity, low cost, direct data analysis, broad range of analytes	On-site source for the sensor, invasive and destructive, sensitive to pH or temperature
Piezoelectric	Mechanical forces in morphogenesis	Converts mechanical vibration into an electric signal	Real-time monitoring of the mechanical environment or plant growth	High cost, no optical readout, labour intensive for fabrication
SERS	Adenine dinucleotide, glucose	Enhances Raman signals of analytes adsorbed in the surface of metal nanoparticles	Ultra-high sensitivity, nonphotobleaching, multiplexing	Limited analytes, instrumentation
FRET	DNA, ATP, glucose, sucrose, metal ions, phytoestrogens	Consists of a recognition element fused to a reporter element, reports a conformational change in the energy transfer between the fluorophores	High sensitivity, low detection limit, high temporal and spatial resolution	Low stability of expressing FP in plant, photobleaching, low photostability, background signal
CoPhMoRe	H_2O_2, NO, glucose, dopamine	Turn off or on fluorescence by molecular recognition mediated by corona phases formed by a surfactant or polymer wrapping of individual fluorescent nanoparticle such as SWNTs	Photostability, non-photobleaching, optical detection *in vivo*	Sensitivity, specificity, stability in vivo rational design of corona phase for each analyte

Source: Kwak et al. (2017).

Due to recent advances in the research of metallic nanoparticles such as gold (Yusoff et al. 2015) and nanocarbon materials including carbon nanotubes and graphene-based materials having unique electronic properties (Geim 2009; Shao et al. 2010; Vilatela and Eder 2012), advances in electrochemical nanosensors has occurred. Generally, the working electrodes substrate is on the mm to um length scale e.g. the most common 1–6 mm diameter glassy carbon electrode or Pt-based ultra-microelectrode with a 5–10 um diameter (Cui and Zhang 2012). Usually, the electrode substrates itself is coated with active nanoparticles ranging from 2 to 100 nm in size (Qureshi et al. 2009). They can be further separated into three categories: amperometric, impedimetric and potentiometric transducers and utilized to detect various targets. These sensors are used for a qualitative study of plant sensing. They have the potential to study redox-active species inside the plant and to indicate its growth and environment conditions by detecting biomolecules such as H_2O_2, ethylene, vitamin C, oxygen, glucose, root auxin, plant thiols and other specific enzymes. Recently, electrochemical nanosensors were developed for online plant analysis. The fabrication of these sensors is based on photoactive nanomaterials such as Fe_2O_3, QDs, Si nanowires. Wang et al. (2015) developed

a photoelectrochemical nanosensor based on Mo-doped bismuth vanadate (BiVO$_4$) nanoparticles, forming a flower-like microsphere about 0.5–1 μm in diameter. This photocatalyst was coated on a glassy carbon electrode with 3 mm diameter or on indium tin oxide (ITO) electrode to form the electrochemical sensors (Wang et al. 2015).

The chemiresistive sensors are another type of electrochemical sensor in which electrochemical resistance is changed due to the adsorption of target molecules. Semiconductors such as carbon nanotubes and metal oxide and conducting polymers such as polypyrrole are generally used as sensing materials. A nanosensor device is fabricated by placing the sensing material in-between two electrodes and monitoring the change in resistance. These nanosensors also include screen-printed electrodes and semiconductors. These types of nanosensors have been applied in several applications. This method has been frequently used for gas sensing such as ethylene (Chauhan et al. 2014; Esser et al. 2012; Krivec et al. 2015; Mirica et al. 2013) and VOCs produced by plants (Degenhardt et al. 2012; Weerakoon et al. 2011).

14.3.2 Optical Nanosensors

These measure the change of the refractive index of the transducer when the target and recognition element produce a complex. They can be classified into two types, namely direct and indirect optic nanosensors. In the former type, signal generation is based on the production of a complex on the transducer surface, whereas the latter type is designed with various labels to detect the binding and extend the signal. Optical nanosensors are nanomaterials that permit high spatial-temporal resolution for monitoring plant signaling molecules through fluorescence signals in spectral regions (Chandra et al. 2015; Reuel et al. 2012; Sun et al. 2016; Wise and Brasuel 2011). These nanosensors can be modified for high stability, specificity for analytes, rapid dynamics, accuracy, and reproducibility (Lichtenstein et al. 2014).

14.3.3 Piezoelectric Nanosensors

These measure mass change and viscoelasticity by recording frequency and modifying a quartz crystal resonator. The piezoelectric effect can convert mechanical vibration into an electric signal and vice versa. Piezoelectricity is a coupling between mechanical and electrical behavior. A unique class of material known as functional oxides including ZnO, SnO$_2$, In$_2$O$_3$, Ga$_2$O$_3$, CdO and PbO$_2$ exhibits semiconducting and piezoelectric properties, and this class is an ideal candidate for fabricating electromechanical-coupled devices (Wang 2007). ZnO is the most commonly studied due to its biocompatible nature and can be easily formed either using a physical method or a chemical approach (Nguyen et al. 2012). Pressure sensors have been demonstrated for measuring forces known as a nm-sized force in the nanonewton range (Wang et al. 2006). Piezoelectric nanosensors are in nanometers scale but the detection process generally goes through a charge voltage amplifier, in centimeter scale. Piezoelectric nanosensors have been well studied in biological systems (Fernández et al. 2012) and can demonstrate a new pathway for real-time monitoring of the mechanical environment in plant and agricultural systems.

14.3.4 Metal or Metalloid Nanoparticles-Based Nanosensors

Inorganic nanomaterials such as silver, gold, silicone, and various metal oxide nanoparticles form a class of nanosensors that use recognizing ligands (e.g. DNA oligos or antibodies) and signal transducers to detect and quantify molecular targets of interest (Darr et al. 2017; Wu et al. 2013). Characteristics sensor responses are recorded and fed through changes in the surface-enhanced optical properties, originating from the localized SPR of nanoparticles. Recently, a myriad of self-assembled nanoparticles with oligonucleotides or proteins have been used to construct sensitive and target-specific nanobiosensors (Cho and Ku 2017; Dubertret et al. 2001; Lau et al. 2017; Li et al. 2020). Silver nanorods have been recently reported as sensor substrates for the identification

of various plant pathogens or produced toxins using SERS (Ahmad et al. 2012; Farber et al. 2020; Farber and Kurouski 2018; Wu et al. 2012).

14.3.5 QUANTUM DOTS

Quantum dots (QDs) are semi-conductor nanocrystals with unique and photophysical properties. QDs have demonstrated tremendous potentials as optical nanosensors and found successful applications as biosensors for plant imaging and disease identifications (Gao et al. 2018; Wang et al. 2016). The ultra-small size of the QDs about 1–10 nm makes them ideal fluorescent imaging contrast agents for rapid detection of plant diseases. The newly emerging carbon dots having low cytotoxicity and high biocompatibility and ease of preparations from fruits and vegetables, are reported valuable in bacterial and fungal imaging (Jin et al. 2015; Kasibabu et al. 2015; Sachdev and Gopinath 2015). Fluorescence resonance energy transfer (FRET) is a well-established mechanism for the construction of QD sensors (Yuan et al. 2013). For FRET sensors, the proximity of the donors (QDs) to acceptors such as Au nanoparticles, carbon nanodots, and organic dyes, leads to an energy transfer which results in quenched fluorescence intensity (Shojaei et al. 2016a; Shojaei et al. 2016b).

14.4 DEVELOPMENTS IN NANOSENSOR TECHNOLOGY

Intracellular sensors for metabolic precursors signaling ligands and nutrients may help elucidate the complex roles of these molecules in plant systems. Some emerging nanosensing platforms as summarized below, have been used or have the potential for use in agriculture. Some emerging nanosensing platforms have been used or have the potential for use in agriculture.

14.4.1 FRET NANOSENSORS

Optical nanosensors based on FRET have been used widely to study cellular concentrations of small molecules, protein dynamics, and biophysical parameters (Zadran et al. 2012). There are two fluorophores-a donor and an acceptor. This type of sensor is based on the observations that photoexcitation of the donor fluorophores will resonantly transfer to the acceptor fluorophore if the distance between the two falls within a nanometer-scale range. These types of nanosensors are either genetically encoded within the plant itself (fluorescence proteins, FPs) or added exogenously as externally synthesized construct (nanoparticles). Both types report conformational changes using proteins, protein domains, nanoparticles, or molecular ligands that modulate the distance between donor and acceptor fluorescent domain of two fluorophores. When the donor is excited, a fraction of the energy is transferred to the acceptor by resonance electron transfer resulting in the fluorescence of the acceptor.

Genetically encoded FRET nanosensors have been used to detect DNA (Maxwell et al. 2002), metal ions (Huang and Murray 2002) and organic compound. So, they can monitor steady-state levels of ions or metabolites. They are typically composed of two spectral variants of fluorophores with spectral overlap such as a reporter unit coupled to a substrate-binding domain or recognition unit. Nanoparticles-based FRET sensors have been developed for the detection of biomolecules. A variety of nanoparticles has been used including gold nanoparticles, semiconductor quantum dots (QDs) and lanthanide-doped up conversion nanoparticles (UCNPs) which act as either a FRET donor or a quencher (Chen et al. 2012). Nanoparticles can provide several merits over traditionally FRET pairs with regard to photostability and emission intensity.

14.4.1.1 Surface-Enhanced Raman Scattering Nanosensors (SERS)

SERS is an ultra-sensitive non-destructive spectroscopic technique that can detect analytes down to the single-molecule level. Raman signals of molecules adsorbed on the surface of metal nanoparticles can be enhanced by a factor of up to 10^{14} that is comparable to ever higher than fluorescent organic dye (Stiles et al. 2008). The SERS effect involves two mechanisms, a long-range

electromagnetic effect (EM) and a short-range chemical effect (CM). The EM mechanism results from the enhancement of the local electromagnetic field due to surface plasmon resonance (SPM) of surface roughness nanoscale features in the 10–200 nm range. The CM mechanism arises from the interaction between adsorbed molecules and the metal surface. Gold and silver nanoparticles have been widely used due to tunable plasmon resonance in the NIR region. Most peaks of the SERS spectra in plants are attributed mainly to adenine-containing materials such as adenine dinucleotide, chlorophylls, flavins, and lipids (Zeiri and Efrima 2006).

14.4.2 CoPhMoRe Nanosensors

This technique is invented by MIT and uses the specific adsorption of a compositionally designed polymer at a nanoparticle interface to enable reorganization in the same way that an amino acid sequence, when folded, can provide a three-dimensional antigen-binding domain. Semiconducting SWNTs have been used in this technique and the surfactant or amphiphilic polymer wrapping. There are two purposes of wrapping - firstly, the wrapping stabilizes the dispersion in an aqueous solution necessary for *in vivo* applications. Secondly, both in plants and other living organisms the wrapping modulates interactions of other molecules with the SWNT surfaces, creating a nonbiological antibody (Zhang et al. 2013). The wrapping molecules have no affinity to the target analyte but when bound to the nanotube surface, a complex is formed that can selectively bind to a specific analyte. SWNTs are an excellent candidate for the underlying florescent nanoparticles because they fluoresce in the NIR range of the spectrum which is the transparency window for *in vivo* optical detection (Barone et al. 2005). Furthermore, the utilization of SWNTs within living plants has been demonstrated for the detection of NO, H_2O_2, and dopamine (Giraldo et al. 2014).

14.4.3 Array-Based Nanosensors

The array-based sensors assemblies composed of several chromophores or synthetic nanomaterials are a format capable of multiplexing and discretion of different analytes. This form of the array-based sensor is also known as the electronic nose (e-nose), which incorporates electronic transducers instead of chemical sensors in the array (Geng et al. 2019; Li et al. 2019; Röck et al. 2008). The array-based approach has advantages in distinguishing highly similar analyte mixtures due to its cross-reactivity and ability to measure molecular fingerprints. The multidimensional output data of sensor arrays can be analyzed with highly precise and predictive results (Cuypers and Lieberzeit 2018). Array-based chemical sensors can also provide optical readout (e.g. fluorescent or colorimetric) which is easier to analyze and interpret compared with e-nose signals (Askim et al. 2013). In addition, chemical sensor arrays are easy to fabricate based on specific chemical interaction. Such nanosensors arrays are extremely cost-effective and more resistant to environmental interference e.g. humidity and temperature) than most e-nose devices. Chemical sensor arrays can be easily prepared by embedding various analytes-responsive dyes such as Lewis acid-base colorants, Bronsted acidic or basic colorants, solvatochromic or vapochromic dyes and redox indicator colorants in hydrophobic nanoporous substrates such as modified silica sol-gels (Li et al. 2019).

An important indirect method for plant disease detection involves the profiling of the volatile chemical signature of diseased plants (Cellini et al. 2017; Cellini et al. 2018). This approach is based on the observations that diseased plants could result in the release or change of the composition of characteristics VOCs that are indicative of the type of biotic or abiotic stresses that host plants have experienced.

14.4.4 Wearable Nanosensors

Wearable sensors integrate flexible substrate materials and nanoelectronics that are advancing the field of sensors for human health monitoring and robotic manipulation, but their application in agriculture

is still unexplored. The field of wearable nanosensors has been widely developed for human skin and clothing application (Zhang et al. 2020). Recently, nanosensors networks based on flexible wearable nanoelectronics circuit implanted on plant surfaces, communicating wirelessly and reporting low concentration of volatile molecules in real-time (Giraldo et al. 2019). The flexible skin due to the elastic properties of wearable nanosensors can be bent on organism with a radius of curvature up to 100 μm. Wearable SWCNT-graphite nanosensors are operated by radio frequency for wireless monitoring with electronic devices without power consumption in response to very concentration of gas molecules. Ethylene, a plant hormone that acts as a key indicator of the onset of fruit ripening, is reported for long-term monitoring at its sub-ppm concentrations by using chemoresistance nanosensors based on SWCNTs equipped with copper complex (Cellini et al. 2017). Carbon-nanotubes-based sensing devices for plant VOCs are now commercially available for agriculture applications, but not for crop health monitoring through plant signaling molecules. Highly stretchable wearable sensors based on graphene carbon nanotubes have been reported to monitor wirelessly a wide range of gas and aqueous phase molecules including glucose (Cellini et al. 2018; Fang et al. 2014). Plant wearables are new frontiers of crop diagnostics which relies on ultrathin and ultra-lightweight nanosensor design to attach flexible sensor devices directly on plant tissues such as leaves for continuous monitoring. With recent advances in the technology of micro-electromechanical system (MEMS) more and more implantable and wearable electronics using conductive nanomaterials have emerged as sensor components for long-term and on-demand monitoring of plant POCs or other biomarkers.

14.4.5 GENETICALLY ENCODED NANOSENSORS

Genetically encoded sensors consisting of a sensory module coupled to fluorescent proteins and are highly selective for the specific analyte can be delivered into plant cells via DNA plasmids. These fluorescent proteins can be detected with high spatial and temporal resolution by florescence imaging devices. A genetically encoded nanobiosensor is a biosensor where 'detector' of biological event is a signaling pathway associated to analyte of interest and 'transducer' for signal transfer is a reported gene that provides a measurable signal (Walsh et al. 2015). Most of the nanobiosensors include biosensors for plant hormones only, that trigger transcriptional regulators degradation and signal transduction leads to nuclear transduction of proteins. Reported genes generally include beta-glucoronidase (GUS), bioluminescent, fluorescent proteins, etc. that allow *in vivo* measurements. Signaling pathways that are biochemically characterized and structurally characterized for plant hormones are chosen for phytohormones signaling that may be transcriptionally based and degradation based or FRET-based. A review on plant genetically encoded nanobiosensors has been reported to explain about various biosensors such as induced biosensors, direct intrinsic biosensors or direct extrinsic biosensors (Oliveira et al. 2015). GFP-based biosensors were used to visualize the spatial and temporal kinetics of cellular regulators such as Ca^{2+}, H^+, H_2O_2, ATP, NH_3^+, Zn^{2+}, NO, peptides, plant hormones, glucose, sucrose, redox, etc. Using such nanobiosensors membrane transport activities could be monitored in real time in living plant cells (Wang et al. 2019). At present developments in such nanobiosensors technologies are still in their primitive stage, but in the future great opportunities for nondestructive measurement will be demonstrated.

14.5 EMERGING NANODIAGNOSTIC TOOLS FOR PLANT DISEASES

Plant disease is a process where living (biotic infection disease due to bacteria, fungi, virus, nematodes, and parasites) or non-living (abiotic non-infectious diseases due to adverse conditions by surrounding environment such as nutrient deficiency, prolonged water/air/soil stress) imposes a change in the normal function of plants. Plant disease may lead to leaf spots, leaf blights, root rots, fruit rots, fruit spots, wild, dieback, result in decline in growth and crop production (Al-Sadi 2017). Plants generally respond to pathogen infection through its defense reactions and various physiological processes such respiration, photosynthesis, nutrient translocation, transpiration, growth

and development, primary carbon metabolism (Berger et al. 2007). Visual detection of change in plant morphology, serological (Shang et al. 2011), electron microscope (Lobert et al. 1987) or PCR (Hongyun et al. 2008) based methods are generally used for the detection of plant diseases. Methods for plant disease detection are classified as direct and indirect methods. Direct methods include serological methods (ELISA, FISH, immune fluorescence, flow cytometry, etc.) and molecular methods (PCR and its variants like RT-PCR, multiplex-PCR). Disease-causing bacteria, viruses, and fungus are accurately detected in a high throughput method. Indirect methods make use of measurement of morphological change, temperature change, transpiration rate change and VOCs in case of infection using thermography, fluorescence imaging, hyperspectral imaging, gas chromatography, etc.

Conventional and contemporary methods of plant disease diagnosis have their individual limitations and features to provide performance parameters such as reliable, low cost, fast (serological/molecular), simple (PCR/ELISA/FC), *in-situ*, specific (imaging/spectral/thermal), sensitive process of measurement, ease of use, continuous measurement and multi-analytes testing, etc. These methods are now supplemented by newer methods such as nanobiosensors that make use of nano- and bio-molecules assemblies to cater to the challenges of routing methods of diagnosis.

Advancement in nanotechnology has conferred highly exciting ingredients to achieve unprecedented levels of performance parameters in nanosensors. The use of a diverse range of nanomaterials such as metals-based, carbon allotropes, polymers, composite, etc. in the form of nanoparticles, nanotubes, nanorods, and nanowires have enabled faster detection and reproducibility in a much better way. Unique tunable and customizable properties of nanomaterials such as optical, high electrical conductivity offer versatile ultra-sensitive responses and detection mechanisms through measuring properties such s thermal, electrical, optical, and piezoelectric. These exciting properties of nanomaterials on the nanometer scale facilitate the congregation of nanomaterials with biomolecules to achieve desired nanosensors for various applications (Arora 2018). Nanosensors can play a major role in combating growing challenges in the form of plant disease detection in agriculture. Hence, timely diagnosis of plant diseases and subsequent preventive measures is one of the crucial elements to sustain the availability of food. Food demand is continuing to increase with the rising human population and is being additionally challenged through environmental challenges such as decreasing agricultural land, pest and pathogen-induced crop losses, etc. Among these pest and pathogen-induced loss range within 12%–80% depending on the type of crop ranging from rice, wheat, maize, barley, soybean groundnut, tomatoes, potatoes, and cotton (Fang and Ramasamy 2015). Hence, nanobiosensors can play an important role in monitoring plant health and timely detection and prevention of plant disease spread.

Microbial infection leads to invoking plant defense (pathogen hypersensitive response) such as local accumulation of phytoalexins, enhancement of several enzymes activates such as chitinase, peroxidase, lipoxygenase, beta-1,3-glucanase, NO, ROS, molecular markers in the form of expression or repression of new genes e.g. pathogen-associated molecular patterns (PAMP) such as activation of kinases (stress/microbial induced), release of chemical in the surroundings such as oxylipins (Zeilinger et al. 2016). Expression of all such biomarkers is used to detect plant pathogenic infection. It has been reported that salt stress (high levels of sodium ions or potassium ions) in the species named *Picea abies* leads to decrease in total phenolics or total carotenoids levels (Schiop et al. 2015). So, levels of phenolics or carotenoids are the most reliable and useful biomarkers for salt stresses.

Detection of VOCs has also been reported to assess in a non-invasive way about the functional information of plant's growth, defense and health status. Generally, electronic nose (e-nose) is used that utilizes nanosensor array, signal conditioning circuit and pattern recognition algorithms in a rapid, cost-effective and in-situ applications. The basic principle of measurement may involve gas sensing, conductivity sensing, gravimetric sensing such as surface acoustic wave, quartz crystal microbalance, optical sensing such as measuring redox potential, reactivity, acid-base reactions, colorimetry, and fluorimetry. An elaborative review (Cui et al. 2018) has explained the collective use of electronic nose for plant pest diagnosis (Figure 14.1) and has reported that plant nanosensors are

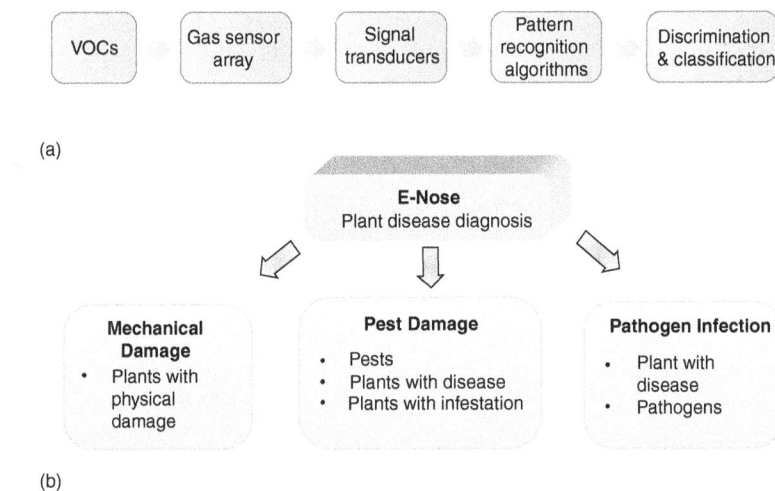

FIGURE 14.1   E-nose: (a). Sensing-interpreting-discriminating process (b). plant disease diagnosis.

still in their early stages and various challenges such as sensor performance, sampling, and detection in open areas and scaling up measurements are yet to be addressed.

Detection and identification of plant disease could be achieved by both direct and indirect methods (Lebeda et al. 2001; Schiop et al. 2015). Direct methods involved analysis of plant pathogens including fungi, bacteria and viruses or biomolecular markers such as carbohydrates, proteins and nucleic acids isolated from infected plant tissues. Indirect methods involve recognition of plant disease through changes in histological or physiological indices such as leaf surface temperature or humidity, spectroscopic features of plant tissues, morphology, growth rate and emission of VOCs. A wide variety of spectroscopic electrochemical or molecular techniques could serve as direct or indirect detection methods (Cui et al. 2018; Singh et al. 2012). There are several categories of nanoscale techniques for agricultural diagnosis in both direct or indirect methods including microneedle patches, nanopore sequencing platform, plant wearables and nanoparticles or array-based sensors (Li et al. 2020).

Several imaging methods such as hyperspectral and thermographic imaging have been used for indirect detection of plant disease. But they have certain demerits such as lack of specificity for disease subtypes or stains and susceptibility to parameter changes of the environment (Jiao et al. 2000; Zhao et al. 2014). Recently, several nanoparticles-based chemo- or bio-sensors are developed and commercialized for agricultural diagnosis (Hongyun et al. 2008; Jarocka et al. 2013; Lobert et al. 1987). The analytes are identified by the characteristics optical or electrical outputs of the nanosensors through transduction mechanisms of the designed sensors interactions. The nanosensor's detection specificity could be enhanced by selective chemical interactions or by the used of biospecific recognition elements such as DNA oligos, aptamers, antibodies and enzymes. The detection sensitivity could be improved by the used of surface-enhanced optical properties e.g. SPR, electron-conductive nanoscale substrates such as carbon-based nanomaterials e.g. graphenes or carbon-nanotubes as transducers. There are several classes of nanosensors or nanobiosensors including metal or metalloid nanoparticles, quantum dots and array-based nanosensors used for plant disease diagnosis.

14.5.1   Detecting Plant Infections

Infections in plants are the major cause of losses in crop production throughout the world. These diseases are caused by pathogens such as fungi, bacteria and viruses (Khater et al. 2017; Tahir et al. 2017). Hence, timely detection of plant infection critically affects the levels of crop productivity. Generally, plant infections are detected by conventional visible symptoms on plant morphology, ELISA, direct

tissue blot immunoassays, molecular techniques such PCR, RT-PCR, Raman spectroscopy, fluorescence spectroscopy, infrared spectroscopy (Lertanantawong et al. 2019; Liu et al. 2016). But recently, nanosensors have been developed for easy and quick methods of plant infection detection.

Most of the bacteria are generally known to be associated with plant are beneficial or saprophytic but one hundred bacterial species are known to cause diseases in plant such as *Agrobacterium*, *Erwinia*, Proteobacteria, *Pseudomonas*, *Xanthomonas*. These bacteria cause infection, leading to loss of crop production and may result in symptoms such as gall and overgrowths, leaf spots, specks and blight, wilds, soft rots, scabs and cankers (Borse and Srivastava 2019; Lertanantawong et al. 2019). A lateral flow immunoassay-based test strip was reported to detect *Ralstonia solanacearum* that causes potato brown rot using enlarging gold nanoparticles (GNPs), antibodies against the pathogen and tetra chloroauric (III) anion reduction used as an amplification solution (Borse and Srivastava 2019; Razo et al. 2019). It is also reported the detection of potato leafroll virus and potato virus X using the silver salt solution as an amplification solution (Drygin et al. 2012; Panferov et al. 2018). An electrochemical method for detection of endophytic bacteria *Agave tequilana* has been reported. Electrochemical method involves the detection of H_2O_2 levels produced by leaves of host plant tissue to destroy the bacterial infection. Bacterial spot disease caused by *Xanthomonas campestris* plant pathogens was identified by a highly sensitive method using fluorescent silica nanoparticles doped with a dinuclear Ruthenium complex and conjugated with a secondary antibody (Luna-Moreno et al. 2019). *Ralstonia solanacearum*, a soil-borne species causing potato bacterial wilt was reported to detect using its genomic DNA with gold nanoparticles functionalized with a specific single-stranded DNA (Nikitin et al. 2018). Fungal infections in plants amount to 80% of infections caused in crops. Fungi are a group of versatile eukaryotic carbon-heterotrophic organisms that can maintain symbiotic/saprophytic or pathogenic relation with plants. Fungi act as decomposers and recyclers of organic materials. Each plant is attacked by fungal pathogens but only a few lead to infections (Knogge 1996). Timely detection of fungal infection is extremely important because damage caused by these pathogens to plant is tremendous. Potato blight, downy mildew of grapes are some samples of fungi infection in crops. A surface plasmon resonance-based immunosensor was reported to detect a fungus *Pseduocerocospora fijiensis* in real leaf extract samples of banana plants using polyclonal antibody against cell wall protein HF1 of *P. fijiensis* immobilized onto gold chip in lateral flow assay (Luna-Moreno et al. 2019). Multiplex field disease diagnostics of six fungal pathogenic of potato crops, namely, *Alternaria solani*, *A. alternate*, *Rhizocotonia solani*, *Collectorichum coccodes*, *Synchytrium subterranean*, *Fusarium* were reported in easy and convenient PCR-based microarray in open microreactors (Nikitin et al. 2018). Late blight in potato caused by *Phytophthora infestans* was visually detected using a novel method that used integration of universal primer mediated asymmetric PCR with gold nanoparticles (AuNP)-based lateral flow biosensor (Zhan et al. 2018). The availability of sensitive methods is important because fungal infections are very severe and can contaminate seed and soil. Yellow rust disease in wheat is caused by fungus *Puccinia striiformis* which impacts the quality and production. So, timely detection of infection at initial stage of crop development is extremely important from agricultural point of view. Hyperspectral analysis was reported to be used as a tool that captured biophysical variations in crops depending on the degree of infestation (Hong and Lee 2018). Similar strategy was adopted to detect rice infested with bacteria leaf blight disease and sugar beet rust disease.

Hong and Lee (2018) published a review on nanosensors for detection plant viruses based on QDs and magnetic NPs through immunomagnetic separation and FRET phenomena through SPR and optical measurement (Hong and Lee 2018) (Table 14.3). Most of the nanobiosensors reported are immune-based that make use of optical modes of detection. Plant viruses such as cucumber *mosiaicirus, Pantoeastewartii, Plum pox virus, Prunus nectrotic ringspot virus, Citrus tristeza virus, Potato virus X* are reported to be detected using immune assays (Drygin et al. 2012; Jarocka et al. 2013; Jarocka et al. 2011; Jiao et al. 2000; Zhao et al. 2014). Optical DNA hybridization nanobiosensors have been reported that provided PCR-independent visual nucleic acid detection, thus

TABLE 14.3

Quantum Dot and Surface Resonance-Based Virus Biosensor

Mode of Detection	Virus	Biomolecular and Nano Conjugate
SPR	PVY	Monoclonal antibody
SPR	MCMV	Anti-MCMV antibody
SPR	BSMV	Specific oligonucleotide from RNA
SPR	ASPB	DNA aptamer from coat protein
SPR	CPMV	Monoclonal antibody
FRET	CPMV	Surface immobilized CPMV CdSc-ZnS core
FRET	CaMV	23 mer derived from CaMV 35S PbS nanoparticle
FRET	CTV	CTV-CP antibody-CdTe
FRET	CTV	AuNPs-CTV-CP/Ads-CTV-CV antibody, AuNP/QD
FRET	GVA	Grapevine virus A type proteins ZnO films
FRET	BPMV, ArMV and ToRSV	Antibody, Fe_2O_3/SiO_2 MNPs and SiO_2/UNCPs
Electrochemical	CTV	Antibody to CTV coat protein-lnP
Electrochemical	CTV	CTV-CP antibody-CdTe

Source: Adapted from Hong and Lee (2018).

offering simplicity, high sensitivity and quick results (Lee et al. 2018; Ortiz-Tena et al. 2018; Ouyang et al. 2018; Shibata et al. 2018).

Recently, a nanobiosensor based on DNA diagnostics using the localized SPR (LSPR)-gold nanoparticles (AuNPs) was developed with a colorimetric nano-biosensing system to detect the unamplified Tomato Yellow Leaf Virus (TYLCV) genome in infected plants (Razmi et al. 2019). Sequential use of restrictions enzymes AcII, Terminal deoxynucleotidyl transferase (TdT) Mg^{2+}-dependent DNAzymes activity, hemin/G-quadruplex DNAzyme was reported for the detection of Cucumber green mottle mosaic virus in visual color change assay using cDNA prepared from total mRNA extracted from seedling of infected crops (Wang et al. 2019). The virus CGMMV belongs to the genus *Tobamovirus* of the Cucurbitaceae family which has a single-stranded, positive sense RNA genome of about 6.4kb and can be transferred mechanically through seeds where infection leads to severe decline in crop productivity. Similarly, exonuclease and polymerase activity of T4 DNA polymerase and Mg^{2+} dependent DNAzymes and hemin/G-quadruplex DNA cascade reaction series was reported to use for detection of watermelon mosaic virus (WMV) through electrochemical differential pulse voltammetric measurement (Wang et al. 2019). In another case, a highly specific and sensitive lateral flow immunochromatography assay (ICA) was developed to detect Grapevine leaf-associated virus 3 (GLRAV-3) causing a loss of quality and productivity of grapes. This method utilized sandwich immunoassay using gold nanoparticles labeled antibody.

E-noses made of metal oxides or conductive polymer coatings have drawn considerable attention as multiplexed gas sensors for tracing VOC biomarkers in plant stress events ranging from pathogen infection, pest invasion, to physical wounding (Fang et al. 2014). Ethylene, one of the most important phytohormones, is detected using chemiresistive or colorimetric method (Esser et al. 2012; Li et al. 2019). A nanoplasmic sensor array comprising gold nanoparticles and a molecularly imprinted sol-gel (Au-NP-S@MISG) for selective detection of terpenes has been designed (Shang et al. 2018). Recently, Li et al (2019) developed a smartphone-integrated VOC sensing platform that utilized plasmonic nanoparticles for the early detection of tomato late blight (Li et al. 2019).

14.5.2 Detecting Abiotic Stress-Induced Plant Disease

Resource deficiencies and plant stress are monitored by assessing plant physical traits through imaging, spectroscopy, and florescence from the visible to the infrared, but these methods are not suitable

for early detection of some types of stresses or resource deficit (Chaerle and Van Der Straeten 2000; Humplík et al. 2015; Li et al. 2014). For example, plant water status parameters are related to multiple stresses including draught, salinity, and pathogenesis. Chlorophyll florescence is not always suitable for early stress detection and is not a specific indicator of stress types (Li et al. 2014).

Nanobiosensors also play vital roles in detecting and controlling the use of pesticides, fertilizers as well as many other growth parameters associated with crops which provide timely information for precise decision-making and agricultural management (Deuschle et al. 2006). Abiotic stress is the positive impact of non-living factors on the plants through their surroundings including salt, water, temperature, pollutants, and heavy metals. Persistent exposure to these non-living variables leads to change in normal biological functions of plant resulting in a diseased state and finally causing retarded growth and development, and death of a plant. Hence, detection of abiotic stress-induced diseases in plants is extremely important. Generally, such stress results have been measured through visual change in the leaves, stem and roots, thermal analysis, expression of some new receptors, proteins, decrease in normal functions, ROS etc. Superoxide (O^{2-}) and other reactive oxygen species (ROS) are generated in response to numerous biotic and abiotic stresses (Xu et al. 2017). Abscisic acid (ABA) is a plant hormone which is involved in plant stress management against environmental stress in the form of heat and drought (Choi and Gilroy 2014). Various studies such as *in vivo* imaging of ROS and redox potential, ATP sensing in tissue gradients and stress, altered gene expression in plants, have successfully demonstrated and plant disease due to abiotic stresses. Vitronectin-like proteins on the surface of plant cells are reported as important biomarkers for monitoring damage of plants under the stress of heavy metals (Wang et al. 2019). Bioprobes such as Abaleons, AAcus and Cameleon-FRET based bioprobe have green fluorescent proteins linked to ABA receptor were developed to measure ABA *in vivo* in plants. These bioprobes change colour of fluorescence or emit green fluorescence upon binging to ABA to receptor (Pandey et al. 2018).

Oren et al (2017) have developed a roll-to-roll fabrication method of a graphene-based wearable sensor that can monitor water evaporations from plant leaves (Oren et al. 2017). The sensing mechanism is based on changes in the electrical resistance of grapheme in different humidity environments. Im et al. (2018) reported a wearable plant drought stress sensor based on a different mechanism where the variation of the capacitance of printed 100 nm-thick gold electrodes on a flexible polyimide (PI) was recorded in real-time to monitor local humidity (Im et al. 2018). Nassar et al. (2018) developed a lightweight and multiplexed plant wearable that integrates temperature (resistance, humidity (capacitance), and strain sensors (resistance) to monitor plant's local microclimate and plant growth (Nassar et al. 2018). Kim et al. (2019) developed a technique to prepare vapor printed polymer electrodes, which can be directly printed on living plant tissues for long-term monitoring of drought and photodamage (Kim et al. 2019).

14.5.3 MONITORING PLANT GROWTH

Phytohormones are molecules that regulate various stages of plant growth and metabolism. Auxin, cytokinin, gibberellin, abscisic acid and ethylene are the five major plant hormones which can work together or independently to influence plant growth at very low concentrations. Indole-3-acetic acid (IAA) is an auxin which is known to be involved in cell growth and cell expansion of stem. It exists in concentration range of nanogram/picogram per gram level and is affected by external environmental factors such as light and temperature. A sensitive electrochemical immunosensor was reported for detection of IAA using GNPs-coated porous grapheme layer onto glassy carbon electrode with detection limit of 0.016 ng mL-1 in extract samples from various plants (Su et al. 2019). A movement of carbohydrate from roots to shoots and leaves give important indication towards plant growth and development. Assimilation of carbon in the form of sugars, movements and processes is generally monitored for various plants to assess the plant growth and production or state of disease. A gel-based grapheme oxide, horse radish peroxidase and dye Ampliflu Red based method reported to study the presence of glucose in roots of various plants by visual assay (Voothuluru et al. 2018).

This method was also utilized to monitor important information towards carbohydrate partitioning under various conditions. For desired crop production, assessment of plant growth is critically important and various methods for this analysis have been developed using nanobiosensors based on grapheme and metallic nanoparticles.

Unique plasmonic nanoprobe has been reported in *Arabidopsis thaliana* for *in vivo* imaging and biosensing of miRNA biotargets (Crawford et al. 2019). This method provided *in vivo* functional imaging of target nucleic acids in plant that have potential for direct imaging within whole plant system, allowing monitoring of plant development and regulations. Such unique example of detecting target nucleic acid in plants *in vivo* possesses immense potential to be developed and can be moulded for various biosensing applications.

Most of the nanobiosensors include biosensors for plant hormones only, that trigger transcriptional regulators degradation and signal transduction leads to nuclear transduction of proteins. Reported genes generally include beta-glucoronidase (GUS), bioluminescent, fluorescent proteins etc. that allow *in vivo* measurements. Signaling pathways which are biochemically characterized and structurally characterized for plant hormones are chosen for phytohormones signaling that may be transcriptional based and degradation based or FRET based. A review on plant genetically encoded nanobiosensors have been reported to explain about various biosensors such as induced biosensors, direct intrinsic biosensors or direct extrinsic biosensors (Walia et al. 2018). GFP based biosensors were used to visualize the spatial and temporal kinetics of cellular regulators such as Ca^{2+}, H^+, H_2O_2, ATP, NH_4^+, Zn^{2+}, NO_3^-, peptides, plant hormones, glucose, sucrose, redox etc. Using such nanobiosensors membrane transport activities could be monitored in real time in living plant cells (Hilleary et al. 2018). At present developments in such nanobiosensors technologies are still in its primitive stage, but in future great opportunities for nondestructive measurement will be demonstrated.

14.5.4 Detecting GM Crops

GM plants or transgenic plants are in use for mass production of crops such as rice, mustard, cotton, tomato for many features such as insect resistance, disease resistance, delayed ripening. There are different regulations imposed by various countries to control the used of GM crops, but still lot of GM crops are available throughout the world. There are less stringent policies imposed by USA and have allowed the cultivation of GM crops and use of their products. According to International Service for Acquisition of Agri-biotech Applications (ISAAA), the global area of GM crops has increased about 112-fold from 1.7 million hectares in 1996 to 190.4 million hectares in 2019 and a total of 71 countries adopted these crops -29 countries planted and 42 additional countries imported including 26EU countries for food, feed and processing indicating future emerging use of GM crops (ISAAA, 2019). There is complete ban over the use of GM crops and food in some countries. Keeping in view the rights of consumers to know which food or crops contain the transgene and in how many copies, more and more countries are issuing qualitative and quantitative labeling regulations for specific insert sequence in transgenic plant.

Routine methods such PCR including multiplex PCR and qPCR, microarray, Sothern blotting, ELISA, western blotting, strip tests are available for the detection of GM crops (Singh et al. 2012). But these methods are time-consuming, labor intensive and costly. Hence, techniques for detection of GM plants are developing to achieve reliable, fast and simple routine detection process throughout the world. A review article by Sánchez-Paniagua Lopez et al. (2018) described almost all nanobiosensors for GMO detection though the use of DNA based biosensors including optical, electrochemical, piezoelectric that mostly relied on the use of 35 CaMV promoter, PEP carboxylase promoter and NOS promoter from *Agrobacterium tumefaciance* only (Sánchez-Paniagua López et al. 2018). Kumar and Arora (2020) reported some recent developments in nanobiosensors based on optical and electrochemical transduction mechanisms for GM plant detection (Kumar and Arora 2020). Various optical methods such as fluorescence based, spectroscopy, surface plasmon

resonance-based biosensor are available for detection of GM plants. Surface-enhanced Raman spectroscopy-barcoded nanobiosensors are reported for detection of *Bacillus thuringensis* (Bt) gene transformed rice, expressing insecticidal proteins. Specific oligonucleotide conjugated silica encapsulated gold nanoparticles were used as 'SERS-barcoded nanoparticles spectroscopic tags' for transgenic rice (Chen et al. 2012).

14.6 SMART SENSING TECHNOLOGY FOR MONITORING CROP HEALTH STATUS

Approaches for monitoring plant signaling molecules associated with early detection of stress or resource deficiencies using nanomaterials-delivered genetic-encode nanosensors, optical nanosensors and wearable sensors can provide a perspective on how smart plant sensors can actuate agricultural devices by communicating with smartphones, hyperspectral imaging cameras, wireless radio frequency devices and metrological stations.

14.6.1 THE POINT-OF-CARE TECHNOLOGY

Development of point-of-care (POC) technology assays has revolutionized the field of diagnostic testing in healthcare sector but in agriculture, it is in the initial stage. This term is generally used to describe diagnosis, monitoring and treatment of a disease at the site-of-need. POC assays have been developed for a huge number of diseases including malaria, HIV and tuberculosis (Cook et al. 2015; Rizzo et al. 2015). The span of POC has broadened growing from its current dominant usage in clinical settings to encompass many sectors such as water, plant and food testing in crop production. This POC technology when applied to plant testing is known as phytodiagnostics with the aim to mitigate crop losses due to various pathogens such as bacterial, viral, and fungal pathogens. These pathogens cause annual crop losses of around 20% worldwide alongside an added 10% post-harvest (Bebber and Gurr 2015). A number of studies have reported that plant pathogens, particularly fungal pathogens, can produce large toxin-associated risks in the crops they infect. Mycotoxins, example of such toxin, are secondary metabolites of pathogenic organisms and can be harmful to animals and human when ingested (Reverberi et al. 2010). At present, the decision to spray crops depends on the farmer's experience and visual methods to determine the presence of pathogens and sometimes based on laboratory-based diagnostics test. But these approaches may allow pathogens to proliferate undetected, particularly those which are asymptomic in early stages of infection or generate spores for survival. POC devices can provide high sensitivity and specificity and inferring that pathogen. Nanosensors for these on-site diagnostic approaches in the field of crop pathology are developed and now available for on-site testing POCT and automated crop monitoring and agrochemical used. Antibody-based diagnostic systems hold great promise in the field of POCT and phytodiagnostics. Even, multiplexing is also now being applied in the field of phytodiagnostics. This approach is beneficial because multiple strains of the same plant pathogens can be determined.

The ever-increasing population and consequent increased global food demands, is leading to mass growing of crops. Automation can improve crop yields, enhance the health of crops and reduce human error such as chemical overuse and seed loss through the precise, routine, remote monitoring of crops for infection. Automation is the use of pre-programmed robots and machinery in the production, monitoring and treatment of goods. Automated monitoring of crops has been successfully demanded the determination of field estimates, crop phenology, drought monitoring, acreage estimation and monitoring of environmental disturbance to crops (Atzberger 2013). Similarly, crop monitoring and automation diagnostics methods for common crop pathogens are routinely carried out across the field. The on-site diagnostics devices would ideally be connected to other automated machinery which would then be relocated to spray the area infected with pathogens with fungicides or insecticides.

14.6.2 Mobile Technology for Crop Diagnostics

Mobile technology is changing very fast and new developments allow information to be relayed to and from a phone without the need for proximity to other connectivity technologies such as Wi-Fi, or Bluetooth. Hence, this is useful in the field of POC when information may need to be gathered and sent remotely. Imaging capabilities of smart phones has been improved rapidly which may be used for on-site colorimetric images (Ozcan 2014). This ability of smart phones to capture and analyse such results is of great interest in the field of POC. Developments of imaging software, particularly mobile apps are undergoing and mobile phones can be linked to wearable sensing devices for new mode of detection using the latest technology. Thus, on-site testing within the field of crop diagnostics has the potential to reduce costs, increase yields and provide farmers with a reliable means to test the quality of their crop.

Advances in POC across all diagnostics discipline are deriving the miniaturization of sensors i.e. nanosensor capable of diagnosis a disease within minutes on-site. The future will see great improvements within the field of POC based on nanosensor technology. Major breakthrough in the development of miniaturized platforms and multiplexed test array will lead to new strategies for crop disease treatment. New strategies used in crop care will increase yield, reduce chemical spraying.

14.6.3 Wireless Sensor Network Technology

The advances in the miniaturization of sensors can work in tandem with the development of aerial and ground robots to produce lightweight machines for monitoring crops. The wireless sensor network (WSN) utilizes hardware and software to connect the sensing, communication and computation components of a remote monitoring circuit. There are several wireless networks available for use in remote monitoring including Wi-Fi, Bluetooth, wibre and zigbee. Drone and unmanned aerial vehicles (UAV) have been reported in the remote monitoring of crops. The UAV fitted with a camera and Global Positioning System (GPS) could simultaneously capture high resolution images of field crops while reporting its position accurately, thereby identifying the location of diseased crops. Other aerial devices which could be beneficial in the field of plant POC are drones. In recent years drone technology has become far cheaper, more robust and more accessible to the public and easier to use. Other methods of remote crop disease detection such as reflectance of spectral wavelengths are also investigated recently. The principle behind this detection method is that healthy plants will exhibit a different reflectance pattern than diseased crops. The pathogenesis infection can present as a number of symptoms including plant dehydration, alternation to chlorophyll levels or changes within other leaf tissues. This system proved useful as the same platform could be used to detect several disease using the same techniques with only the optical wavelengths reading changing.

14.7 MONITORING OF CROP HEALTH STATUS IN REAL-TIME

Recently, novel sensing tools have been used to complement existing remote sensing tools for continuous monitoring of crops with high reliability and improved signal-to-noise ratios. Nanosensors based approaches are providing a pathway to transducer the invisible crop stress-related chemical signal into optical, electrical or wireless signals that can be recorded. Electronic devices such as smartphones, hyperspectral imaging cameras, can establish direct communications with these nanosensors in plants.

Nanomaterials are promising candidates for plant diseases detection due to the remarkable biospecificity of engineered molecule as recognitions at the nanoscale. Owing to the rapid advances in nanotechnology and modern nanofabrication techniques, a great progress in a variety of useful nanosensors and nanostructured platforms has been continuously emerging for plant disease analysis. Newly developed nanodiagnostics tools can be used for the precision plant disease detection. Nucleic acid amplification, sequencing and VOC analysis now can be potentially performed directly

in the crop field in a much faster and cost-effective manner. This has become possible due to the recent innovation of rapid plant DNA extraction technology enabled by microneedles, miniature DNA sequencing chips, and smart phones-based VOC sensors. In future, it is expected that more and more powerful nanosensor and probes integrated with multimodal detection mechanisms will be developed for quick detection and determination of infections caused by pant pathogens as well as many other biotic and abiotic stresses. However, there are several challenges such as toxicity and environmental impacts of ENMs, promptness of data sharing and disease forecasting and long-term sensor stability in extreme conditions such as hot or cold weather, intensive sun exposure and heavy wear. So, safety concerns much be addressed before any nanosensors can be commercialized and deployed to the crop field.

Since the foremost prerequisite of disease diagnosis is always the timely report and forecast of infecting events on-site. The next generation of nanosensors is expected to be more wirelessly connected that can provide near real-time measurement. Continuous monitoring of VOC emission of plants is expected to provide more time dynamic information and therefore enable more accurate monitoring of plant stresses. In order to support continuous measurement, sensor miniaturization, wireless data transmission, and integration with computational date processing pipelines such as artificial intelligence and machine learning will be among the critical areas to be addressed further. Lastly, more durable and robust nanosensors that can withstand various environmental conditions such as temperature, humidity in the crop field are anticipated before any nanosensors can be deployed to the real field. Despite the challenges, the recent development of miniature and cost-effective nanodiagnostics tools and nanosensors has shown tremendous potentials in improving plant disease diagnosis management and crop health monitoring in the long run. The future of nanosensor is indeed very bright in the coming era of digital farm and precision agriculture. Smart plant-sensing devices and nanosensors have the potential to provide capabilities in real-time monitoring of crops in response to stresses by reporting health status and then control actuation of electronic devices for improving crop productivity (Figure 14.2).

14.8   NANOSENSOR COMMUNICATION AND ACTUATION SYSTEM WITH MACHINES

Nanotechnology is on the verge of generating the tools for establishing real-time two-way communications channels between nanosensors embedded in plants and electronic devices merged with crops. Wong et al. (2017) reported that SWCNT sensors embedded in leaves have converted plants into self-powered chemical detectors that report the presence of groundwater analytes through nIR optical signals (Wong et al. 2017). These SWCNT nanosensors merged within plants communicate the presence of the analyte through optical signals that could trigger e-mails and text messages or communicate directly to a smartphone using Bluetooth technology. Nanosensors are, thus, a promising tool to create smart crops that communicate their health-status to agricultural devices. Nanosensors that communicate through optical signals, wireless or wired channels have the capacity to integrate with existing agricultural electronic devices, including smartphones, hyperspectal imaging cameras, high-throughput phenotyping instrumentation, radio frequency devices and metrological stations (Bai et al. 2016; Lee et al. 2014; Lelong et al. 2008; White et al. 2012; Zarco-Tejada et al. 2012). Smartphones are already used to monitor crops with accurate GPS information and active communication through Bluetooth and the Internet of Things (García-Tejero et al. 2018; Gubbi et al. 2013). Unmanned aerial vehicle are also already involved in crop management by monitoring vegetation indexes through multispectral bands from the visible to the nIR and detecting water stress with a 40 cm resolution (Lelong et al. 2008; Zarco-Tejada et al. 2012). Nanosensors could be used for high-throughput chemical phenotyping of plants through terrestrial platform-based multi-sensors system with vegetation index sensors, thermal infrared radiometers, spectrometers and visible cameras (Wolfert et al. 2017).Wireless or wired nanosensors data can be collected and transmitted to

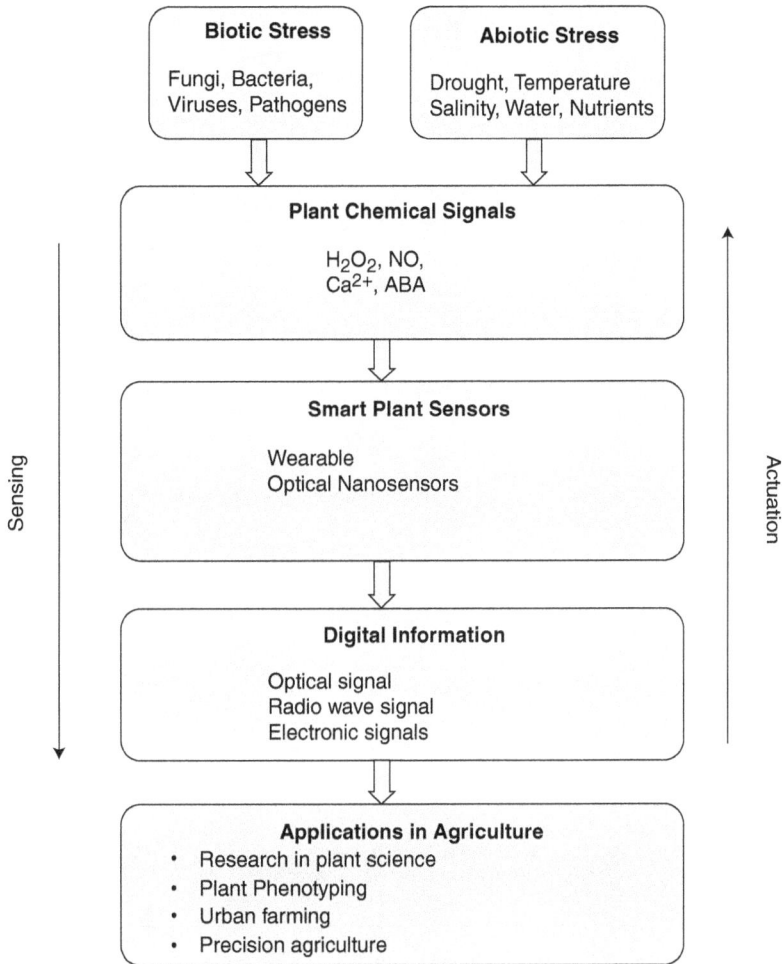

FIGURE 14.2 Smart plant sensors: process of communication via chemical signals produced by stress and electronic devices to equipment.

field-deployable environmental and phenotyping stations. These stations using internet connectivity are capable of cloud-based storage and data analysis. Remote sensing technologies with the help of computer nanosensors for stress management (Baret et al. 2007). Hence, nanosensors will allow both the crop health status in real-time and the actuation of agricultural equipments with the help of electronic devices (Figure 14.3). Automated wireless control systems have been applied to regulate water and nutrient supply in hydroponics and do not suffer from the interference of environmentally related factors (Ibayashi et al. 2016). So, research on automatic actuation of agricultural equipment for fine-tuning crop health status should be focused and explored in coming years.

14.9   CHALLENGES AND FUTURE PERSPECTIVES

Nanobiosensors have already applications in clinical diagnosis of disease, food quality control and environmental monitoring. Clinical diagnosis involves estimation of analytes (such as sucrose, glucose, cholesterol, urea, uric acid), immunological biomarkers, molecular markers, diseases causing agents (such as toxins, microbes, bacterial, fungal, viral), drugs etc. Food quality control including microbes, toxins constituents of foods or GM food for maintaining desired quality of food item.

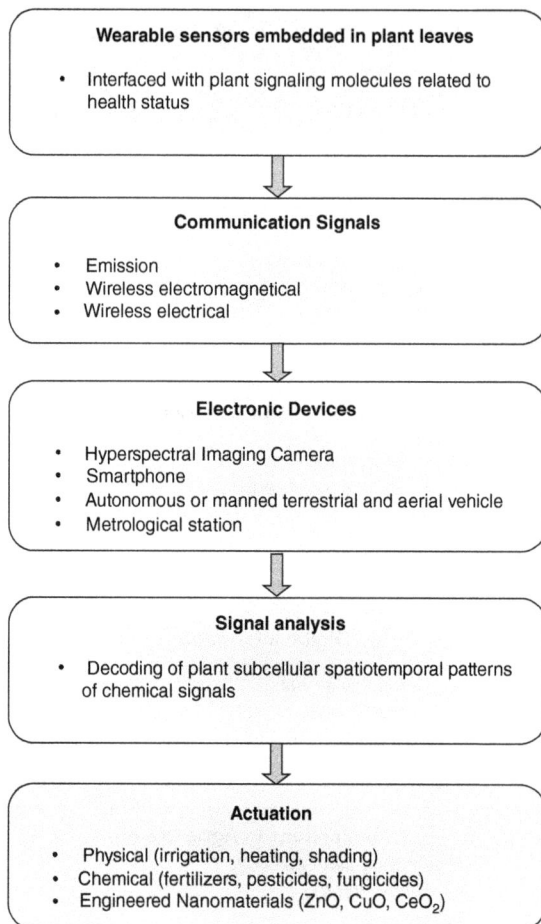

FIGURE 14.3 Wearable sensors and actuation of equipment through electronic devices.

Environmental monitoring includes assessing environmental pollutants in flora and fauna, water, air and soil. Generally, monitoring plant growth and disease have been categorized either as part of food quality control or environmental monitoring. Very less attention has been paid towards developing a desired diagnostic method for plants.

Nanobiosensors that have been established for other applications can be easily translated to the needs of crop production. New research should be focused on new composites made of nano-assemblies of metal-organic framework with properties for vast range of customized applications featuring different working phenomena/principles ranging from chemiluminesce source, quenching abilities, super conducting to super capacitive, utilizing FRET phenomenon, electrochemical/optical labeling and intrinsic catalytic abilities, while imparting stability, biocompatibility near transparency, super-elasticity and mechanical strength to nano-and biomolecular assemblies for fabrication of nanosensors for detection of various analytes related to plant and crop production.

Chimeric methods combining molecular and immunological methods can be used to improve the performance of exiting nanosensors including PCR-based lateral flow immunobiosensors. Genetically encoded biosensors, transcriptional nanobiosensors that would facilitate *in vivo* applications in monitoring crop health status in real-time can be further improved. Use of metabolomics including proteomics and lipidomics, Next-Gen sequencing (NGS) are some disciplines that still need to be trapped for biosensing application in agricultural sector. Nanobiosensors can be coupled

with GPS and robotics system to create smart delivery systems that detect, map and treat specific area in a field prior to or during the onset of symptoms of a particular disease. Transcriptional or transcriptomic nanobiosensors are new concepts being developed at nano-scale levels for tracking metabolites *in vivo* for metabolic engineering and synthetic biology-based application. Such type nanobiosensors have ability to sense specific metabolites. Till date, there are only primitive studies but in future such studies will provide insight towards the biological events and metabolite occurring *in vivo* which are till date not understandable or yet to resolved.

In future, large data-set technologies that transmit process and actuate devices will synergize the integration of smart plant sensors with agricultural devices. There are several challenges such as accuracy, applicability and durability under crop field conditions for nanosensors-based monitoring of crop health in real-time. Multiple chemicals signaling and environmental parameters are likely needed to detect and analyze for an accurate assessment of plant health status. Multiplexing is crucial to further improve the sensing approaches for signaling molecules or a combination of these molecules. Additionally, integrating microarray technologies, MEMS, microfluidics can facilitate wide range of modulation to achieve small sample size, online, automated measurements of multiple samples.

Considering the importance of monitoring crop health status in real-time, researchers should be encouraged to divert their attention focusing on research in the field of nanobiosensors technologies using innovative nanomaterials and novel biomarkers for monitoring crop health status in real-time.

14.10 CONCLUSIONS

Detection of plant disease, metabolic parameters within cellular microenvironment is extremely important for monitoring crop growth and productivity. Nanobiosensors have already established their presence for diagnosis various parameters of clinical, environmental and food quality control. But nanobiosensors for plant disease diagnosis are limited. This review is an attempt to throw spotlight on recent development in this direction. These mainly include use of gold nanoparticles, magnetic nanoparticles, grapheme and its variants, QDs or their combinations for detection of bacterial, fungal, and viral pathogens for plant disease detection using both optical and electrochemical measurements. These nanobiosensors demonstrated the detection of various parameters such as carbohydrate, nutrients, proteins, receptors, phytohormones, miRNA. Use of transcriptomic biosensor and genetic encoded biosensors is another arena that has shown enormous potential for studying *in vivo* microenvironment of plant dynamics. In the coming years, various metallic and carbon-based nanomaterials will be used for plant diagnostics. Various novel techniques and unique feature/flexibility of nanomaterials and opportunities to nano-tune various properties will soon realize fabrication of Next-Gen Nanobiosensors that will provide additional features of portability, real-time detection, accuracy, and simultaneous analysis of different analytes in a single device. Achievements made so far suggest that nanobiosensors are the pioneers for the future disease diagnosis devices for monitoring crop health status in real-time that offer unlimited opportunities to be tapped.

Nanotechnology innovation is running fast in many fields of life science, smart application in agricultural science still lag behind, particularly delivery of agrochemicals and biosensing. Nanomaterials have been used as sensing material to develop nanosensors in the field of agriculture and food sector. Although the use of nanosensors in agriculture is at an initial stage, but nanomaterials are reported to use as tools for detection and quantification of plant metabolic flux, residual of pesticides in food and viral, bacterial and fungal pathogens. Nanomaterials-based nanobiosensors are very promising because of rapid detection and precise quantification of bacteria, viruses, and fungi in plants. Various types of nanosensors have been used in plants and reviewed including fluorescence resonance energy transfer-based nanosensors, carbon-based electrochemical nanosensors, nanowires nanosensors, plasmonic nanosensors, and antibody nanosensors.

Although nanosensors have been used in human and animals for several decades to address a broad number of applications, their use in plant system is emerging, particularly in precision

agriculture and urban farming. Through a better understanding of plant molecular biology and signaling pathways, innovative therapeutic uses of existing molecules within plants have been discovered. The installation of nanosensors or nanoscale wireless nanosensors in living plants is currently applied to enable the real-time monitoring and early detection of potential problems related to biochemistry and metabolism. Intracellular sensors for metabolic precursors signaling ligands and nutrients may help elucidate the complex roles of these molecules in the plant system.

REFERENCES

Ahmad, F., Siddiqui, M. A., Babalola, O. O., & Wu, H. (2012). Biofunctionalization of nanoparticle assisted mass spectrometry as biosensors for rapid detection of plant associated bacteria. *Biosensors and Bioelectronics, 35*(1), 235–242. https://doi.org/10.1016/j.bios.2012.02.055.

Al-Sadi, A. (2017). Impact of plant diseases on human health. *International Journal of Nutrition, Pharmacology, Neurological Diseases, 7*(2), 21. https://doi.org/10.4103/ijnpnd.ijnpnd_24_17.

Arora, K. (2018) Advances in nano based biosensors for food and agriculture. In: Gothandam KM, Ranjan S, Dasgupta N, Ramalingam C, Lichtfouse E (eds) *Nanotechnology, Food Security and Water Treatment.* Springer International Publishing, Cham, pp. 1–52. https://doi.org/10.1007/978-3-319-70166-0_1.

Askim, J. R., Mahmoudi, M., & Suslick, K. S. (2013). Optical sensor arrays for chemical sensing: The optoelectronic nose. *Chemical Society Reviews, 42*(22), 8649. https://doi.org/10.1039/c3cs60179j.

Atzberger, C. (2013). Advances in remote sensing of agriculture: Context description, existing operational monitoring systems and major information needs. *Remote Sensing, 5*(2), 949–981. https://doi.org/10.3390/rs5020949.

Bagal-Kestwal, D., Kestwal, R. M., & Chiang, B. (2015). Invertase-nanogold clusters decorated plant membranes for fluorescence-based sucrose sensor. *Journal of Nanobiotechnology, 13*(1). https://doi.org/10.1186/s12951-015-0089-1.

Bai, G., Ge, Y., Hussain, W., Baenziger, P. S., & Graef, G. (2016). A multi-sensor system for high throughput field phenotyping in soybean and wheat breeding. *Computers and Electronics in Agriculture, 128,* 181–192. https://doi.org/10.1016/j.compag.2016.08.021.

Baret, F., Houles, V., & Guerif, M. (2007). Quantification of plant stress using remote sensing observations and crop models: The case of nitrogen management. *Journal of Experimental Botany, 58*(4), 869–880. https://doi.org/10.1093/jxb/erl231.

Barone, P. W., Baik, S., Heller, D. A., & Strano, M. S. (2005). Near-infrared optical sensors based on single-walled carbon nanotubes. *Nature Materials, 4*(1), 86–92. https://doi.org/10.1038/nmat1276.

Bebber, D. P., & Gurr, S. J. (2015). Crop-destroying fungal and oomycete pathogens challenge food security. *Fungal Genetics and Biology, 74,* 62–64. https://doi.org/10.1016/j.fgb.2014.10.012.

Berger, S., Sinha, A. K., & Roitsch, T. (2007). Plant physiology meets phytopathology: Plant primary metabolism and plant pathogen interactions. *Journal of Experimental Botany, 58*(15–16), 4019–4026. https://doi.org/10.1093/jxb/erm298.

Borse, V., & Srivastava, R. (2019). Process parameter optimization for lateral flow immunosensing. *Materials Science for Energy Technologies, 2*(3), 434–441. https://doi.org/10.1016/j.mset.2019.04.003.

Cederroth, C. R., & Nef, S. (2009). Soy, phytoestrogens and metabolism: A review. *Molecular and Cellular Endocrinology, 304*(1–2), 30–42. https://doi.org/10.1016/j.mce.2009.02.027.

Cellini, A., Blasioli, S., Biondi, E., Bertaccini, A., Braschi, I., & Spinelli, F. (2017). Potential applications and limitations of electronic nose devices for plant disease diagnosis. *Sensors, 17*(11), 2596. https://doi.org/10.3390/s17112596.

Cellini, A., Buriani, G., Rocchi, L., Rondelli, E., Savioli, S., Rodriguez Estrada, M. T., Cristescu, S. M., Costa, G., & Spinelli, F. (2018). Biological relevance of volatile organic compounds emitted during the pathogenic interactions between apple plants and Erwinia amylovora. *Molecular Plant Pathology, 19*(1), 158–168. https://doi.org/10.1111/mpp.12509.

Chaerle, L., & Van Der Straeten, D. (2000). Imaging techniques and the early detection of plant stress. *Trends in Plant Science, 5*(11), 495–501. https://doi.org/10.1016/s1360-1385(00)01781-7.

Chakraborty, S., & Newton, A. C. (2011). Climate change, plant diseases and food security: An overview. *Plant Pathology, 60*(1), 2–14. https://doi.org/10.1111/j.1365-3059.2010.02411.x.

Chandra, S., Chakraborty, N., Dasgupta, A., Sarkar, J., Panda, K., & Acharya, K. (2015). Chitosan nanoparticles: A positive modulator of innate immune responses in plants. *Scientific Reports, 5*(1). https://doi.org/10.1038/srep15195.

Chaudhuri, B., Hörmann, F., Lalonde, S., Brady, S. M., Orlando, D. A., Benfey, P., & Frommer, W. B. (2008). Protonophore- and pH-insensitive glucose and sucrose accumulation detected by FRET nanosensors in arabidopsis root tips. *The Plant Journal*, *56*(6), 948–962. https://doi.org/10.1111/j.1365-313x.2008.03652.x.

Chauhan, R., Moreno, M., Banda, D. M., Zamborini, F. P., & Grapperhaus, C. A. (2014). Chemiresistive metal-stabilized thiyl radical films as highly selective ethylene sensors. *RSC Advances*, *4*(87), 46787–46790. https://doi.org/10.1039/c4ra07560a.

Chen, G., Song, F., Xiong, X., & Peng, X. (2013). Fluorescent Nanosensors based on fluorescence resonance energy transfer (FRET). *Industrial & Engineering Chemistry Research*, *52*(33), 11228–11245. https://doi.org/10.1021/ie303485n.

Chen, K., Han, H., Luo, Z., Wang, Y., & Wang, X. (2012). A practicable detection system for genetically modified rice by SERS-barcoded nanosensors. *Biosensors and Bioelectronics*, *34*(1), 118–124. https://doi.org/10.1016/j.bios.2012.01.029.

Cho, I., & Ku, S. (2017). Current technical approaches for the early detection of foodborne pathogens: Challenges and opportunities. *International Journal of Molecular Sciences*, *18*(10), 2078. https://doi.org/10.3390/ijms18102078.

Choi, W., & Gilroy, S. (2014). Plant biologists FRET over stress. *eLife*, *3*. https://doi.org/10.7554/elife.02763.

Cognet, L., Tsyboulski, D. A., Rocha, J. R., Doyle, C. D., Tour, J. M., & Weisman, R. B. (2007). Stepwise quenching of exciton fluorescence in carbon nanotubes by single-molecule reactions. *Science*, *316*(5830), 1465–1468. https://doi.org/10.1126/science.1141316.

Cook, J., Aydin-Schmidt, B., González, I. J., Bell, D., Edlund, E., Nassor, M. H., Msellem, M., Ali, A., Abass, A. K., Mårtensson, A., & Björkman, A. (2015). Loop-mediated isothermal amplification (LAMP) for point-of-care detection of asymptomatic low-density malaria parasite carriers in Zanzibar. *Malaria Journal*, *14*(1). https://doi.org/10.1186/s12936-015-0573-y.

Cos, P., De Bruyne, T., Apers, S., Berghe, D. V., Pieters, L., & Vlietinck, A. J. (2003). Phytoestrogens: Recent developments. *Planta Medica*, *69*(7), 589–599. https://doi.org/10.1055/s-2003-41122.

Crawford, B. M., Strobbia, P., Wang, H., Zentella, R., Boyanov, M. I., Pei, Z., Sun, T., Kemner, K. M., & Vo-Dinh, T. (2019). Plasmonic nanoprobes for in vivo multimodal sensing and bioimaging of microRNA within plants. *ACS Applied Materials & Interfaces*, *11*(8), 7743–7754. https://doi.org/10.1021/acsami.8b19977.

Cui, F., & Zhang, X. (2012). Electrochemical sensor for epinephrine based on a glassy carbon electrode modified with graphene/gold nanocomposites. *Journal of Electroanalytical Chemistry*, *669*, 35–41. https://doi.org/10.1016/j.jelechem.2012.01.021.

Cui, S., Ling, P., Zhu, H., & Keener, H. (2018). Plant pest detection using an artificial nose system: A review. *Sensors*, *18*(2), 378. https://doi.org/10.3390/s18020378.

Cuypers, W., & Lieberzeit, P. A. (2018). Combining two selection principles: Sensor arrays based on both biomimetic recognition and chemometrics. *Frontiers in Chemistry*, *6*. https://doi.org/10.3389/fchem.2018.00268.

Darr, J. A., Zhang, J., Makwana, N. M., & Weng, X. (2017). Continuous hydrothermal synthesis of inorganic nanoparticles: Applications and future directions. *Chemical Reviews*, *117*(17), 11125–11238. https://doi.org/10.1021/acs.chemrev.6b00417.

de San Celedonio, R. P., Abeledo, L. G., & Miralles, D. J. (2018). Physiological traits associated with reductions in grain number in wheat and Barley under waterlogging. *Plant and Soil*, *429*(1–2), 469–481. https://doi.org/10.1007/s11104-018-3708-4.

Degenhardt, D. C., Greene, J. K., & Khalilian, A. (2012). Temporal dynamics and electronic nose detection of stink bug-induced volatile emissions from cotton bolls. *Psyche: A Journal of Entomology*, *2012*, 1–9. https://doi.org/10.1155/2012/236762

Delledonne, M., Xia, Y., Dixon, R. A., & Lamb, C. (1998). Nitric oxide functions as a signal in plant disease resistance. *Nature*, *394*(6693), 585–588. https://doi.org/10.1038/29087.

Demas, J. N., DeGraff, B. A., & Coleman, P. B. (1999). Peer reviewed: Oxygen sensors based on luminescence quenching. *Analytical Chemistry*, *71*(23), 793A–800A. https://doi.org/10.1021/ac9908546.

Deuschle, K., Chaudhuri, B., Okumoto, S., Lager, I., Lalonde, S., & Frommer, W. B. (2006). Rapid metabolism of glucose detected with FRET glucose nanosensors in epidermal cells and intact roots of arabidopsis RNA-silencing mutants. *The Plant Cell*, *18*(9), 2314–2325. https://doi.org/10.1105/tpc.106.044073.

Drygin, Y. F., Blintsov, A. N., Grigorenko, V. G., Andreeva, I. P., Osipov, A. P., Varitzev, Y. A., Uskov, A. I., Kravchenko, D. V., & Atabekov, J. G. (2012). Highly sensitive field test lateral flow immunodiagnostics of PVX infection. *Applied Microbiology and Biotechnology*, *93*(1), 179–189. https://doi.org/10.1007/s00253-011-3522-x.

Dubertret, B., Calame, M., & Libchaber, A. J. (2001). Single-mismatch detection using gold-quenched fluorescent oligonucleotides. *Nature Biotechnology*, *19*(4), 365–370. https://doi.org/10.1038/86762.

Dumbrepatil, A. B., Lee, S., Chung, S. J., Lee, M. G., Park, B. C., Kim, T. J., & Woo, E. (2010). Development of a nanoparticle-based FRET sensor for ultrasensitive detection of phytoestrogen compounds. *The Analyst, 135*(11), 2879. https://doi.org/10.1039/c0an00385a

Esser, B., Schnorr, J. M., & Swager, T. M. (2012). Selective detection of ethylene gas using carbon nanotube-based devices: Utility in determination of fruit ripeness. *Angewandte Chemie International Edition, 51*(23), 5752–5756. https://doi.org/10.1002/anie.201201042.

Fahad, S., Bajwa, A. A., Nazir, U., Anjum, S. A., Farooq, A., Zohaib, A., Sadia, S., Nasim, W., Adkins, S., Saud, S., Ihsan, M. Z., Alharby, H., Wu, C., Wang, D., & Huang, J. (2017). Crop production under drought and heat stress: Plant responses and management options. *Frontiers in Plant Science, 8*. https://doi.org/10.3389/fpls.2017.01147.

Fang, Y., & Ramasamy, R. (2015). Current and prospective methods for plant disease detection. *Biosensors, 5*(3), 537–561. https://doi.org/10.3390/bios5030537.

Fang, Y., Umasankar, Y., & Ramasamy, R. P. (2014). Electrochemical detection of P-ethylguaiacol, a fungi infected fruit volatile using metal oxide nanoparticles. *The Analyst, 139*(15), 3804–3810. https://doi.org/10.1039/c4an00384e.

Farber, C., Bryan, R., Paetzold, L., Rush, C., & Kurouski, D. (2020). Non-invasive characterization of single-, double- and triple-viral diseases of wheat with a hand-held Raman spectrometer. *Frontiers in Plant Science, 11*. https://doi.org/10.3389/fpls.2020.01300.

Farber, C., & Kurouski, D. (2018). Detection and identification of plant pathogens on maize kernels with a hand-held Raman spectrometer. *Analytical Chemistry, 90*(5), 3009–3012. https://doi.org/10.1021/acs.analchem.8b00222.

Fernández, J., García-Aznar, J., & Martínez, R. (2012). Piezoelectricity could predict sites of formation/resorption in bone remodelling and modelling. *Journal of Theoretical Biology, 292*, 86–92. https://doi.org/10.1016/j.jtbi.2011.09.032.

Fisher, M. C., Henk, D. A., Briggs, C. J., Brownstein, J. S., Madoff, L. C., McCraw, S. L., & Gurr, S. J. (2012). Emerging fungal threats to animal, plant and ecosystem health. *Nature, 484*(7393), 186–194. https://doi.org/10.1038/nature10947.

Flavel, B. S., Moore, K. E., Pfohl, M., Kappes, M. M., & Hennrich, F. (2014). Separation of single-walled carbon nanotubes with a gel permeation chromatography system. *ACS Nano, 8*(2), 1817–1826. https://doi.org/10.1021/nn4062116.

Gao, G., Jiang, Y., Sun, W., & Wu, F. (2018). Fluorescent quantum dots for microbial imaging. *Chinese Chemical Letters, 29*(10), 1475–1485. https://doi.org/10.1016/j.cclet.2018.07.004.

García-Tejero, I., Ortega-Arévalo, C., Iglesias-Contreras, M., Moreno, J., Souza, L., Tavira, S., & Durán-Zuazo, V. (2018). Assessing the crop-water status in almond (Prunus dulcis mill.) trees via thermal imaging camera connected to smartphone. *Sensors, 18*(4), 1050. https://doi.org/10.3390/s18041050.

Geim, A. K. (2009). Graphene: Status and prospects. *Science, 324*(5934), 1530–1534. https://doi.org/10.1126/science.1158877.

Geng, Y., Peveler, W. J., & Rotello, V. M. (2019). Array-based "Chemical nose" sensing in diagnostics and drug discovery. *Angewandte Chemie International Edition, 58*(16), 5190–5200. https://doi.org/10.1002/anie.201809607.

Gilroy, S., Suzuki, N., Miller, G., Choi, W., Toyota, M., Devireddy, A. R., & Mittler, R. (2014). A tidal wave of signals: Calcium and ROS at the forefront of rapid systemic signaling. *Trends in Plant Science, 19*(10), 623–630. https://doi.org/10.1016/j.tplants.2014.06.013.

Giraldo, J. P., Landry, M. P., Faltermeier, S. M., McNicholas, T. P., Iverson, N. M., Boghossian, A. A., Reuel, N. F., Hilmer, A. J., Sen, F., Brew, J. A., & Strano, M. S. (2014). Plant nanobionics approach to augment photosynthesis and biochemical sensing. *Nature Materials, 13*(4), 400–408. https://doi.org/10.1038/nmat3890.

Giraldo, J. P., Wu, H., Newkirk, G. M., & Kruss, S. (2019). Nanobiotechnology approaches for engineering smart plant sensors. *Nature Nanotechnology, 14*(6), 541–553. https://doi.org/10.1038/s41565-019-0470-6.

Gomez-Roldan, V., Fermas, S., Brewer, P. B., Puech-Pagès, V., Dun, E. A., Pillot, J., Letisse, F., Matusova, R., Danoun, S., Portais, J., Bouwmeester, H., Bécard, G., Beveridge, C. A., Rameau, C., & Rochange, S. F. (2008). Strigolactone inhibition of shoot branching. *Nature, 455*(7210), 189–194. https://doi.org/10.1038/nature07271.

Gubbi, J., Buyya, R., Marusic, S., & Palaniswami, M. (2013). Internet of things (IoT): A vision, architectural elements, and future directions. *Future Generation Computer Systems, 29*(7), 1645–1660. https://doi.org/10.1016/j.future.2013.01.010.

Guillaume, C., Isabelle, C., Marc, B., & Thierry, A. (2018). Assessing frost damages using dynamic models in walnut trees: Exposure rather than vulnerability controls frost risks. *Plant, Cell & Environment, 41*(5), 1008–1021. https://doi.org/10.1111/pce.12935.

Hatfield, J. L., Gitelson, A. A., Schepers, J. S., & Walthall, C. L. (2008). Application of spectral remote sensing for agronomic decisions. *Agronomy Journal, 100*(S3). https://doi.org/10.2134/agronj2006.0370c.

Hilleary, R., Choi, W., Kim, S., Lim, S. D., & Gilroy, S. (2018). Sense and sensibility: The use of fluorescent protein-based genetically encoded biosensors in plants. *Current Opinion in Plant Biology, 46*, 32–38. https://doi.org/10.1016/j.pbi.2018.07.004.

Hong, S., & Lee, C. (2018). The current status and future outlook of quantum dot-based biosensors for plant virus detection. *The Plant Pathology Journal, 34*(2), 85–92. https://doi.org/10.5423/ppj.rw.08.2017.0184.

Hongyun, C., Wenjun, Z., Qinsheng, G., Qing, C., Shiming, L., & Shuifang, Z. (2008). Real time TaqMan RT-PCR assay for the detection of cucumber green mottle mosaic virus. *Journal of Virological Methods, 149*(2), 326–329. https://doi.org/10.1016/j.jviromet.2008.02.006.

Howe, G. A., Major, I. T., & Koo, A. J. (2018). Modularity in Jasmonate signaling for Multistress resilience. *Annual Review of Plant Biology, 69*(1), 387–415. https://doi.org/10.1146/annurev-arplant-042817-040047.

Huang, T., & Murray, R. W. (2002). Quenching of [Ru(bpy)3]2+Fluorescence by binding to AU nanoparticles. *Langmuir, 18*(18), 7077–7081. https://doi.org/10.1021/la025948g.

Humplík, J. F., Lazár, D., Husičková, A., & Spíchal, L. (2015). Automated phenotyping of plant shoots using imaging methods for analysis of plant stress responses – A review. *Plant Methods, 11*(1). https://doi.org/10.1186/s13007-015-0072-8.

Ibayashi, H., Kaneda, Y., Imahara, J., Oishi, N., Kuroda, M., & Mineno, H. (2016). A reliable wireless control system for tomato hydroponics. *Sensors, 16*(5), 644. https://doi.org/10.3390/s16050644.

Im, H., Lee, S., Naqi, M., Lee, C., & Kim, S. (2018). Flexible PI-based plant drought stress sensor for real-time monitoring system in smart farm. *Electronics, 7*(7), 114. https://doi.org/10.3390/electronics7070114.

ISAAA's GM Approval Database. Brief 55: Global Status of Commercialized Biotech/GM Crops: 2019 http://www.isaaa.org/gmapprovaldatabase/.

Jarocka, U., Radecka, H., Malinowski, T., Michalczuk, L., & Radecki, J. (2013). Detection of prunus necrotic Ringspot virus in plant extracts with Impedimetric Immunosensor based on glassy carbon electrode. *Electroanalysis, 25*(2), 433–438. https://doi.org/10.1002/elan.201200470.

Jarocka, U., Wąsowicz, M., Radecka, H., Malinowski, T., Michalczuk, L., & Radecki, J. (2011). Impedimetric Immunosensor for detection of plum pox virus in plant extracts. *Electroanalysis, 23*(9), 2197–2204. https://doi.org/10.1002/elan.201100152.

Jiao, K., Sun, W., & Zhang, S. (2000). Sensitive detection of a plant virus by electrochemical enzyme-linked immunoassay. *Fresenius' Journal of Analytical Chemistry, 367*(7), 667–671. https://doi.org/10.1007/s002160000423.

Jin, X., Sun, X., Chen, G., Ding, L., Li, Y., Liu, Z., Wang, Z., Pan, W., Hu, C., & Wang, J. (2015). PH-sensitive carbon dots for the visualization of regulation of intracellular pH inside living pathogenic fungal cells. *Carbon, 81*, 388–395. https://doi.org/10.1016/j.carbon.2014.09.071.

Joshi, R., Singla-Pareek, S. L., & Pareek, A. (2018). Engineering abiotic stress response in plants for biomass production. *Journal of Biological Chemistry, 293*(14), 5035–5043. https://doi.org/10.1074/jbc.tm117.000232.

Kasibabu, B. S., D'souza, S. L., Jha, S., & Kailasa, S. K. (2015). Imaging of bacterial and fungal cells using fluorescent carbon dots prepared from carica papaya juice. *Journal of Fluorescence, 25*(4), 803–810. https://doi.org/10.1007/s10895-015-1595-0.

Kaushik, S., & Djiwanti, S. R. (2019). Nanofertilizers: Smart delivery of plant nutrients. In: Panpatte D, Jhala Y (eds) *Nanotechnology for Agriculture: Crop Production & Protection*. Springer, Singapore. https://doi.org/10.1007/978-981-32-9374-8_3.

Khater, M., De la Escosura-Muñiz, A., & Merkoçi, A. (2017). Biosensors for plant pathogen detection. *Biosensors and Bioelectronics, 93*, 72–86. https://doi.org/10.1016/j.bios.2016.09.091.

Kiegle, E., Moore, C. A., Haseloff, J., Tester, M. A., & Knight, M. R. (2000). Cell-type-specific calcium responses to drought, salt and cold in the Arabidopsis root. *The Plant Journal, 23*(2), 267–278. https://doi.org/10.1046/j.1365-313x.2000.00786.x.

Kim, J. J., Allison, L. K., & Andrew, T. L. (2019). Vapor-printed polymer electrodes for long-term, on-demand health monitoring. *Science Advances, 5*(3), eaaw0463. https://doi.org/10.1126/sciadv.aaw0463.

Kim, T., Böhmer, M., Hu, H., Nishimura, N., & Schroeder, J. I. (2010). Guard cell signal transduction network: Advances in understanding Abscisic acid, CO_2, and Ca^{2+} signaling. *Annual Review of Plant Biology, 61*(1), 561–591. https://doi.org/10.1146/annurev-arplant-042809-112226.

Klessig, D. F., Durner, J., Noad, R., Navarre, D. A., Wendehenne, D., Kumar, D., Zhou, J. M., Shah, J., Zhang, S., Kachroo, P., Trifa, Y., Pontier, D., Lam, E., & Silva, H. (2000). Nitric oxide and salicylic acid signaling in plant defense. *Proceedings of the National Academy of Sciences, 97*(16), 8849–8855. https://doi.org/10.1073/pnas.97.16.8849.

Knogge, W. (1996). Fungal infection of plants. *The Plant Cell*, *8*(10), 1711. https://doi.org/10.2307/3870224.

Krivec, M., Gunnigle, G., Abram, A., Maier, D., Waldner, R., Gostner, J., Überall, F., & Leitner, R. (2015). Quantitative ethylene measurements with MOx chemiresistive sensors at different relative air humidities. *Sensors*, *15*(11), 28088–28098. https://doi.org/10.3390/s151128088.

Kruss, S., Landry, M. P., Vander Ende, E., Lima, B. M., Reuel, N. F., Zhang, J., Nelson, J., Mu, B., Hilmer, A., & Strano, M. (2014). Neurotransmitter detection using corona phase molecular recognition on fluorescent single-walled carbon nanotube sensors. *Journal of the American Chemical Society*, *136*(2), 713–724. https://doi.org/10.1021/ja410433b.

Kulma, A., & Szopa, J. (2007). Catecholamines are active compounds in plants. *Plant Science*, *172*(3), 433–440. https://doi.org/10.1016/j.plantsci.2006.10.013.

Kumar, V., & Arora, K. (2020). Trends in nano-inspired biosensors for plants. *Materials Science for Energy Technologies*, *3*, 255–273. https://doi.org/10.1016/j.mset.2019.10.004.

Kwak, S., Wong, M. H., Lew, T. T., Bisker, G., Lee, M. A., Kaplan, A., Dong, J., Liu, A. T., Koman, V. B., Sinclair, R., Hamann, C., & Strano, M. S. (2017). Nanosensor technology applied to living plant systems. *Annual Review of Analytical Chemistry*, *10*(1), 113–140. https://doi.org/10.1146/annurev-anchem-061516-045310.

Lager, I., Looger, L. L., Hilpert, M., Lalonde, S., & Frommer, W. B. (2006). Conversion of a putative Agrobacterium sugar-binding protein into a FRET sensor with high selectivity for sucrose. *Journal of Biological Chemistry*, *281*(41), 30875–30883. https://doi.org/10.1074/jbc.m605257200.

Lakowicz, J. R., Chowdury, M. H., Ray, K., Zhang, J., Fu, Y., Badugu, R., Sabanayagam, C. R., Nowaczyk, K., Szmacinski, H., Aslan, K., & Geddes, C. D. (2006). Plasmon-controlled fluorescence: A new detection technology. *Plasmonics in Biology and Medicine III*. https://doi.org/10.1117/12.673106.

Lau, H. Y., Wu, H., Wee, E. J., Trau, M., Wang, Y., & Botella, J. R. (2017). Specific and sensitive isothermal electrochemical biosensor for plant pathogen DNA detection with colloidal gold nanoparticles as probes. *Scientific Reports*, *7*(1). https://doi.org/10.1038/srep38896.

Lebeda, A., Luhová, L., Sedlářová, M., & Jančová, D. (2001). The role of enzymes in plant-fungal pathogens interactions/Die Rolle der Enzyme in den Beziehungen zwischen Pflanzen und pilzlichen Erregern. *Zeitschrift für Pflanzenkrankheiten und Pflanzenschutz/Journal of Plant Diseases and Protection*, 89–111.

Lee, K., Park, J., Lee, M., Kim, J., Hyun, B. G., Kang, D. J., Na, K., Lee, C. Y., Bien, F., & Park, J. (2014). In-situ synthesis of carbon nanotube–graphite electronic devices and their integrations onto surfaces of live plants and insects. *Nano Letters*, *14*(5), 2647–2654. https://doi.org/10.1021/nl500513n.

Lee, S., An, K., Shin, S., Jun, K., Naveen, M., & Son, Y. (2018). "Turn-on" fluorescent and colorimetric detection of Zn^{2+} ions by rhodamine-cinnamaldehyde derivative. *Journal of Nanoscience and Nanotechnology*, *18*(8), 5333–5340. https://doi.org/10.1166/jnn.2018.15380.

Lelong, C., Burger, P., Jubelin, G., Roux, B., Labbé, S., & Baret, F. (2008). Assessment of unmanned aerial vehicles imagery for quantitative monitoring of wheat crop in small plots. *Sensors*, *8*(5), 3557–3585. https://doi.org/10.3390/s8053557.

Lertanantawong, B., Krissanaprasit, A., Chaibun, T., Gothelf, K. V., & Surareungchai, W. (2019). Multiplexed DNA detection with DNA tweezers in a one-pot reaction. *Materials Science for Energy Technologies*, *2*(3), 503–508. https://doi.org/10.1016/j.mset.2019.05.001.

Lew, T. T., Koman, V. B., Silmore, K. S., Seo, J. S., Gordiichuk, P., Kwak, S., Park, M., Ang, M. C., Khong, D. T., Lee, M. A., Chan-Park, M. B., Chua, N., & Strano, M. S. (2020). Real-time detection of wound-induced H_2O_2 signalling waves in plants with optical nanosensors. *Nature Plants*, *6*(4), 404–415. https://doi.org/10.1038/s41477-020-0632-4.

Li, L., Zhang, Q., & Huang, D. (2014). A review of imaging techniques for plant phenotyping. *Sensors*, *14*(11), 20078–20111. https://doi.org/10.3390/s141120078.

Li, Z., Paul, R., Ba Tis, T., Saville, A. C., Hansel, J. C., Yu, T., Ristaino, J. B., & Wei, Q. (2019). Non-invasive plant disease diagnostics enabled by smartphone-based fingerprinting of leaf volatiles. *Nature Plants*, *5*(8), 856–866. https://doi.org/10.1038/s41477-019-0476-y.

Li, Z., Yu, T., Paul, R., Fan, J., Yang, Y., & Wei, Q. (2020). Agricultural nanodiagnostics for plant diseases: Recent advances and challenges. *Nanoscale Advances*, *2*(8), 3083–3094. https://doi.org/10.1039/c9na00724e.

Lichtenstein, A., Havivi, E., Shacham, R., Hahamy, E., Leibovich, R., Pevzner, A., Krivitsky, V., Davivi, G., Presman, I., Elnathan, R., Engel, Y., Flaxer, E., & Patolsky, F. (2014). Supersensitive fingerprinting of explosives by chemically modified nanosensors arrays. *Nature Communications*, *5*(1). https://doi.org/10.1038/ncomms5195.

Lin, Y., Qasim, M., Hussain, M., Akutse, K. S., Avery, P. B., Dash, C. K., & Wang, L. (2017). The herbivore-induced plant volatiles methyl salicylate and menthol positively affect growth and pathogenicity of Entomopathogenic fungi. *Scientific Reports*, *7*(1). https://doi.org/10.1038/srep40494.

Liu, H., Nishide, D., Tanaka, T., & Kataura, H. (2011). Large-scale single-chirality separation of single-wall carbon nanotubes by simple gel chromatography. *Nature Communications*, 2(1). https://doi.org/10.1038/ncomms1313.

Liu, J., Cui, M., Zhou, H., & Zhang, S. (2016). Efficient double-quenching of electrochemiluminescence from CdS:Eu QDs by hemin-graphene-Au nanorods ternary composite for ultrasensitive immunoassay. *Scientific Reports*, 6(1). https://doi.org/10.1038/srep30577.

Lobert, S., Heil, P. D., Namba, K., & Stubbs, G. (1987). Preliminary X-ray fiber diffraction studies of cucumber green mottle mosaic virus, watermelon strain. *Journal of Molecular Biology*, 196(4), 935–938. https://doi.org/10.1016/0022-2836(87)90415-3.

Luna-Moreno, D., Sánchez-Álvarez, A., Islas-Flores, I., Canto-Canche, B., Carrillo-Pech, M., Villarreal-Chiu, J., & Rodríguez-Delgado, M. (2019). Early detection of the fungal banana black Sigatoka pathogen Pseudocercospora fijiensis by an SPR immunosensor method. *Sensors*, 19(3), 465. https://doi.org/10.3390/s19030465.

Mandal, N., Adhikary, S., & Rakshit, R. (2020). Nanobiosensors: Recent developments in soil health assessment. *Soil Analysis: Recent Trends and Applications*, 285–304. https://doi.org/10.1007/978-981-15-2039-6_15.

Maxwell, D. J., Taylor, J. R., & Nie, S. (2002). Self-assembled nanoparticle probes for recognition and detection of biomolecule s. *Journal of the American Chemical Society*, 124(32), 9606–9612. https://doi.org/10.1021/ja025814p.

Mickelbart, M. V., Hasegawa, P. M., & Bailey-Serres, J. (2015). Genetic mechanisms of abiotic stress tolerance that translate to crop yield stability. *Nature Reviews Genetics*, 16(4), 237–251. https://doi.org/10.1038/nrg3901.

Mirica, K. A., Azzarelli, J. M., Weis, J. G., Schnorr, J. M., & Swager, T. M. (2013). Rapid prototyping of carbon-based chemiresistive gas sensors on paper. *Proceedings of the National Academy of Sciences*, 110(35), E3265–E3270. https://doi.org/10.1073/pnas.1307251110.

Mittler, R. (2017). ROS are good. *Trends in Plant Science*, 22(1), 11–19. https://doi.org/10.1016/j.tplants.2016.08.002.

Nassar, J. M., Khan, S. M., Villalva, D. R., Nour, M. M., Almuslem, A. S., & Hussain, M. M. (2018). Compliant plant wearables for localized microclimate and plant growth monitoring. *NPJ Flexible Electronics*, 2(1). https://doi.org/10.1038/s41528-018-0039-8.

Neill, S. J. (2002). Hydrogen peroxide and nitric oxide as signalling molecules in plants. *Journal of Experimental Botany*, 53(372), 1237–1247. https://doi.org/10.1093/jexbot/53.372.1237.

Nguyen, T. D., Deshmukh, N., Nagarah, J. M., Kramer, T., Purohit, P. K., Berry, M. J., & McAlpine, M. C. (2012). Piezoelectric nanoribbons for monitoring cellular deformations. *Nature Nanotechnology*, 7(9), 587–593. https://doi.org/10.1038/nnano.2012.112.

Nikitin, M., Deych, K., Grevtseva, I., Girsova, N., Kuznetsova, M., Pridannikov, M., Dzhavakhiya, V., Statsyuk, N., & Golikov, A. (2018). Preserved microarrays for simultaneous detection and identification of six fungal potato pathogens with the use of real-time PCR in matrix format. *Biosensors*, 8(4), 129. https://doi.org/10.3390/bios8040129.

Oliveira, S. F., Bisker, G., Bakh, N. A., Gibbs, S. L., Landry, M. P., & Strano, M. S. (2015). Protein functionalized carbon nanomaterials for biomedical applications. *Carbon*, 95, 767–779. https://doi.org/10.1016/j.carbon.2015.08.076.

Oren, S., Ceylan, H., & Dong, L. (2017). Helical-shaped graphene tubular spring formed within microchannel for wearable strain sensor with wide dynamic range. *IEEE Sensors Letters*, 1(6), 1–4. https://doi.org/10.1109/lsens.2017.2764046.

Ortiz-Tena, J. G., Rühmann, B., & Sieber, V. (2018). Colorimetric determination of sulfate via an enzyme cascade for high-throughput detection of sulfatase activity. *Analytical Chemistry*, 90(4), 2526–2533. https://doi.org/10.1021/acs.analchem.7b03719.

Ouyang, H., Tu, X., Fu, Z., Wang, W., Fu, S., Zhu, C., Du, D., & Lin, Y. (2018). Colorimetric and chemiluminescent dual-readout immunochromatographic assay for detection of pesticide residues utilizing g-C_3N_4/BiFeO$_3$ nanocomposites. *Biosensors and Bioelectronics*, 106, 43–49. https://doi.org/10.1016/j.bios.2018.01.033.

Ozcan, A. (2014). Mobile phones democratize and cultivate next-generation imaging, diagnostics and measurement tools. *Lab Chip*, 14(17), 3187–3194. https://doi.org/10.1039/c4lc00010b.

Padilla, F. M., Gallardo, M., Peña-Fleitas, M. T., De Souza, R., & Thompson, R. B. (2018). Proximal optical sensors for nitrogen management of vegetable crops: A review. *Sensors*, 18(7), 2083. https://doi.org/10.3390/s18072083.

Pandey, R., Teig-Sussholz, O., Schuster, S., Avni, A., & Shacham-Diamand, Y. (2018). Integrated electrochemical chip-on-Plant functional sensor for monitoring gene expression under stress. *Biosensors and Bioelectronics*, 117, 493–500. https://doi.org/10.1016/j.bios.2018.06.045.

Panferov, V. G., Safenkova, I. V., Byzova, N. A., Varitsev, Y. A., Zherdev, A. V., & Dzantiev, B. B. (2018). Silver-enhanced lateral flow immunoassay for highly-sensitive detection of potato leafroll virus. *Food and Agricultural Immunology, 29*(1), 445–457. https://doi.org/10.1080/09540105.2017.1401044.

Patolsky, F., & Lieber, C. M. (2005). Nanowire nanosensors. *Materials Today, 8*(4), 20–28. https://doi.org/10.1016/s1369-7021(05)00791-1.

Quan, L., Zhang, B., Shi, W., & Li, H. (2008). Hydrogen peroxide in plants: A versatile molecule of the reactive oxygen species network. *Journal of Integrative Plant Biology, 50*(1), 2–18. https://doi.org/10.1111/j.1744-7909.2007.00599.x.

Qureshi, A., Kang, W. P., Davidson, J. L., & Gurbuz, Y. (2009). Review on carbon-derived, solid-state, micro and nano sensors for electrochemical sensing applications. *Diamond and Related Materials, 18*(12), 1401–1420. https://doi.org/10.1016/j.diamond.2009.09.008.

Razmi, A., Golestanipour, A., Nikkhah, M., Bagheri, A., Shamsbakhsh, M., & Malekzadeh-Shafaroudi, S. (2019). Localized surface plasmon resonance biosensing of tomato yellow leaf curl virus. *Journal of Virological Methods, 267*, 1–7. https://doi.org/10.1016/j.jviromet.2019.02.004.

Razo, S. C., Panferova, N. A., Panferov, V. G., Safenkova, I. V., Drenova, N. V., Varitsev, Y. A., Zherdev, A. V., Pakina, E. N., & Dzantiev, B. B. (2019). Enlargement of gold nanoparticles for sensitive immunochromatographic diagnostics of potato brown rot. *Sensors, 19*(1), 153. https://doi.org/10.3390/s19010153.

Reuel, N. F., Mu, B., Zhang, J., Hinckley, A., & Strano, M. S. (2012). Nanoengineered glycan sensors enabling native glycoprofiling for medicinal applications: Towards profiling glycoproteins without labeling or liberation steps. *Chemical Society Reviews, 41*(17), 5744. https://doi.org/10.1039/c2cs35142k.

Reverberi, M., Ricelli, A., Zjalic, S., Fabbri, A. A., & Fanelli, C. (2010). Natural functions of mycotoxins and control of their biosynthesis in fungi. *Applied Microbiology and Biotechnology, 87*(3), 899–911. https://doi.org/10.1007/s00253-010-2657-5.

Rizzo, J. M., Shi, S., Li, Y., Semple, A., Esposito, J. J., Yu, S., Richardson, D., Antochshuk, V., & Shameem, M. (2015). Application of a high-throughput relative chemical stability assay to screen therapeutic protein formulations by assessment of conformational stability and correlation to aggregation propensity. *Journal of Pharmaceutical Sciences, 104*(5), 1632–1640. https://doi.org/10.1002/jps.24408.

Röck, F., Barsan, N., & Weimar, U. (2008). Electronic nose: Current status and future trends. *Chemical Reviews, 108*(2), 705–725. https://doi.org/10.1021/cr068121q.

Rolland, F., Baena-Gonzalez, E., & Sheen, J. (2006). Sugar sensing and signaling in plants: Conserved and novel mechanisms. *Annual Review of Plant Biology, 57*(1), 675–709. https://doi.org/10.1146/annurev.arplant.57.032905.105441.

Sachdev, A., & Gopinath, P. (2015). Green synthesis of multifunctional carbon dots from coriander leaves and their potential application as antioxidants, sensors and bioimaging agents. *The Analyst, 140*(12), 4260–4269. https://doi.org/10.1039/c5an00454c.

Sánchez-Paniagua López, M., Manzanares-Palenzuela, C. L., & López-Ruiz, B. (2018). Biosensors for GMO testing: Nearly 25 years of research. *Critical Reviews in Analytical Chemistry, 48*(5), 391–405. https://doi.org/10.1080/10408347.2018.1442708.

Schiop, S. T., Al Hassan, M., Sestras, A. F., Boscaiu, M., Sestras, R. E., & Vicente, O. (2015). Identification of salt stress biomarkers in Romanian Carpathian populations of picea abies (L.) Karst. *PLoS One, 10*(8), e0135419. https://doi.org/10.1371/journal.pone.0135419.

Schmälzlin, E., Van Dongen, J. T., Klimant, I., Marmodée, B., Steup, M., Fisahn, J., Geigenberger, P., & Löhmannsröben, H. (2005). An optical multifrequency phase-modulation method using microbeads for measuring intracellular oxygen concentrations in plants. *Biophysical Journal, 89*(2), 1339–1345. https://doi.org/10.1529/biophysj.105.063453.

Scholthof, K. G., Adkins, S., Czosnek, H., Palukaitis, P., Jacquot, E., Hohn, T., Hohn, B., Saunders, K., Candresse, T., Ahlquist, P., Hemenway, C., & Foster, G. D. (2011). Top 10 plant viruses in molecular plant pathology. *Molecular Plant Pathology, 12*(9), 938–954. https://doi.org/10.1111/j.1364-3703.2011.00752.x.

Shang, H., Xie, Y., Zhou, X., Qian, Y., & Wu, J. (2011). Monoclonal antibody-based serological methods for detection of cucumber green mottle mosaic virus. *Virology Journal, 8*(1). https://doi.org/10.1186/1743-422x-8-228.

Shang, L., Liu, C., Chen, B., & Hayashi, K. (2018). Plant biomarker recognition by molecular imprinting based localized surface Plasmon resonance sensor array: Performance improvement by enhanced hotspot of AU Nanostructure. *ACS Sensors, 3*(8), 1531–1538. https://doi.org/10.1021/acssensors.8b00329.

Shao, Y., Wang, J., Wu, H., Liu, J., Aksay, I., & Lin, Y. (2010). Graphene based electrochemical sensors and biosensors: A review. *Electroanalysis, 22*(10), 1027–1036. https://doi.org/10.1002/elan.200900571.

Shibata, H., Henares, T. G., Yamada, K., Suzuki, K., & Citterio, D. (2018). Implementation of a plasticized PVC-based cation-selective optode system into a paper-based analytical device for colorimetric sodium detection. *The Analyst*, *143*(3), 678–686. https://doi.org/10.1039/c7an01952a.

Shojaei, T. R., Salleh, M. A., Sijam, K., Rahim, R. A., Mohsenifar, A., Safarnejad, R., & Tabatabaei, M. (2016a). Detection of citrus tristeza virus by using fluorescence resonance energy transfer-based biosensor. *Spectrochimica Acta Part A: Molecular and Biomolecular Spectroscopy*, *169*, 216–222. https://doi.org/10.1016/j.saa.2016.06.052.

Shojaei, T. R., Salleh, M. A., Sijam, K., Rahim, R. A., Mohsenifar, A., Safarnejad, R., & Tabatabaei, M. (2016b). Fluorometric immunoassay for detecting the plant virus citrus tristeza using carbon nanoparticles acting as quenchers and antibodies labeled with CdTe quantum dots. *Microchimica Acta*, *183*(7), 2277–2287. https://doi.org/10.1007/s00604-016-1867-7.

Singh, A., Singh, M. P., Sharma, V., Verma, H., & Arora, K. (2012). Molecular techniques. *Chemical Analysis of Food: Techniques and Applications*, 407–461. https://doi.org/10.1016/b978-0-12-384862-8.00013-3.

Singsaas, E. L., & Sharkey, T. D. (1998). The regulation of isoprene emission responses to rapid leaf temperature fluctuations. *Plant, Cell and Environment*, *21*(11), 1181–1188. https://doi.org/10.1046/j.1365-3040.1998.00380.x.

Stiles, P. L., Dieringer, J. A., Shah, N. C., & Van Duyne, R. P. (2008). Surface-enhanced Raman spectroscopy. *Annual Review of Analytical Chemistry*, *1*(1), 601–626. https://doi.org/10.1146/annurev.anchem.1.031207.112814.

Su, Z., Xu, X., Cheng, Y., Tan, Y., Xiao, L., Tang, D., Jiang, H., Qin, X., & Wang, H. (2019). Chemical pre-reduction and electro-reduction guided preparation of a porous graphene bionanocomposite for indole-3-acetic acid detection. *Nanoscale*, *11*(3), 962–967. https://doi.org/10.1039/c8nr06913a.

Sun, M., Sun, B., Liu, Y., Shen, Q., & Jiang, S. (2016). Dual-color fluorescence imaging of magnetic nanoparticles in live cancer cells using conjugated polymer probes. *Scientific Reports*, *6*(1). https://doi.org/10.1038/srep22368.

Suzuki, N., Miller, G., Salazar, C., Mondal, H. A., Shulaev, E., Cortes, D. F., Shuman, J. L., Luo, X., Shah, J., Schlauch, K., Shulaev, V., & Mittler, R. (2013). Temporal-spatial interaction between reactive oxygen species and abscisic acid regulates rapid systemic acclimation in plants. *The Plant Cell*, *25*(9), 3553–3569. https://doi.org/10.1105/tpc.113.114595.

Suzuki, N., Rivero, R. M., Shulaev, V., Blumwald, E., & Mittler, R. (2014). Abiotic and biotic stress combinations. *New Phytologist*, *203*(1), 32–43. https://doi.org/10.1111/nph.12797.

Swanson, S. J., Choi, W., Chanoca, A., & Gilroy, S. (2011). In vivo imaging of Ca^{2+}, pH, and reactive oxygen species using fluorescent probes in plants. *Annual Review of Plant Biology*, *62*(1), 273–297. https://doi.org/10.1146/annurev-arplant-042110-103832.

Tahir, M. A., Hameed, S., Munawar, A., Amin, I., Mansoor, S., Khan, W. S., & Bajwa, S. Z. (2017). Investigating the potential of multiwalled carbon nanotubes based zinc nanocomposite as a recognition interface towards plant pathogen detection. *Journal of Virological Methods*, *249*, 130–136. https://doi.org/10.1016/j.jviromet.2017.09.004.

Tsuchiya, Y., Yoshimura, M., Sato, Y., Kuwata, K., Toh, S., Holbrook-Smith, D., Zhang, H., McCourt, P., Itami, K., Kinoshita, T., & Hagihara, S. (2015). Probing strigolactone receptors in striga hermonthica with fluorescence. *Science*, *349*(6250), 864–868. https://doi.org/10.1126/science.aab3831.

Tvrdy, K., Jain, R. M., Han, R., Hilmer, A. J., McNicholas, T. P., & Strano, M. S. (2013). A kinetic model for the deterministic prediction of gel-based single-chirality single-walled carbon nanotube separation. *ACS Nano*, *7*(2), 1779–1789. https://doi.org/10.1021/nn305939k.

Umehara, M., Hanada, A., Yoshida, S., Akiyama, K., Arite, T., Takeda-Kamiya, N., Magome, H., Kamiya, Y., Shirasu, K., Yoneyama, K., Kyozuka, J., & Yamaguchi, S. (2008). Inhibition of shoot branching by new terpenoid plant hormones. *Nature*, *455*(7210), 195–200. https://doi.org/10.1038/nature07272.

Van Loon, L. C., Geraats, B. P., & Linthorst, H. J. (2006). Ethylene as a modulator of disease resistance in plants. *Trends in Plant Science*, *11*(4), 184–191. https://doi.org/10.1016/j.tplants.2006.02.005.

Vilatela, J. J., & Eder, D. (2012). Nanocarbon composites and hybrids in sustainability: A review. *ChemSusChem*, *5*(3), 456–478. https://doi.org/10.1002/cssc.201100536.

Voothuluru, P., Braun, D. M., & Boyer, J. S. (2018). An in vivo imaging assay detects spatial variability in glucose release from plant roots. *Plant Physiology*, *178*(3), 1002–1010. https://doi.org/10.1104/pp.18.00614.

Walia, A., Waadt, R., & Jones, A. M. (2018). Genetically encoded biosensors in plants: Pathways to discovery. *Annual Review of Plant Biology*, *69*(1), 497–524. https://doi.org/10.1146/annurev-arplant-042817-040104.

Walsh, R., Morales, J. M., Skipwith, C. G., Ruckh, T. T., & Clark, H. A. (2015). Enzyme-linked DNA dendrimer nanosensors for acetylcholine. *Scientific Reports*, *5*(1). https://doi.org/10.1038/srep14832.

Wang, L., Han, D., Ni, S., Ma, W., Wang, W., & Niu, L. (2015). Photoelectrochemical device based on Mo-doped BiVO4 enables smart analysis of the global antioxidant capacity in food. *Chemical Science*, 6(11), 6632–6638. https://doi.org/10.1039/c5sc02277k.

Wang, P., Lombi, E., Zhao, F., & Kopittke, P. M. (2016). Nanotechnology: A new opportunity in plant sciences. *Trends in Plant Science*, 21(8), 699–712. https://doi.org/10.1016/j.tplants.2016.04.005.

Wang, X., Cheng, M., Yang, Q., Wei, H., Xia, A., Wang, L., Ben, Y., Zhou, Q., Yang, Z., & Huang, X. (2019). A living plant cell-based biosensor for real-time monitoring invisible damage of plant cells under heavy metal stress. *Science of the Total Environment*, 697, 134097. https://doi.org/10.1016/j.scitotenv.2019.134097.

Wang, X., Zhou, J., Song, J., Liu, J., Xu, N., & Wang, Z. L. (2006). Piezoelectric field effect transistor and nanoforce sensor based on a single ZnO nanowire. *Nano Letters*, 6(12), 2768–2772. https://doi.org/10.1021/nl061802g.

Wang, Y., Liu, J., & Zhou, H. (2019). Visual detection of cucumber green mottle mosaic virus based on terminal deoxynucleotidyl transferase coupled with DNAzymes amplification. *Sensors*, 19(6), 1298. https://doi.org/10.3390/s19061298.

Wang, Z. L. (2007). The new field of nanopiezotronics. *Materials Today*, 10(5), 20–28. https://doi.org/10.1016/s1369-7021(07)70076-7.

Weerakoon, K. A., Shu, J. H., & Chin, B. A. (2011). A chemiresistor sensor with a poly3-hexylthiophene active layer for the detection of insect infestation at early stages. *IEEE Sensors Journal*, 11(7), 1617–1622. https://doi.org/10.1109/jsen.2010.2103359.

White, J. W., Andrade-Sanchez, P., Gore, M. A., Bronson, K. F., Coffelt, T. A., Conley, M. M., Feldmann, K. A., French, A. N., Heun, J. T., Hunsaker, D. J., Jenks, M. A., Kimball, B. A., Roth, R. L., Strand, R. J., Thorp, K. R., Wall, G. W., & Wang, G. (2012). Field-based phenomics for plant genetics research. *Field Crops Research*, 133, 101–112. https://doi.org/10.1016/j.fcr.2012.04.003.

Wise, K., & Brasuel, M. (2011). The current state of engineered nanomaterials in consumer goods and waste streams: The need to develop nanoproperty-quantifiable sensors for monitoring engineered nanomaterials. *Nanotechnology, Science and Applications*, 73. https://doi.org/10.2147/nsa.s9039.

Wolfert, S., Ge, L., Verdouw, C., & Bogaardt, M. (2017). Big data in smart farming – A review. *Agricultural Systems*, 153, 69–80. https://doi.org/10.1016/j.agsy.2017.01.023.

Wong, M. H., Giraldo, J. P., Kwak, S., Koman, V. B., Sinclair, R., Lew, T. T., Bisker, G., Liu, P., & Strano, M. S. (2017). Nitroaromatic detection and infrared communication from wild-type plants using plant nanobionics. *Nature Materials*, 16(2), 264–272. https://doi.org/10.1038/nmat4771.

Wu, S., Mou, C., & Lin, H. (2013). Synthesis of mesoporous silica nanoparticles. *Chemical Society Reviews*, 42(9), 3862. https://doi.org/10.1039/c3cs35405a.

Wu, X., Gao, S., Wang, J., Wang, H., Huang, Y., & Zhao, Y. (2012). The surface-enhanced Raman spectra of aflatoxins: Spectral analysis, density functional theory calculation, detection and differentiation. *The Analyst*, 137(18), 4226. https://doi.org/10.1039/c2an35378d.

Wujcik, E. K., Wei, H., Zhang, X., Guo, J., Yan, X., Sutrave, N., Wei, S., & Guo, Z. (2014). Antibody nanosensors: A detailed review. *RSC Advances*, 4(82), 43725–43745. https://doi.org/10.1039/c4ra07119k.

Xia, Y., Song, L., & Zhu, C. (2011). Turn-on and near-infrared fluorescent sensing for 2,4,6-Trinitrotoluene based on hybrid (Gold Nanorod)–(Quantum dots) assembly. *Analytical Chemistry*, 83(4), 1401–1407. https://doi.org/10.1021/ac1028825.

Xu, J., Tran, T., Padilla Marcia, C. S., Braun, D. M., & Goggin, F. L. (2017). Superoxide-responsive gene expression in arabidopsis thaliana and zea mays. *Plant Physiology and Biochemistry*, 117, 51–60. https://doi.org/10.1016/j.plaphy.2017.05.018.

Yoshida, T., Mogami, J., & Yamaguchi-Shinozaki, K. (2014). ABA-dependent and ABA-independent signaling in response to osmotic stress in plants. *Current Opinion in Plant Biology*, 21, 133–139. https://doi.org/10.1016/j.pbi.2014.07.009.

Yuan, L., Lin, W., Zheng, K., & Zhu, S. (2013). FRET-based small-molecule fluorescent probes: Rational design and Bioimaging applications. *Accounts of Chemical Research*, 46(7), 1462–1473. https://doi.org/10.1021/ar300273v.

Yusoff, N., Pandikumar, A., Ramaraj, R., Lim, H. N., & Huang, N. M. (2015). Gold nanoparticle based optical and electrochemical sensing of dopamine. *Microchimica Acta*, 182(13–14), 2091–2114. https://doi.org/10.1007/s00604-015-1609-2.

Zadran, S., Standley, S., Wong, K., Otiniano, E., Amighi, A., & Baudry, M. (2012). Fluorescence resonance energy transfer (fret)-based biosensors: Visualizing cellular dynamics and bioenergetics. *Applied Microbiology and Biotechnology*, 96(4), 895–902. https://doi.org/10.1007/s00253-012-4449-6.

Zarco-Tejada, P., González-Dugo, V., & Berni, J. (2012). Fluorescence, temperature and narrow-band indices acquired from a UAV platform for water stress detection using a micro-hyperspectral imager and a thermal camera. *Remote Sensing of Environment*, *117*, 322–337. https://doi.org/10.1016/j.rse.2011.10.007.

Zeilinger, S., Gupta, V. K., Dahms, T. E., Silva, R. N., Singh, H. B., Upadhyay, R. S., Gomes, E. V., Tsui, C. K., & Nayak, S. C. (2016). Friends or foes? Emerging insights from fungal interactions with plants. *FEMS Microbiology Reviews*, *40*(2), 182–207. https://doi.org/10.1093/femsre/fuv045.

Zeiri, L., & Efrima, S. (2006). Surface-enhanced Raman scattering (SERS) of microorganisms. *Israel Journal of Chemistry*, *46*(3), 337–346. https://doi.org/10.1560/u792-l827-5511-8520.

Zhan, F., Wang, T., Iradukunda, L., & Zhan, J. (2018). A gold nanoparticle-based lateral flow biosensor for sensitive visual detection of the potato late blight pathogen, Phytophthora infestans. *Analytica Chimica Acta*, *1036*, 153–161. https://doi.org/10.1016/j.aca.2018.06.083.

Zhang, J., Landry, M. P., Barone, P. W., Kim, J., Lin, S., Ulissi, Z. W., Lin, D., Mu, B., Boghossian, A. A., Hilmer, A. J., Rwei, A., Hinckley, A. C., Kruss, S., Shandell, M. A., Nair, N., Blake, S., Şen, F., Şen, S., Croy, R. G., … Strano, M. S. (2013). Molecular recognition using corona phase complexes made of synthetic polymers adsorbed on carbon nanotubes. *Nature Nanotechnology*, *8*(12), 959–968. https://doi.org/10.1038/nnano.2013.236.

Zhang, Y., Haghighi, P. D., Burstein, F., Yap, L. W., Cheng, W., Yao, L., & Cicuttini, F. (2020). Electronic skin wearable sensors for detecting lumbar–pelvic movements. *Sensors*, *20*(5), 1510. https://doi.org/10.3390/s20051510.

Zhao, Y., Liu, L., Kong, D., Kuang, H., Wang, L., & Xu, C. (2014). Dual amplified electrochemical immunosensor for highly sensitive detection of Pantoea stewartii sbusp. stewartii. *ACS Applied Materials & Interfaces*, *6*(23), 21178–21183. https://doi.org/10.1021/am506104r.

Zhu, Q., Wang, L., Dong, Q., Chang, S., Wen, K., Jia, S., Chu, Z., Wang, H., Gao, P., Zhao, H., Han, S., & Wang, Y. (2017). FRET-based glucose imaging identifies glucose signalling in response to biotic and abiotic stresses in rice roots. *Journal of Plant Physiology*, *215*, 65–72. https://doi.org/10.1016/j.jplph.2017.05.007.

15 Artificial Intelligence-aided Bioengineering of Eco-friendly Microbes for Food Production
Policy and Security Issues in a Developing Society

Wilson Nwankwo, Charles Oluwaseun Adetunji, Kingsley Eghonghon Ukhurebor and Ayodeji Samuel Makinde
Edo State University Uzairue

CONTENTS

15.1 INTRODUCTION

The issues of food production, availability, security and safety have been an utmost apprehension to researchers as well as all the relevant stakeholders globally. This is a result of the incessant upsurge in the population of the world; in affirmation of this fact, the United Nations (UN, 2019) has projected that the world population would reach 9.7 billion by 2050 from the recent estimate of about 7.7 billion. Indisputably, this incessant upsurge in the population of the world, would logically, increase and correspondingly influence the demand for food globally if urgent stages are not taken to address the incessant population growth as well as the advancement of food production, availability, security and safety.

According to "the Food and Agriculture Organization of the United Nations, (FOA, 2019)", about half of the estimated 821 million persons that are believed to have insufficient food (that is those living in poverty) are those persons that devote most of their time and lives for the production of food (farmers or agriculturalists) for the benefits of others These farmers or agriculturalists are mostly those in a developing society of the world. This is basically ascribed to the susceptibility of these sets of farmers or agriculturalists to the numerous agricultural hazards resulting naturally or ensuing as a result of human activities that have continually threatened human and other living organisms' existence as well as the entire environment. Such agricultural hazards as well as other

DOI: 10.1201/9781003268468-15

environmental hazards could result in extreme or unfavourable weather conditions, conflict, market shocks, etc (Nwankwo et al., 2020a–d; FOA, 2015; FOA, 2019).

According to "the Artificial Intelligence, Emerging Trends, SDG2, SDG6 (2019)"; these farmers or agriculturalists in developing society of the world are mostly not commercial, meaning that they are smallholder farmers or agriculturalists, whose agricultural produce account for about 70% of the global food consumption, are predominantly susceptible to several hazards and food insecurity. However, according to them, evolving machinery and technologies such as Artificial Intelligence (AI), have been predominantly auspicious in tackling and mitigating these challenges and menaces such as deficiency of expertise, climate change, resource enhancement and user confidence.

The foremost of these challenges and menaces is the climate change issues. According to the "Intergovernmental Panel on Climate Change (IPCC, 2014), climate change is a statistically substantial change in either the average state of the climate or in the inconsistencies in the average state of the climate, taking place over a long period". It has to do with a modification in the climate resulting either from natural internal processes or from external factors by human direct or indirect activities (IPCC, 2014; Field et al., 2014). These changes are constituent of the atmosphere, coupled with the usual variation in the climate observed over a longer period and this could influence agriculture, radio waves, etc (Field et al., 2014; FOA, 2017; Ukhurebor and Umukoro, 2018; Ukhurebor et al., 2019; Ukhurebor and Azi, 2019; Ukhurebor and Nwankwo, 2020). Similarly, it has been reported that in several regions of the world the release of Greenhouse gasses (GHGs) such as Carbon (IV) Oxide (CO_2), Methane (CH_4) and Nitrous Oxide (N_2O) have increased tremendously as a result of human activities such as agricultural and industrial actions (Field et al., 2014). Expectedly, in the next coming decades numerous individuals most especially those living in the developing economy would encounter a deficit of water and food with degenerative health conditions due to climate change influences (Field et al., 2014; FOA, 2015; Ukhurebor and Abiodun, 2018). The issue of the health and other effects resulting from these environmental effluences is now a cause of serious concern to environmental scientists globally (Izuogu, 2015), and if we are to combat these challenges, we should look toward technology and innovation; no wonder "the United Nations Conference on Trade and Development (UNCTAD, 2017)", reported that technology and innovation would contribute significantly in ensuring food production, availability and security in the upcoming years. However, at the moment, there are several ongoing scientific deliberations among environmental scientists concerning the core source of global warming which consistently results in climate change (Nwankwo and Ukhurebor, 2019; Herndon and Whiteside, 2019). The apparent disagreement that the main source of global warming is not only CO_2 heat retention and other GHGs, but particulate pollution that engrosses radiation, heats the troposphere and upsurges the efficiency of atmospheric-convective heat elimination from the earth's surface (Nwankwo and Ukhurebor, 2019; Herndon and Whiteside, 2019). Nevertheless, there have been continued research studies aimed at unravelling other tendencies which possibly cause environmental effluences that perhaps result in global warming (Nwankwo and Ukhurebor, 2019; Herndon and Whiteside, 2019). As reported by Nwankwo et al. (2020b), the continuous-increasing progression in communication and information technology via the use of computers and enormous data centres had also been recognized as possible sources of environmental effluence, dilapidation, and core contributing factor to climate change issues globally.

Attaining food production, availability and security in the upcoming years, by means of water, eco-friendly microbes and all other essential resources in a maintainable way is one of the foremost challenges we presently face and if appropriate procedures are not applied, the upcoming generations would suffer more of the impacts ensuing therein. Since water, eco-friendly microbes and all other essential resources are essential ingredients for agriculture, it is therefore recommended that cautious monitoring and management of eco-friendly microbes and all other essential resources efficient in agriculture vis-à-vis food production, availability and security, as well as exploring opportunities to manage these essential resources are crucial.

The question on how can we manage the performance of water, eco-friendly microbes and all other essential resources use in agriculture, should come to mind. However, some relevant agencies are working hard in this regard. The FAO recently developed a publicly available near real-time database by means of satellite data that would allow the monitoring and management of agricultural water productivity (FOA, 2017; FOA, 2019).

Innovative developments in earth observation machinery specify that currently, it is now possible to monitor crucial data for sustainable agricultural production as well as other natural resources management by means of satellite remote sensing. This is an indication that there need for a more all-inclusive operational procedure such as Artificial Intelligence, that would be both systematically vigorous and effective at several scales.

Artificial Intelligence could be of great assistance in aiding subsistence agriculturalists/farmers in developing society such as in African countries in addressing the menaces and challenges that could ensue during their agricultural activities more efficiently. Such menaces and challenges like diseases or viruses as well as other adverse organisms have beleaguered these regions over some decades now, regardless of the all-embracing investment from some relevant supporting agencies such as; "the Organization for Food and Agriculture of the United Nations (FAO), the United Nations Environment Programme (UNEP), the United Nations Economic Commission for Africa (UNECA), the United Nations Environment Programme (UNEP), the United Nations Industrial Development Organization (UNIDO), the United Nations Development Programme (UNDP), the New Partnership for Africa's Development (NEPAD), the World Health Organization (WHO), the International Monetary Funds (IMF), the African Development Bank (ADB), the World Bank", etc (Serdeczny et al., 2017; FOA, 2019).

As reported by "the Artificial Intelligence, Emerging Trends, SDG2, SDG6 (2019)", the combination of satellite imagery with Artificial Intelligence machinery has been predominantly supportive in assisting individuals, the governments and organizations such as the "World Food Programme (WFP)" in addressing the susceptibility by providing a more granulated understanding of the influence of precise shocks on farming or agriculture, as well as in the prediction of agricultural yields and future proceedings. Additionally, they also reported that Artificial Intelligence in agriculture could be also useful in developing "credit risk scoring systems" that would be of assistance to farmers or agriculturalists who cannot assess credit antiquity or collateral in acquiring access to loans from the relevant monetary/financial establishments. "The Artificial Intelligence, Emerging Trends, SDG2, SDG6 (2019)", also reported that Artificial Intelligence is also presently being applied in remote sensing technology. The remote sensing technology was used for the empowerment of "the Organization for Food and Agriculture of the United Nations (FAO), Water Productivity through Open access of Remotely sensed derived data (WaPOR) project (the FAO's WaPOR project)" in providing more understanding about the quantity of food that could be yielded per unit of water. This is mostly beneficial for the relevant stakeholders as well as policy-makers who would be well prepared to boost good agricultural activities and prioritize aspects for optimum benefits. By means of the application of these machinery, a vast quantity of data is as well being generated; and this is an aspect that could be habitually sensitive for farmers or agriculturalists cautious to menace divulging their imminent plans or pricing statistics to participants.

There are presently indications that at Microsoft, "FarmBeats" developed a "data-driven, precision agriculture" by means of sensors and drones in targeting pesticides and water to where they are mostly required. In the same way, the importance of amenities, infrastructure and precision agriculture devices has also assisted smallholder farmers or agriculturalists in reducing the used pesticide, encourage agricultural multiplicity as well as counter monocropping "(Artificial Intelligence, Emerging Trends, SDG2, SDG6, 2019)".

There is no doubt that with the combination of working with smart algorithms or an Artificial Intelligence with other conventional procedures, there is going to be a great advancement in attaining one of the major transformations for increasing food availability, safety, security and sustainability

globally. However, the issue of how individual farmers or agriculturists in most remote regions of the world like those of the developing society could benefit from this Artificial Intelligence technology in the form of Artificial Intelligence -aided bioengineering of eco-friendly microbes for food production is another call for concern.

From creation onwards, there has been quest for modernization and innovation on how to improve food production. According to Vitorino and Bessa (2017), modernization and innovation can be projected for the use of eco-friendly molecules and microbes in the production of foods and other food-related resources for both individuals and industrial applications. Nevertheless, the present accelerated steps of innovative food production are a result of the swift combination of biotechnological or bioengineering procedures and methods that have assisted us in identifying innovative eco-friendly molecules and microbes or even the genetic enhancement of identified species of organisms (Vázquez and Villaverde, 2013; Vitorino and Bessa, 2017; Waditee-Sirisattha et al., 2016).

Even before now, history has it that the influences of microbes were not explored in agricultural and medical sciences, apart from the fact that they were basically recognized and known villains (Vázquez and Villaverde, 2013; Vitorino and Bessa, 2017; Waditee-Sirisattha et al., 2016). However, as the moment, beneficial eco-friendly microbes such as plant development agents or promoters as well "phytopathogen regulators or controllers are now being required by several agricultural plants and numerous species of organisms are being applied as "biofactories" of significant pharmacological particles or molecules (Vázquez and Villaverde, 2013; Vitorino and Bessa, 2017; Waditee-Sirisattha et al., 2016). The applications of microbes to biofactories is presently not the only aspect where we presently explore microbes' relevance; microbes have also been discovered for the synthesis of various fuel molecules (biofuel), chemicals (biochemicals) and numerous industrial polymers as well as strains environmental significant as a result of their biodecomposition, bioremediation, biodegradation biomonitoring or biosorption attributes and capacity are now continuously be of great interest to both researchers and industrial activities (Vázquez and Villaverde, 2013; Vitorino and Bessa, 2017; Waditee-Sirisattha et al., 2016).

Hence, this chapter intends to make available an all-inclusive detail as to the application of Artificial Intelligence-aided bioengineering of eco-friendly microbes would be of assistance in food production, availability, security and safety in developing societies. The chapter would also attempt evaluation policies and security issues in developing societies as they relate to food production, availability and safety, so as to proffer appropriate recommendations that would assist in the advancement of agricultural and environmental sustainability.

15.2 ARTIFICIAL INTELLIGENCE-AIDED BIOENGINEERING OF ECO-FRIENDLY MICROBES FOR FOOD PRODUCTION

The last decade has witnessed consistent and ground-breaking research and industrial developments geared at harnessing the various innovative attributes and capacity of microbes in many socioeconomic domains. A completely integrative area or what is now known in different parlance as technological microbiology, bioengineering, biotechnology, etc. has evolved as a mainstay of modern food and agricultural revolution. With AI becoming a science with multidisciplinary applications, biotechnology/bioengineering processes are being re-engineered and the era of developing powerful strains of eco-friendly microbes for food production is here.

Agreeably, the integration of AI with Biotechnology is quite complex. However, it is believed that these multifaceted procedures and techniques that have been applied in heterologous and metabolomics engineering (biotechnology/bioengineering), could be gradually incorporated into microbial reengineering and production processes leading to the emergence of innovative and enhanced food production and economic sufficiency. This advancement adjudged as remarkable in some quarters does raise issues bordering on socioeconomic, legitimacy, and policy.

Accordingly, in this chapter, we would attempt to bring to bear recent developments in the classical study of microbes especially the re-discovery of novel species of such, selection and their

incorporation into various food production activities vis-à-vis the vital policy and security issues associated with such developments.

Biotechnology/bioengineering is very broad (Vázquez and Villaverde, 2013; Vitorino and Bessa, 2017; Waditee-Sirisattha et al., 2016), and its subdomainsorscope could be grouped as follows:

- Food biotechnology/bioengineering
- Agricultural biotechnology/bioengineering
- Chemical and fuel biotechnology/bioengineering
- Environmental biotechnology/bioengineering
- Medical biotechnology/bioengineering
- Materials biotechnology/bioengineering

We would however try to narrow attention to those aspects/areas that relate to food production and thereafter an attempt would be made to assess the nature and relevance of artificial intelligence-aided bioengineering of eco-friendly microbes for food production.

15.2.1   BIOTECHNOLOGY/BIOENGINEERING IN FOOD PRODUCTION AND PROCESSING

Notwithstanding the application of biotechnological/bioengineering procedures in the areas of industrial food-processing and industrial agronomy which happened to precede the technological developments in the 1970s (Vázquez and Villaverde, 2013; Vitorino and Bessa, 2017; Waditee-Sirisattha et al., 2016) the present trend integrates the use of genetically improved microbes as well as the use of enzymes and some other chemicals and biochemicals composites that are gotten as a result of the chemical reaction in microbes (microbial metabolism), with the intention of refining efficiency, enhancing organoleptic physiognomies as well as ascribing innovative nutritional roles to some foods (Vázquez and Villaverde, 2013; Vitorino and Bessa, 2017; Waditee-Sirisattha et al., 2016). Consequently, Vitorino and Bessa (2017) reported that the roles of microbes in contemporary food production could be grouped into two, viz:

- Microbes serve as starters in fermentations (genetically improved microbes or engineered microbes are not acceptable in this circumstance).
- Microbes are used as a place of work in the invention and production of food constituents (genetically improved microbes be may acceptable, however, would hardly take part unswervingly in the food fermentation procedure in this circumstance), where the metabolite is disinfected from biotechnology/bioengineering fermentation and included as a clean additive to the food medium. According to Vitorino and Bessa (2017), the contribution of the microbes in this circumstance is not direct.

Genetic engineering has been proven to be useful in the modification of the physiognomies of yeasts; refining their attributes and performance in the fermentation procedure (Vitorino and Bessa, 2017). Presently, as a result of this procedure breads and pastas of improved quality are now be obtained within a small period of time. Yeasts have been enhanced to endure temperature and pH variations as well as to develop with the greater yield on an array of substrates (Vitorino and Bessa, 2017). Composites such as "trehalose and proline" that are involved in the stress tolerance tendency in yeasts are auspicious for the growth of resilient strains (Vitorino and Bessa, 2017; Takagi and Shima, 2015). Consequently, yeasts that are exposed to innovative procedures such as ultra violet (UV) radiation allow foods with innovative nutritional qualities to be established, such foods are with amplified vitamin D contents (Lipkie et al., 2016). According to Belda et al. (2016), the collection of β-lyase-creating yeasts enhances "aromatic thiol" release and, subsequently, the sensory physiognomies of wines, whereas in the study carried out by Tofalo et al. (2016), they reported that the collection of yeasts focusing in some procedures such as "flocculation" could enhance the fermentation of distinct wines like sparkling wines.

Presently, the use of "non-saccharomyces strains", that were previously seen as yeasts of lesser standing or yeasts that formed unwanted changes, has absolutely impacted the "vinification procedure", which specified the capability of the strains to generate enzymes, minor metabolites, ethanol, glycerol as well as other composites which could upsurge the organoleptic intricacy of wines (Vitorino and Bessa, 2017; Padilla et al., 2016).

According to reports from some studies (Mokoena et al., 2016; Satish Kumar et al., 2013; Vitorino and Bessa, 2017); the prospection of "lactic acid bacteria" that are existing in products which are fermented with diverse cultures has produced innovative bases of "probiotics" as well as the detection of strains which could enhance the eminence of fermented products. According to Gawkowski and Chikindas (2013), these "probiotics" that are existing in microbes that have been allied to the host well-being assistance. Prasad et al. (1998) reported that the foremost known "probiotic microbes" are those that belong to the "genera lactobacillus and Bifidobacterium". For history, they are reportedly used in producing fermented dairy products, some strains of both "genera" are gradually being applied in the formulation of useful foods (Vitorino and Bessa, 2017). As a result of this, Enujiugha and Badejo (2017) reported that there is now an increasing quantity of probiotic foods readily accessible, as well as a swiftly evolving diversity of non-dairy probiotic brewery foods.

Several enzyme preparations of the source of microbes have been assessed in food processing. Reports from several studies have shown that amylases which are gotten from cultures of "Aspergillusniger" as reported in the study of Omemu et al. (2005); Adejuwon et al. (2015) as well as some other studies or "Bacillus subtilis" as in reported in the study of Salman et al. (2016); Ploss et al. (2016) as well as some other studies, for instance, have been applied as a substitute for chemical extracts in the dealing and treatment of wheat flour (Bueno et al., 2016), they assist in the improvement in the preparation of bread for baking as well as tolerating the attainment of some pre-cooked foods. "Aspergillusniger" and "Rhizomucormiehei "strains have been reported to be very auspicious for the manufacture of "extracellular lipases", which enable enzyme recovery (Messias et al., 2011; Vitorino and Bessa, 2017).

According to Sharma et al. (2001), these "microbial lipases" are actually used in the "hydrolysis" procedure of milk fat; refining and enhancing the "aromatization of dairy products". As reported by Sharma et al. (2001), they could also boost the aroma or odour of beverage foods as well as the eminence of "margarine and mayonnaise".

Reports from some studies such as Andersen et al. (2016); Ismail et al. (2016); Mateo et al. (2007); Sheldon (2007) and several others, have shown that "cellulases and pectinases", are used specifically in juice elucidation and the reduction of the fluid frictional force (viscosity), and have are also been without difficulty in the recovery from cultures of "filamentous fungi" which are effective in the breaking down (degradation) of plant "biomass", such as *Cladosporiumsphaerospermum, Penicilliumchrysogenum*and *Trichodermaviride"*.

Furthermore, the machinery for the "immobilization" of these enzymes in "pre-fabricated supports or polymer matrices" advances their steadiness, action and selectivity, supporting their benefits and application as well as recycle process for elongated periods in industrial reactors (Mateo et al., 2007; Vitorino and Bessa, 2017; Sheldon, 2007). According to Carroll et al. (2016), enzymes from microbes could also acquire natural aromas and flavours for foods, even though these composites can habitually be unswervingly acquired from the general biochemical reaction of filamentous fungi (metabolism) such as "Aspergillusniger" and "Pycnoporuscinnabarinus", that could serve in the procedure which can bring about the synthesis of vanillin (a vital food additive, from autoclaved maize bran. In addition, yeasts of the "genus Pichia" have been reported as an additive for the fermentation of coffee in order to enhance its eminence and taste. According to Saerens and Swiegers (2016), it is so because they upsurge the manufacture of the natural additives "isoamyl acetate".

Microbial biosynthetic trails have been discovered primarily as a result of their enzymatic convert low-cost precursors, like glucose or glycerol to form high-cost aromatic composites. This was demonstrated by Nielsen et al. (2010), where they synthesized "E. coli of acetoin", which is in charge of part of buttery fragrance, using glucose as a substrate by means of glucose as a substrate.

Increase in the human population globally has amplified the quest for innovative means of producing foods. Protein that is removed from cultured microbial "biomass" also known as "single-cell protein" are now been used as supplementary protein in normal diets, according to Anupama and Ravindra (2000) now used in replacing the high-cost expensive conventional or traditional sources as such easing the challenges of insufficient or deficiencies of protein. The "single-cell protein" is now commonly used as a means of nutritional protein in both animal and human nourishment.

Several microbes (bacteria) strains such as "bacillus, Hydrogenomonas, Methanomonas, Methylomonas, and Pseudomonas" have been reportedly used as a substrate for manufacturing "single-cell proteins" industrially (Vitorino and Bessa, 2017), this is as a result fact that these bacteria contain around 80% basic protein in the entire dry bulk. The furthermost used yeasts for locating "single-cell proteins" are "saccharomyces, candida, and rhodotorula". According to Patelski et al. (2015), the production of these yeasts is mostly practical since these microbes are could use widespread diversity of substrates. however, the "single-cell proteins" gotten are inadequate in sulphur-comprising amino acids.

According to Nalage et al. (2016), the furthermost frequently used "filamentous fungi" are "fusarium, aspergillus and penicillium" and amongst the "prokaryotic algae", the furthermost used are those that belong to the "genus spirulina", with roughly 65% of their dehydrated mass comprising of protein. The likelihood of using microbes to attain food, food essences as well as "microbial biomass" for food safety and sustainability has strengthened the food industries especially the processing ones, which have incessantly exploited innovative possibilities for traditional or conventional foods, such as flavours, consistencies and scents as well as the detection of innovative food products.

Notwithstanding the countless spring advancing made by biotechnological or bioengineering in the aspect of food production as well as microbial biosensor expansion, there are still some eminent challenges that still require attention. Innovative microbes studies need to be carried out, as such artificial intelligence-aided biotechnological or bioengineering of eco-friendly microbes especially for food production so as to evaluate the efficiency as well procedures for immobilizing cells of microbes that still require some level of development.

15.3 APPLICATION OF ARTIFICIAL INTELLIGENCE-AIDED BIOENGINEERING OF ECO-FRIENDLY MICROBES FOR FOOD PRODUCTION

There is no doubt as to the importance of AI in the contemporary trend and schemes in both socio-economic, biological, biochemical, medical and political aspects of our endeavours. Several nations in the world especially the developed ones are beginning to place more emphasis on the development of AI research, as it is evident that the future control of several events is possibly affected in all aspects by the influential technology and machinery of Artificial Intelligence. Presently, Artificial Intelligence-aided bioengineering of eco-friendly microbes are now been used for food production processes. Existing procedures are being redefined and extensively reinforced. Examining some research developments would help shed more light.

The combined effect of pH, time, NaCl and ethanol concentration on biofilm formation of *Staphylococcus aureus* was described in the study by Vaezi et al. (2020). The authors of this study analysed and evaluated the impact of pH, ethanol and NaCl concentrations at 37°C, and modelled the results from the artificial neural network, after 24 and 48 hours of incubation time. Due to its strong correlation between experimental and modelling results, the artificial neural network was observed with strong modelling efficiency. It was therefore concluded that pH, ethanol, NaCl and time were effective parameters for the formation of biofilm and that the ANN can model them, without being linear.

Harfouche et al. (2019) use the artificial intelligence of the next generation to improve climate-resilient plant breeding. In terms of genomics and phenomics, the authors proposed an insight into the complex biological processes that underlie plant functions in the sense of ecological

disturbances. The combination of genomics and phenomics would also speed up the production of climate-sensitive plants; however, these omics technologies generate massive, heterogeneous and complex data more rapidly than is currently possible to evaluate them. Consequently, artificial intelligence of the next generation is used in the survey, classification, analysis and interpretation in multiple modalities of integrated omic data.

The use of the fermentation and microorganisms in various areas, including food, climate and human health, was discussed in the study carried out by Feng et al. (2018). It also claimed from the obtained results that the use of modernization, automation and artificial intelligence technology could open up new room for fermentation technology. This would be of great benefit and assistance for the regulation of the fermentation rate as well as the monitoring and management of the fermentation path.

Singh et al. (2008) in their study present two separate artificial intelligence strategies, namely an artificial neural network and genetic algorithms. These artificial intelligence strategies are reportedly beneficial for optimizing the fermentation medium used in gelling the production of glucan sucrase. The concentrations of three medium components include Tween 80, Saccharose and K_2HPO_4 as inputs to the model of the neural network and the function of the enzyme as output. To optimize the input space in the neural network model, a genetic algorithm was used to find optimal configurations for maximum enzyme activity. The combination of the artificial neural network with genetic algorithm predicts a maximum of 6.92 U / mL operation at medium 0.54% (v / v) of 80, 5.98% (w / v) saccharose and 1.01% (w / v) of K_2HPO_4. The projected model for artificial intelligence provided a 6.0% increase in enzyme activity, this show and indicates more improved and beneficial results.

15.4   POLICY AND SECURITY ISSUES ON AI-BIOENGINEERED MICROBES

Synthetic Biology has been identified as a sustainable subject of biotechnology that utilizes novel biological systems or the application of re-designing existing ones for the aim of achieving numerous goals. This could be referred to as a disruptive technology that gave a supportive ideal to the Bioeconomy which possess several ideas of giving novel solutions to several challenges faced by the whole world in several sectors such as environmental, global healthcare, manufacturing, and agriculture (Cameron et al., 2014; Bueso and Tangney, 2017; French, 2019). However, it has been identified that the generation of numerous products through the application of synthetic biology there are some limitations that are being reported most especially which hampers the delivery of adequate products.

However, several governments have raised numerous concerns that synthetic biology could increase the amount of agents that might be generated in form of a contaminant or hazards. Therefore, there is a need to intensify more effort towards the development of, identification and evaluation as well as detection techniques for the identification of biological and chemical threats (Wang and Zhang, 2019). Therefore, there is a need for the government of different countries to put investment most especially towards the development of science and technology and the introduction of numerous regulatory processes as well as the amendment of regulatory rules. The application of artificial intelligence will play a lot of role in this regard. This will go a long way most especially towards the development of critical investment, especially in artificial intelligence for rapid development of national security, medicine, remediation, manufacturing, food and energy production. Their application in the maintenance environment could not be over-emphasized (Adetunji et al., 2020; Adetunji et al 2019a, b; Adetunji and Adejumo 2019; Adetunji et al., 2018; Adetunji et al., 2017; Adetunji and Adejumo, 2017).

The application of synthetic biology has been recognized as a typical illustration of dual-use technology which uncountable benefits but there are so many risks that are associated with their application. These could create several fear which could unintentionally or intentionally constitute damage to the environment or human beings. Atypical examples are the capability to develop

engineered virus-specific shuttles and more effective production of gene therapies for overwhelming inherited disorders.; hence, engineering viruses which could lead to the development of more dangerous pathogens most especially those that could constitute harm.

There are numerous arguments that synthetic biology possesses the capability to impose several risks which has instigated numerous precautions that could mitigate against its several hazards.

Moreover, there are several uncertainty as well as the secluded possibility of such risks that could prevent the growth of necessary technology.

Therefore, there is a need that researchers as well as their host institution and various funding bodies need to take into consideration so as to evaluate may be the planned research will be misused or not. There is a need to introduce necessary measures that could mitigate the possibility of misuse as well as its significances should be executed and evidently communicated.

Moreover, most synthetic biology community needs to be cognized of these problems by participating and utilizing adequate open dialogue with regulatory bodies as well as engaging in horizon scanning exercises most especially with the media. The identification of these risk will go a long way towards the application of synthetic biology or artificial intelligence for resolving global challenges such as adequate provision of food through proper implementation of food security strategy, generation of biofuels, adequate production of more effective medicines will go a long way towards the acceptance by the general public.

Also, there is a need for countries to establish international agreement and mutual understanding on the application of synthetic biology and their application for effective development of products and towards the achievement of sustainable development goals. Provision of adequate training to the researcher should also be encouraged most especially on the application of research ethics and on the best way to perform such research. The introduction of training to the student will go a long way towards highlighting their function as science ambassadors and influencers. It must be noted that their training should not only introduce them to adequate knowledge and skills but must be able to point out necessary awareness as well as the application of artificial intelligence and synthetic biology. Therefore, it has become paramount that every scientist must be cognizant of the various means that could be used for the identification of risks well as their misuse

A typical example of this situation involves the application of synthetic biology research which has several applications in the multiple utilization of gene drive technology which could be utilized to effectively disseminate a specific suite of genes during the course of a population. The advantage of utilizing gene drive technology entails the removal of disease-carrying insect populations as well as the prevention of intruding pests' species which has led to the development of several concerns about the unplanned ecological influences of decreasing or eradicating a population (Callaway, 2018; Collins, 2018).

Also, there is a need to develop some certain research on the potential of the pathogens to target certain tissues most especially in the body or certain chemicals in the environment that could play a crucial function towards the delivery of necessary clean-up polluted environment or targeted therapies. There is a need to release some natural techniques that could lead to the liberation of environmental bioremediation interventions most especially when released on a larger scale.

Moreover, there is place necessary measures that could bridge the gaps that exist between research and development scale up as well as their communication. Also, there is a need to adopt a regulation that could double up with the speed of these new innovative technologies while more emphasis should be placed on the product in place of the process used in their manufacturing (Tait et al., 2017). Inappropriate regulatory agendas could dishearten private sector investment in artificial intelligence or synthetic biology.

15.5 CONCLUSION AND FUTURE RECOMMENDATIONS

AI has emerged as a science and machinery with multifaceted and multi-domain applications including biotechnology and bioengineering. The intent of chapter is to examine two things: first,

the applicability and relevance of AI to experimental and industrial production of eco-friendly microbes and in particular compatible strains of microbes that could improve food production in all respects without compromising the health of consumers. Second, since the deployment of bio-engineering techniques in the genetic engineering of new streams of improved crops (genetically modified organisms) in agricultural production, vital policy and security issues had been raised and with AI, issues would be raised on security and safety of consumers of such food. In consistent with the aforementioned, we conclude that in any society, the answers to issues bordering on the security and safety of technology or the products of such technology(food production in this context) must be well-articulated and discussed and accepted if any meaningful progress is to be made not only during the implementation of such projects but also in the distribution, marketing and consumption circles since discrimination or non-acceptance of the technology or products realized through such would ultimately lead to wasted efforts and economic losses on the part of the promoters and producers alike.

Thus, it is our submission that whereas the adoption of AI-aided Bioengineering microbes for food production is a vital socioeconomic strategy, efforts should be made to unravel and address every potential issue that may pose a threat to the acceptance of the technology. It therefore follows that while food production could be boosted immensely by integrating AI into bioengineering of useful microbes and subsequent utilization in specific food production operations, there should be a synergy between researchers, the organized private sector, farmers/food producers, and public agencies in order to create a policy that addresses the concerns of all parties.

REFERENCES

Adejuwon. A. O., Oluduro, A. O., Agboola, F. K., Olutiola, P. O., Burkhardt, B. A., & Segal, S. J. (2015). Expression of α-amylase by *Aspergillus niger*: effect of nitrogen source of growth medium. *Advances in Bioscience and Bioengineering*, 3, 12–19.

Adetunji, C. O, & Adejumo, I. O (2017). Nutritional assessment of mycomeat produced from different agricultural substrates using wild and mutant strains from Pleurotus sajor-caju during solid state fermentation. *Animal Feed Science and Technology*, 224, 14–19. https://doi.org/10.1016/j.anifeedsci.2016.12.004.

Adetunji, C. O., & Adejumo, I. O. (2018a). Efficacy of crude and immobilizedenzymes from bacillus licheniformis for production of biodegraded feather meal and their assessment on chickens. *Environmental Technology & Innovation*, 11, 116–124. https://doi.org/10.1016/j.eti.2018.05.002.

Adetunji, C. O., & Adejumo, I. O. (2018b). Potency of agricultural wastes in mushroom (Pleurotus sajor-caju) biotechnology for feeding broiler chicks (Arbor acre). *International Journal of Recycling of Organic Waste in Agriculture*, 8(1), 37–45. https://doi.org/10.1007/s40093-018-0226-6.

Adetunji, C. O., Adejumo, I. O., Afolabi, I. S., Adetunji, J. B., & Ajisejiri, E. S. (2018). Prolonging the shelf life of 'Agege sweet' orange with chitosan–rhamnolipid coating. *Horticulture, Environment, and Biotechnology*, 59(5), 687–697. https://doi.org/10.1007/s13580-018-0083-2.

Adetunji, C. O., Afolabi, I. S., & Adetunji, J. B. (2019a). Effect of rhamnolipid-aloe Vera gel edible coating on post-harvest control of rot and quality parameters of 'Agege sweet' orange. *Agriculture and Natural Resources*. https://doi.org/10.34044/j.anres.2019.53.4.06.

Adetunji, C. O., Oloke, J. K., Bello, O. M., Pradeep, M., & Jolly, R. S. (2019b). Isolation, structural elucidation and bioherbicidal activity of an eco-friendly bioactive 2-(hydroxymethyl) phenol, from pseudomonas aeruginosa (C1501) and its ecotoxicological evaluation on soil. *Environmental Technology & Innovation*, 13, 304–317. https://doi.org/10.1016/j.eti.2018.12.006.

Adetunji, C. O., Oloke, J. K., Phazang, P., & Sarin, N. B. (2020). Influence of eco-friendly phytotoxic metabolites from Lasiodiplodia pseudotheobromae C1136 on physiological, biochemical, and ultrastructural changes on tested weeds. *Environmental Science and Pollution Research*, 27(9), 9919–9934. https://doi.org/10.1007/s11356-020-07677-9.

Adetunji, C. O., Oloke, J. K., Prasad, G., & Akpor, O. B. (2017). Environmental influence of cultural medium on bioherbicidal activities of *Pseudomonasaeruginosa*C1501 on mono and dico weeds. *Polish Journal of Natural Sciences*, 32(4), 659–670.

Andersen, B., Poulsen, R., & Hansen, G. H. (2016). Cellulolytic and xylanolytic activities of common indoor fungi. *International Biodeterioration & Biodegradation*, *107*, 111–116. https://doi.org/10.1016/j.ibiod.2015.11.012.

Anupama, & Ravindra, P. (2000). Value-added food. *Biotechnology Advances*, 18(6), 459–479. https://doi. org/10.1016/s0734-9750(00)00045-8.

Artificial Intelligence, Emerging Trends, SDG2, SDG6 (2019). How AL can improve agriculture for better food security. https://news.itu.int/ai-for-food-security/.

Belda, I., Ruiz, J., Navascués, E., Marquina, D., & Santos, A. (2016). Improvement of aromatic thiol release through the selection of yeasts with increased β-lyase activity. *International Journal of Food Microbiology*, 225, 1–8. https://doi.org/10.1016/j.ijfoodmicro.2016.03.001.

Bueno, M. M., Thys, R. C., & Rodrigues, R. C. (2016). Microbial enzymes as substitutes of chemical additives in baking wheat flour—Part II: Combined effects of nine enzymes on dough rheology. *Food and Bioprocess Technology*, 9(9), 1598–1611. https://doi.org/10.1007/s11947-016-1744-8.

Callaway, E. (2018). Ban on 'gene drives' is back on the UN's agenda — worrying scientists. *Nature*, 563(-7732), 454–455. https://doi.org/10.1038/d41586-018-07436-4.

Cameron, D. E., Bashor, C. J., & Collins, J. J. (2014). A brief history of synthetic biology. *Nature Reviews Microbiology*, 12(5), 381–390. https://doi.org/10.1038/nrmicro3239.

Carroll, A. L., Desai, S. H., & Atsumi, S. (2016). Microbial production of scent and flavor compounds. *Current Opinion in Biotechnology*, 37, 8–15. https://doi.org/10.1016/j.copbio.2015.09.003.

Collins, J. P. (2018). Gene drives in our future: Challenges of and opportunities for using a self-sustaining technology in pest and vector management. *BMC Proceedings*, 12(S8). https://doi.org/10.1186/s12919-018-0110-4.

Enujiugha, V. N., & Badejo, A. A. (2017). Probiotic potentials of cereal-based beverages. *Critical Reviews in Food Science and Nutrition*, 57(4), 790–804. https://doi.org/10.1080/10408398.2014.930018.

Feng, R., Chen, L., & Chen, K. (2018). Fermentation trip: Amazing microbes, amazing metabolisms. *Annals of Microbiology*, 68(11), 717–729. https://doi.org/10.1007/s13213-018-1384-5.

Field, C.B., Barros, V.R., Dokken, D., Mach, K., Mastrandrea, M., Bilir, T., Chatterjee, M., Ebi, K.L., Estrada, Y.O., Genova, R.C., et al. (2014). Climate change 2014: Impacts, adaptation, and vulnerability. Part A: Global and sectoral aspects. Contribution of Working Group II to the Fifth Assessment Report of the Intergovernmental Panel on Climate Change; Cambridge University Press: Cambridge, UK; New York, NY, USA.

Flores Bueso, Y., & Tangney, M. (2017). Synthetic biology in the driving seat of the bioeconomy. *Trends in Biotechnology*, 35(5), 373–378. https://doi.org/10.1016/j.tibtech.2017.02.002.

Food and Agriculture Organization of the United Nations (FOA, 2015). The impact of natural hazards and disasters on agriculture and food security and nutrition. Rome, Italy. https://reliefweb.int/sites/reliefweb. int/files/resources/a-i4434e_0.pdf.

Food and Agriculture Organization of the United Nations (FOA, 2017). The future of food and agriculture – Trends and challenges. Rome, Italy. http://www.fao.org/3/a-i6583e.pdf.

Food and Agriculture Organization of the United Nations (FOA, 2019). The state of food and agriculture 2019. Moving forward on food loss and waste reduction. Rome, Italy. http://www.fao.org/3/ca6030en/ ca6030en.pdf.

French, K. E. (2019). Harnessing synthetic biology for sustainable development. *Nature Sustainability*, 2(4), 250–252. https://doi.org/10.1038/s41893-019-0270-x.

Gawkowski, D., & Chikindas, M. (2013). Non-dairy probiotic beverages: The next step into human health. *Beneficial Microbes*, 4(2), 127–142. https://doi.org/10.3920/bm2012.0030.

Harfouche, A. L., Jacobson, D. A., Kainer, D., Romero, J. C., Harfouche, A. H., Scarascia Mugnozza, G., Moshelion, M., Tuskan, G. A., Keurentjes, J. J., & Altman, A. (2019). Accelerating climate resilient plant breeding by applying next-generation artificial intelligence. *Trends in Biotechnology*, 37(11), 1217–1235. https://doi.org/10.1016/j.tibtech.2019.05.007.

Herndon, J. M., & Whiteside, M. (2019). Further evidence that particulate pollution is the principal cause of global warming: Humanitarian considerations. *Journal of Geography, Environment and Earth Science International*, 1–11. https://doi.org/10.9734/jgeesi/2019/v21i130117.

IPCC. (2014). Impacts, adaptation, and vulnerability. Part A: Global and sectoral aspects. Contribution of Working Group II to the Fifth Assessment Report of the Intergovernmental Panel on Climate Change; Cambridge University Press: Cambridge, UK; New York, NY, USA.

Ismail, A. S., Abo-Elmagd, H. I., & Housseiny, M. M. (2016). A safe potential juice clarifying pectinase from Trichoderma viride EF-8 utilizing Egyptian onion skins. *Journal of Genetic Engineering and Biotechnology*, 14(1), 153–159. https://doi.org/10.1016/j.jgeb.2016.05.001.

Izuogu, C. U. (2015). Environmental hazard effects on agricultural production among rural households in IMO state, Nigeria. *Universal Journal of Public Health*, 3(6), 229–233. https://doi.org/10.13189/ujph. 2015.030601.

Lipkie, T. E., Ferruzzi, M., & Weaver, C. M. (2016). Bioaccessibility of vitamin D from bread fortified with UV-treated yeast is lower than bread fortified with crystalline vitamin D2 and bovine milk. *The FASEB Journal*, 30(S1). https://doi.org/10.1096/fasebj.30.1_supplement.918.6.

Mateo, C., Palomo, J. M., Fernandez-Lorente, G., Guisan, J. M., & Fernandez-Lafuente, R. (2007). Improvement of enzyme activity, stability and selectivity via immobilization techniques. *Enzyme and Microbial Technology*, 40(6), 1451–1463. https://doi.org/10.1016/j.enzmictec.2007.01.018.

Messias, J. M., Costa, B. Z., Lima, V. M., Giese, C., Dekker, R. F., & Barbosa, A. D. (2011). Lipases microbianas: Produção, propriedades E aplicações biotecnológicas. *Semina: Ciências Exatas e Tecnológicas*, 32(2), 213–234. https://doi.org/10.5433/1679-0375.2011v32n2p213.

Mokoena, M. P., Mutanda, T., & Olaniran, A. O. (2016). Perspectives on the probiotic potential of lactic acid bacteria from African traditional fermented foods and beverages. *Food & Nutrition Research*, 60(1), 29630. https://doi.org/10.3402/fnr.v60.29630.

Nalage, D., Khedkar, G., Kalyankar, A., Sarkate, A., Ghodke, S., Bedre, V., & Khedkar, C. (2016). Single cell proteins. *Encyclopedia of Food and Health*, 790–794. https://doi.org/10.1016/b978-0-12-384947-2.00628-0.

Nielsen, D. R., Yoon, S., Yuan, C. J., & Prather, K. L. (2010). Metabolic engineering of acetoin and meso-2, 3-butanediol biosynthesis in E. coli. *Biotechnology Journal*, 5(3), 274–284. https://doi.org/10.1002/biot.200900279.

Nwankwo, W., Olayinka, A. S., & Ukhurebor, K. E. (2020a). Nanoinformatics: Why design of projects on Nanomedicine development and clinical applications may fail? *2020 International Conference in Mathematics, Computer Engineering and Computer Science (ICMCECS)*. https://doi.org/10.1109/icmcecs 47690.2020.246992.

Nwankwo, W., Olayinka, S. A., & Ukhurebor, K. E. (2020b). Green computing policies and regulations: a necessity? *International Journal of Scientific & Technology Research*, 9(1), 4378–4383.

Nwankwo, W., & Ukhurebor, K.E. (2019). An X-ray of connectivity between climate change and particulate pollutions. *Journal of Advanced Research in Dynamical and Control Systems*, 11(8) Special Issue, 3002–3011.

Nwankwo, W., Ukhurebor, K.E., & Aigbe, U.O. (2020c). Climate change and innovation technology: a synopsis. *Technology Reports of Kansai University*, 63(3), 383–391.

Nwankwo, W., Ukhurebor, K., & Ukaoha, K. (2020d). Knowledge discovery and analytics in process reengineering: A study of Port clearance processes. *2020 International Conference in Mathematics, Computer Engineering and Computer Science (ICMCECS)*. https://doi.org/10.1109/icmcecs47690.2020.246989.

Omemu, A. M., Akpan, I., Bankole, M. O., & Teniola, O. D. (2005). Hydrolysis of raw tuber starches by amylase of Aspergillusniger AM07 isolated from the soil. *African Journal of Biotechnology*, 4, 19–25.

Padilla, B., Gil, J. V., & Manzanares, P. (2016). Past and future of non-saccharomyces yeasts: From spoilage microorganisms to biotechnological tools for improving wine aroma complexity. *Frontiers in Microbiology*, 7. https://doi.org/10.3389/fmicb.2016.00411.

Patelski, P., Berlowska, J., Dziugan, P., Pielech-Przybylska, K., Balcerek, M., Dziekonska, U., & Kalinowska, H. (2015). Utilisation of sugar beet bagasse for the biosynthesis of yeast SCP. *Journal of Food Engineering*, 167, 32–37. https://doi.org/10.1016/j.jfoodeng.2015.03.031.

Ploss, T. N., Reilman, E., Monteferrante, C. G., Denham, E. L., Piersma, S., Lingner, A., Vehmaanperä, J., Lorenz, P., & Van Dijl, J. M. (2016). Homogeneity and heterogeneity in amylase production by bacillus subtilis under different growth conditions. *Microbial Cell Factories*, 15(1). https://doi.org/10.1186/s12934-016-0455-1.

Prasad, J., Gill, H., Smart, J., & Gopal, P. K. (1998). Selection and characterisation of lactobacillus and Bifidobacterium strains for use as probiotics. *International Dairy Journal*, 8(12), 993–1002. https://doi.org/10.1016/s0958-6946(99)00024-2.

Saerens, S., & Swiegers, J. H. (2016). Enhancement of coffee quality and flavor by using Pichiakluyveri yeast starter culture for coffee fermentation. US 20160058028 A1.

Salman, T., Kamal, M., Ahmed, M., Siddiqa, S. M., Khan, R. A., & Hassan A. (2016). Medium optimization for the production of amylase by Bacillus subtilis RM16 in Shake-flask fermentation. *Pakistan Journal of Pharmaceutical Science*, 29, 439–444.

Satish Kumar, R., Kanmani, P., Yuvaraj, N., Paari, K. A., Pattukumar, V., & Arul, V. (2013). Traditional Indian fermented foods: A rich source of lactic acid bacteria. *International Journal of Food Sciences and Nutrition*, 64(4), 415–428. https://doi.org/10.3109/09637486.2012.746288.

Serdeczny, O., Adams, S., Baarsch, F., Coumou, D., Robinson, A., Hare, W., Schaeffer, M., Perrette, M., & Reinhardt, J. (2017). Climate change impacts in sub-Saharan Africa: From physical changes to their social repercussions. *Regional Environmental Change*, 17(6), 1585–1600. https://doi.org/10.1007/s10113-015-0910-2.

Sharma, R., Chisti, Y., & Banerjee, U. C. (2001). Production, purification, characterization, and applications of lipases. *Biotechnology Advances*, 19(8), 627–662. https://doi.org/10.1016/s0734-9750(01)00086-6.

Sheldon, R. (2007). Enzyme immobilization: The quest for optimum performance. *Advanced Synthesis & Catalysis*, 349(8–9), 1289–1307. https://doi.org/10.1002/adsc.200700082.

Singh, A., Majumder, A., & Goyal, A. (2008). Artificial intelligence based optimization of exocellular glucansucrase production from Leuconostoc dextranicum NRRL B-1146. *Bioresource Technology*, 99(17), 8201–8206. https://doi.org/10.1016/j.biortech.2008.03.038.

Tait, G., Banda, A., & Watkins, A. (2017). Proportionate and Adaptive Governance of Innovative Technologies (PAGIT): A framework to guide policy and regulatory decision making. Innogen Institute. Available online at: https://www.innogen.ac.uk/reports/1222.

Takagi, H., & Shima, J. (2015). "Stress tolerance of baker's yeast during bread-making processes," in *Stress Biology of Yeasts and Fungi* eds Takagi H., Kitagaki H. (Tokyo: Springer) pp. 23–42. https://doi.org/10.1007/978-4-431-55248-2-2.

Tofalo, R., Perpetuini, G., Di Gianvito, P., Arfelli, G., Schirone, M., Corsetti, A., & Suzzi, G. (2016). Characterization of specialized flocculent yeasts to improve sparkling wine fermentation. *Journal of Applied Microbiology*, 120(6), 1574–1584. https://doi.org/10.1111/jam.13113.

Ukhurebor, K., & Abiodun, I. (2018). Variation in annual rainfall data of forty years (1978–2017) for south-south, Nigeria. *Journal of Applied Sciences and Environmental Management*, 22(4), 511. https://doi.org/10.4314/jasem.v22i4.13.

Ukhurebor, K., & Umukoro, O. (2018). Influence of meteorological variables on UHF radio signal: Recent findings for EBS, Benin City, south-south, Nigeria. *IOP Conference Series: Earth and Environmental Science*, 173, 012017. https://doi.org/10.1088/1755-1315/173/1/012017.

Ukhurebor, K., Olayinka, S., Nwankwo, W., & Alhasan, C. (2019). Evaluation of the effects of some weather variables on UHF and VHF receivers within Benin City, south-south region of Nigeria. *Journal of Physics: Conference Series*, 1299, 012052. https://doi.org/10.1088/1742-6596/1299/1/012052.

Ukhurebor, K. E., & Azi, S. O. (2019). Review of methodology to obtain parameters for radio wave propagation at low altitudes from meteorological data: New results for Auchi area in Edo state, Nigeria. *Journal of King Saud University - Science*, 31(4), 1445–1451. https://doi.org/10.1016/j.jksus.2018.03.001.

Ukhurebor, K. E., & Nwankwo, W. (2020). Estimation of the refractivity gradient from measured essential climate variables in iyamho-auchi, Edo state, south-south region of Nigeria. *Indonesian Journal of Electrical Engineering and Computer Science*, 19(1), 276. https://doi.org/10.11591/ijeecs.v19.i1.pp276-284.

United Nations (UN, 2019). https://www.un.org/development/desa/en/news/population/world-population-prospects-2019.html.

United Nations Conference on Trade and Development (UNCTAD, 2017). The role of science, technology and innovation in ensuring food security by 2030. New York and Geneva. https://unctad.org/en/PublicationsLibrary/dtlstict2017d5_en.pdf.

Vaezi, S. S., Poorazizi, E., Tahmourespour, A., & Aminsharei, F. (2020). Application of artificial neural networks to describe the combined effect of pH, time, NaCl and ethanol concentrations on the biofilm formation of staphylococcus aureus. *Microbial Pathogenesis*, 141, 103986. https://doi.org/10.1016/j.micpath.2020.103986.

Vázquez, E., & Villaverde, A. (2013). Microbial biofabrication for nanomedicine: Biomaterials, nanoparticles and beyond. *Nanomedicine*, 8(12), 1895–1898. https://doi.org/10.2217/nnm.13.164.

Vitorino, L. C., & Bessa, L. A. (2017). Technological microbiology: Development and applications. *Frontiers in Microbiology*, 8. https://doi.org/10.3389/fmicb.2017.00827.

Waditee-Sirisattha, R., Kageyama, H., & Takabe, T. (2016). Halophilic microorganism resources and their applications in industrial and environmental biotechnology. *AIMS Microbiology*, 2(1), 42–54. https://doi.org/10.3934/microbiol.2016.1.42.

Wang, F., & Zhang, W. (2019). Synthetic biology: Recent progress, biosafety and biosecurity concerns, and possible solutions. *Journal of Biosafety and Biosecurity*, 1(1), 22–30. https://doi.org/10.1016/j.jobb.2018.12.003.

16 Recent Developments and Application of Potato Plant-Based Polymers

Omorefosa Osarenkhoe Osemwegie, Abiola Folakemi Olaniran,
Clinton Emeka Okonkwo, A. Adewumi,
Oluwakemi Christianah Erinle, Godshelp Osas Egharevba
and Adejumoke Abosede Inyinbor
Landmark University

Charles Oluwaseun Adetunji
Edo State University Uzairue

CONTENTS

DOI: 10.1201/9781003268468-16

16.1 INTRODUCTION

Potato ranks as one of the topmost important tuber/vegetable crops, it is arbitrarily the fifth food crop in the world behind cereals. Moreso, it was selected as the food security crop option by the United Nations for many disadvantaged populations to restore dietary stability in their already under-nutritious food supply chain and socioeconomic life (Olaniran et al., 2020a). But for the unanticipated encumbrance of the COVID-19 pandemic, the global production volume of potato crops (*Solanum* spp) e.g. Irish, Orange-fleshed, Purple-fleshed, etc could have risen beyond 370 million tons by 2022 according to the FAO (Matharu et al., 2018). This increase is hypothetically driven by the crops' production, applications, derived products, markets, economic and industrial versatilities, coupled with their expanded species(180) and cultivar varieties (Chandrasekara and Kumar, 2016). Also, the changing paradigm in the traditional consumption form of fresh potatoes (boiled, baked, roasted, fried) to functional processed food products is associated with urbanization, husbandry, and a staggering range of appetite, and recipe diversities. These factors facilitate the rising global potato production volume (Wang et al., 2018; Olaniran et al., 2020a).

The waste fractions from potato processing and starch extractions constitute good sources of fibres, proteins, phenolic compounds and antioxidants that can be valorized for further uses. Furthermore, animal feeds, formulated human food products in form of dehydrated, frozen, fermented, or canned foods as well as other industrial feedstock or by-products from potatoes have been reported (Priedniece et al., 2017; Vithu et al., 2020). While its food functionality may undermine orientating it for applications in product development and bio-industrialization, the struggle for a safer environment with zero toxicological consequences from antropogenics and other human activities is rather a more compelling concern enforcing research investigations into eco-friendlier possibilities. It is hypothetical to assume that the food vacuum created by the over-exploitation of a crop for other benefits (product developments, industrialization, research) is recompensed naturally by another crop of corresponding biological values thereby mitigating possible consequences. There is therefore no gainsaying that potato biopolymers are becoming substantially recognized in biopolymer prospecting and businesses. Potatoes' amenability to propagation, exploitable traits abundance, gene editing, and introgressive hybridization made them one of the best crops of choice for biotechnology as well as novel economic products' explorations (Halterman et al., 2016).

Potato, depending on the species, cultivar, and variety, are composed of diverse starch as well as non-starch polysaccharides like hemicelluloses, cellulose, pectin, lignin, xyloglucan, heteroxylans, and heteromannans (Hao et al., 2013). Studies have already related the functionality, composition and diversity of these heterogeneous polysaccharides with processing technologies, nutritiveness of food products, and potential industrial applications. This is with particular focus on the maneuverability of their starch-sugar ratio characteristic for unequivocal functionalities (Kita, 2002). However, Dupuis and Liu (2019) reported that starch forms the bulk of potato polysaccharides with amylopectin (60%–80%, branched), amylose (20%–40%, straight chain) and approximately 800 ppm phosphate bond composites. Torres and Dominguez (2020) noted the economic prospect in potato starch and its other biopolymer derivatives and estimated a rise in their global market demands to 4.4 million tons by 2024 *ceteris paribus*. In more recent times, different genetic maneuverability technologies have led to the emergence of potato varieties with either only amylopectin or a ranging amylase-amylopectin content ratio. This exerts a concomitant influence on the overall

starch content, properties (viscosity, adhesiveness, digestibility, gelatinization temperature), and application versatility (Zhao et al., 2018a). The unique influence of environmental factors on the physicochemical characteristic of potato starch as well as biopolymers derived from the tubers, particularly those associated with the extensive phosphorylation of the amylopectin molecules and the synthesis of starch, has been reported (Yusuf et al., 2003; Wang et al., 2020a). However, this is without a clear understanding on how the environment influences starch quantity, configuration, solubility, crystallinity, gelatinizability and purity variations among potato varieties. Literature is however scanty on the techniques amenable for both small and large-scale production of starch and other polysaccharides from potato tubers. It is likely at this point to suggest that there is a remote correlation between potato-starch quality and application versatility with the type of extraction technique (Colussi et al., 2020; Oluwasina et al., 2019). Although the phytochemical comparisons of potato varieties with other root crops exist in the literature, there is insignificant data on the quantification and physicochemical comparison of potato starch with those of many other crops' starches (Molavi et al., 2015; Leonel et al., 2017; Olaniran et al., 2020b). It is equally likely that many developing nations would rather rally their overall potato harvests to mitigate hunger endemism rather than exploit it for bio-industrial benefits. This lends credence to the seeming dearth of research publications from many developing nations on potato application for non-food, industrial or technological benefits.

Potato starch ranks as one of the most versatile in the world and is slowly gathering recognition as an important natural resource tool in driving bioeconomic development in many developed nations. In recent decades, this biopolymer has been subjected to diverse modifications (manipulations). This is purposely to overcome its native use (ethnomedicine, food, animal feeds from peelings, biofertilizers) limitations, and offer succor to environmental safety drives as well as the management of mounting agrowastes disposal challenges like greenhouse gas emissions, landfills stench and toxic emissions from wastes' incineration. Additionally, the medication is to derive benign end use benefit(s) in food production systems, pharmaceuticals, biorefinery processes, degradable plastic manufacturing, and as prebiotic in animal feeds (Juarez-Arellano et al., 2020; Yusoff et al., 2021). According to Do Val Siqueira et al. (2020) heat, fermentation, chemicals, and even microbes have been used to hydrolyze starch feedstock, including potato starch, into building blocks of polylactic acid (PAL), hydroxymethylfurfural (HMF) and poly-3-hydroxybutyrate (P3HB) that are key chemicals in the creation of bioplastics, particularly for food packaging. This potentially validates the tractability and versatility of potato prime polymer (starch), placing it at a similar level as those of cassava and corn. Also, this forms the basis for further investigation of other potato polysaccharides for different utility possibilities relating to engineering design, industries, and biotechnology (Shafqat et al., 2020). Similarly, the wastewater from the processing of potato starch is reportedly valuable due to its rich content of protein, potassium salt, chemical oxygen demand (COD) and organic substances. According to Li et al. (2020), this can be harnessed, harmlessly purified using biologically safe processes (biological, thermo-chemical precipitation, membrane separation technology) and transformed into high value-adding food products as well as low molecular weight inhibitors for medicines using the Expanded Bed Absorption (EBA) technology (Waglay et al., 2014). Deliberate attempts using diverse technologies including gene encoding enzyme activities to strategically promote the endogenous production of other useful polymers like elastin, collagen, cyanophycin and polyhydroxyalkanoates (PHAs) were reported by Snell et al. (2015). However, the growing research attention accorded potato polymers and their successful manipulations for a diverse range of economic benefits may, in some developing nations like Nigeria, be encumbered by social, infrastructural, agricultural, economic and political challenges. It is also inferable that technology influences potato products' bioactivity, functionality, versatility, and acceptability. Hence, this article reviews the knowledge, advancement and the current range of applications of potato biopolymers with a special bias for potato starch. In addition, it aims at providing insight into their role in mitigating the challenges of environmental safety, and comparing them to those of other crops.

16.2   COMPARISON WITH OTHER BIOEQUIVALENT MATERIAL

Several studies have investigated the development of starch-based biopolymers using various sources of starch like potato (Zhang et al., 2018), maize (Luchese et al., 2017), wheat (Domene-López et al., 2019), cassava (Zhang et al., 2018), and Jicama (*Pachyrhizuserosus*) (Abral et al., 2019). Starch-based bioplastics have been exploited to prolong food's shelf life and promising marketability potentials for food products (Zhong et al., 2020). The shear/tensile strength and the rupture rate at elongation observed for starch films can be similar to those from artificial polymers such as LDPE (low-density polyethylene) which are presently sold as food containers even after concerns were raised about their recyclability and biodegradability (Montero et al., 2017). Starched-based bioplastics represent about 85%–90% of the Bio-plastics industries (Ebrahimian, 2022) with corn starch being the lead feedstock for the manufacture of decomposable plastics and edible biopolymer films in recent years (Luchese et al., 2017). Souza et al., (2010), found corn preponderance to be the possible reason for being the global preference as a source of consumable starch (65%) ahead of potato (13%) and cassava (11%). Owing to its large starch content (6.9%–30.7%) on wet basis and high availability, the potato has been equally identified as a major bioplastic source by Jabeen et al. (2015). Potato is ~38% higher in amylose, which makes it more appropriate for the development of films with potentially laudable technical characteristics, sturdier and versatile mechanical functions. This suggests that potato is relatively competitive with corn, cassava, rice and wheat-based starches as bioresource feedstock for industrial purposes.

While literature is scanty on the comparative advantages of starch homopolymers from different plant sources, the amylose/amylopectin ratio may be the main determinant of possible physico-chemical variability separating starch from different sources. de Azevedo et al. (2020) compared the decomposable films obtained from corn starch with that of potato starch and observed a more uniform compound mixture by the corn starch compared with potato starch bioplastics. This significantly improved the translucent, physical cohesiveness, and gelation properties of corn starch. Furthermore, Basiak et al. (2017), attributed the slight advantage of the innate physicochemical properties of corn starch over potato starch to its amylose content (linear structure) and the higher amylopectin content (highly branched structure) of the latter. In addition, potato starch is hypothesized to biodegrade faster (5 days) than corn starch (40 days), a characteristic that affects their individual end use preference in food preservation and storage. The retrogradation, mechanical, and thermal properties of unreinforced corn, and possibly cassava starch were better than those of potato-based starch (Versino and García, 2014; Molavi et al., 2015).

Further comparative investigation of the physicochemical characteristics of starch sourced from maize, wheat and potato respectively was carried out by Basiak et al. (2017) in biodegradable films. Their study affirmed maize starch films as the thickest (112 µm) whilst potato starch films were the thinnest and least heterogeneous (55 µm). This observation validated the findings of Jabeen et al. (2015) on lesser amylose composition of potato starch which underscored the reduction in film thickness, and their mechanical variability as well as barrier properties against water vapour (Bonilla et al., 2013). The potato starch film showed more translucence than the opalescence of maize and wheat indicating probable higher lipid content (Jiménez et al., 2012, 2013; Sun et al., 2013). Furthermore, the deformability (E%) of wheat starch film was greater than that of potato starch films but had less rigidity (TS and YM) than potato starch film. While the hardness (E percent), elongation rupture rate and viscidness properties of wheat and corn starches are higher than those of potato starch, it's very low amylose content of potato starch film confers the least permeability and makes it interesting amenable to blend with other biopolymers of different molecular weight as well as plasticizers, physical and chemical modification for functional versatility (Jiménez et al., 2012; Muscat et al., 2012; Acevedo-Morantes et al., 2021). Several other intrinsic physical characteristics of thermoplastic starch sources such as viscosity-average molecular weights (potato -3.97×10^{17}, corn -2.14×10^{17}, cassava -3.3×10^{8}, yam -1.01×10^{17}), young modulus, tensile strength, and pH (sorghum -4.57, potato -6.92) have been compared (Molavi et al., 2015; Omotoso et al., 2015;

Richard et al., 2015; Ismail et al., 2016; Moongngarm, 2013). All these properties directly correlate with the starch source selection, film forming property, functionality, and end-use preferences are fundamental considerations in the decisive application of biopolymers. This is also possible implication for the choice of method that optimizes the anticipated qualities of starch and quantity of extraction. Therefore, the physicochemical versatility, retrogradation, amenability, availability, marketability and affordability of potato starch make it a strong competitive amorphous equivalent for biopolymers as well as replacement of artificial/synthetic polymers productions.

16.3 DYNAMICS OF POTATO BIOPOLYMER IN REGIONAL BIOECONOMIC GROWTH

Although the value chain of potato biopolymers in bioeconomic growth in still insignificant in most developing nations due to rooted attention on resolving crisis of hunger. Consequently, the slow adoption of innovative agrowastes conversion technologies that affords value-adding benefits to humanity in some developing nations like Nigeria has remotely caused agro-wastes overload in the environment. This condition has given impetus to smart preservation techniques and rise in exploitation of biopolymers as preservatives.

Plastics have found great relevance in various products' packaging and their indiscriminate disposals constitute serious environmental and waste management challenges that have resulted into the search for alternate biodegradable polymers. Hence biopolymers from natural sources such as carbohydrate, protein and lipid are currently being explored in diverse research investigations, and for verse functionalities as well as applications. The pressures from the accelerated demand for benign products, agricultural expansion due to urbanization, global trading profiteering, and dietary diversifications have oriented attention to preserving humanity's healthiness and environment with the use of biologically safer natural sources including biopolymers. These natural resources are generally deemed to be environmentally friendly due to their biodegradability, renewable, readily available, valorizeable, and economical (Wang et al., 2020a). Consequently, potato being a natural resource is a globally popular food is found to be an excellent agro-industrial feedstock and sustainable resource for bioproducts production for different possibilities of application due to its significant tendency for wastage due to colour change, infections, size and quality failure.

The relative economic importance of natural biopolymers has been underexplored before now causing huge wastes at high disposal costs on systems. For instance, about 4.5 million in tonnes of waste is generated from various potato processing activities in the United States (US) annually. This imposes huge disposal cost on the US systems (Kang and Min, 2010). Negligible economic advantage of potato peels in compounding animal feeds was not adequate in the significant depletion of the tons of wastes from agriculture and potato processing systems until their more recent application in biopolymers. Low-valued agro wastes such as potato peels have drastically reduced the cost of edible biopolymer thereby positively enhancing the economy (Kang and Min, 2010).

16.3.1 APPLICATIONS OF POTATO BIOPOLYMER AND THEIR CONSEQUENTIAL ECONOMIC ADVANTAGES

Potato biopolymer has found use in food, feed, pharmaceutical, and medical industries as well as composite materials in constructions (Karaky et al., 2019; Udayakumar et al., 2021). The ability to impart stiffness, strength and stability in materials has made its use most compelling (Verma et al., 2022).

16.3.1.1 Potato Biopolymer in Food Industries

Biopolymers are currently being explored in some operations of the food industry. Various food packing employing synthetic plastics are now considering biopolymers as cleaner and environmentally

friendlier alternatives. Carbohydrate-based biopolymers have found great attraction as industrial feedstock or products because of their huge availability, ease of processing, and low cost (Moreno et al., 2015; Borah et al., 2017; Kabir et al., 2020). The adoption of seemingly useless material in such great application afford strategic waste management solutions and consequently enhances regional economy. For instance, waste management has been established to present a positive economic impact on Nola in Southern Italy (Feo et al., 2017). Biopolymers of whole and waste of potato are valued for their minor anionic, non-toxic and biocompatibility properties (Dong and Cui, 2021). Other properties such as amino acid profile and techno-functional properties unique to potatoes make potato biopolymer highly attractive to the food industries (Li et al., 2021). Edible films/biopolymers obtained from potato and other starch give mechanical resistance, halt unwanted oxidation and deal with moisture loss (Dash et al., 2019).

Edible biopolymers have given the food industries huge success in areas where the synthetic polymer has been limited. Such vivid examples are in cased, coated or complex foods with various components and layers which are not removed before cooking or consumption (Kowalczyk, 2016). Furthermore, waste containing pectins, flavonoids and essential oil has been blended with potato peels in biopolymer production hence developing active packaging materials. Such active packaging materials may have anti-inflammatory, antioxidant, anti-microbial as well as anxiolytic properties that confer preservative and additive roles in food products (Kowalczyk, 2016; Borah et al., 2017). The inherent antioxidants, proteins and fibers contents of potato are revered in cost-effective production of biopolymer packaging materials with a direct positive impact on the economy (Torres et al., 2020).

In a bid to maximize the economic gains of meat industries, potato proteins in blend with linseed oil, chitosan and ZnO nanoparticles have been developed into a unique biopolymer for meat preservation (Wang et al., 2020b). This potato-based biopolymer preserved the freshness of the meat and repressed microbial proliferation. Potato blended with chitosan and tapioca starch has also indicated great antibacterial activities (Verma et al., 2022).

The various applications of biopolymer in food industries will greatly reduce production cost, maximize profit and impact the economy via products generation as well as export trade. These economic gains have been well captured by Stoica et al. (2020).

16.3.1.2   Potato Biopolymer in Constructions

The production of construction material such as cement has also had great impact on the environment. Four part of a hundredth of global CO_2 release is from cement production hence reducing cement consumption will make for a greener environment (Shanmugavel et al., 2020). Recently, the harmful impact of cement being controlled as well as improved using plant residues such as potato peels and extracts to enhance the properties of cement (Shanmugavel et al., 2020; Torres and Dominguez, 2020). Biomodification of construction material goes a long way in reducing its damaging effect on the environment. Plant residues and biopolymers may therefore potentially serve as sustainable material for the modification of construction materials for more benign impact (Shanmugavel et al., 2020).

Various plant residues and biopolymers are becoming attractive to increase the strength and rigidity of construction materials. Potato peel powder has found use in this regard as its addition to ordinary clay may enhance the sound insulation, water absorption, dry shrinkage, water shrinkage, porosity and soluble salts properties of the construction material (Ghorbani et al., 2021). Interestingly, potato biopolymer blended with sugar beet pulp equally gave excellent mechanical properties as well as thermal strength to building material (Karaky et al., 2019).

Potato biopolymers have potential economic viability with excellent applications as composite to strengthen polyurethane (PUR) foams (Członka et al., 2018), biocomposite in polylactide production (Chen et al., 2020), biomaterials and nutraceutical (Dong and Cui, 2021) amongst others.

16.4   RECENT ADVANCES IN POTATO STARCH AND BIOPOLYMER APPLICATIONS

Potato starch is a fine, white powdery substance derived from potatoes. Like other starches, it is commonly used to thicken, gel, texturize and increase crispiness in baked goods. It is also used in clear soup, confections, fillings, and/or as excipient, dilent, disintegrant, and binder in drug production.

Biopolymers are organic-based molecules which comprise repeating units called monomers and produced by living organisms or synthesized from/with organically derived materials.

Potato starch is a biopolymer that is derived from potato tubers. The root tubers of potato plant are made up of leucoplasts (starch grains). The leucoplast is extracted from the tubers by crushing the potato tuber cells, thus releasing the starch grains. The application of potato starch depends on the type (variety, cultivar, species) of potato starch in question, chemical architecture or structure and composition of the potato, broad interchangeability of the starch, and technological method used in its extraction.

Application of Potato starch in the 21st century is unprecedented due to rapidity in advancement of technology and biotechnology, and their integration into different facets of the economic value chain for bioeconomic emergence. Some of the areas of application, particularly in many developing worlds including Nigeria, are as follows.

16.4.1   FILM MAKING

Potato starch has been used to make eco-friendly starch films void of pollutants (Balakrishnan et al., 2018; Janik et al., 2018). It does present diverse advantages that include minimal cost of production (MITCH, 1984; Zhang et al., 2012), high biodegradability (Wilpiszewska and Czech, 2014; Arvanitoyannis et al., 1998), and easier processing methods (Kraak, 1992; Zhang et al., 2018). Potato starch films are optimistic viable commercial preservative films for elongating the shelf-life and nutritiveness of foods (Ferreira Saraiva et al., 2016). Several kinds of starch have been employed in producing films, potato starch presents better film-forming properties than that of other cereal starch used for film production (Vargas et al., 2017). Nano-SiO_2/potato starch films of different sizes were synthesized to improve the film properties of potato starch to further dispel possible incidence of microorganisms and fortify their impermeability (Zhang et al., 2018).

16.4.2   FOOD AND BEVERAGE INDUSTRY

Starch in form of polysaccharides is abundant in nature, and it is universally used in its gelatinized form because it digests very quickly. Enzyme-defiance dextrins of potato starch were produced using microwave-assisted technology, which in turn increased its dietary fibre contents (Jochym and Nebesny, 2017). Potato starch is used in food processing industry for anti-caking, as thickener, gelling agent, texturizer and binder. It also consumed as snack foods, baked goods, gravies, noodles, sauces, shredded cheese, soups, doughnuts, and yeast in filtration as well as fermented beverages (Odebode et al., 2008; Zhang et al., 2012; Castanha et al., 2019; Jagadeesan et al., 2020).

16.4.3   PAPER INDUSTRY

The paper industry has longer history of starch use in the production of paper (Kraak, 1992; Grommers and van der Krogt, 2009). Potato starch act as a flocculant and an adhesive in paper production processing (Maurer, 2009). It is used in preference to corn starch (Qasem, 2020). This is because cationic potato starch in paper enhanced the retention of fine fibres and fillers, thus improving its mechanical properties (MITCH, 1984; Kraak, 1992; Maurer, 2009). Furthermore, cationic

amylopectin starch is a relatively abundant, highly degradable/compostable, and less expensive natural polymer compared with plant-based cellulose or fossil-based polymers. Hence, it is equally amenable to produce medical bandages, orthopaedic cast, paper napkins toilet papers and other medical dressings by various electrospinning devices and technologies (Li et al., 2018; Costa et al., 2020).

16.4.4   ADHESIVE

The starch manufacturing industry sells to consumers potato starch for thickening and as adhesive (Kraak, 1992). The adhesive is applied in fastening paper together due to its nontoxic nature to the environment (Liu et al., 2014; Schwartz and Whistler, 2009). This is in addition to its consumption as swallow by several natives in some west African regions, and routine use in laundry process to enhance the crispness, structure as well as wrinkling resistance of clothes.

16.4.5   STABILIZER IN YOGURT MANUFACTURING

Stabilizers are very necessary components in dairy products making because of their added advantages. Extracted potato starch has been used as stabilizers in yogurt manufacturing (Altemimi, 2018). The addition of sufficient concentrations of potato starch in the production line would provide appreciable results texturization (Jimoh and Kolapo, 2007; Altemimi, 2018).

16.4.6   TEXTILE MANUFACTURING

Stiffening of fabric materials can be done using potato starch instead of wax (Zhang et al., 2012). This is done by spaying the fabric with a colloidal solution of potato starch (Grommers and van der Krogt, 2009; Maurer, 2009). Potato starch gives the fabric or cotton a fine finishing, thus improving its appearance (Kraak, 1992; Bergthaller et al., 1999). It is an effective sizing material that is still valuable in textile production line (Clays and Clays, 2009; Maurer, 2009).

16.4.7   NANOPARTICLES SYNTHESIS

Several nano particles have been synthesized from potato starch due to its aqueous ability, non-toxicity, environmentally friendlily nature, biocompatibility, improved film properties, and effectiveness (MITCH, 1984; Jiang et al., 2016; Wu et al., 2019).

16.4.8   OIL DRILLING AND MINING

Potato starch lowers the rate of filtration of fluid and enhances the properties of mud cake consistency which is a primary trait, shows effectiveness in fluid rheology and acts as a good viscosifier. It is also used to improve the gel strength of the fluid (Nmegbu et al., 2015; Heo et al., 2017; Assi, 2018; Horstmann et al., 2018).

16.4.9   WATER TREATMENT

Potato starch waste water has been reported to be used in fermentation to get bioflocculant for coagulation *Rhodococcus erythropolis* (Guo et al., 2015)due to its biodegradable and harmless impact on the environment (Pu et al., 2014; Bergthaller et al., 1999; Dash et al., 2019). Additionally, potato starch wastewater has been exploited for the isolation of potato protease inhibitors (PPIs) which is a major recovery protein that is attracting attention as a promising bioactive substance (Liu et al., 2020).

16.4.10 PHARMACEUTICAL INDUSTRIES

Starches are used in the pharma-drug industry for diverse purposes that include as binder, excipient, glidant, diluent, granulation, disintegrant, tableting, and capsule filling (Maurer, 2009; Szepes, 2009).

16.4.11 THERMOPLASTIC STARCH

Mościcki et al. (2012) reported the use of potato starch as means of making thermoplastic starch. He employed variant blends of potato starch combined alongside with glycerol to form various ranges of thermoplastic starch pellets which can then be processed (Janik et al., 2018).

16.4.12 CERAMIC PROCESSING

Potato starch is used in the processing of ceramic due to its large size (46–49 μm) and its chemical constituents, mainly carbon, hydrogen and oxygen, which enhances its combustibility without any trace of residue after burning. Starch is employed as a pore formation agent in the technology employed for making ceramic (Gregorová et al., 2006).

16.4.13 PACKAGING MATERIAL

The rising use of synthetic-polymer for packaging has led to serious ecological burdens as a result of their non-biodegradability and food safety concerns from seeming infiltration of the food chain by extraneous toxicants (Janik et al., 2018). This therefore necessitated the need for the use of bio-materials as substitutes to conventional synthetic polymers, potato-starch polylactide, and polyvinyl alcohol in food packaging production (Arvanitoyannis et al., 1998; Mościcki et al., 2012; Ferreira Saraiva et al., 2016; Balakrishnan et al., 2018; Janik et al., 2018). Scientific reports have shown that these potato starch packaging material disintegrates in *in situ* marine environment as well as in the laboratory (Franz, 2015; Janik et al., 2018; Dash et al., 2019).

16.5 TECHNIQUES ADVANCED FOR THE EXPLORATION OF RECYCLABLE POTATO-BASED

The recalcitrant nature of synthetic polymers or petroleum-derived polymers (e.g. synthetic plastic packaging bags) used in agro-allied industries, their environmental pollution and biosystems' health hazards have attracted polymer scientists across the globe to benign biopolymers (like bioplastics) as possible alternatives (Cruz-Gálvez et al., 2018; Shafqat et al., 2020). Furthermore, the inability of petroleum and coal-based polymers to seamlessly feedstock the natural biogeochemical recycling system and preserved environmental resources is also a raising concern to environmentalists, toxicologists and physical health experts in most part of the world. Biopolymers are natural polymers derived from chemically synthesized biomaterials or completely biosynthesized biomaterials by living organisms (Mahalakshmi et al., 2019; Maraveas, 2020). Aside from the facts that natural obtained biopolymers are degradable, eco-friendly, user-friendly, recyclable and cost-effective, they also offer biological benefits that may include biocompatibility, antioxidant, antimicrobial and pesticidal when activated bioactive substances like essential oil (Varghese et al., 2020; Syafiq et al., 2020). Their application according to Gowthaman et al. (2021) mediates social, economic and environmental repercussions of ecosystems' service disbalance. Biopolymers are hydrophilic (i.e. they swell due to adsorption of moisture) and have poor thermal, physic-mechanical and barrier properties (Vinod et al., 2020) whose limitations have been drastically mitigated in recent years by some specific modification techniques. Some of the products produced with biopolymers are edible films, aerogels, and electrolyte, foams and plastics (Zhao et al., 2018b). Some of the biopolymers used in the industrial space to produce most biodegradable plastics are polylactic acid (PLA),

polyhydroxyalkanoates (PHA), and polyhydroxybutyrate (PHB) (Ranganathan et al., 2020). Most potential biomaterials sources are classified into three groups; protein (e.g. soy protein, casein, gelatin and wheat gluten), polysaccharides (e.g. starch, lignin, cellulose, pectin, chitin, fungal and algae) and by-products (e.g. peels, leaves and bran). Although polysaccharides like starch are more explored since they are available in large quantities, versatile and cheap and their byproducts are normally protein-polysaccharide composites (Ilyas and Sapuan, 2020). Consequently, research in the use of potato-based polymer is of high interest and recently trending due to its ranking as the fourth most important food crop in the world, after rice, wheat and corn (Torres et al., 2020). The physic-mechanical properties of most polymer-based composites are improved with plasticizing/ compatibilization agents, and nano-fillers (Islam et al., 2020). Industries in the world release large tonnage of potato byproducts of which 33.3% accounts for potato protein (PP) which are valuable fillers in many polymer composites (Członka et al., 2018). Potato-based polymers have been used in diverse forms in recent years. For instance, it is used as reinforcement in the production of rigid polyurethane foams (Członka et al., 2018). Rigid polyurethane foam (RPF) is a widely used plastics produced from the reaction between polyol and an isocyanate. In other to reduce the environmental impact caused by the use of these plastics, potato-based additives like PP was integrated to its total mass content in the range of 0.1–5% wt. Incorporation of PP to the RPF improved the compressive strength, reduced the thermal conductivity and water absorption properties even though 0.1% wt of PP was recommended as adequate.

Bio-based edible films have also been receiving research attention as a replacement for non-biodegradable/compostable conventional packaging films and partial solution to non-recyclable waste-disposal management problems (Ncube et al., 2020). Consequently, biopolymer film has been produced from potato peels even though studies showed that the product exerts lower mechanical properties than biopolymer films from some other sources. The improvement of the mechanical, along with tensile, moisture barrier, microstructure, storage, antimicrobial, optical and colour qualities of potato-derived biopolymer films have been studied using different combinations of reinforcement composites (e.g. chitosan, oil, pomace, bacterial cellulose, curcumin, glycerol, sorbitol, ZnO NPs, etc.), homogenization, ultrasound and irradiation techniques (Kang and Min, 2010; Farajpour et al., 2020; Wang et al., 2020b; Xie et al., 2020; Liu et al., 2021). These methods of biofilm production also significantly improved the antioxidant characteristics and reduced the lipid oxidation tendency of products (Xie et al., 2020). Dash et al. (2019) developed sweet potato starch-lemon waste pectin film (3:1) impregnated with titanium oxide (TiO_2) particles for improved the water barrier and mechanical properties due to good intra- and intermolecular interactions of the composites. The addition of lavender essential oil significantly enhanced the antimicrobial and antioxidant properties of potato starch-furcellaran-gelatin films but lowered the mechanical strength (Jamróz et al., 2018). Similarly, biofilms produced from 2.4% wt potato-starch containing 1% glycerol was plasticized with acetone and methanol extracts from *Hibiscus sabdariffa* (Cruz-Gálvez et al., 2018). The antimicrobial assay investigation of the film further showed higher sensitivity by pathogens like *Escherichia coli* to the acetone extract plasticized film compared the methanolic modified film. This suggests a likely differential level of alteration of the physicochemical properties of the potato-based film due to the addition methanolic and acetonic extracts. Significantly reduced permeability of water vapor is a unique and important characteristic considered in the production of packaging materials in the food industries and therefore one of the basic essential properties of biopolymer films (Nandi and Guha, 2018). Reinforcement of potato polysaccharides with composites, the nature of composites, composition and concentration variability are optimized based on the desired physical, optical, mechanical and barrier properties. Nandi and Guha (2018) used an optimized combination of 3.7% (potato starch), 0.4% (guar gum), and 15% (glycerol) respectively in their biofilm experiment. Zhao and Saldaña (2019) applied Subcritical fluid technology to produce of potato-based bioactive films composited with and without the gallic acid. The process which involved the filling of a reactor of subcritical fluid system with blended samples of potato cull, potato peel, glycerol and water, with or without gallic acid was homogenized to obtain the desired temperature and pressure as well as starch gelatinization and reaction. This method

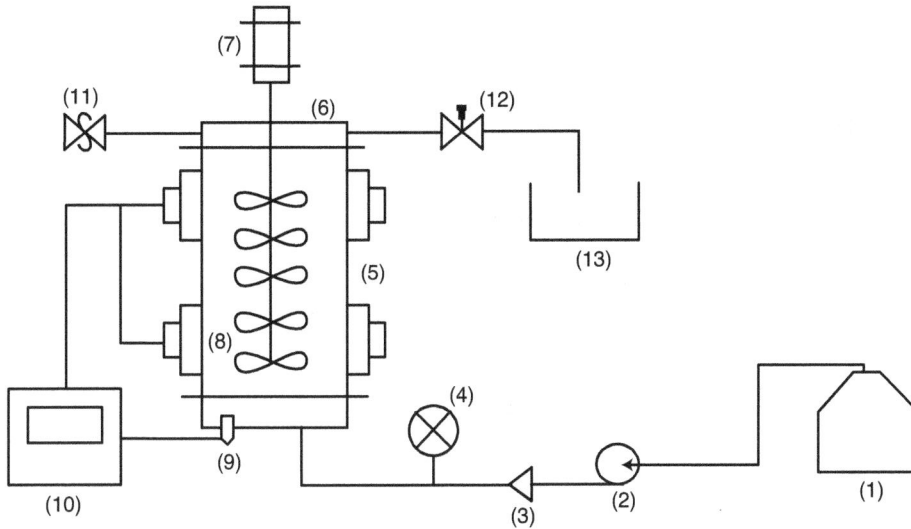

FIGURE 16.1 Subcritical fluid technology system: (1) Solvent reservoir, (2) Pump, (3) One-way valve, (4) Pressure gauge, (5) Band heaters, (6) Pressurized fluid reaction vessel, (7) Motor stirrer controlled by the control panel, (8) Stirrer, (9) Thermocouple, (10) Temperature controller, (11) Safety valve, (12) Back pressure regulator, and (13) Sample collection. (Zhao and Saldaña, 2019).

produced potato-based film with optimal tensile strength with increasing concentration of the glycerol (plasticizer) and higher antioxidant capacity at 0.3 g gallic/g starch combination. It is therefore likely that the Subcritical fluid technology is a hypothetical technique for optimizing potato-based bioactive films' production (Figure 16.1). Furthermore, Sharma et al. (2020) have associated the assured benefits of food safety, food quality retention, extension of shelf life, prevention of physical, environmental, chemical and microbiological alterations of both nutritional and sensory properties during storage and transport from food packages of biological sources. Glycerol and sorbitol plasticizers showed capacity in improving the mechanical, optical, and barrier properties of sweet potato-based starch film (Ballesteros-Mártinez, et al., 2020; Miller et al., 2020) while also suggesting a concentration of 50% sorbitol composite for edible film production. Reinforcement techniques have also helped optimized antibacterial potential as evident in the work of Gopi et al. (2019) in which a potato-starch/tapioca-starch/chitosan-based biopolymer combined with turmeric nanofibers inhibits *Escherichia coli*, *Bacillus cereus*, *Salmonella typhimurium* and *Staphylococcus aureus*. Moreso, nanofillers do not only act as reinforcement agent in biofilm production, but they also serve as an active ingredient for biopolymers (Jamróz et al., 2018).

Ultrasound technology (UT) is amenable to biofilms production (Figure 16.2). Increase in the ultrasound treatment time causes better film qualities such as lower water vapor permeability, moisture adsorption, and water solubility; better film strength; higher elongation capacity; and lighter colour (Borah et al., 2017). It was however noted in a study by Coezee et al. (2020) that ultrasound technique may be unsuitable for potato-based polymer production involving nanocomposites. Biocomposite of poly (lactic acid) and potato pulp powder (20% wt) was alsoreported in literature to help improve the mechanical strength bio-based and petroleum-based waxes (Righetti et al., 2019). Melting technology was applied to produce poly (3-hydroxybutyrate-co-3-hydroxyvalerate) biocomposite using coffee silver skin, an agro-based material whose addition to the matrix enhanced the stiffness, and heat deflection temperature properties, and crystalline nature of the composite (Gigante et al., 2021). In addition, the production of solid-based polymer electrolyte casted with Lithium trifluoromethanesulfonate ($LiCF_3SO_3$) as well as having potato-starch/graphene oxide composite has lend credence to the use of nanotechnology in potato-based polymer-matrix nano-composite explorations. Therefore, the presence of the nanomaterial in the polymer matrix may have implications for the physical and conductivity characteristics of polymer materials (Azli

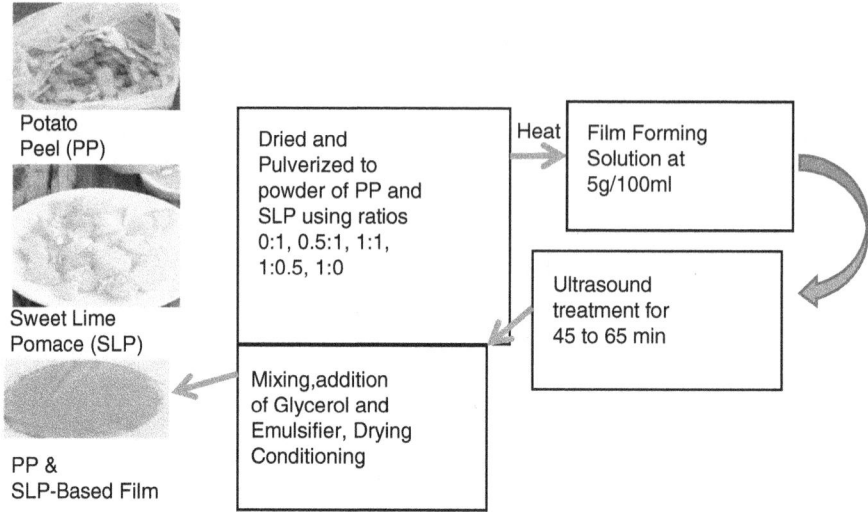

FIGURE 16.2 Illustration of processes involved in using ultrasound technique potato peel and sweet lime pomace Biofilm production (Borah et al., 2017 modified).

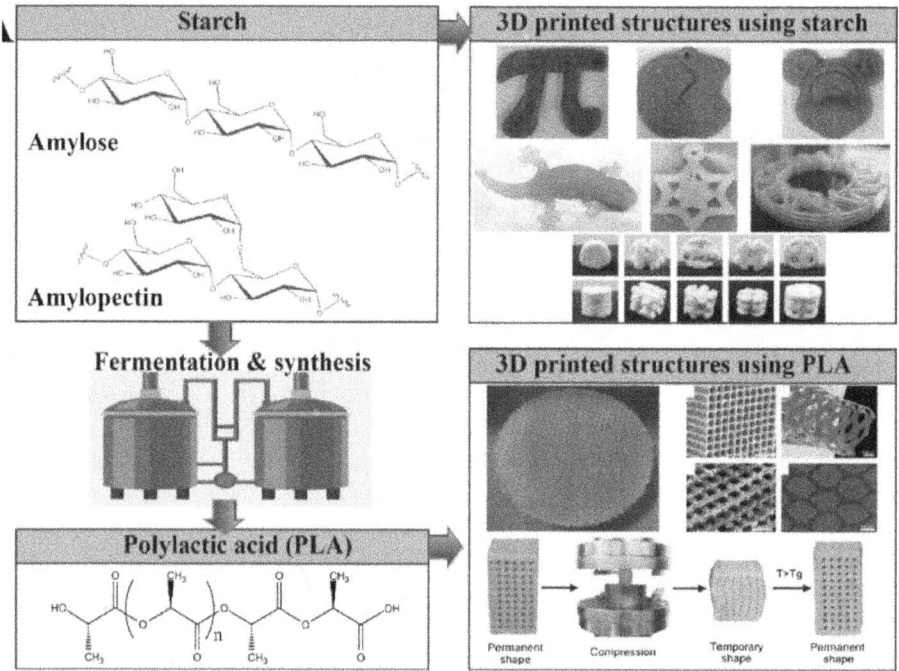

FIGURE 16.3 3D printable materials from starch and polylactic acid

et al., 2020). The rising popularity of nanotechnology in food biopolymer products is still a concern and requires better understanding of their toxicological effects, recyclability, disposal, heat and electrical conductivities effect for acceptability in the food industries. More recently, biopolymers have sufficed in the production of 3D printable materials using starch-based polymers (Figure 16.3). This phenomenon has helped to reduce the use of fossil-based materials whose unregulated exploit has caused unprecedented environmental and health oddity (Liu et al., 2019).

16.6 FACTORS AFFECTING MEDIUM-LARGE SCALE PRODUCTION OF POTATO POLYMERS

The recent global trend enforced by increasing intoxication of the environment, and corresponding rise in health hazard concerns from chronic use of synthetic substances, have given impetus to human attention shift to safer food crops as well as plants as sources of natural product or raw materials for byproduct needs' generation (Nogueira et al., 2020). This new paradigm might exert unprecedented pressure on global food balance and forest resources, create apparent competition for food, and undermine food security strategies. Many tuber (cassava, yam, potatoes) and cereal (corn, maize, sorghum, wheat) crops are becoming more attractive as acceptable alternative raw materials/feedstock for the production of biopolymers (biofilms, exopolysaccharides, bioplastics, etc.), animal feeds, drug delivery/gelatinizing agents, industrial starch, biofuel production and biofertilzers (Vithu et al., 2020). These secondary uses of food crops however have implications on the social, economic, foreign exchange, industrial and agricultural perspectives of many nations all over the world.

Although many food crops now enjoyed usage in the production of a diverse range of biopolymers, potato is fast gaining acceptances as one of the best sources of thermoplastic starch in many food and pharmaceutical industries due to their starch quality versatility. The persistent use of potato as a veritable source of starch in biopolymer production might be marred by a number of factors popular among which is their recent declared status, based on their nutritiveness, by the United Nations as a hunger food for solving global malnutrition challenges (Wang et al., 2018). Other factors, particularly as evident in Nigeria militating medium to large scale use of potato as a sustainable source of starch for degradable bioplastic packaging materials or biopolymer films by the food industries and domestic applications are as follows.

16.6.1 COMPETITION WITH FOOD NEED

If current trend persists, the rising demands for food crops and other natural resources as feedstock of several industrial systems will over time conflict with food need purposes in all ramifications. This competition is likely affect on the availability and cost of these industrial feedstock for scalar production of potato-based polymers (Wool and Sun, 2011). Although conceptualized use of these natural resources is to partially reduce and replace fossil biomass dependence in the production of polymers. The present dichotomy, a bid to finding lasting supply balance between reorienting industrial values for sustainable environmental safety and hunger eradication, has become the subject of scientific discussion among the community of global scientists. While there is apparently no literature that has evaluated economic impact of food crops' biomass as raw material for industrial productions, it is logical to assume the trend might unbalance supply-demand dynamics and pose a raw material challenge to biochemical, biopolymer and food industries (Babu et al., 2013). Globally, there is a need to systematically roll back the use of food-based commodities for eco-friendly technologies that are uses cellulose-based feedstock like wood, parts of non-edible plants (stems and leaves) and wastes (domestic, municipal, agricultural and industrial) (Pillai, 2010).

16.6.2 RISING COST OF FOOD

The geometric population rate increase has doubled the world food need by over 60% while the constant fight against food loss is still undercutting supply. Food needs are encumbered by environmental, disease and economic vagaries coupled with their demand for non-food product generation by the industries. Potatoes along with other popular tuber and cereal crops are staple food and constitute regular household diets in many nations which suggests that they might be less available or scarce for other non-food-based applications. This coupled with poor government agricultural support initiatives might impact negatively on the price and slow down potato-based polymer production (Glenn et al., 2014).

16.6.3 HIGH-QUALITY FEEDSTOCK

According to literature, corn in comparison to other crops including potato is richer in starch, produces higher yields, and provides high-quality feedstock for the production of high-quality lactic acid, which is needed for an efficient synthetic process in polylactic acid (PLA) production. According to Babu et al. (2013), most biopolymer firms' preference for corn and other food crops in producing biopolymers adversely contest with their food values and accessibility as feedstock for compounding animal feeds.

16.6.4 PERFORMANCE OF POTATO POLYMERS

The key disadvantages of unmodified starch-based materials like potato is poor mechanical behaviour and high-water vapour permeability (WVP), which can be due to starch's highly hydrophilic nature. Potato starch polymer also exhibits poor performance properties when subjected to severe environmental conditions capable of limiting its broad application (Shit and Shah, 2014). Except for strain at break, thermoplastic starch-based materials' mechanical properties (tensile ability, elastic modulus, and strain at break) are weak when compared to other thermoplastics, restricting their use in soft plastics applications. While they may undermine their wide use and acceptability, diverse affordable modification technologies now abound in the industries to activate and tinkered with the desired properties required for the end-use functionality.

16.6.5 COST OF PRODUCTION

Although there is a pressing need for bioplastic and bio-composite materials, the cost of their production must be effective. The current affordability of degradable and recalcitrant plastics products allowed them to dominate a large portion of today's local and international polymer markets. This may be as a result of the low cost input of raw materials, energy requirement and production space in the production system (Mohanty et al., 2002). Irrespective of the cost effectiveness, large-scale production development is still undermined by the cost of acquiring the optimization technologies for potato starch-based polymer s production. Similarly, the time to develop efficient potato polymer is high (Mooney, 2009) while the currency of potato in the biopolymer production system and market is juvenile and still being received with scepticism in Nigeria even though it has a promising prospect of rapid adoption in the near future (Al-Oqla and Sapuan, 2020).

16.6.6 PROCESS ABILITY

Starch-based polymers like the potato polymer production are common in laboratory using the casting technique. This technique is apparently less cost-effective for even medium to industrial scale production due to the longer drying period of the film-forming mixture. Addressing this challenge often requires scientific expertise in the management of thermoplastic treatments that include sheet/film extrusion, foaming extrusion, reactive extrusion and injection moulding (Ribba et al., 2017). The need for and timing of hydrothermal treatments during the process is however of observed limitation to the adoption of the technique for large-scale production. Furthermore, the use of carefully monitored processing conditions and the astute combination of complex additives also militate against the friendly application of the technique and apparently complicate the processing is being investigated as a way to improve these processing techniques (Fink, 2019).

16.6.7 BIODIVERSITY AND GLOBAL WARMING

Mono-cropping of energy crops such as potatoes is a booster to biodiversity with a tendency to increase beneficial energy crop production (intensify energy crop cultivation, irrigation, and pesticide

use). Agro-allied and agrochemical application in agricultural systems along with growing global biofuel, and biofeedstock trades endangers ecosystem services' sustainability. The surge in demand for biofuels and bio-feedstocks is likely to hasten the exploitation of bioresources and cultivation expansion of natural feedstock to service safe industrial systems' activities. It may also facilitate the deforestation of rainforests and grassland agriculture both of which are highly efficient carbon sinks (Jiang et al., 2020). Emissions of the highly potent N_2O greenhouse gas are expected to surge as a result of increased fertilizer use. Consequently, planned conversion of rainforest and grassland into new cropland for the production of biopolymers could release 200 times more carbon dioxide into the atmosphere for decades, if not centuries, while transitioning from fossil to bio-based energy sources (Sanyang and Jawaid, 2019). Technically, the intention behind the use of biological-based starch sources to feed biopolymer productions is also likely to contribute undetectably to environmental stress.

16.6.8 RESEARCH AND TECHNOLOGY

Building large-scale plants can be daunting due to the absence of expertise with modern technology, and there is a rarity of scientific literature on large-scale potato polymer processing (Nayak, 1999). As previously mentioned, potato polymers have some limitations, particularly in the areas of processability, non-moisture resistance, and performance. Therefore, further research may be needed to effectively improve their storage, value, and microbial resistance in order to overcome these limitations and encourage their wider industrial applications (Zhong et al., 2020). Additionally, there is the need to develop novel polymers made entirely from bio-based compounds that meet end-user specifications, functionalities, and intrinsically have complete biodegradable as well as recyclable characteristics (Madbouly et al., 2015). Europe, unlike Nigeria is one of the few developed continents leading in world bioplastics research and innovation, while Asia is atop in its production and consumption (Ramesh Kumar et al., 2020). However, only a few other countries have joined this league to advance biopolymer research. In order to make these technologies commercially viable, and accessible, it is important to develop and create (i) modern manufacturing pathways that can modify conventional systems for increased productivity, (ii) effective management of biomass resources, new biological strains/enzymes, and productive processing techniques for recovery of bio-based products. Research firms are not motivated to advance on improved bio-based composite because of lack of funding, which wasn't the case during the polymer science and technology boom after World II (Gandini, 2008). Hence, the government legislature has to provide funding stimulus, improve technological growth and form of regulations facilitate the industrial production of different bio-based polymers relying on food and non-food-based sources.

16.7 CONCLUSION

Relative to other tuber crops, potatoes as well as its derivable polymers have proven to be most versatile, tractable and functionally stable raw materials with applications that stretch beyond its traditional use in the production biofertifilizers, biogas, feeds, and as food to more novel use as feedstock for bioproducts, by-products and biocomposites production. Potato biopolymers constitute one of the value-adding derivatives from potato tubers that have been found to be toxicologically safe, biocompatible, physic-chemically stable, and amenable to a wide range of modification and optimization technologies that ascertain end use functionality. In many developing West African nations, potato is found to be relatively the least consumed compared to other tubers like cassava and yam which are more recurrent dietary preferences. The reason for this trend may require further socio-economic impact investigations even though the sizes, rate of spoilage, short shelf life, and prices of potato could have accounted for it. This is aiding the enlargement of the potato supply chain for the fulfilment of bioeconomy interests that include the inventing of new value chains that could afford product groups' shift in raw materials from fossil to renewable energy (bioresources)

and carbon sources taking into account planetary limits. Hence, research, technological, and business investments by developing nations in biopolymers, particularly from potatoes, would facilitate industrial growth and a new generation of safer product groups with a diverse range of value-added bioactivities as well as end use functionalities. The inherent thermo-pressure and mechanical versatility of potato qualify it as a competitive candidate for application in biotechnology, nanotechnology, tissue engineering, renewable energy processes, food systems, pharmaceuticals, and other areas of human endeavours. Its utility benefits, some of which yet defy scientific detection, and mass valorization have also drastically reduce agrowastes build-up and mediate other attendant wastes related crises in the environment. Furthermore, it indirectly contributes to the United Nation's Sustainable Development Goals (SDGs) 2030 attainment by assuring the protection of life in water, clean environment, affordable and clean energy coupled with growth.

REFERENCES

Abral, H., SoniSatria, R., Mahardika, M., Hafizulhaq, F., Affi, J., Asrofi, M., Handayani, D., Sapuan, S. M., Stephane, I., Sugiarti, E., & Muslimin, A. N. (2019). Comparative study of the physical and tensile properties of Jicama (Pachyrhizuserosus) starch film prepared using three different methods. *Starch/Stärke*, 1800224. doi: 10.1002/star.201800224.

Acevedo-Morantes, M. T., Pineros-Guerrero, N., & Ortega-Toro, R. (2021). Advances in thermoplastic starch-based biopolymers: Fabrication and improvement. In: Ahmed, S. (ed.) *Advanced Green Materials: Fabrication, Characterization and Applications of Biopolymers and Biocomposites*. Woodhead Publishing in Materials: Sawston, Cambridge, 205–255 pp.

Al-Oqla, F. M., & Sapuan, S. M. (2020). *Advanced Processing, Properties, and Applications of Starch and other Bio-Based Polymers*. Elsevier: Amsterdam, Netherlands.

Altemimi, A. B. (2018). Extraction and optimization of potato starch and its application as a stabilizer in yogurt manufacturing. *Foods*, 7(2). doi: 10.3390/foods7020014.

Arvanitoyannis, I., Biliaderis, C. G., Ogawa, H., & Kawasaki, N. (1998). Biodegradable films made from low-density polyethylene (LDPE), rice starch and potato starch for food packaging applications: Part 1. *Carbohydrate Polymers*, 36(2–3), 89–104. doi: 10.1016/S0144-8617(98)00016-2.

Assi, A. H. (2018). Potato starch for enhancing the properties of the drilling fluids. *Iranian Journal of Chemical and Petroleum Engineering*, 19(3), 33–40.

Azli, A. A., Manan, N. S. A., Aziz, S. B., & Kadir, M. F. Z. (2020). Structural, impedance and electrochemical double-layer capacitor characteristics of improved number density of charge carrier electrolytes employing potato starch blend polymers. *Ionics*, 26(11), 5773–5804. doi: 10.1007/s11581-020-03688-1.

Babu, R. P., O'Connor, K., & Seeram, R. (2013). Current progress on bio-based polymers and their future trends. *Progress in Biomaterials*, 2(1), 1–16.

Balakrishnan, P., Gopi, S., Sreekala, M. S., & Thomas, S. (2018). UV resistant transparent bionanocomposite films based on potato starch/cellulose for sustainable packaging. *Starch/Staerke*, 70(1–2), 1–34. doi: 10.1002/star.201700139.

Ballesteros-Mártinez, L., Pérez-Cervera, C., & Andrade-Pizarro, R. (2020). Effect of glycerol and sorbitol concentrations on mechanical, optical, and barrier properties of sweet potato starch film. *NFS Journal*, 20, 1–16. doi: 10.1016/j.nfs.2020.06.002.

Basiak, E., Lenart, A., & Debeaufort, F. (2017). Effect of starch type on the physico-chemical properties of edible films. *International Journal of Biological Macromolecules*, 98, 348–356. doi: 10.1016/j.ijbiomac.2017.01.122.

Bergthaller, W., Witt, W., & Goldau, H. P. (1999). Potato starch technology. *Starch/Staerke*, 51(7), 235–242. doi: 10.1002/(sici)1521-379x(199907)51:7<235::aid-star235>3.0.co;2-7.

Bonilla, J., Atarés, L., Vargas, M., & Chiralt, A. (2013). Properties of wheat starch film-forming dispersions and films as affected by chitosan addition. *Journal of Food Engineering*, 114(3), 303–312. doi: 10.1016/j.jfoodeng.2012.08.005.

Borah, P. P., Das, P., & Badwaik, L. S. (2017). Ultrasound treated potato peel and sweet lime pomace-based biopolymer film development. *Ultrasonics Sonochemistry*, 36, 11–116. doi: 10.1016/j.ultsonch.2016.11.010.

Castanha, N., Santos, D. N., Cunha, R. L., & Augusto, P. E. D. (2019). Properties and possible applications of ozone-modified potato starch. *Food Research International*, 116, 1192–1201. doi: 10.1016/j.foodres.2018.016.064.

Chandrasekara, A., & Kumar, T.J. (2016). Roots and tuber crops as functional foods: A review on phytochemical constituents and their potential health benefits. *International Journal of Food Science*, 2016, 3631647. doi: 10.1155/2016/3631647.

Chen, W., Oldfield, T. L., Cinelli, P., Righetti, M. C., & Holden, N. M. (2020). Hybrid life cycle assessment of potato pulp valorisation in biocomposite production. *Journal of Cleaner Production*, 269, 122366.

Coezee, D., Venkataraman, M., Militky, J. and Petru, M. (2020). Influence of nanoparticles on thermal and electrical conductivity of composites. *Polymer*, 12, 742. doi: 10.3390/polym12040742.

Colussi, R., Kringel, D., Kaur, L., da Rosa Zavareze, E., Renato Guerra Dias, A., & Singh, J. (2020). Dual modification of potato starch: Effects of Heat-moisture and high pressure treatments on starch structure and functionalities. *Food Chemistry*, 126475. doi: 10.1016/j.foodchem.2020.126475.

Costa, N. N., de Faria Lopes, L., Ferreira, D. F., de Prado, E. M. L., Severi, J. A., Resende, J. A., de Paula Careta, F., Ferreira, M. C. P., Carreira, L. G., de Souza, S. O. L., Cotrim, M. A. P., Boeing, T., de Andrade, S. F., Orefice, R. L., & Villanova, J. C. O. (2020). Polymeric films containing pomegranate peel extract based on PVA/starch/PAA blends for use in wound dressing: In vitro analysis and phytochemical evaluation. *Materials Science & Engineering C*, 109, 110643.

Cruz-Gálvez, A. M., Castro-Rosas, J., Rodríguez-Marín, M. L., Cadena-Ramírez, A., Tellez-Jurado, A., Tovar-Jiménez, X., Chavez-Urbiola, E. A., Abreu-Corona, A., & Gómez-Aldapa, C. A. (2018). Antimicrobial activity and physicochemical characterization of a potato starch-based film containing acetonic and methanolic extracts of Hibiscus sabdariffa for use in sausage. *LWT*, 93, 300–305. doi: 10.1016/j.lwt.2018.02.064.

Członka, S., Bertino, M. F., & Strzelec, K. (2018). Rigid polyurethane foams reinforced with industrial potato protein. *Polymer Testing*, 68, 135–145. doi: 10.1016/j.polymertesting.2018.04.006.

Dash, K. K., Ali, N. A., Das, D., & Mohanta, D. (2019). Thorough evaluation of sweet potato starch and lemon-waste pectin based-edible films with nano-titania inclusions for food packaging applications. *International Journal of Biological Macromolecules*, 139, 449–458. doi: 10.1016/j.ijbiomac.20116.07.193.

de Azevedo, L. C., Rovani, S., Santos, J. J., Dias, D. B., Nascimento, S. S., Oliveira, F. F., Silva, L. G. A., & Fungaro, D. A. (2020). Biodegradable films derived from corn and potato starch and study of the effect of silicate extracted from sugarcane waste ash. *ACS Applied Polymer Materials*, 2(6), 2160–21616. doi: 10.1021/acsapm.0c00124.

Do Val Siqueira, L., Arias, C. I. L. F., Maniglia, B. C., & Tadini, C. C. (2020). Starch-based biodegradable plastics: Methods of production, challenges and future perspectives. *Current Opinion in Food Science*. doi: 10.1016/j.cofs.2020.10.020.

Domene-López, D., Delgado-Marín, J. J., Martin-Gullon, I., García-Quesada, J. C., & Montalbán, M. G. (2019). Comparative study on properties of starch films obtained from potato, corn and wheat using 1-ethyl-3-methylimidazolium acetate as plasticizer. *International Journal of Biological Macromolecules*, 135, 845–854. doi: 10.1016/j.ijbiomac.20116.06.004.

Dong, D., & Cui, B. (2021). Fabrication, characterization and emulsifying properties of potato starch/soy protein complexes in acidic conditions. *Food Hydrocolloids*, 115, 106600.

Ebrahimian, F., Denayer, J. F. M., & Karimi, K. (2022). Potato peel waste biorefinery for the sustainable production of biofuels, bioplastics, and biosorbents. *Bioresource Technology*, 360, 127609. https://doi.org/10.1016/j.biortech.2022.127609.

Farajpour, R., Djomeh, Z. E., Moeini, S., Tavakolipour, H., & Safayan, S. (2020). Structural and physic-mechanical properties of potato starch-olive oil edible films reinforced with zein nanoparticles. *International Journal of Biological Molecules*, 149, 941–950.

Feo, G. D., Ferrara, C., Finelli, A., & Grosso, A. (2017). Environmental and economic benefits of the recovery of materials in a municipal solid waste management system. *Environmental Technology*, 2017. doi: 10.1080/09593330.2017.1411395.

Ferreira Saraiva, L. E., de Naponucena, L. O. M., da Silva Santos, V., Silva, R. P. D., de Souza, C. O., Evelyn Gomes Lima Souza, I., de Oliveira Mamede, M. E., & Druzian, J. I. (2016). Development and application of edible film of active potato starch to extend mini panettone shelf life. *LWT: Food Science and Technology*, 73, 311–3116. doi: 10.1016/j.lwt.2016.05.047.

Fink, J. K. (2019). *The Chemistry of Bio-Based Polymers*. John Wiley & Sons: Hoboken, NJ.

Franz, P. (2015). Verification Report ETV01/2015 Aerobic Biodegradation of Third generation Mater Bi under marine condition By NOVAMONT Spa Table of Contents. 1–116. https://ec.europa.eu/environment/ecoap/sites/ecoap_stayconnected/files/etv/vn20150004_verification_report_novamont.pdf.

Gandini, A. (2008). Polymers from renewable resources: A challenge for the future of macromolecular materials. *Macromolecules*, 41(24), 9491–9504. doi: 10.1021/ma801735u.

Ghorbani, M., Dahrazma, B., Saghravani, S. F., & Yousofizinsaz, G. (2021). A comparative study on physico-chemical propertiesofenvironmentally-friendly lightweight bricks having potato peel powder and sour orange leaf. *Construction and Building Materials*, 276, 121937.

Gigante, V., Seggiani, M., Cinelli, P., Signori, F., Vania, A., Navarini, L., Amato, G., & Lazzeri, A. (2021). Utilization of coffee silverskin in the production of poly(3-hydroxybutyrate-co-3-hydroxyvalerate) bio-polymer-based thermoplastic biocomposites for food contact applications. *Composites Part A: Applied Science and Manufacturing*, 140, 106172. doi: 10.1016/j.compositesa.2020.106172.

Glenn, G. M., Orts, W., Imam, S., Chiou, B.-S., & Wood, D. F. (2014). Starch plastic packaging and agri-culture applications. In: Halley, J., & Avérous, L., (eds.) *Starch Polymers*. Elsevier: Amsterdam, The Netherlands, pp. 421–452.

Gopi, S., Amalraj, A., Jude, S., Thomas, S., & Guo, Q. (2019). Bionanocomposite films based on potato, tapi-oca starch and chitosan reinforced with cellulose nanofiber isolated from turmeric spent. *Journal of the Taiwan Institute of Chemical Engineers*, 96, 664–671. doi: 10.1016/j.jtice.20116.01.003.

Gowthaman, N. S. K., Lim, H. N., Sreeraj, T. R., Amalraj, A., & Gopi, S. (2021). Advantages of biopoly-mers over synthetic biopolymers: Social, economic and environmental aspects. In: Thomas, S., Gopi, S., & Amalray, A. (eds.) *Biopolymers and Their Industrial Applications: From Plant, Animal and Marine Sources, to Functional Products*. Elservier: Amsterdam, Netherlands, 351–372 pp. doi: 10.1016/B978-0-12-819240-5.00015-8.

Gregorová, E., Pabst, W., & Bohačenko, I. (2006). Characterization of different starch types for their applica-tion in ceramic processing. *Journal of the European Ceramic Society*, 26(8), 1301–13016. doi: 10.1016/j.jeurceramsoc.2005.02.015.

Grommers, H. E., & van der Krogt, D. A. (2009). Potato starch: Production, modifications and uses. In: *Starch* (Third Edition). Elsevier: Amsterdam, Netherlands. doi: 10.1016/B978-0-12-746275-2.00011-2.

Guo, J., Zhang, Y., Zhao, J., Zhang, Y., Xiao, X., Wang, B., & Shu, B. (2015). Characterization of a bioflocu-lant from potato starch wastewater and its application in sludge dewatering. *Applied Microbiology and Biotechnology*, 99(13), 5429–5437. doi: 10.1007/s00253-015-6567-4.

Halterman, D., Guenthner, J., Collinge, S., Butler, N., & Douches, D. (2016). Biotech potatoes in the 21st cen-tury: 20 years since the first biotech potato. *American Journal of Potato Research*, 93, 1–20.

Hao, L., Zang, J., Lu, J., Ba, J., & Yu, J. (2013). Characterization of a new polysaccharide from potato starch. *Journal of Food Processing and Preservation*, 38(4), 1409–1415.

Heo, H., Lee, Y. K., & Chang, Y. H. (2017). Rheological, pasting, and structural properties of potato starch by cross-linking. *International Journal of Food Properties*, 20(2), 2138–2150. doi: 10.1080/10942912.2017.1368549.

Horstmann, S. W., Axel, C., & Arendt, E. K. (2018). Water absorption as a prediction tool for the applica-tion of hydrocolloids in potato starch-based bread. *Food Hydrocolloids*, 81, 129–138. doi: 10.1016/j.foodhyd.2018.02.045.

Ilyas, R. A., & Sapuan, S. M. (2020). Biopolymers and biocomposites: Chemistry and technology. *Current Analytical Chemistry*, 16(5), 500–503. doi: 10.2174/1573411016052000603095311.

Islam, H. B. M. Z., Susan, Md. A. B. H., & Imran, A. B. (2020). Effects of plasticizers and clays on the physi-cal, chemical, mechanical, thermal, and morphological properties of potato starch-based nanocompos-ite films. *ACS Omega*, 5(28), 17543–17552. doi: 10.1021/acsomega.0c02012.

Ismail, N. A., Mohd Tahir, S., Norihan, Y., Abdul Wahid, M. F., Khairuddin, N. E., Hashim, I., Rosli, N., & Abdullah, M. A. (2016). Synthesis and characterization of biodegradable starch-based bioplastics. *Materials Science Forum*, 846, 673–678. doi: 10.4028/www.scientific.net/msf.846.673.

Jabeen, N., Majid, I., & Nayik, G. A. (2015). Bioplastics and food packaging: A review. *Cogent Food & Agriculture*, 1(1). doi: 10.1080/23311932.2015.1117749.

Jagadeesan, S., Govindaraju, I., & Mazumder, N. (2020). An insight into the ultrastructural and physiochemi-cal characterization of potato starch: A review. *American Journal of Potato Research*, 97(5), 464–476. doi: 10.1007/s12230-020-09798-w.

Jamróz, E., Juszczak, L., & Kucharek, M. (2018). Investigation of the physical properties, antioxidant and antimicrobial activity of ternary potato starch-furcellaran-gelatin films incorporated with lavender essential oil. *International Journal of Biological Macromolecules*, 114, 1094–1101. doi: 10.1016/j.ijbiomac.2018.04.014.

Janik, H., Sienkiewicz, M., Przybytek, A., Guzman, A., Kucinska-Lipka, J., & Kosakowska, A. (2018). Novel biodegradable potato starch-based compositions as candidates in packaging industry, safe for marine environment. *Fibers and Polymers*, 19(6), 1166–1174. doi: 10.1007/s12221-018-7872-1.

Jiang, S., Liu, C., Wang, X., Xiong, L., & Sun, Q. (2016). Physicochemical properties of starch nanocomposite films enhanced by self-assembled potato starch nanoparticles. *LWT - Food Science and Technology*, 69, 251–257. doi: 10.1016/j.lwt.2016.01.053.

Jiang, T., Duan, Q., Zhu, J., Liu, H., & Yu, L. (2020). Starch-based biodegradable materials: Challenges and opportunities. *Advanced Industrial and Engineering Polymer Research,* 3(1), 8–18. doi:doi: 10.1016/j. aiepr.20116.11.003.

Jiménez, A., Fabra, M. J., Talens, P., & Chiralt, A. (2012). Effect of sodium caseinate on properties and ageing behaviour of corn starch based films. *Food Hydrocolloids,* 29(2), 265–271. doi: 10.1016/j. foodhyd.2012.03.014.

Jiménez, A., Fabra, M. J., Talens, P., & Chiralt, A. (2013). Physical properties and antioxidant capacity of starch–sodium caseinate films containing lipids. *Journal of Food Engineering,* 116(3), 695–702. doi: 10.1016/j.jfoodeng.2013.01.010.

Jimoh, K. O., & Kolapo, A. L. (2007). Effect of different stabilizers on acceptability and shelf-stability of soy-yoghurt. *African Journal of Biotechnology,* 6(8), 1000–1003. doi: 10.4314/ajb.v6i8.57031.

Jochym, K. K., & Nebesny, E. (2017). Enzyme-resistant dextrins from potato starch for potential application in the beverage industry. *Carbohydrate Polymers,* 172, 152–158. doi: 10.1016/j.carbpol.2017.05.041.

Juarez-Arellano, E. A., Urzua-Valenzuela, M., Peña-Rico, M. A., Aparicio-Saguilan, A., Valera-Zaragoza, M., Huerta-Heredia, A. A., & Navarro-Mtz, A. K. (2020). Planetary ball-mill as a versatile tool to controlled potato starch modification to broaden its industrial applications. *Food Research International,* 109870. doi: 10.1016/j.foodres.2020.109870.

Kabir, E., Kaur, R., Lee, J., Kim, K., & Kwon, E. E. (2020). Prospects of biopolymer technology as an alternative option for non-degradable plastics and sustainable management of plastic wastes. *Journal of Cleaner Production,* 258, 120536.

Kang, H. J., & Min, S. C. (2010). Potato peel-based biopolymer film development using high-pressure homogenization, irradiation, and ultrasound. *LWT - Food Science and Technology,* 43(6), 903–9016. doi: 10.1016/j.lwt.2010.01.025.

Karaky, H., Maalouf, C., Bliard, C., Gacoina, A., El Wakila, M. L. N., & Polidori, G. (2019). Characterization of beet-pulp fiber reinforced potato starch biopolymercomposites for building applications. *Construction and Building Materials,* 203, 711–721.

Kita, A. (2002). The influence of potato chemical composition on crisp texture. *Food Chemistry,* 76(2), 173–1716. doi: 10.1016/s0308-8146(01)00260-6.

Kowalczyk, D. (2016). Biopolymer/candelilla wax emulsion films as carriers of ascorbic acid: A comparative study. *Food Hydrocolloids,* 52, 543–553.

Kraak, A. (1992). Industrial applications of potato starch products. *Industrial Crops and Products,* 1(2–4), 107–112. doi: 10.1016/0926-6690(92)90007-I.

Li, Y.-D., Xu, T.-C., Xiao, J.-X., Zong, A.-Z., Qiu, B., Jia, M., Liu, L.-N., & Liu, W. (2018). Efficacy of potato resistant starch prepared by microwave-toughening treatment. *Carbohydrate Polymers,* 192, 299–307.

Li, H., Shi, W., Du, Q., Zhou, R., Zeng, X., Zhang, H., & Qin, X. (2020). Recovery and purification of potato proteins from potato starch wastewater by hollow fiber separation membrane integrated process. *Innovative Food Science & Emerging Technologies,* 102380. doi: 10.1016/j.ifset.2020.102380.

Li, M., Blecker, C., & Karboune, S. (2021). Molecular and air-water interfacial properties of potato protein upon modification via laccase-catalyzed cross-linking and conjugation with sugar beet pectin. *Food Hydrocolloids,* 112, 106236.

Liu, J., Guo, L., Yang, L., Liu, Z., & He, C. (2014). The rheological property of potato starch adhesives. *Advance Journal of Food Science and Technology,* 6(2), 275–2716. doi: 10.19026/ajfst.6.24.

Liu, J., Sun, L., Xu, W., Wang, Q., Yu, S., & Sun, J. (2019). Current advances and future perspectives of 3D printing natural-derived biopolymers. *Carbohydrate Polymers,* 207, 297–316. doi: 10.1016/j. carbpol.2018.11.077.

Liu, Y.-T., Hu, X.-P., Bai, Y., Zhao, Q.-Y., Yu, S.-Q., Tian, Y.-X., Bian, Y.-Y., Li, S.-H., Li, T.-P. (2020). Preparation and antioxidative stability of the potato protease inhibitors (PPIs) from potato starch wastewater.

Liu, J., Huang, J., Ying, Y., Hu, L., & Hu, Y. (2021). pH-sensitive and antibacterial films developed by incorporating anthocyanins extracted from polyvinylalcohol/nano-ZnO matrix: Comparative study. *International Journal of Biological Molecules,* 178, 104–112.

Luchese, C. L., Spada, J. C., & Tessaro, I. C. (2017). Starch content affects physicochemical properties of corn and cassava starch-based films. *Industrial Crops and Products,* 109, 619–626. doi: 10.1016/j. indcrop.2017.016.020.

Luchese, C. L., Benelli, P., Spada, J. C., & Tessaro, I. C. (2018). Impact of the starch source on the physico-chemical properties and biodegradability of different starch-based films. *Journal of Applied Polymer Science,* 135(33), 46564. doi: 10.1002/app.46564.

Madbouly, S., Zhang, C., & Kessler, M. R. (2015). *Bio-Based Plant Oil Polymers and Composites.* Elsevier: Amsterdam, Netherlands; William Andrew: Norwich, NY.

Mahalakshmi, M., Selvanayagam, S., Selvasekarapandian, S., Moniha, V., Manjuladevi, R., & Sangeetha, P. (2019). Characterization of biopolymer electrolytes based on cellulose acetate with magnesium perchlorate $(Mg(ClO_4)_2)$ for energy storage devices. *Journal of Science: Advanced Materials and Devices*, 4(2), 276–284. doi: 10.1016/j.jsamd.20116.04.006.

Maraveas, C. (2020). Production of sustainable and biodegradable polymers from agricultural waste. *Polymers*, 12(5), 1127. doi: 10.3390/polym12051127.

Matharu, A. S., de Melo, E. M., & Houghton, J. A. (2018). Food supply chain waste: A functional periodic table of bio-based resources. In: Bhaskar, T., Pandey, A., Mohan, S. V., Lee, D.-J., & Khanal, S. K. (eds.), *Waste Biorefinery: Potential and Perspectives*. Elsevier: Amsterdam, Netherlands. pp. 219–236. doi: 10.1016/b978-0-444-63992-16.00007-0.

Maurer, H. W. (2009). Starch in the paper industry. In: *Starch* (Third Edition). Elsevier: Amsterdam, Netherlands. doi: 10.1016/B978-0-12-746275-2.00018-5.

Miller, K., Silcher, C., Lindner, M., & Schmid, M. (2020). Effects of glycerol and sorbitol on optical, mechanical, and gas barrier properties of potato peel-based films. *Packaging Technology and Science*, 34(1), 11–23. doi: 10.1002/pts.2536.

Mitch, E. L. (1984). Potato starch: Production and uses. *Starch: Chemistry and Technology*, 479–490. doi: 10.1016/b978-0-12-746270-7.50020-3.

Mohanty, A. K., Misra, M., & Drzal, L. T. (2002). Sustainable bio-composites from renewable resources: Opportunities and challenges in the green materials world. *Journal of Polymers and the Environment*, 10(1), 19–26. doi: 10.1023/A:1021013921916.

Molavi, H., Behfar, S., Shariati, M. A., Kviani, M., & Atarod, S. (2015). A review on biodegradable starch films. *Journal of Microbiology, Biotechnology and Food Sciences*, 4(5), 456–461. doi: 10:15414/jmbfs.2015.4.5.456-461.

Montero, B., Rico, M., Rodríguez-Llamazares, S., Barral, L., & Bouza, R. (2017). Effect of nanocellulose as a filler on biodegradable thermoplastic starch films from tuber cereal and legume. *Carbohydrate Polymer*, 157, 1094–1104.

Mooney, B. P. (2009). The second green revolution? Production of plant-based biodegradable plastics. *Biochemical Journal*, 418(2), 219–232. doi: 10.1042/BJ20081769%.

Moongngarm, A. (2013). Chemical compositions and resistant starch content in starchy foods. *American Journal of Agricultural and Biological Sciences*, 8(2), 107–113. doi: 10.3844/ajabssp.2013.107.113.

Moreno, O., Atarés, L., & Chiralt, A. (2015). Effect of the incorporation of antimicrobial/antioxidant proteins on the properties of potato starch films. *Carbohydrate Polymers*, 133, 353–364.

Mościcki, L., Mitrus, M., Wójtowicz, A., Oniszczuk, T., Rejak, A., & Janssen, L. (2012). Application of extrusion-cooking for processing of thermoplastic starch (TPS). *Food Research International*, 47(2), 291–2916. doi: 10.1016/j.foodres.2011.07.017.

Muscat, D., Adhikari, B., Adhikari, R., & Chaudhary, D. S. (2012). Comparative study of film forming behaviour of low and high amylose starches using glycerol and xylitol as plasticizers. *Journal of Food Engineering*, 109(2), 189–201. doi: 10.1016/j.jfoodeng.2011.10.019.

Nandi, S., & Guha, P. (2018). Modelling the effect of guar gum on physical, optical, barrier and mechanical properties of potato starch based composite film. *Carbohydrate Polymers*, 200, 498–507. doi: 10.1016/j.carbpol.2018.08.028.

Nayak, P. L. (1999). Biodegradable polymers: Opportunities and challenges. *Journal of Macromolecular Science, Part C*, 39(3), 481–505. doi: 10.1081/MC-100101425.

Ncube, L. K., Ude, A. U., Ogunmuyiwa, E. N., Zulkifli, R., & Beas, I. N. (2020). Environmental impact of food packaging materials: A review of contemporary development from conventional plastics to polylactic acid based materials. *Materials*, 13, 4994. doi: 10.3390/ma13214994.

Nmegbu, J., Nyeche, W., & Jane, P. (2015). Drilling mud formulation using potato starch (*Ipomoea batatas*). *International Journal of Engineering Research and Applications (IJERA)*, 5(9), 48–54.

Odebode, S. O., Egeonu, N., & Akoroda, M. O. (2008). Promotion of sweetpotato for the food industry in Nigeria. *Bulgarian Journal of Agricultural Science*, 14(3), 300–308.

Olaniran, A. F., Okonkwo, C. E., Osemwegie, O. O, Iranloye, Y. M., Afolabi, Y. T. Alejolowo, O. O., Nwonuma, C. O., & Badejo, T. E. (2020a). Production of a complementary food: Influence of cowpea soaking time on the nutritional, anti-nutritional and antioxidant properties of the cassava-cowpea-orange fleshed potato blends. *International Journal of Food Science*, 2020, 1–10.

Olaniran, A. F., Okonkwo, C. E., Owolabi, A. O, Osemwegie, O. O., & Badejo, T. E. (2020b) Proximate composition and physicochemical properties of formulated cassava, cowpea and potato flour blends. *IOP Conference Series: Earth and Environmental Science*, 445(012042), 1–7. doi: 10.1088/1755-1315/445/1/01204.

Oluwasina, O. O., Olaleye, F. K., Olusegun, S. J., Oluwasina, O. O., & Mohallem, N. D. S. (2019). Influence of oxidized starch on physico-mechanical, thermal properties, and atomic force micrographs of cassava starch bioplastic film. *International Journal of Biological Macromolecules*, 135, 282–293. doi: 10.1016/j.ijbiomac.20116.05.150.

Omotoso, M. A., Adeyefa, O. S., Animashaun, E. A., & Osibanjo, O. O. (2015). Biogradable starch film from cassava, corn, potato and yam. *Chemistry and Material Research*, 7(12), 15–24.

Pillai, C. K. S. (2010). Challenges for natural monomers and polymers: Novel design strategies and engineering to develop advanced polymers. *Designed Monomers and Polymers*, 13(2), 87–121. doi: 10.1163/138 577210X12634696333190.

Priedniece, V., Spalvins, K., Ivanovs, K., Pubule, J., & Blumberga, D. (2017). Bioproducts from potatoes a review. *Environmental and Climate Technologies*, 21, 18–27.

Pu, S. Y., Qin, L. L., Che, J. P., Zhang, B. R., & Xu, M. (2014). Preparation and application of a novel bioflocculant by two strains of Rhizopus sp. using potato starch wastewater as nutrilite. *Bioresource Technology*, 162, 184–191. doi: 10.1016/j.biortech.2014.03.124.

Qasem, E. A. (2020). United States Patent. 2(12).

Ramesh Kumar, S., Shaiju, P., & O'Connor, K. E. (2020). Bio-based and biodegradable polymers: State-of-the-art, challenges and emerging trends. *Current Opinion in Green and Sustainable Chemistry*, 21, 75–81. doi: 10.1016/j.cogsc.20116.12.005.

Ranganathan, S., Dutta, S., Moses, J. A., & Anandharamakrishnan, C. (2020). Utilization of food waste streams for the production of biopolymers. *Heliyon*, 6(9), e04891. doi: 10.1016/j.heliyon.2020.e04891.

Ribba, L., Garcia, N. L., D'Accorso, N., & Goyanes, S. (2017). Disadvantages of starch-based materials, feasible alternatives in order to overcome these limitations. In: *Starch-Based Materials in Food Packaging*. Elsevier, Amsterdam, Netherlands, pp. 37–76.

Richard, M., Emarold, M., Robert, M., & Amon, M. (2015). Characterisation of botanical starches as potential substitutes of agar in tissue culture media. *African Journal of Biotechnology*, 14(8), 702–713. doi: 10.5897/ajb2014.142816.

Righetti, M., Cinelli, P., Mallegni, N., Massa, C., Aliotta, L., & Lazzeri, A. (2019). Thermal, mechanical, viscoelastic and morphological properties of poly(lactic acid) based biocomposites with potato pulp powder treated with waxes. *Materials*, 12(6), 990. doi: 10.3390/ma12060990.

Sanyang, M. L., & Jawaid, M. (2019). *Bio-Based Polymers and Nanocomposites: Preparation, Processing, Properties & Performance*. Springer: New York.

Schwartz, D., & Whistler, R. L. (2009). History and future of starch. In: *Starch* (Third Edition). Elsevier: Amsterdam, Netherlands. doi: 10.1016/B978-0-12-746275-2.00001-X.

Shafqat, A., Tahir, A., Mahmood, A., & Pugazhendhi, A. (2020). A review on environmental significance carbon foot prints of starch based bio-plastic: A substitute of conventional plastics. *Biocatalysis and Agricultural Biotechnology*, 101540. doi: 10.1016/j.bcab.2020.101540.

Shanmugavel, D., Selvaraj, T., Ramadoss, R., & Raneri, S. (2020). Interaction of a viscous biopolymer from cactus extract with cement paste to produce sustainable concrete. *Construction and Building Materials*, 257, 119585.

Sharma, R., Jafari, S. M., & Sharma, S. (2020). Antimicrobial bio-nanocomposites and their potential applications in food packaging. *Food Control*, 112, 107086. doi: 10.1016/j.foodcont.2020.107086.

Shit, S. C., & Shah, P. M. (2014). Edible polymers: Challenges and opportunities. *Journal of Polymers*, 2014, 1–13.

Snell, K. D., Singh, V., & Brumbley, S. M. (2015). Production of novel biopolymers in plants: Recent technological advances and future prospects. *Current Options in Biotechnology*, 32, 68–75.

Souza, A. C., Ditchfield, C., & Tadini, C. C. (2010). Biodegradable films based on biopolymer for food industries. In: Passos, M. L., & Ribeiro, C. P. (eds.) *Innovation in Food Engineering: New Techniques and Products*. CRC Press: Boca Raton, FL, pp. 511–537.

Stoica, M., Antohi, V. M., Zlati, M. Z., & Stoica, D. (2020). The financial impact of replacing plastic packaging by biodegradable biopolymers: A smart solution for the food industry. *Journal of Cleaner Production*, 277, 124013.

Sun, Q., Sun, C., & Xiong, L. (2013). Mechanical, barrier and morphological properties of pea starch and peanut protein isolate blend films. *Carbohydrate Polymers*, 98(1), 630–637. doi: 10.1016/j.carbpol.2013.06.040.

Syafiq, R., Sapuan, S. M., Zuhri, M. Y. M., Ilyas, R. A., Nazrin, A., Sherwani, S. F. K., & Khalina, A. (2020). Antimicrobial activities of starch-based biopolymers and biocomposites incorporated with plant essential oils: A review. *Polymers*, 12(10), 2403. doi: 10.3390/polym12102403.

Szepes, P. S. A. (2009). Potato starch in pharmaceutical technology: A review.

Torres, M. D., & Dominguez, H. (2020). Vaorisation of potato wastes. *International Journal of Food Science and Technology*, 55, 2296–2304.

Torres, M. D., Fradinho, P., Rodríguez, P., Falqué, E., Santos, V., & Domínguez, H. (2020). Biorefinery concept for discarded potatoes: Recovery of starch and bioactive compounds. *Journal of Food Engineering*, 275, 109886. doi: 10.1016/j.jfoodeng.20116.109886.

Udayakumar, G. P., Muthusamy, S., Selvaganesh, B., Sivarajasekar, N., Rambabu, K., Banat, F., Sivamani, S., Sivakumar, N., Hosseini-Bandegharaei, A., & Show, P. L. (2021). Biopolymers and composites: Properties, characterization and their applications in food, medical and pharmaceutical industries. *Journal of Environmental Chemical Engineering*, 9, 105322.

Vargas, C. G., Costa, T. M. H., de Rios, A. O., & Flôres, S. H. (2017). Comparative study on the properties of films based on red rice (Oryza glaberrima) flour and starch. *Food Hydrocolloids*, 65, 96–106. doi: 10.1016/j.foodhyd.2016.11.006.

Varghese, S. A., Siengchin, S., & Parameswaranpillai, J. (2020). Essential oils as antimicrobial agents in biopolymer-based food packaging: A comprehensive review. *Food Bioscience*, 38, 100785. doi: 10.1016/j.fbio.2020.100785.

Verma, D., Dogra, V., Chaudhary, A. A., & Mordia, R. (2022). 5 - Advanced biopolymer-based composites: construction and structural applications. In Deepak Verma, Mohit Sharma, Kheng Lim Goh, Siddharth Jain, Himani Sharma (eds.) *Woodhead Publishing Series in Composites Science and Engineering, Sustainable Biopolymer Composites*. Sawston, UK: Woodhead Publishing, pp. 113–128, https://doi.org/10.1016/B978-0-12-822291- 1.00010-5.

Versino, F., & García, M. A. (2014). Cassava (Manihot esculenta) starch films reinforced with natural fibrous filler. *Industrial Crops and Products*, 58, 305–314. doi: 10.1016/j.indcrop.2014.04.040.

Vinod, A., Sanjay, M. R., Suchart, S., & Jyotishkumar, P. (2020). Renewable and sustainable biobased materials: An assessment on biofibers, biofilms, biopolymers and biocomposites. *Journal of Cleaner Production*, 258, 120978. doi: 10.1016/j.jclepro.2020.120978.

Vithu, P., Dash, S. K., Rayaguru, K., Panda, M. K., & Nedunchezhiyan, M. (2020). Optimization of starch isolation process for sweet potato and characterization of the prepared starch. *Journal of Food Measurement and Characterization*. doi: 10.1007/s11694-020-00401-8.

Waglay, A., Karboune, S., & Alli, I. (2014). Potato protein isolates: Recovery and characterizationof their properties. *Food Chemistry*, 142, 373–382.

Wang, A., Li, R., Ren, L., Gao, X., Zhang, Y., Ma, Z., & Luo, Y. (2018). A comparative metabolomics study of flavonoids in sweet potato with different flesh colors (Ipomoea batatas (L.) Lam). *Food Chemistry*, 260, 124–134. doi: 10.1016/j.foodchem.2018.03.125.

Wang, C., Chang, T., Dong, S., Zhang, D., Ma, C., Chen, S., & Li, H. (2020a). Biopolymer films based on chitosan/potato protein/linseed oil/ZnO NPs to maintain the storage quality of raw meat. *Food Chemistry*, 332, 127375. doi: 10.1016/j.foodchem.2020.127375.

Wang, H., Yang, Q., Ferdinand, U., Gong, X., Qu, Y., Gao, W., & Liu, M. (2020b). Isolation and characterization of starch from light yellow, orange, and purple sweet potatoes. *International Journal of Biological Macromolecules*. doi: 10.1016/j.ijbiomac.2020.05.2516.

Wilpiszewska, K., & Czech, Z. (2014). Citric acid modified potato starch films containing microcrystalline cellulose reinforcement: Properties and application. *Starch/Staerke*, 66(7–8), 660–667. doi: 10.1002/star.201300093.

Wool, R., & Sun, X. S. (2011). *Bio-Based Polymers and Composites*. Elsevier: Amsterdam, Netherlands.

Wu, J., Huang, Y., Yao, R., Deng, S., Li, F., & Bian, X. (2019). Preparation and characterization of starch nanoparticles from potato starch by combined solid-state acid-catalyzed hydrolysis and nanoprecipitation. *Starch/Staerke*, 71(9–10), 1–8. doi: 10.1002/star.201900095.

Xie, Y., Niu, X., Yang, J., Fan, R., Shi, J., Ullah, N., Feng, X., & Chen, L. (2020). Active biodegradable films based on the whole potato peel incorporated with bacterial cellulose and curcumin. *International Journal of Biological Macromolecules*, 150, 480–491. doi: 10.1016/j.ijbiomac.2020.01.291.

Yusoff, N. H., Pal, K., Narayanan, T., & de Souza, F. G. (2021). Characterizations: Polylactic acid (PLA) incorporated with tapioca starch for packaging applications. *Journal of Molecular Structure*, In Press. doi: 10.1016/j.molstruc.2021.129954.

Yusuf, M., Tester, R. F., Ansell, R., & Snap, C. E. (2003). Comparison and properties of starches extracted from tubers of different potato varieties grown under the same environmental conditions. *Food Chemistry*, 82, 283–2816.

Zhang, R., Wang, X., & Cheng, M. (2018). Preparation and characterization of potato starch film with various size of nano-SiO$_2$. *Polymers*, 10(10), 1172. doi: 10.3390/polym10101172.

Zhang, B., Gong, H., Lü, S., Ni, B., Liu, M., Gao, C., Huang, Y., & Han, F. (2012). Synthesis and characterization of carboxymethyl potato starch and its application in reactive dye printing. *International Journal of Biological Macromolecules*, 51(4), 668–674. doi: 10.1016/j.ijbiomac.2012.07.003.

Zhao, Y., & Saldaña, M. D. A. (2019). Use of potato by-products and gallic acid for development of bioactive film packaging by subcritical water technology. *The Journal of Supercritical Fluids*, 143, 97–106. doi: 10.1016/j.supflu.2018.07.025.

Zhao, S., Malfait, W. J., Guerrero-Alburquerque, N., Koebel, M. M., & Nyström, G. (2018a). Biopolymer aerogels and foams: Chemistry, properties, and applications. *Angewandte Chemie International Edition*, 57(26), 7580–7608. doi: 10.1002/anie.201709014.

Zhao, X., Andersson, M., & Andersson, R. (2018b). Resistant starch and other dietary fiber components in tubers from a high-amylose potato. *Food Chemistry*, 251, 58–63.

Zhong, Y., Godwin, P., Jin, Y., & Xiao, H. (2020). Biodegradable polymers and green-based antimicrobial packaging materials: A mini-review. *Advanced Industrial and Engineering Polymer Research*, 3(1), 27–35. doi: 10.1016/j.aiepr.20116.11.002.

17 The Introduction of Biotechnology into Food Engineering

Daniel Ingo Hefft
University Centre Reaseheath

CONTENTS

17.1 INTRODUCTION

Food production systems are complex as we live in an interconnected world where food has become international and supply chains are expanding across the globe. Food manufacturing and production is one of the few industries that serve one of humanity's most essential needs.

Food and water are the foundation for human survival as they satisfy our most basic and physiological needs (Maslow, 1943).

The idea of Maslow's hierarchy of human motivation, which clearly points out how essential food and water are (the foundation of the pyramid), has also been recaptured in the United Nations 2030 Sustainable Development Goals (SDGs) (United Nations, 2015). In fact, 2 of the 16 Sustainable Development Goals are clearly linked to food and water.

These are namely:

- **Goal 2**: End hunger, achieve food security and improved nutrition and promote sustainable agriculture.
- **Goal 6**: Ensure availability and sustainable management of water and sanitation for all.

With predictions forecasting a massive increase in population by 2050 – 7.7 billion in 2019 to 9.7 billion in 2050 (UN Department of Economic and Social Affairs, 2019) food manufacturing and agriculture are under increasing strain to feed this ever-increasing number of humans on a planet of obviously limited resources.

Traditional agriculture and food production looked at a simple expansion model, i.e., for agriculture this would be the exploitation of new land for farming usage to increase production. Since the post-World War II era, this model has shifted towards focusing on practices based on yield increases and a peak of a yield-focused agriculture has been reached in the 1950s/1960s with the so-called Green Revolution (Altieri et al., 2018).

This chapter will explore the introduction of biotechnology into food engineering. After a short introduction to the key concepts, the main focus will be laid on the role of biotechnology in modern

DOI: 10.1201/9781003268468-17

food engineering disciplines. This will also cover insights into the educational landscape with recommendations on why biotechnology should play a bigger role in the training of food engineers worldwide.

17.1.1 Biotechnology

Biotechnology is a composition word of the ancient Greek words (bíos, English "life" and technología, the "art-appropriate treatise on an art or science" (Pape, 1914)).

Biotechnology is an interdisciplinary science that deals with the use of enzymes, cells, and entire organisms in technical applications with goals including the development of new and/or more efficient processes for the production of chemical compounds and of diagnostic methods.

With biotechnology being a very broad term covering numerous areas to research, it is therefore commonly divided into different branches according to the respective areas of application (Table 17.1). Some of these branches overlap so this subdivision is not always clear. In some cases, the border of distinction between biotechnology and other industries becomes blunder (i.e., food engineering, biosystems engineering).

In some cases, the terms are not yet established or are defined differently by different groups of scientists. However, each branch is usually associated with a colour.

Biotechnology is multidisciplinary and combines findings from many subject areas, such as microbiology, molecular biology, genetics, (bio)chemistry, bioinformatics bioprocess engineering.

At its heart, biotechnology is based on chemical reactions that are catalysed by free enzymes or enzymes present in cells (bio-catalysis or bio-conversion). In a traditional sense, this would be the use of yeasts and fungi to process dairy or to ferment beer and wine (Qiao, 2020). Modern biotechnology emerged in the 19th century and moved from a traded knowledge of application-effect practices (i.e., yeasts ferment sugary solutions) into a microbiology-based discipline (i.e., the production of antibiotics from microorganisms) (Barrels-Shallow, 2012). In the middle of the 20th century, this has been expanded by the use of molecular and genetic biological tools as well as genetic engineering capabilities. This latest development has made it possible to use industrial scale and industry competitive manufacturing processes for chemical compounds, allowing large-quantity production of active ingredients for the food, personal care, and pharmaceutical industry. But biotechnology also delivers basic chemicals required for the chemical industry to develop novel diagnostic methods, biosensors, and new plant cultivars (Buchholz and Collins, 2013).

TABLE 17.1

Overview of the Different Branches of Biotechnology and Their Key Application Fields

Branch of Biotechnology	Application Area
Green biotechnology	Agriculture, plant biotechnology (Yashveer et al., 2014)
Blue biotechnology	Biotechnological use of marine resources (DaSilva, 2004)
Red biotechnology	Medicine, pharmaceutical drug development and discovery, medical biotechnology (Gartland et al., 2013)
Violet biotechnology	Law and ethics (Kafarski, 2012)
White biotechnology	Industry, industrial biotechnology (Ulber and Sell, 2007)
Brown biotechnology	Environmental sciences, nature conservation, soil preservation (Rodríguez-Núñez et al., 2020)
Dark biotechnology	Warfare, bio-terrorism (Kafarski, 2012)
Gold biotechnology	Bioinformatics (Kafarski, 2012)
Yellow biotechnology	Food industry(Vilcinskas, 2013)
Grey biotechnology	Waste management, environmental engineering (Siddhartha, 2016)

Considering the education of biotechnologists, biotechnology suffers some issues that are similar to those engineering disciplines are facing with educator's inability to scope the discipline and, in consequence, giving career advice (Atmojo et al., 2018).

17.1.2 FOOD ENGINEERING

Whilst the discipline of biotechnology is well established, the opposite applies for food engineering. Food Engineering as a discipline is fairly young. The discipline roots back to primary production and biosystems engineering (more commonly known in Europe as *agricultural engineering*).

Food Engineering suffers unlike many other engineering disciplines such as mechanical or electrical engineering a lack of definition (Kostaropoulos, 2012), hence the overall profession lacks recognition; despite it playing such an important role in modern society.

A definition of food engineering has been given by Hefft (2019), stating:

> Food Engineering is a technical multidisciplinary profession that deals with the system and structures of food, production processes as well as physical, (bio)chemical, and biological transformation processes. It is based on scientific laws and economical, ecological and social, cultural and religious norms.

This lack of recognition can be showcased impressively by a web search looking at the distribution of various food-related degrees in the UK within a scientific context (undergraduate and postgraduate degrees combined) (Figure 17.1).

It becomes apparent that technology and engineering degrees relating to food are not proportionally represented across food disciplines. This falls in trend with overall issues other engineering disciplines are facing, however, food engineering has been in particular affected (Saguy et al., 2013). A 2018 report by *Engineering UK* highlights this crisis by showcasing that 70% of parents cannot tell what the role of an engineer entails and just over 50% of educators feel confident to give career advice in engineering (Engineering UK, 2018).

This does not only put strain onto an industry crying out for years that there is a shortage of engineering skills in the food community – even at the point where current predictions talk of a permanent gap – (Mayes, 2021; Pendrous, 2016), but also reflects on the following:

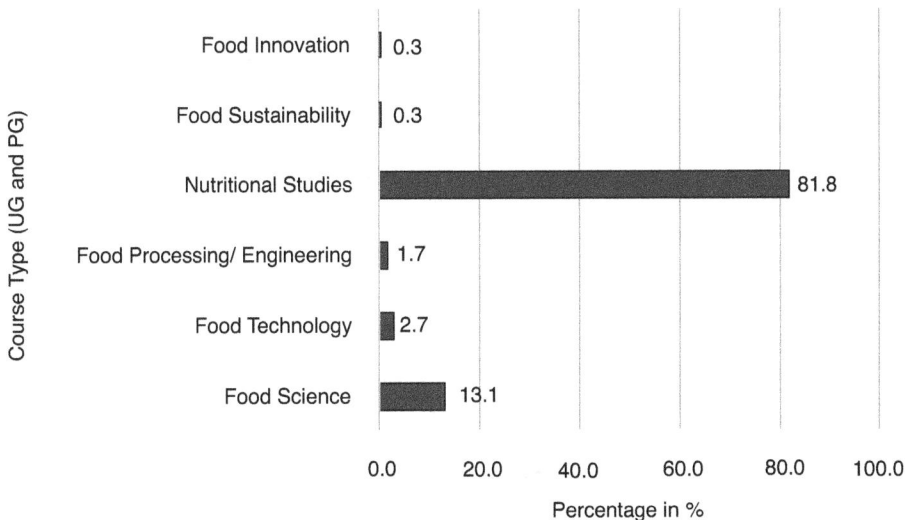

FIGURE 17.1 University degrees of food degrees (undergraduate (UG) and postgraduate (PG) combined) based on a web search (https://www.hotcoursesabroad.com/).

- Decades of outsourcing of engineering activities for cost-saving purposes (Grant Thornton UK LLP, 2017).
- Young people not opting for a career choice in engineering (Becker, 2010).
- Gender and identity crises within engineering (Faulkner, 2007; Hodgkinson and Hamill, 2006).
- Decline in students mathematics and science performance skills (Gottfried et al., 2009; Stokke, 2015; Vigdor, 2013).
- Massive investment shortfalls in the food industry compared to other sectors for Research and Development activities (Department for Business Innovation and Skills, 2010).
- Asynchrony between industry needs and educational offering (Figure 17.1).
- A lack of differentiation of food engineering from other disciplines (Hefft and Higgins, 2021).
- A lack of representation and independence from other engineering disciplines (Hefft and Higgins, 2021).

Given the close relationship between food engineering and biosystems engineering, Figure 17.2 shows a brief historic timeline on the key agricultural revolutions, including an outlook what is to be predicted to be the fifth Agricultural Revolution.

The *British Agricultural Revolution* is a pivotal moment in the history of food production as it marks the time when humanity's agricultural activity moved from subsistence farming to an agricultural system that was able to feed people in cities. Farming has evolved from a mean of survival into a business opportunity (Swaminathan and Philippines, 1987).

With this pivotal moment in primary production, farming and food manufacturing industries fuelled the Industrial Revolution. When comparing this timeline (Figure 17.2) with development in

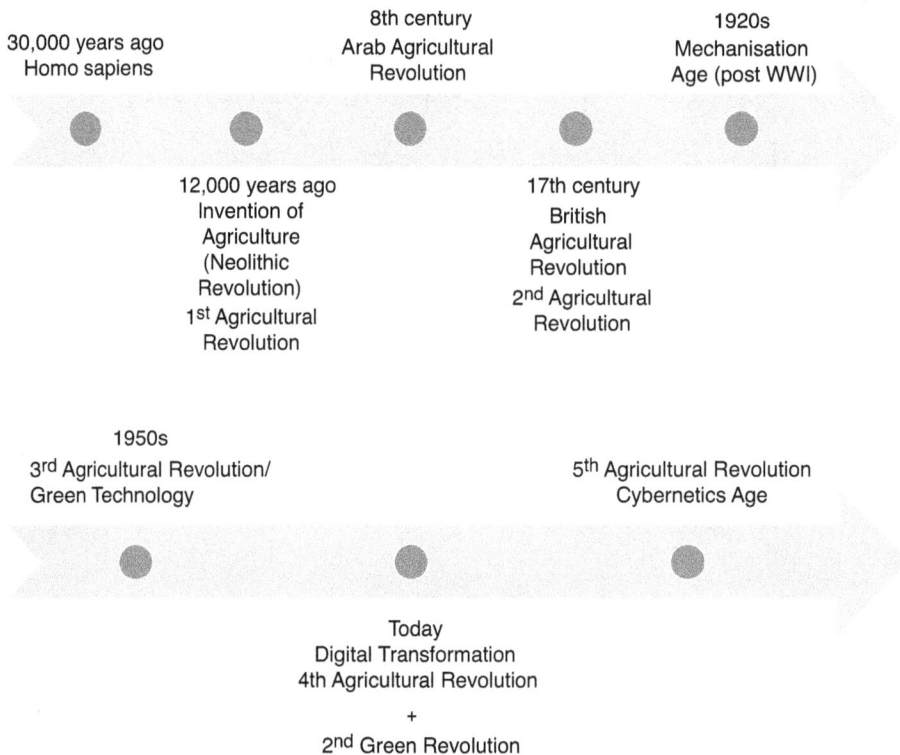

FIGURE 17.2   Agricultural revolutions.

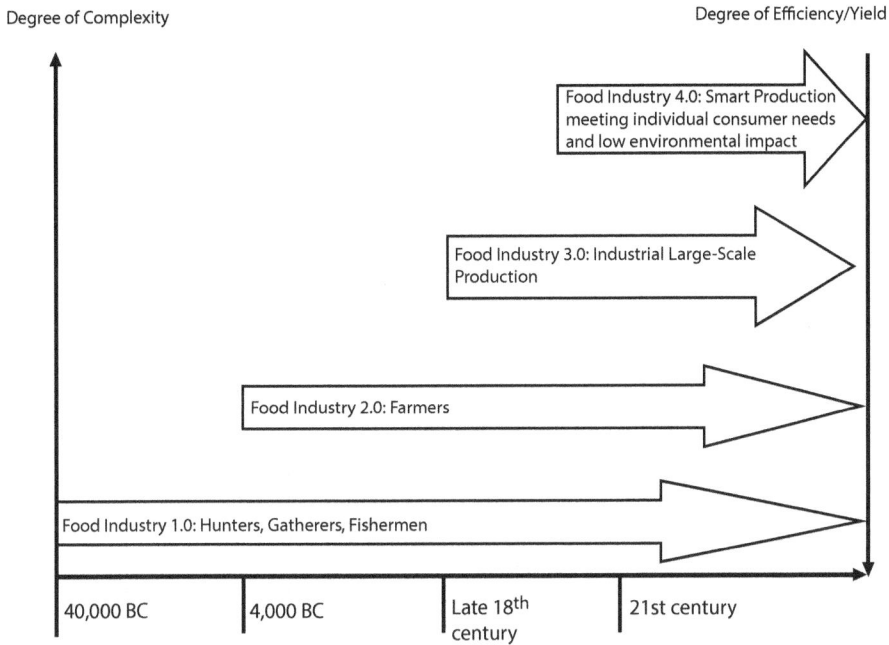

FIGURE 17.3 Evolution of the food industry.

the food industry (Figure 17.3), a clear relationship between evolution in the agricultural sector its impact on food manufacturing.

Again, synergies are apparent between philosophies in the agricultural sector (yield focus at its peak attention in the 1950s) and food manufacturing (introducing drive for cheap, large-scale production of food based on lean manufacturing principles over the past 50 years) (Hefft and Higgins, 2021).

And even looking at the most current developments, food and agriculture follow similar trends with industrial food production/agriculture moving towards diversified individual food production/agro-ecological systems. To meet this individualised trend towards individualised nutrition and products, food engineers will be tasked to develop new processes that can deliver on such agenda. Biotechnology and technologies such as 3D-printing are just a few techniques that are currently under investigation (Rosenthal et al., 2021).

Distinction must be drawn between the application of biotechnological tools and the use of biotechnology in the food industry.

Biotechnology has been present within the food industry for centuries, be it for the fermentation of dough, beer, wine, and dairy, however, this has not followed a disciplined approach as in of a biotechnologist's doing to assist a food handler to manufacture products.

17.2   THE ROLE OF BIOTECHNOLOGY IN MODERN FOOD PRODUCTION

- Baked Goods
- Meats
- Beverages – non-alcoholic
- Beverages – alcoholic

Traditional biotechnology for new foods and beverages - ScienceDirect
The roots—a short history of industrial microbiology and biotechnology | SpringerLink

17.3 CHALLENGES

- Power struggle between Food Eng and Biotechnology
- Educators education
- Ensuring food safety and quality
- Skill updating of current Food Engs

REFERENCES

Altieri, M. A., Farrell, J. G., Hecht, S. B., Liebman, M., Magdoff, F., Murphy, B., Norgaard, R. B., & Sikor, T. O. (2018). *Agroecology*. CRC Press. doi: 10.1201/9780429495465.

Atmojo, I. R. W., Sajidan, S., Ashadi, W., & Nugraha, D. A. (2018). The profiles of pre-service elementary teachers (PETs) in biotechnology using RCB. *International Conference on Science and Applied Science (ICSAS)*, 020063. doi: 10.1063/1.5054467.

Barrels-Shallow, V. (2012). *History of Industrial Microbiology*. New Age Publishers: New Delhi.

Becker, F. S. (2010). Why don't young people want to become engineers? Rational reasons for disappointing decisions. *European Journal of Engineering Education*, 35(4), 349–366. doi: 10.1080/03043797.2010. 489941.

Buchholz, K., & Collins, J. (2013). The roots: A short history of industrial microbiology and biotechnology. *Applied Microbiology and Biotechnology*, 97(9), 3747–3762. doi: 10.1007/s00253-013-4768-2.

DaSilva, E. J. (2004). The colours of biotechnology: Science, development and humankind. *Electronic Journal of Biotechnology*, 7(3), 1–2.

Department for Business Innovation and Skills. (2010). The 2010 R & D Scoreboard.

Engineering UK. (2018). Engineering UK 2018: Synopsis and recommendations.

Faulkner, W. (2007). Nuts and bolts and people. *Social Studies of Science*, 37(3), 331–356. doi: 10.1177/0306312706072175.

Gartland, K., Bruschi, F., Dundar, M., Gahan, P., Viola Magni, M., & Akbarova, Y. (2013). Progress towards the 'Golden Age' of biotechnology. *Current Opinion in Biotechnology*, 24, S6–S13. doi: 10.1016/j.copbio.2013.05.011.

Gottfried, A. E., Marcoulides, G. A., Gottfried, A. W., & Oliver, P. H. (2009). A latent curve model of parental motivational practices and developmental decline in math and science academic intrinsic motivation. *Journal of Educational Psychology*, 101(3), 729–739. doi: 10.1037/a0015084.

Grant Thornton UK LLP. (2017). FDF Economic contribution and growth opportunities. https://www.fdf.org.uk/publicgeneral/FDF-Economic-contribution-Full-report.pdf.

Hefft, D. I. (2019). Engineering for Food. Institution of Agricultural Engineers. https://iagre.org/food-engineering-group.

Hefft, D. I., & Higgins, Ś. (2021). Food industry and engineering: Quo vadis? *Journal of Food Process Engineering*, 44(8). doi: 10.1111/jfpe.13766.

Hodgkinson, L., & Hamill, L. (2006). Engineering careers in the UK: Still not what women want? *Industry and Higher Education*, 20(6), 403–412. doi: 10.5367/000000006779882986.

Kafarski, P. (2012). Rainbow code of biotechnology. *Chemik*, 66(8), 814–816.

Kostaropoulos, A. E. (2012). Food engineering within sciences of food. *International Journal of Food Studies*, 1(2), 109–113.

Maslow, A. H. (1943). A theory of human motivation. *Psychological Review*, 50(4), 370–396. doi: 10.1037/h0054346.

Mayes, J. (2021). U.K. faces 'permanent' gaps in food supply, industry chief warns. *Bloomberg*. https://www.bloomberg.com/news/articles/2021-09-10/u-k-faces-permanent-gaps-in-food-supply-industry-chief-warns.

Pape, W. (1914). *Handwörterbuch der griechischen Sprache* (3rd ed.). Vieweg & Sohn: Braunschweig.

Pendrous, R. (2016). Food sector needs more engineers urgently. Food Manufacture. https://www.foodmanufacture.co.uk/Article/2016/10/14/More-food-and-drink-engineers-needed-urgently.

Qiao, W. (2020). Realization and practical application analysis of lactic acid bacteria fermentation bioengineering. *IOP Conference Series: Earth and Environmental Science*, 615, 012111. doi: 10.1088/1755-1315/615/1/012111.

Rodríguez-Núñez, K., Rodríguez-Ramos, F., Leiva-Portilla, D., & Ibáñez, C. (2020). Brown biotechnology: A powerful toolbox for resolving current and future challenges in the development of arid lands. *SN Applied Sciences*, 2(7), 1187. doi: 10.1007/s42452-020-2980-0.

Rosenthal, A., Maciel Guedes, A. M., dos Santos, K. M. O., & Deliza, R. (2021). Healthy food innovation in sustainable food system 4.0: Integration of entrepreneurship, research, and education. *Current Opinion in Food Science*, 42, 215–223. doi: 10.1016/j.cofs.2021.07.002.

Saguy, I. S., Singh, R. P., Johnson, T., Fryer, P. J., & Sastry, S. K. (2013). Challenges facing food engineering. *Journal of Food Engineering*, 119(2), 332–342. doi: 10.1016/j.jfoodeng.2013.05.031.

Siddhartha, M. (2016). Innovating beyond medical sciences with (Internet of Things) IoT. *GJESR*, 3(9).

Stokke, A. (2015). What to do about Canada's declining math scores? *SSRN Electronic Journal*. doi: 10.2139/ssrn.2613146.

Swaminathan, M. S., & Philippines, L. B. (1987). The green revolution. In *The Future Development of Maize and Wheat in the Third World*. International Maize and Wheat Improvement Center: Mexico, pp. 1–27.

Ulber, R., & Sell, D. (Eds.). (2007). *White Biotechnology* (Vol. 105). Springer: Berlin Heidelberg. doi: 10.1007/978-3-540-45696-4.

UN Department of Economic and Social Affairs. (2019). World Population Prospects 2019.

United Nations. (2015). Transforming our world: The 2030 Agenda for Sustainable Development.

Vigdor, J. L. (2013). Solving America's math problem: Tailor instruction to the varying needs of the students. *Education Next*, 13, 42. https://link.gale.com/apps/doc/A313012647/AONE?u=anon~7578e6ee&sid=googleScholar&xid=a7be9c79.

Vilcinskas, A. (Ed.). (2013). *Yellow Biotechnology I* (Vol. 135). Springer: Berlin Heidelberg. doi: 10.1007/978-3-642-39863-6.

Yashveer, S., Singh, V., Kaswan, V., Kaushik, A., & Tokas, J. (2014). Green biotechnology, nanotechnology and bio-fortification: Perspectives on novel environment-friendly crop improvement strategies. *Biotechnology and Genetic Engineering Reviews*, 30(2), 113–126. doi: 10.1080/02648725.2014.992622.

18 Plant Resident Microorganisms

A Boon for Plant Disease Management

Ajit Kumar Savani
Assam Agricultural University

E. Rajeswari
PJTSAU

Bandana Saikia and Ashok Bhattacharyya
Assam Agricultural University

K. Dinesh
University of Agricultural Sciences

CONTENTS

18.1 INTRODUCTION

Endophyte is derived from the Greek word 'endon' (within) and 'phyte' (plant). The term endophyte refers to interior colonization of plants by bacterial or fungal microorganisms. Petrini (1991) first defined endophyte as microorganism living in the plant organization for a certain stage of its life and would not cause disease. Anton de Bary coined the term endophyte in 1886 to describe microorganisms that colonize internal tissues of stems and leaves. Perotti (1926), Hallmann et al. (1997) and Azevedo et al. (2000) reported that bacteria on roots and in the rhizosphere benefit from root exudates, but some bacteria and fungi are capable of entering the plant as endophytes that do not cause harm and could establish a mutualistic association. Wagenaar and Clardy (2001) identified endophytes as microorganisms growing in the intercellular spaces of higher plants and they are recognized as one of the most

DOI: 10.1201/9781003268468-18

chemically promising groups of microorganisms in terms of diversity and pharmaceutical potential. James and Olivares (1998) stated that all bacteria that colonize the interior of plants, including active and latent pathogens, can be considered to be as endophytes. Kado (1992) and Quispel (1992) suggested that endophytic bacteria establish endosymbiosis with the plant, whereby the plant receives an ecological benefit from the presence of the symbiont. It is now commonly accepted that for each of the nearly 300,000 existing plant species host, at least one or even several hundred strains of endophytes have been reported (Strobel and Daisy, 2003). Endophytic fungi belong to the phylum Ascomycota and they were often closely related to fungi known to cause diseases, either in healthy tissue or as secondary invaders of damaged tissue (Schardl and Tsai, 1993). This suggests that endophytes may have evolved from pathogens or vice versa. It has proven impacts on plant growth promotion, biological control and alleviation of abiotic stresses. Exploring the huge diversity of endophytes can bring potential strains which add to the valuable genetic stock of biological agents for sustainable agriculture.

The size of the tissue fragment taken for isolation also affects the number of endophytic fungi that appeared on the media. Reducing the size and increasing the number of leaf fragments will increase the number of fungal species isolated (Gamboa et al., 2003). Almost all studies relating to endophytes of different plant species are based on, one-time sampling of the plant tissue (mainly, the leaf) resulting in a snapshot record of its endophyte community. Although such sampling may be useful for comparing endophyte assemblages of plants of different lineages or geographic locations, such single screenings do not reveal the dynamics of endophyte communities. Suryanarayanan and Thennarasan (2004) showed that the species composition of the foliar endophyte assemblage of a tropical tree changes significantly with time. For example, colonization by foliar endophytes in *Fraxinusexcelsi* is heavily dependent on microclimatic factors and leaf characteristics resulting in highly variable spatial and temporal distribution patterns of them in an individual tree (Scholtysik et al., 2013). Thus, periodic sampling is essential to obtain complete information on the endophyte community status of a plant. Existence of balanced antagonism between the virulence of colonizing endophytes and the plant defense response has been reported to be a key factor involved in the endophytic relationship without producing any apparent disease symptom (Maciá-Vicente et al., 2009).

18.2   ECOLOGY OF ENDOPHYTES

Geographical and seasonal factors affect the assemblages of endophytic fungi as they are subjected to different selection pressure at each ecological niche (Petrini, 1991). Saxena et al. (2019) critically noticed that plants growing in greater biodiversity and special habitats like deteriorated ecological environment conditions, (*Piriformosporaindica* isolated from cactus in the deteriorated environment) plants surrounded by pathogen-infected plants showing no symptoms, plants that have been exploited for human use as traditional medicines in some places should be considered and plants occupied under certain ancient landmass were also likely to lodge good endophytes.

18.2.1   POPULATION DYNAMICS OF ENDOPHYTES IN DIFFERENT PLANT PARTS

The endophytic microbial community inhabits different plant structures, such as leaves, petioles, reproductive structures, twigs, bark and roots (Rodriguez et al., 2009). Investigations were carried out on different organs of a wide variety of plant species of economic interest, including food crops such as common bean (*Phaseolus vulgaris* L.), Cocoa (*Theobroma cacao* L.), soyabean (*Glycine max* (L.) Merr.) and wheat (*Triticum aestivum*) had revealed the diversity of the microbial population. Endophytes are present in different plant parts but greater diversity was reported from root tissues in some of the studies (Saxena et al., 2019). Specificity and variability in the distribution of endophytic microorganisms in different tissues of the medicinal plants have clearly indicated the variation in diversity of microbes is influenced by different chemical compositions in plant tissue (Jalgaonwala et al., 2017).

However, the composition of endophytic microorganisms may be greatly influenced by the habitat of the plants (Suryanarayanan, 2013). Host defence metabolites exert a selection pressure in

the evolution of endophytism. A comparison of the endophyte diversity of varieties of plant species indicated a difference in antifungal compound composition may influence the microbial load (Mueller-Harvey and Dohana, 1991), and their tolerance to such chemicals would be of heuristic value. Maize tissues are rich in antifungal benzoxazinoids but harbour endophytes that are tolerant to those compounds (Saunders and Kohn, 2009). Recent studies have dispelled that horizontal gene transfer events from plant hosts to endophytes are responsible for the heightened biosynthetic ability of endophytes (Heinig et al., 2013; Sachin et al., 2013).

18.2.2 INFLUENCE OF CLIMATE AND TOPOGRAPHY ON POPULATION DIVERSITY OF ENDOPHYTES

Symbiotic plant-fungal interactions are of widespread interest to ecological research as they influence important ecosystem processes including plant productivity, plant diversity, and plant pathogen interactions (Van Bael et al., 2012). Many environmental factors influence the plant endophyte interactions; however host plant response to endophyte infection is mainly mediated by the host genotype, endophytic strain, resource availability (Hesse et al., 2003; Singh et al., 2011; Qawasmeh et al., 2012; Wani et al., 2015). Tolerance to abiotic factors such as thermo and salt tolerance is observed in certain plants colonized with endophytes (Redman, 2002; Waller et al., 2005).

Plants growing in different geographical regions are confronted with different environmental challenges (Arnold and Lutzoni, 2007). The diversity of endophytes associated with the plant varies not only temporally but spatially as well (Herera et al., 2010; Ek-ramos et al., 2013). For instance, studies showed that endophytes may increase in incidence, diversity, and host breadth as a function of latitude (Arnold and Lutzoni, 2007). Furthermore, endophyte communities from high latitude were characterized by relatively few fungal species representing several classes of Ascomycota, whereas tropical endophytes assemblages were dominated by a small number of classes. Different plants growing in similar environmental conditions do not harbor the same endophytes. Endophyte diversity is expected to be high in tropical forests as they support a high diversity of plant hosts (Zimmerman and Vitousek, 2012).

Few studies have analyzed the effect of different environment variables on endophyte diversity. Plant endophyte population also depends on many variables such as plant growth stage, the plant tissue analyzed; the health of the plant; the nutritional state of the plant; the type of soil and its condition (including pH and moisture content); the altitude; the temperature; *etc.* (Hardoim et al., 2008). It has been evident that the abundance and hyper diversity of endophytes in tropical forests, rainy slopes of mountains and a positive correlation with increasing annual precipitation (Carroll and Carroll, 1978; Rodrigues and Samuels, 1999; Holmes et al., 2004; Bulgari et al., 2011).

18.3 METHOD OF ISOLATION

Isolation of endophytes and confirming the identity of the isolated fungi or bacteria as exact endophytes or some surface contaminants is becoming an important matter of discussion, whenever any researchers are engaged in working with endophytes. Endophytes are isolated by removing surface contamination and plating the content of plants' inner parts on a suitable agar medium. Isolation of endophytes is one of the most important steps in endophyte research as it ensures that no surface contamination is isolated as endophytes. Various methods are being used for endophyte isolation. The method of surface sterilization should be strong enough to remove the entire surface microflora; at the same time, it should be gentle on endophytes (i.e. surface sterilants should not kill the endophytes). This may be a little difficult to achieve because if we use very strong surface sterilization to ensure complete removal of surface microflora, it kills some of the endophytes near to surface. On the other hand, if we use relatively dilute sterilants in order to "not kill" any endophyte near to surface, the surface microflora may not be eradicated. The culture-dependent surface sterilization methods, may not eradicate surface organisms. Therefore, the method used for surface sterilization should be standardized according to different types of tissue and sterilants used, so that eradication of surface flora can be achieved with minimum harm to the endophyte population (Figure 18.1).

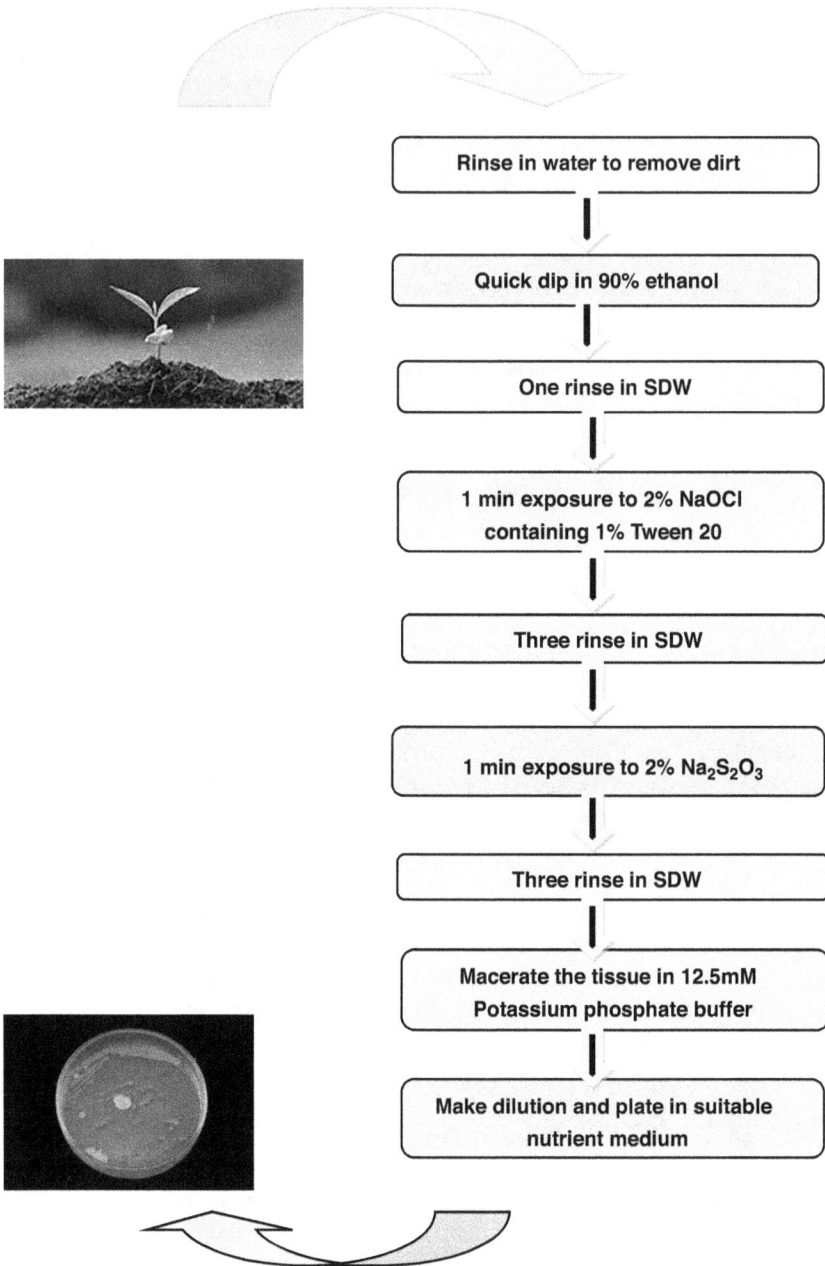

FIGURE 18.1　Isolation of bacterial endophytes (conceptualized from Upreti and Thomas, 2015 and Zinniel et al., 2002; the protocol has been illustrated by taking tomato as model plant, it can vary with plant species; SDW = sterile distilled water, NaOCl = Sodium hypochloride, $Na_2S_2O_3$ = Sodium thiosulphate)

Plantlets were washed to harvest roots and shoot with the least damage. Plantlets were surface sterilized using the standard protocol, root and shoots are separated and grounded in phosphate buffer. The filtrate is plated on nutrient agar (NA) plates (strictly monitored for up to a week to ensure freedom from all accidental contaminants) in appropriate dilutions. Tissue imprints of surface-sterilized roots and shoots were also plated to ensure the elimination of surface microflora (Sessitsch et al., 2002). Here, care should be taken to judgment sterilization, as the tissue imprint of a surface sterile tissue may also give microbial colony from cut edges of tissue because sap coming

out from tissue contains some endophytes. These should not be considered as contaminants. Plates after incubation at 30°C were observed for distinct colony types for 2–7 days. All colonies should be categorized based on surface morphology. The most common method used for the isolation of endophytes *i.e.* fungi (McInroy and Kloepper, 1995; Hallmann et al., 1997) and bacteria (Zinniel et al., 2002; Upreti and Thomas, 2015).

18.4   MODE OF ENTRY

Endophyte gets the benefit of the niche close to the host which bypasses the complex competition for food and space with other microbes as it happens in the rhizosphere. Research has been conducted to confer the mode of entry of endophytes to the plants. Apart from seed-transmitted endophytes which are already present in plants, root is the major source of entry. However, there are other possible sites like in fruits endophytes can enter through flowers, natural openings of leaves (stomata or lenticels).

Entry of *Burkholderia* sp. Strain PsJN in *Vitisvinifera* L. cv. Chardonnay was determined under gnotobiotic conditions. The visualization of the mode of entry and colonization was done by tagging with gfp (PsJN::gfp2x) or gusA (PsJN::gusA11) gene. Secretion of cell wall degrading endoglucanase and endo-polygalacturonase was observed as a mode of entry (Compant et al., 2005). Sharma et al. (2008) studied the colonization pattern of one endophyte isolate in wheat by tagging it with gusA/gfp genes. They found that cracks developed near lateral root emergence are major sites from where the endophyte enters and spreads to intercellular spaces as well as vascular bundles. Similar mode of endophyte entry was also observed in rice roots; however, external application of endophytes tagged with radioisotopes active P^{32} endophytic bacteria inoculated on cocoa pods and roots has clearly indicated that endophytic bacteria colonize in the plant system if applied externally (Kurian et al., 2012).

18.5   APPLICATIONS IN AGRICULTURE

Microbes with tremendous capacities for plant growth promotion, biocontrol and abiotic stress alleviation are being explored in rhizosphere, endosphere, phyllosphere and unique ecological niches. In past decades, promoting eco-friendly agriculture had explored a lot on rhizospheric microflora for plant growth and health promotion and recently the focus is shifting to explore endophytes to enhance agriculture production.

18.5.1   Biotic Stress Alleviation

It is not evident that all endophytes are involved in protection against pathogens, but reports indicate strongly that the endophytes are having huge potential to be used as excellent biocontrol agents as they endorse disease tolerance against a wide array of plant pathogens. Induced systemic resistance (ISR), production of antifungal compounds- surfectin, fungicin, proteases, chitinases, siderophore production, competition for nutrients, volatile organic compounds, etc. are few of the major mechanisms of pathogen suppression by endophytes.

Since induction of systemic resistance was reported first time in bacteria in *Pseudomonas fluorescens* strain G8-4 against anthracnose disease of cucumber, massive efforts have been poured to harness it for suppressing plant pathogens. Induction of systemic resistance against pathogens involves the interaction of plants with microbial-associated molecular patterns (MAMPs) present in beneficial microbes. Endophytes are reported to suppress plant pathogens by various mechanisms like production of ammonia, hydrogen cyanide (HCN), volatile organic compounds (VOCs), antibiotics, biosurfactants, oxidative enzymes, hydrolytic enzymes, antioxidants, secondary metabolites, siderophore, quorum sensing degraders, antimicrobial allelochemicals, etc. In the process of making more effective biological inoculants, it is imperative to explore the hidden potential of endophytes. Endosphere of plants is one of the potential niches to be explored for novel secondary metabolites and biologically active compounds. This could be a useful source of biological inputs

in agriculture. As a biocontrol agent endophytes could form an extra line of defense against plant pathogens attacking their host. Endophytes also obtain the benefit of being close to plants and not deprived of rhizosphere microflora (Sturz et al., 2000). Bacterial endophytes provide additional mechanisms like niche exclusion, production of novel secondary metabolites, direct antagonism inside the plants, barrier effects for vascular pathogens, etc., than rhizospheric microorganisms (Rosenblueth and Martínez-Romero, 2006). All such effects make endophytes a suitable candidate to be used as biological inoculants in the agriculture production system for biological control. The effect of endophytes is always referred to as a combined defect of the interaction of microorganisms and its host plant; therefore, it could be more suitable for sustainable agriculture.

18.5.2 ABIOTIC STRESS ALLEVIATION

Plants have the internal capability to withstand a certain amount of salt stress. Some of the beneficial microbes also enhance the plant tolerance to such stresses by multiple times. Mayak et al. (2004) have reported the use of plant growth promoting bacteria for salinity alleviation, by studying ACC-deaminase producing *Achromobcterpiechaudii* strain enhanced growth and production of tomato seedlings up to an extent of 172 mM concentration of sodium chloride. Palaniyandi et al. (2014) have studied salinity stress –alleviating activity of streptomyces strain PGPA39 was evaluated using "Micro Tom" tomato plants with 180 mmol 1-1 Nacl stress under gnotobiotic condition. A significant increase in plant biomass and chlorophyll content and a reduction in leaf proline content were observed in PGPA39-inoculated tomato plants under salt stress compared with control and salt stress non-inoculated plants. Endophytes have great potential to become successful inoculants for agriculture production by having novel secondary metabolites and bioactive compounds. The use of *Rhizobium* in legume production is an established practice for dinitrogen fixation. There are several other reports of dinitrogen fixing endophytes from cereal crops like rice, wheat, sugarcane, etc. Endophytes like *Glucanoacetobacter, Azoarcus*spp., *Herbaspirillum* spp., *Pseudomonas stutzeri, Azorhizobiumcaulinodans,* etc are reported from cereals. Endophytic inoculants are reported to solubilize mineral nutrients in the rhizosphere. This enhances the performance of crops as well as reduces the necessity of adding a higher amount of chemical fertilizers. Phytoharmones control the physiology and performance of plants and endophytes enhances plant growth by modulating these phytoharmone-producing bacterial endophyte for enhanced growth of plants.

18.6 CONCLUSION

The changes that occurred during the green revolution have increased the usage of pesticides and chemicals in agriculture production. This continuous usage of chemicals over a long period had lead to deleterious effects on soil and brought concerns regarding ecology and environment. On other hand, pesticides are toxic to health and also increase resistance to the pathogen. Hence, to maintain sustainable agriculture production and to overcome biotic and abiotic stress, it is essential to implement environmentally safe approaches in plant health management. Among these, the usage of endophytes will be a highly potential weapon for maintaining plant health management. Endophytes are very potential in controlling diseases, preventing emerging diseases, abitotic stress and improving plant growth and health when used wisely at proper time. The practical usage of endophytes in plant protection is yet to gain momentum. Still there is a need for a better understanding of the mechanism of endophytes in defense signaling in host plant and about host plant microbiome. Rachel Carson has rightly stated in her book "Silent spring" which has made to think about the nature by many business tycoons all over the world as "In nature nothing exists alone". Anthropogenic interventions should exploit the relations for the benefit of mankind instead of ruining the human existence by disturbing the relations, this might be most probably a great gift that we can give to our future generations, instead of a barren planet.

REFERENCES

Arnold, A. E., & Lutzoni, F. (2007). Diversity and host range of foliar fungal endophytes: Are tropical leaves biodiversity hotspots? *Ecology, 88*(3), 541–549. doi: 10.1890/05-1459.

Azevedo, J. L., Maccheroni Jr., W., Pereira, J. O., & De Araújo, W. L. (2000). Endophytic microorganisms: A review on insect control and recent advances on tropical plants. *Electronic Journal of Biotechnology, 3*(1), 40–46. doi: 10.2225/vol3-issue1-fulltext-4.

Bulgari, D., Casati, P., Crepaldi, P., Daffonchio, D., Quaglino, F., Brusetti, L., & Bianco, P. A. (2011). Restructuring of endophytic bacterial communities in grapevine yellows-diseased and recovered vitis-vinifera L. plants. *Applied and Environmental Microbiology, 77*(14), 5018–5022. doi: 10.1128/aem.00051-11.

Carroll, G. C., & Carroll, F. E. (1978). Studies on the incidence of coniferous needle endophytes in the Pacific Northwest. *Canadian Journal of Botany, 56*(24), 3034–3043. doi: 10.1139/b78-367.

Compant, S., Reiter, B., Sessitsch, A., Nowak, J., Clément, C., & Ait Barka, E. (2005). Endophytic colonization of vitisvinifera L. by plant growth-promoting bacterium Burkholderia Sp. Strain PsJN. *Applied and Environmental Microbiology, 71*(4), 1685–1693. doi: 10.1128/aem.71.4.1685-1693.2005.

Ek-Ramos, M. J., Zhou, W., Valencia, C. U., Antwi, J. B., Kalns, L. L., Morgan, G. D., Kerns, D. L., & Sword, G. A. (2013). Spatial and temporal variation in fungal endophyte communities isolated from cultivated cotton (Gossypiumhirsutum). *PLoS One, 8*(6), e66049. doi: 10.1371/journal.pone.0066049.

Gamboa, M. A., Laureano, S., & Bayman, P. (2003). Measuring diversity of endophytic fungi in leaf fragments: Does size matter? *Mycopathologia, 156*(1), 41–45. doi: 10.1023/a:1021362217723.

Hallmann, J., Quadt-Hallmann, A., Mahaffee, W. F., & Kloepper, J. W. (1997). Bacterial endophytes in agricultural crops. *Canadian Journal of Microbiology, 43*(10), 895–914. doi: 10.1139/m97-131.

Hardoim, P. R., Van Overbeek, L. S., & Elsas, J. D. (2008). Properties of bacterial endophytes and their proposed role in plant growth. *Trends in Microbiology, 16*(10), 463–471. doi: 10.1016/j.tim.2008.07.008.

Heinig, U., Scholz, S., & Jennewein, S. (2013). Getting to the bottom of taxol biosynthesis by fungi. *Fungal Diversity, 60*(1), 161–170. doi: 10.1007/s13225-013-0228-7.

Herrera, J., Khidir, H. H., Eudy, D. M., Porras-Alfaro, A., Natvig, D. O., & Sinsabaugh, R. L. (2010). Shifting fungal endophyte communities colonize Bouteloua gracilis: effect of host tissue and geographical distribution. *Mycologia, 102*(5), 1012–1026.

Hesse, U., Schöberlein, W., Wittenmayer, L., Förster, K., Warnstorff, K., Diepenbrock, W., & Merbach, W. (2003). Effects of neotyphodiumendophytes on growth, reproduction and drought-stress tolerance of three Loliumperenne L. genotypes. *Grass and Forage Science, 58*(4), 407–415. doi: 10.1111/j.1365-2494.2003.00393.x.

Holmes, K. A., Schroers, H., Thomas, S. E., Evans, H. C., & Samuels, G. J. (2004). Taxonomy and biocontrol potential of a new species of Trichoderma from the Amazon basin of South America. *Mycological Progress, 3*(3), 199–210. doi: 10.1007/s11557-006-0090-z.

Jalgaonwala, R. E., Mohite, B. V., & Mahajan, R. T. (2017). A review: Natural products from plant associated endophytic fungi. *Journal of Microbiology and Biotechnology Research, 1*(2), 21–32.

James, E. K., & Olivares, F. L. (1998). Infection and colonization of sugar cane and other graminaceous plants by endophyticDiazotrophs. *CriticalReviewsinPlantSciences,17*(1),77–119.doi:10.1080/07352689891304195.

Kado, C. I. (1992). Plant pathogenic bacteria. In: Ballows, A., Trüper, G. G., Dworkin, M., Harder, W., & Schleifer, K. H. (Eds.), *The Prokaryotes.* Springer-Verlag: NewYork, pp. 660–662.

Kurian, P. S., Abraham, K., & Kumar, P. S. (2012). Endophytic bacteria–do they colonize within the plant tissues if applied externally?. *Current Science, 103*(6), 626–628.

Maciá-Vicente, J., Rosso, L., Ciancio, A., Jansson, H., & Lopez-Llorca, L. (2009). Colonisation of Barley roots by endophytic Fusariumequiseti and Pochoniachlamy dosporia: Effects on plant growth and disease. *Annals of Applied Biology, 155*(3), 391–401. doi: 10.1111/j.1744-7348.2009.00352.x.

Mayak, S., Tirosh, T., & Glick, B. R. (2004). Plant growth-promoting bacteria that confer resistance to water stress in tomatoes and peppers. *Plant Science, 166*(2), 525–530. doi: 10.1016/j.plantsci.2003.10.025.

McInroy, J. A., & Kloepper, J. W. (1995). Survey of Indigenous bacterial endophytes from cotton and sweet corn. *Plant and Soil, 173*(2), 337–342. doi: 10.1007/bf00011472.

Mueller-Harvey, I., & Dhanoa, M. S. (1991). Varietal differences among sorghum crop residues in relation to their phenolic HPLC fingerprints and responses to different environments. *Journal of the Science of Food and Agriculture, 57*(2), 199–216. doi: 10.1002/jsfa.2740570206.

Palaniyandi, S., Damodharan, K., Yang, S., & Suh, J. (2014). Streptomyces Sp. strain PGPA39 alleviates salt stress and promotes growth of 'Micro Tom' tomato plants. *Journal of Applied Microbiology, 117*(3), 766–773. doi: 10.1111/jam.12563.

Perotti, R. (1926). On the limits of biological enquiry in soil science. *Proceedings of International Society of Soil Science*, *1*(2), 146–161.

Petrini, O. (1991). Fungal endophyte of three leaves. In: Andrews, J. & Hirano, S. S., (Eds.), *Microbial Ecology of Leaves*. Spring-Verlag: New York, pp. 179–197.

Qawasmeh, A., Obied, H. K., Raman, A., & Wheatley, W. (2012). Influence of fungal endophyte infection on phenolic content and antioxidant activity in grasses: Interaction between loliumperenne and different strains of Neotyphodiumlolii. *Journal of Agricultural and Food Chemistry*, *60*(13), 3381–3388. doi: 10.1021/jf204105k.

Quispel, A. (1992). A search of signals in endophytic microorganisms. In: Verma, D. P. S. (Ed.), *Molecular Signals in Plant-Microbe Communications*. Boca Raton, FL: CRC Press, Inc, pp. 471–491.

Redman, R. S. (2002). Thermotolerance generated by plant/Fungal symbiosis. *Science*, *298*(5598), 1581. doi: 10.1126/science.1072191.

Rodrigues, K. F., & Samuels, G. J. (1999). Fungal endophytes of Spondias mombin leaves in Brazil. *Journal of Basic Microbiology: An International Journal on Biochemistry, Physiology, Genetics, Morphology, and Ecology of Microorganisms*, *39*(2), 131–135.

Rodriguez, R. J., White Jr, J. F., Arnold, A. E., & Redman, R. S. (2009). Fungal endophytes: Diversity and functional roles. *New Phytologist*, *182*(2), 314–330. doi: 10.1111/j.1469-8137.2009.02773.x.

Rosenblueth, M., & Martínez-Romero, E. (2006). Bacterial endophytes and their interactions with hosts. *Molecular Plant-Microbe Interactions*, *19*(8), 827–837. doi: 10.1094/mpmi-19-0827.

Sachin, N., Manjunatha, B.L.,P. MohanaKumara, G., Ravikanth, S., Shweta, T.S., Suryanarayanan, K.N., Ga neshaiah, R., & Uma Shaanker (2013). Do endophytic fungi possess pathway genes for plant secondary metabolites? *Current Science*, *104*, 178–182.

Saunders, M., & Kohn, L. M. (2009). Evidence for alteration of fungal endophyte community assembly by host defense compounds. *New Phytologist*, *182*(1), 229–238. doi: 10.1111/j.1469-8137.2008.02746.x.

Saxena, A. K., Sahu, P. K., Singh, U. B., Chakdar, H., & Bagul, S. Y. (2019). Bacterial endophytes in agriculture: Concepts to application-A training manual. Published by Director, ICAR-National Bureau of Agriculturally Important Microorganisms, Kushmaur, Maunath Bhanjan, Uttar Pradesh (India). 2009, 1–123.

Schardl, C. L., & Tsai, H. (1993). Molecular biology and evolution of the grass endophytes. *Natural Toxins*, *1*(3), 171–184. doi: 10.1002/nt.2620010305.

Scholtysik, A., Unterseher, M., Otto, P., & Wirth, C. (2013). Spatio-temporal dynamics of endophyte diversity in the canopy of European ash (Fraxinus excelsior). *Mycological Progress*, *12*(2), 291–304. doi: 10.1007/s11557-012-0835-9.

Sessitsch, A., Reiter, B., Pfeifer, U., & Wilhelm, E. (2002). Cultivation-independent population analysis of bacterial endophytes in three potato varieties based on eubacterial and Actinomycetes-specific PCR of 16S rRNA genes. *Fems Microbiology Ecology*, *39*(1), 23–32.

Sharma, M., Schmid, M., Rothballer, M., Hause, G., Zuccaro, A., Imani, J., Kämpfer, P., Domann, E., Schäfer, P., Hartmann, A., & Kogel, K. (2008). Detection and identification of bacteria intimately associated with fungi of the order Sebacinales. *Cellular Microbiology*, *10*(11), 2235–2246. doi: 10.1111/j.1462-5822.2008.01202.x.

Singh, L. P., Gill, S. S., & Tuteja, N. (2011). Unraveling the role of fungal symbionts in plant abiotic stress tolerance. *Plant Signaling & Behavior*, *6*(2), 175–191. doi: 10.4161/psb.6.2.14146.

Strobel, G., & Daisy, B. (2003). Bioprospecting for microbial endophytes and their natural products. *Microbiology and Molecular Biology Reviews*, *67*(4), 491–502.

Sturz, A. V., Christie, B. R., & Nowak, J. (2000). Bacterial endophytes: Potential role in developing sustainable systems of crop production. *Critical Reviews in Plant Sciences*, *19*(1), 1–30. doi: 10.1080/07352680091139169.

Suryanarayanan, T. S. (2013). Endophyte research: Going beyond isolation and metabolite documentation. *Fungal Ecology*, *6*(6), 561–568. doi: 10.1016/j.funeco.2013.09.007.

Suryanarayanan, T. S., & Thennarasan, S. (2004). Temporal variation in endophyte assemblages of Plumeriarubra leaves. *Fungal Diversity*, *15*, 197–204.

Upreti, R., & Thomas, P. (2015). Root-associated bacterial endophytes from Ralstoniasolanacearum resistant and susceptible tomato cultivars and their pathogen antagonistic effects. *Frontiers in Microbiology*, *6*, 655. doi: 10.3389/fmicb.2015.00255.

Van Bael, S. A., Estrada, C., Rehner, S. A., Santos, J. F., & Wcislo, W. T. (2012). Leaf endophyte load influences fungal garden development in leaf-cutting ants. *BMC Ecology*, *12*(1), 12–23. doi: 10.1186/1472-6785-12-23.

Wagenaar, M. M., & Clardy, J. (2001). Dicerandrols, new antibiotic and cytotoxic dimers produced by the fungus Phomopsislongicolla isolated from an endangered mint. *Journal of Natural Products*, *64*(8), 1006–1009. doi: 10.1021/np010020u.

Waller, F., Achatz, B., Baltruschat, H., Fodor, J., Becker, K., Fischer, M., Heier, T., Huckelhoven, R., Neumann, C., von Wettstein, D., Franken, P., & Kogel, K. (2005). The endophytic fungus Piriformosporaindica reprograms Barley to salt-stress tolerance, disease resistance, and higher yield. *Proceedings of the National Academy of Sciences, 102*(38), 13386–13391. doi: 10.1073/pnas.0504423102.

Wani, Z. A., Ashraf, N., Mohiuddin, T., & Riyaz-Ul-Hassan, S. (2015). Plant-endophyte symbiosis, an ecological perspective. *Applied Microbiology and Biotechnology, 99*(7), 2955–2965. doi: 10.1007/s00253-015-6487-3.

Zimmerman, N. B., & Vitousek, P. M. (2012). Fungal endophyte communities reflect environmental structuring across a Hawaiian landscape. *Proceedings of the National Academy of Sciences, 109*(32), 13022–13027. doi: 10.1073/pnas.1209872109.

Zinniel, D. K., Lambrecht, P., Harris, N. B., Feng, Z., Kuczmarski, D., Higley, P., Ishimaru, C. A., Arunakumari, A., Barletta, R. G., & Vidaver, A. K. (2002). Isolation and characterization of endophytic colonizing bacteria from agronomic crops and prairie plants. *Applied and Environmental Microbiology, 68*(5), 2198–2208. doi: 10.1128/aem.68.5.2198-2ef208.2002.

19 Application of Remote Sensing in Smart Agriculture Using Artificial Intelligence

Krishi Godhani, Adit Patel, Devdutt Thakkar and Manan Shah
Pandit Deendayal Petroleum University

CONTENTS

19.1 INTRODUCTION

Agriculture plays an important role in the lives of man. It serves as a medium for survival and also raises the economy of the country. It can be referred to as the backbone of the economic system in some countries. It provides employment opportunities to the people who are illiterate. Agriculture not only provides food but also provides raw materials for industries like textile, sugar, etc. It ensures a constant food supply and food security for the population. The World Summit on Food Security declared that in 2050, "The world's population is expected to grow to almost 10 billion by 2050, increasing agricultural demand - in a scenario of modest economic growth - by some 50 percent

DOI: 10.1201/9781003268468-19

357

compared to 2013" (Weiss, 2020). Due to the rise in the global population, the food demand is more than the food supply. Not only the increase in food production, but also we need to take care about the quality and quantity of land, climatic conditions, crop growth, etc. The annual shifts of the crops in a particular field should be taken care of. Nowadays, modern people lack awareness about the cultivation of crops at the right time and at the right place. Because of these cultivating habits, the seasonal climatic conditions are also being changed against the fundamental assets like soil, water and air which lead to insecurity of food. It would be easier to increase the production if farmers can identify which crop is best for the season.

This can be achieved by using Artificial Intelligence (Shah et al., 2020; Patel et al., 2020a; Ahir et al., 2020). Artificial Intelligence plays a great role in various applications in the industry (Pandya et al., 2020; Sukhadia et al., 2020; Patel et al., 2020b; Kundalia et al., 2020). Artificial Intelligence (AI) is the ability of an algorithmic digital computer or computer-controlled machine or a robot to perform tasks commonly related with intelligent beings (Jani et al., 2020; Parekh et al., 2020; Gandhi et al., 2020; Panchiwala and Shah, 2020; Shah et al., 2020; Patel et al., 2020b). With the help of AI, we can analyze various things such as climatic conditions, crop yield, etc. Artificial Intelligence can also detect weeds and suggests herbicides to be sprayed. AI has the potential to solve all the problems faced by farmers. It can be applied to farming in many ways (Talaviya et al., 2020; Jha et al., 2020; Kakkad et al., 2019; Pathan et al., 2020). One such application is remote sensing. Remote sensing basically means to acquire information about the farm and the crops without making any physical contact with it. It uses methods like photography, surveying, etc (Jha et al., 2020; Pathan et al., 2020; Talaviya et al., 2020). Remote sensing data can greatly contribute to the monitoring of earth's surface features by providing timely, synoptic, cost-efficient and repetitive information about the earth's surface. The advantage of remote sensing is its ability to provide repeated information without destroying the sample of the crop, which can be used for providing valuable information for precision agricultural applications. Remote sensing along with GIS is highly beneficial for creating spatio-temporal basic informative layers which can be successfully applied to diverse fields including flood plain mapping, hydrological modelling, surface energy flux, urban development, land use changes, crop growth monitoring, drought monitoring and stress detection (Shanmugapriya et al., 2019,). Hence, remote sensing is an important key to the development of an effective and economical agriculture in the near future.

19.2 APPLICATION OF REMOTE SENSING IN AGRICULTURE

Remote sensing has gained a lot of attention these days. It is widely addressed based on its specific applications, specific platforms, specific locations and sensors. There are many advances in the field of remote sensing.

The basic principle on which remote sensing works is "Different objects based on their physical, chemical and structural properties reflect or emit different amounts of energy in different wavelength ranges of the electromagnetic spectrum." Remote-sensing techniques record data in both: the visible region and in the invisible region of the electromagnetic spectrum. The essential components of remote sensing include a signal from a source/light, sensors on a plate form and sensing components that performs signal reception, storage, processing, information extraction and decision-making. The data products like air photos and digital data are interpreted in visual form and digital form. Decision-making is done by maps and statistics.

Remote sensing can be mainly divided into three categories: ground based, air-borne and satellite.

1. **Ground-based remote sensing**: It is useful for small-scale operational field monitoring. This technology has better temporal, spectral, and spatial resolutions as compared to air-borne and satellite remote sensing. A disadvantage of ground-based remote sensing is one of efficiency and often time reduced to evaluating small areas when compared with aircraft and satellite-mounted sensors, which can be used to evaluate much larger areas at a time.

2. **Airborne remote sensing**: Airborne remote sensing was mainly carried out with the use of piloted aircrafts. However, this is replaced by Unmanned Aerial Vehicles (UAVs) which are aircrafts remotely piloted from the ground. Its low cost, lightweight and low airspeed aircrafts make it well suited for this job. Sensors are mounted such that they face downward and sideward. One of the advantages of airborne remote sensing over satellite remote sensing is that airborne offers very high spatial resolution. However, it has a low coverage area and it is not cost-effective to map a larger area.

3. **Satellite remote sensing**: The remote-sensing satellites are equipped with sensors facing the earth. They are like the "eyes in the sky" continuously observing the earth while revolving in their orbits. A major disadvantage of this method is that the image quality gets degraded because of the atmospheric effects like absorption and scattering of radiations.

19.3   PRECISION FARMING

Precision farming, also called satellite farming or site-specific crop management, is a method in which farmers optimise input such as water and fertilisers to enhance the quality, quantity and yield. This also involves minimising pests and diseases through spatially targeted application of a precise amount of pesticide. It focuses on observation, measurement and responses to variability in crops, fields and animals. Precision practices are facilitated by satellite and communication technologies mainly by Global Navigation Satellite System (GNSS) coordination networks. Precision Farming systems can be considered as a superset of precision farming which includes a combination of soil sampling, soil scanning data, fertilisers and chemicals with the combination of auto-steer and yield mapping technologies remotely monitored using smartphones. Auto-steer is automated steering and positioning of the vehicles on the landscape. Its advantage is that it reduces the overlap between passes of the machine, thus cutting down energy and the expenses on the chemical fertilisers. It makes farming vehicles to operate in accordance with a pre-set planning path. Such type of ideal systems can regulate the input of seeds and agrochemicals according to the needs (Figure 19.1).

19.3.1   Tools and Equipment

1. **Global positioning system (GPS)**: GPS is a set of satellites that is used to determine the ground positions of different objects. It uses the transmission of microwave signals from a network of satellites. The importance of knowing the precise location is:
   • The soil samples and the laboratory results can be linked to the soil maps.
   • Fertilisers and pesticides can be prearranged to fit soil properties.
   • Monitoring and recording field data can be done.

2. **Geographical Information System (GIS)**: GIS is a system designed for capturing, storing, analysing and managing all types of geographical data. It links information in one place so that it can be inferred whenever required. GIS helps individuals and organizations better understand spatial patterns and relationships. This system is cost saving which results in greater efficiency. GIS also analyses rock information characteristics and identifies the best site location.

3. **Remote sensors**: Remote sciences in agricultural terms refer to viewing crops from a low flying aircraft without actually coming in contact and providing a map to identify the field problems early and more effectively. Transfer of information is achieved by the use of electromagnetic radiation (EMR). They indicate variations in field colours corresponding to changes in soil types, field boundaries, etc.

While working with Remote sensed images for agricultural decision making, there are many issues that must be taken care of.

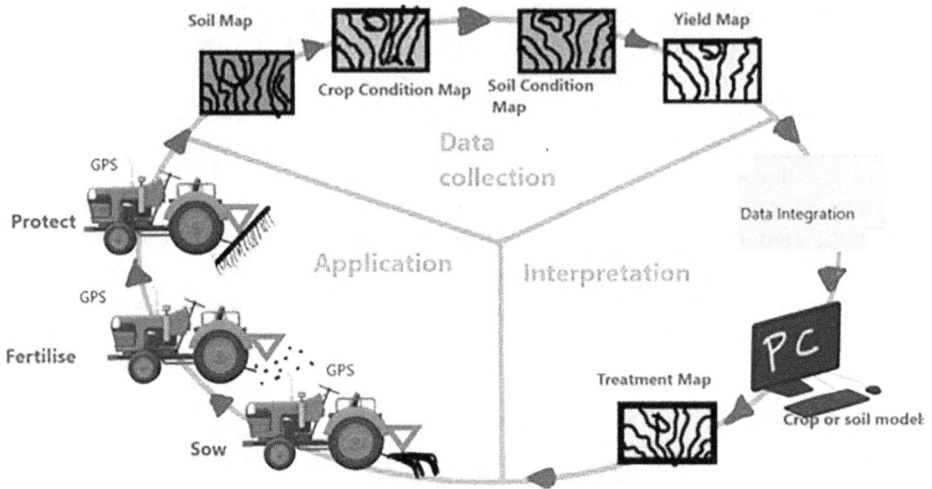

FIGURE 19.1   Precision farming cycle.

19.3.2 IMAGE RELATED ISSUES

1. **Geometric precision**: The sensors mounted on the unmanned aerial vehicles (UAV), satellite or aircraft, while capturing images, are influenced by various unavoidable factors like earth revolution, topographic relief, and the dynamic state of the platform. This results in images which are geometrically distorted which do not exactly correspond to the ground location. Correction of the geometrically distorted image is the first step to be done before the remote-sensed images are used for interpretation and analysis. This rectification is commonly called ortho-rectification. The points on the earth with known locations are called ground control points (GCPs). They are commonly used to ortho-rectify and as checkpoints for validation and quality assessment of those rectified images. Some of the easily recognizable GCPs are road intersections and landmarks. The traditional approach involves manual feeding of information related to camera systems such as lens distortion and focal length, ground elevation and the altitude of the sensor which creates relationships between the sensor, image and target surface. In situations when this information is not available, GCPs are found manually with prior ortho-rectified images. This is done using commercial software packages like ENVI and ERDAS Imagine.
2. **Image resolution**: While analysing images, four types of resolutions are taken into consideration. These are spatial, temporal, spectral and radiometric. Among these, spatial and spectral are significant as they influence the ability to extract information from an image.
   - **Spatial resolution**: Spatial resolution refers to the pixel size. It determines the size of the smallest distinguishable features in an image. In images with high spatial resolution, small objects can be detected, which displays features in detail. When the size of the pixel is high, multiple features are represented by a single pixel. Pixel size is determined by the distance between the sensor platform and the target and the field view of the sensor. Images obtained by UAVs have higher spatial resolution as compared to those obtained by satellites. The process of converting an image with a pixel size different from the original is known as re-scaling.
   - **Spectral resolution**: It refers to the ability of a sensor to distinguish between wavelength intervals in the electromagnetic spectrum. This is the main factor that differentiates multispectral images from hyperspectral. Hyperspectral images deal with thousands of fine wavelength intervals whereas multispectral images deal with fewer

bands. Hyperspectral images provide detailed information. Sensors with high spectral resolution are useful in distinguishing features that are not easily detectable by sensors with broad wavelength ranges. Although hyperspectral images have the ability to capture variability in crop and soil conditions, there are some limitations associated with it. These include high cost of sensor and high image storage requirement.

- **Temporal resolution**: Temporal resolution refers to the frequency at which the images are collected over the same field. Images from UAVs and manned aircrafts have a high temporal resolution as compared to satellite images due to flexibility in scheduling flight plans. During the crop growing season, it is important to obtain images at frequent intervals so that agricultural decision-making such as nutrient application and irrigation becomes much easier. Timely monitoring of crop growth during the critical growth stages helps farmers to identify problems.

- **Radiometric resolution**: Radiometric resolution refers to a sensor's ability to identify and differentiate slight differences in reflected or emitted energy. Data is digitised and recorded as a positive digital number which varies from zero to the selected resolution power of 2. For example, if the resolution of an image is 4-bit, there would be 16 (2^4) digital values ranging from zero to 15. Higher radiometric resolution shows that it is more sensitive in detecting small differences.

19.4   CROP YIELD FORECASTING

The joining of RS and CSM for crop yield forecasting has been investigated for right around three decades, and is defended by the way that RS can evaluate crop status at some random time during the developing season, while CSM can portray crop development consistently all through the season . RS by implication can give measure to shelter state factors utilised by the CSM just as both spatial and transient data about those factors which would then be able to be utilised to change the model recreation. Since the principal satellite data opened up to researchers, they have created calculations to gauge overhang state factors, for example, LAI, vegetation portion and part of APAR. One of the principle mix strategies among RS and CSM centers around modifying the LAI reenacted with the harvest models against the one evaluated through RS. LAI it is a significant agronomic parameter since leaves are the place water and CO_2 are traded between the plant and the climate; moreover, the LAI is utilised to show crop evapotranspiration, biomass gathering and last yield . Analysts chipping away at such mixes commonly embraced three fundamental advances: (i) gauge shade factors with RS; (ii) run the CSM; (iii) utilise an appropriate mix technique to change model runs. The initial step can influence the ensuing aftereffects of the incorporation in light of the fact that, in the event that the yield variable isn't appropriately assessed, at that point altering the model with a one-sided variable will prompt an off-base model assessment. There are two different ways of utilizing RS for the estimation of covering factors: using measurable/exact connections; and the Physical Reflectance Models that recreate the communications between sun-powered shafts (which is a meaning of the concentrated stream of molecule, for example, the light motion) and the different shelter segments using physical laws in which the LAI can be entered as info acquired from the yield model.

There are many methods to estimate crop yield.

For the interpretation and analysis of what is being sensed, calibration of remote-sensing data has to be done. The information collected on location is known as the ground truth and the collection of ground truth data allows calibration of remote-sensing data.

The identification of the farm, the identification of the year, the identification of the crop fields like detailed crop-map, crop species, crop yield data, the acreage in the given farm, dates of sowing and harvesting are the key data that can be provided by Ground truth data. Other chief remarks are the stress, change in technology, etc.

It is seen that this ground truth information is sufficient for classification of the high-resolution remote-sensing data.

Now this high-resolution remote-sensing data allows good assistance in the identification of the low-resolution pixels, which are the features for a given crop species in the research and development phase and in the routine application.

The land use maps and the ground truth crop field map both are combinedly used for the identification of the Advanced Very-High-Resolution Radiometer (AVHRR pixel groups) during the calibration and development (C&D) phase. The known reference crop fields with known conditions and yields are shared by the AVHRR pixel groups. Clean pixels can be obtained when the AVHRR pixel group contains only one crop species.

In a definite region, experimental - measurable models consider crop yield for a long time and powerful factors on crop yield are found. At that point, crop yield is identified with a successful parameter by an observational condition and the coefficient of each factor is found. Presently by these coefficients, crop yield is assessed. Each arrangement of observational models relates crop respect to one set of elements. In most relations, compelling components are natural.

Resource optimization significantly increases the performance of the production system for two reasons: firstly, the production increases in a particular area and secondly, the quantity of inputs such as fertilisers and herbicides decreases as well as consequent reduction in cost.

19.4.1 OPTICAL REMOTE SENSING-BASED MAPPING METHOD

Optical remote detecting sensors were broadly utilised for mapping rice regions around the world. It was utilised in segregating land use/land spread and estimating crop regions because of their capacity to see the Earth surface in the otherworldly range 0.4–2.5 μm. The most ordinarily applied optical sensors include: Landsat (for the most part MSS, TM and ETM+), SPOT-VGT, NOAA/AVHRR, MODIS, and so on. These satellite sensors have the capability of getting multi-worldly and multi-ghastly reflectance information over croplands that can be utilised for determining time arrangement of vegetation lists (VIs), determined as an element of red, blue, and infrared phantom bands. A number of studies have investigated the handiness of optical remote detecting sensors to distinguish rice territories.

19.4.2 MICROWAVE REMOTE SENSING-BASED MAPPING METHOD

One of the prime focal points of microwave remote detecting is related with its capacity of obtaining pictures hypothetically under any climate conditions, for example, overcast spread, downpour, day off, and sun-based irradiance. In this manner, the radar pictures gathered from microwave sensors give a magnificent symbolism source to mapping rice regions, where rice development happens during blustery seasons with prevailing shady conditions. Since the 1990s, analysts have investigated the helpfulness of microwave information recovered from various satellites (e.g., ERS-1 and 2, RADARSAT-1 and 2, ENVISAT ASAR, and so on.). When all is said in done, the transient variety of radar backscatter over the developing season was the key factor in outlining rice zones.

19.5 CLIMATE CHANGE MONITORING

There are a lot of ways in which climate change can be monitored today, and each way has its own extent of effectiveness for a certain type of terrain. For example, a permafrost region, simplification of surface energy balance is taken into account to reduce the number of local variables needed. Index methods which use minimum climatic and soils information to outline the broad regional features of permafrost distribution are not used for they are not well suited to the assessments of climate change impacts, since they do not contain explicit relations between climate and the temperature conditions for permafrost. A common approach taken to monitor the climate change of a permafrost region uses a combination of simplified coupling between the climate and ground

together with an analytical model for the thermal offset effect which results from the seasonal variation in ground thermal properties. The model used will then be able to predict the permafrost temperature under any combination of climatic, surface and lithologic conditions. For the model, the temperature regime is divided into four levels:

1. The temperature at standard screen height;
2. The temperature at the snow surface;
3. The temperature at the ground surface;
4. The temperature at the top of permafrost.

On a mean annual basis, the temperatures at each of these levels differ instantaneously because of the change in heat transfer with time. If the mean annual air temperature (MAAT) and mean annual surface temperature (MAST) are to be compared as illustrated by the ground temperature data for Fort Simpson region one might infer that the regions with MAATs greater then about −5°C should be permafrost free since MAST is warmer then MAAT in most cases. However, that is not the case and is also believed that around 30% of the Fort Simpson region lies under the permafrost region. Hence clearly stating the presence of some sort of mechanism that produces mean ground temperatures that are considerably lower than the surface mean, which can also be explained by the phenomenon of thermal offset. Hence the model executes two types of processes which are surface processes and subsurface processes.

19.5.1 Surface Processes

Local factors generally cancel the influence of larger-scale macro climatic factors on ground thermal conditions. The principal influences of the vegetation canopy are the reduction of solar radiation reaching the ground surface and the variable effects on the accretion and persistence of snow cover (Rouse, 1984). Additionally, the convergence of precipitation and transpiration by the canopy influences the ground thermal regime through the water balance. In general, a thicker snowpack usually restricts heat loss from the ground to a greater extent. As a result, during the winter months, the ground surface temperature is varied by the local and regional spatial variability of the persistence of snow cover (due to the vegetation and topography) to a great extent. And additionally, inter annual variability of snow packs leads to an inter annual variation in the temperature of the ground and beyond that simply due to the variation in air temperatures. Hence N-factor is applied as the transfer function between the air and the temperature at the ground surface, to account for the local influence of the vegetation and snow cover. Here the N factor incorporates all the microclimatic effects due to vegetation (radiation, convection, evapotranspiration etc.) during the summer and is dominated by the snow covers during the winters. Hence to obtain the values applicable to ground surfaces seasonal N-factors are applied as multipliers to air temperature indices.

$$DDT_S = N_T\, DDT_A \tag{19.1}$$

$$DDF_S = N_F\, DDF_A \tag{19.2}$$

where DDT_S=surface thawing degree days (seasonal thawing index)
DDT_A=air freezing degree days (seasonal freezing index)
DDF_A=air freezing degree days (seasonal freezing index)
DDF_S=Surface freezing degree days (seasonal freezing index)
(Note that DDF is assigned a positive value when the temperature is below 0°C, so that (DDT_A-DDF_A) 365 yields the mean annual air temperature while $(N_T\, DDT_A - N_F\, DDF_A)/365$ yields the mean annual ground surface temperature.)

19.5.2 SUBSURFACE PROCESSES

The correspondence between the ground surface temperature and the temperature at the top of permafrost (TTOP) is determined fundamentally by conductive heat flow, although non-conductive heat transfer processes may be important within the active layer under some circumstance. Hence the subsurface process deals with the thermal offset which is caused due to the thermal conductivity of the soil which becomes greater when frozen than when thawed. Seasonal variations in the moisture content of the soil near the ground surface can augment the effect (i.e. drier surface conditions in summer, wetter in rainy, spring, autumn and winter). For example, variations in the ground thermal regime due to the thermal influence of the organic soil were much more important than the vegetation canopy. We have determined that TTOP is related to ground surface temperature as follows:

$$\text{TTOP} = \left(k_T \ \text{DDTs} - k_F \ \text{DDFs} \right) / \ k_F \ P \tag{19.3}$$

Where k_T=thermal conductivity of ground (thawed)
  k_F=thermal conductivity of ground (frozen)
  P=period (365 days)

19.5.2.1 A Functional Climate-Permafrost Model
Combining Eqs. (19.1)–(19.3), we get the complete model as:

$$\text{TTOP} = \left(k_T N_T \ \text{DDT}_A - k_T N_F \ \text{DDF}_A \right) / \ k_F P \tag{19.4}$$

Hence providing a general formula of the climate and permafrost relationship taking all the discussed factors into consideration. And hence various case scenarios about the climate can also be studied by altering the values in Eq. (19.4) hence allowing us to monitor the climate based on the information available to us in this region. Similarly, this model can also be used in monitoring climate at the sites of exposed bedrock as they will produce the most direct signal of climate change on ground thermal regime.

19.5.2.2 A General Approach on Climate Monitoring over All Regions
After going through the above examples and a detailed explanation of one such example, it indicated to a certain extent that the climate is also dependent on the solar radiation emitted by the surface of the ground and based on this indication a model was created which through remote sensing would be able to monitor climate with the help of solar surface radiations emitted by the ground over all the regions i.e. of the whole earth. For this, a satellite is used which retrieves surface radiation accounts with a high structural and profane resolution and a large areal coverage. Through radiometric measurements, satellite sensors provide information on the interaction of solar radioactive fluctuations with the atmosphere and the earth's surface and become the basis for retrieval of surface radiation datasets. There is kept a 23-year-long (1993–2005) continuous and validated climate data record of surface solar irradiance, direct irradiance and effective cloud albedo recently generated by the Satellite Application Facility on Climate Monitoring (CM SAF) which is a part of the European Organisation for the Exploitation of Meteorological Satellites (EUMETSAT) Satellite Application Facilities (SAFs) network. Within the CMSAF, the generation of satellite-derived data records for climate monitoring is placed special emphasis on (Schulz et al., 2009). Using the instruments on-board of the METEOSAT First Generation Satellites (MFGS) The surface solar irradiance, direct irradiance and effective cloud albedo are presented by MVII (METEOSAT Visible and Infrared Imager) based on its visible channel (0.45–1 μm). The model created for the monitoring of climate involves the process of the climate version of the Heliosat algorithm (Cano et al., 1986) which has a self-calibration method which automatically accounts for the degradation of the individual satellite instruments during their lifetime and the breaks induced by changes between

different satellite instruments in the generation of climate data record; and also has an improved algorithm for the determination of clear-sky reflectivity which is based on a recently developed algorithm for the retrieval of the clear sky reflection used to calculate *pcs* (clear sky reflection) (Dürr and Zelenograd, 2009; Zelenka, 2001) where a 7 day running mean of *pcs* is used instead of using a monthly field of clear sky normalised counts (*p*) because of it being a comparatively faster method to find the temporal evolution of clear sky reflectivity if *p* exceeds *pcs* by eup 2 tests are conducted, the first test Which involves Application of a slow adaptation of *pcs* to the current minimum counts is applied if *p* is to a certain extent greater than *pcs* and is given an assumed adaptation time of 7 days Which correspond to the timescale of involved processes (example changes in vegetation, soil properties). Additionally, a sudden strong decrease in *p* below pcs-elbow Also leads to a slow change of p_{cs} and should Prevent the contamination of *pcs* with cloud shadows. If *p* falls below *pcs* by up to Elow then test two is conducted Which executes a fast adaptation of *pcs*. This latter scheme is applied to events with relatively fast processes such as, e.g. melting of snow. If both the tests fail i.e. if *p* is rather large as occurring in case of clouds and snow, *pcs* is not changed. If this happens for a span of time of $t_s = 28$ days subsequent days (i.e. the pixel is very bright at all times) then the pixel is assigned to be snow-covered. This algorithm is based on low temporal variability of snow compared to clouds and thus detecting very high reflectivity in one-time slot over a certain number of days indicates snow. This strategy can fail in the cases of persistent cloud decks (e.g. marine stratocumulus or fog). In the case of positive snow detection, p_{cs} subsequently evolve according to Equation (19.4) so that *pcs* increases to the range of the current minimum normalised counts and the basic scheme can be applied in the subsequent time steps.

Test 1: The CM SAF, solar surface irradiance, direct irradiance and effective cloud albedo datasets are presented below in Figure 19.1 as seasonal means on the full circle with highest radiation values in the regions with highest sun elevation and lowest values in winter hemisphere (lowest sun elevation). In the given Figure 19.1 the stratocumulus region close to western is very well depicted by the shadowing effect of clouds on radiation (especially for SID), and in the tropics and South African coast with large amount of cumulus clouds (Figure 19.2). There exists a number of similar data sets for solar surface irradiance. The HelioClim project also applies a Heliosat algorithm to the MFG satellite data to obtain solar surface irradiance on the METEOSAT disk from 1985 to 2005 and furthermore The flux dataset from the international satellite cloud climatology project and surface radiation budget dataset from the global energy and water cycle experiment provide global solar surface irradiance in this particular time gap which is from 1983 to 2005. And finally the model based on the analysis dataset ERA-interim from the European Centre for medium-range weather forecasts does also provide global Solar surface irradiance fields starting in 1989. These are the four data sets that are used for Inter comparison for the CM SAF solar surface irradiance dataset.

Overall, it was shown that the target accuracy is achieved for monthly and daily means of the global and direct surface solar irradiance in the CM SAF CDR. The effective cloud albedo is a central input quantity for the calculation of SIS and SID. The high accuracy of these data sets demonstrates in turn the high quality of the effective cloud albedo. In general, the validation demonstrates the outstanding quality of the CM-SAF data set for climate monitoring and analysis, also in inter-comparison with other data sets. As a consequence of the high spatial resolution, the CM-SAF data can be also applied to regional climate monitoring and analysis. Moreover, the successful retrieval of high-quality data from the non-calibrated Meteosat First Generation satellite enables the generation of long-term data sets. Heliosat in combination with MAGIC can be also applied to the Meteosat Second Generation satellites, which enables the prolongation and continuation of this high-quality data set. Thus, the analysis of climate trends and extremes becomes possible. The self-calibration approach enables the application of Heliosat to other geostationary satellites, e.g. GOES, as well as to the Meteosat East satellites. Within this scope Heliosat has been implemented at the Joint Research Center for the processing of Meteosat East data. The first validation results show a similar accuracy (T. Huld, Personal communication). The application of the method to other geostationary satellites enables the extension of the data set toward a

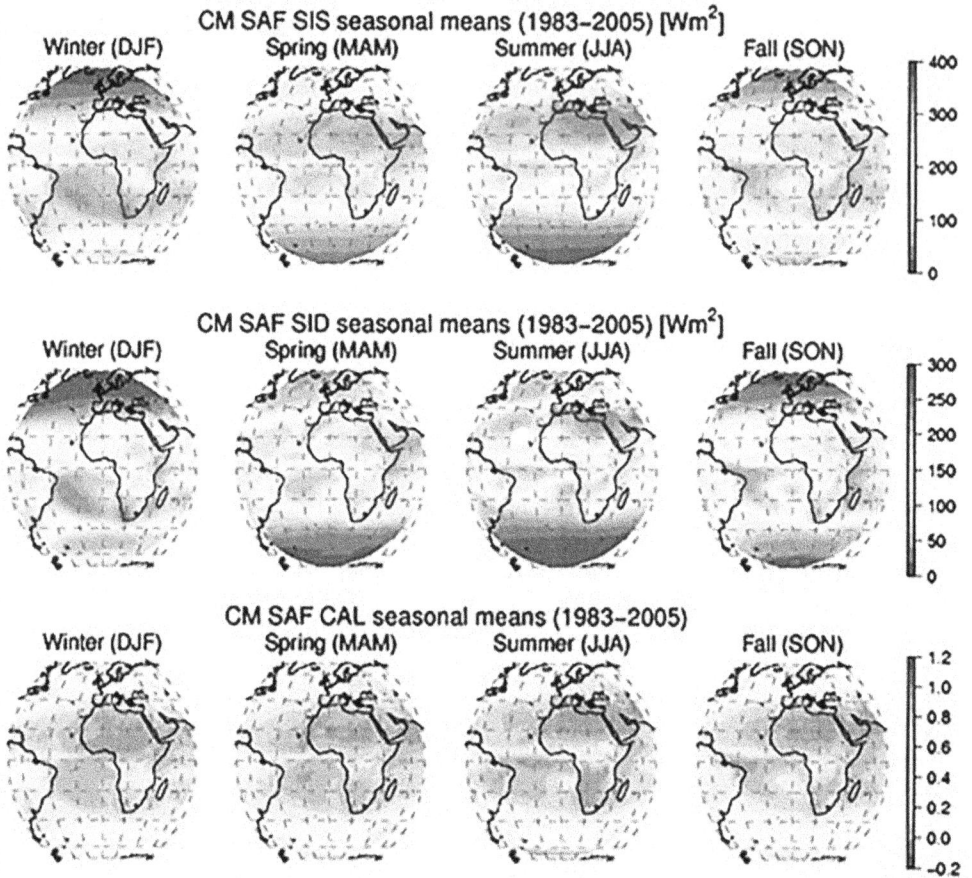

CM SAF SIS seasonal means (1983–2005) [Wm²]
Winter (DJF)   Spring (MAM)   Summer (JJA)   Fall (SON)

CM SAF SID seasonal means (1983–2005) [Wm²]
Winter (DJF)   Spring (MAM)   Summer (JJA)   Fall (SON)

CM SAF CAL seasonal means (1983–2005)
Winter (DJF)   Spring (MAM)   Summer (JJA)   Fall (SON)

FIGURE 19.2   SIS means Solar Surface Irradiance; SID means Direct Irradiance; Cal means Effective Cloud Albedo.

geostationary ring, which increases the attractivity of the data set for further applications, e.g. the evaluation of global climate models.

Test 1 : ρcs;t ≤ρt≤ρcs;t þup or ρtbρcs;t–low →ρcs;tþl ¼ 6 7 ρcs;t þ 1 7 ρt slow evolution ðÞ ð4Þ
Test 2 : ρcs;t–low≤ρtbρcs;t →ρcs;tþl ¼ 1 2 ρcst þ 1 2 ρt fast evolution ðÞ ð5Þ else : ρcs;tþl ¼ρcs;t ð6Þ The subscript t denotes the current and t+1 the subsequent time step. We use variable bandwidths up and low depending on ρmax and ρcs given in Dürr and Zelenka (2009) instead of the fixed ones used in Zelenka (2001).

up ¼0:125ρmax þ
8 ρcs−0:15 ρmax ðÞ 0:25ρmax ð7Þ
low ¼0:0875ρmax þ
6 ρcs−0:15ρmax ðÞ 0:25ρmax ð8Þ

In order to better account for snow events, a reduced t_s=7 is applied in case the considered pixel already experienced a snow event in the past (Zelenka, 2001). The shorter time range is chosen to still avoid that persistent fog is mistaken for snow and that elevated sites had the time to adapt ρcs

after a sudden snowfall. A past snow event is, on one hand, determined by a large amplitude of ρcs (given by Δρcs ¼ρcs; max–ρcs; min > 0:55 ρmax) reflecting the range of the clear sky reflectivity between snow and soil/vegetation. Further, a higher probability of snow events is also given if ρcs > ρsnow with ρsnow being an empirically derived snow threshold. It includes a term ρsnow, min that is high enough to exclude oceans or other dark surfaces and it includes a term ρsnow, range that accounts for the higher probability of a former snow event for pixels with a high amplitude ρcs. ρsnow; min ¼0:48ρmax ð9Þ ρsnow; range ¼max 0:15ρmax;0:6ρcsð Þ ð 10Þ ρsnow ¼max ρsnow; min; ρcs; min þρsnow; range ð11Þ The calculation of the effective cloud albedo CAL for non-snow pixels follows the standard Heliosat method (see Eq. 19.1). In case snow is detected the calculation of the cloud albedo CAL is modified according to Dürr and Zelenka (2009) in which ρcs is substituted by ρsnow, min and ρmax is increased by a factor of 1.41 in order to artificially increase the count range and, thus, enhance the contrast. This empirically developed modified cloud albedo formulation generally gives lower cloud albedo values in order to account for the radiative properties of snow. This includes mainly the reflectivity of the bright current surface which leads to higher surface irradiance values.

CALs ¼
ρ−0: 48ρmax 1:41ρmax−0:48ρmax ¼
ρ−0: 48ρmax 0:93ρmax ð12Þ

2.3: Satellite data

Data from EUMETSAT's geostationary Meteosat satellites of the First Generation (Meteosats 2–7) are used. They were in operation from 1982 to 2005 at a location directly over the equator at a longitude of 0° at an altitude of about 36,000 km. The resulting visible disk reaches up to 80°N/S and 80°E/W, respectively. Meteosat 7 is still operational above the Indian Ocean.

19.6 CROP IDENTIFICATION

Object-based feature selection and crop identification at different phenological stages and field conditions in Yolo County agro ecosystem. The OCIM methodology involves several consecutive steps and provides three levels of crop identification. After segmentation of remote images into realistic and homogeneous areas, cropland was firstly discriminated between permanent crops, summer crops, winter cereals and meadows. Then, the model describes which crop is growing in every parcel and, simultaneously, the intra-class variations attributed to specific crop management operations. Several spectral (vegetation indices), textural and hierarchical features from three different crop-growing periods were selected in a decision tree structure, revealing the physical background affected by every kind of crop, as well as their relationships to field crop stages and crop calendar. The spectral features formed the decision tree framework and contributed around 90% to the model. Some textural features were also necessary to discriminate between crop fields with similar spectral responses, mainly permanent crops (orchards, vineyard, alfalfa and meadow). Out of 336 object-based features studied, the 3-period model included 24 features that made relevant contributions, divided them up into 16 spectral, 6 textural and 2 hierarchical features, demonstrating the potential of this technique to optimise the feature selection for crop classification and, consequently, reducing the computational time of image analysis procedure. NDVI was the index that made the chief contribution to OCIM by identifying the main groups of crops based on the presence and vigor of green vegetation within the fields. In addition, other vegetation indices incorporating SWIR bands (NDSVI, NDI7, NDTI and LCA) were also important to crop identification performance because of their relationship to field properties such as moisture, vegetation vigor, non-photosynthetic vegetation and bare soil. Studying the evolution of field and crop status by using several images in different growing stages was essential for crop identification. The features calculated in the late summer period contributed around 60% to the models produced, followed by mid-spring (30%) and early-summer (10%). Intra-class variations were detected and grouped to different local crop calendars (mainly, planting schedule), orchard

structure (tree size or separation) and land management operations (cover crops, residue or tillage after harvest).

OCIM was built using rules based on the physical properties of the studied crops and so, most of these rules can be extrapolated to other Mediterranean climate regions, although future analyses are necessary to evaluate its strength. Moreover, ASTER spatial, spectral and temporal resolution fulfilled requirements concerning general field size and crop calendars and, therefore, similar results are expected using other calibrated multior hyper-spectral imagery with similar resolutions such as Landsat or Hyperion. In a subsequent companion investigation, researchers are developing an extended rule-set version of OCIM methodology for cropland classification of multi-year satellite imagery, so as to monitor detailed crop parameters at a field scale for the integration of remote sensing and modelling to assess temporal and spatial variations of greenhouse gas emissions in agro-ecosystems.

19.7 LIMITATIONS

Remote sensing has been helping farmers for many years. Agriculture sector contributes around 6.4% of the world's economic production. From 3 billion people living in rural areas, roughly 2.5 billion people are dependent on agriculture for their livelihood (FAO). But this method has some drawbacks:

- Remote sensing is fairly an expensive method especially when it comes to analyzing smaller areas.
- When there is a need to analyze different aspects of the photographs, taking repetitive photographs becomes expensive.
- Collection of data, specifying the resolution and calibrating the sensors is done by humans. This gives rise to human error.
- Remote-sensing systems such as radar emit their own electromagnetic radiation which affects the phenomenon.
- Large scale maps cannot be prepared from the satellites which makes remote-sensing data collection incomplete.
- Atmospheric phenomena may affect the images being analyzed.

19.8 CONCLUSION

This review paper gives an overview of how remote sensing can be applied in agriculture. In recent years, the agricultural industry has marked significant advancements which rendered the use of remote-sensing data. In the last decade, studies mainly focused on showing the feasibility and the operationality of techniques and methods previously developed. A large part of the effort was dedicated to deep learning and machine learning that answers the operational needs, taking advantage of the unique amount of data available, while the fundamental research on remote sensing progresses slower.

Yield models, remote detecting and satellite remote sensing have had equal improvement courses. Yield models are relied upon to adjust to encourage environmental change and look into just the prerequisites of quality phenotype modeling while satellites are used to detect the climate change giving an even more precision insight for the sustenance of each model in the course of the future. Thinking about the favorable circumstances and the wide materialism of the blend of the three, it is sure that the joint effort of remote detecting information with crop models will increase, hopefully by consolidating computerised methodology that will improve their presentation. The alleged large information upset is the setting where this cooperation will certainly unfurl. Procedures for information absorption and quality checking should be at the core of this promising road of research on the off chance that it is to convey solid and dexterous forecasts.

Remote sensing also includes describing and modelling variation in plant and soil species. It aims for increasing economic returns, as well as reducing the energy input and the environmental impact of agriculture. Hence, remote sensing is an important key to the development of an effective and economical agriculture in the near future.

DECLARATION

AUTHORS CONTRIBUTION

All the authors make a substantial contribution to this manuscript. KG, AP, DT and MS participated in drafting the manuscript. KG, AP, DT and MS wrote the main manuscript. All the authors discussed the results and implication on the manuscript at all stages.

ACKNOWLEDGEMENTS

The authors are grateful to Department of Information and Communication Technology, and Department of Chemical Engineering School of Technology, Pandit Deendayal Petroleum University for the permission to publish this research.

AVAILABILITY OF DATA AND MATERIAL

All relevant data and material are presented in the main paper.

COMPETING INTERESTS

The authors declare that they have no competing interests.

FUNDING

Not Applicable

CONSENT FOR PUBLICATION

Not applicable.

ETHICS APPROVAL AND CONSENT TO PARTICIPATE

Not applicable.

REFERENCES

Ahir, K., Govani, K., Gajera, R., & Shah, M. (2020). Application on virtual reality for enhanced education learning, military training and sports. *Augmented Human Research*, 5(1), 7. doi: 10.1007/s41133-019-0025-2.

Cano, D., Monget, J., Albuisson, M., Guillard, H., Regas, N., & Wald, L. (1986). A method for the determination of the global solar radiation from meteorological satellite data. *Solar Energy*, 37(1), 31–39. doi: 10.1016/0038-092x(86)90104-0.

Dürr, B., & Zelenka, A. (2009). Deriving surface global irradiance over the Alpine region from METEOSAT second generation data by supplementing the HELIOSAT method. *International Journal of Remote Sensing*, 30(22), 5821–5841. doi: 10.1080/01431160902744829.

Gandhi, M., Kamdar, J., & Shah, M. (2020). Preprocessing of non-symmetrical images for edge detection. *Augmented Human Research*, 5(1). doi: 10.1007/s41133-019-0030-5.

Jani, K., Chaudhuri, M., Patel, H., & Shah, M. (2020). Machine learning in films: An approach towards automation in film censoring. *Journal of Data, Information and Management, 2*(1), 55–64. doi: 10.1007/s42488-019-00016-9.

Jha, K., Doshi, A., Patel, P., & Shah, M. (2020). A comprehensive review on automation in agriculture using artificial intelligence. *Artificial Intelligence in Agriculture, 2*, 1–12. doi: 10.1016/j.aiia.2019.05.004.

Kakkad, V., Patel, M., & Shah, M. (2019). Biometric authentication and image encryption for image security in cloud framework. *Multiscale and Multidisciplinary Modeling, Experiments and Design, 2*(4), 233–248. doi: 10.1007/s41939-019-00049-y.

Kundalia, K., Patel, Y., & Shah, M. (2020). Multi-label movie genre detection from a movie poster using knowledge transfer learning. *Augmented Human Research, 5*(1). doi: 10.1007/s41133-019-0029-y.

Lefèvre, M., Wald, L., & Diabaté, L. (2007). Using reduced data sets ISCCP-B2 from the Meteosat satellites to assess surface solar irradiance. *Solar Energy, 81*(2), 240–253. doi: 10.1016/j.solener.2006.03.008.

Panchiwala, S., & Shah, M. (2020). A comprehensive study on critical security issues and challenges of the IoT world. *Journal of Data, Information and Management, 2*(4), 257–278. doi: 10.1007/s42488-020-00030-2.

Pandya, R., Nadiadwala, S., Shah, R., & Shah, M. (2020). Buildout of methodology for meticulous diagnosis of K-complex in EEG for aiding the detection of Alzheimer's by artificial intelligence. *Augmented Human Research, 5*(1), 3. doi: 10.1007/s41133-019-0021-6.

Parekh, P., Patel, S., Patel, N., & Shah, M. (2020). Systematic review and meta-analysis of augmented reality in medicine, retail, and games. *Visual Computing for Industry, Biomedicine, and Art, 3*(1). doi: 10.1186/s42492-020-00057-7.

Patel, H., Prajapati, D., Mahida, D., & Shah, M. (2020a). Transforming petroleum downstream sector through big data: A holistic review. *Journal of Petroleum Exploration and Production Technology, 10*(6), 2601–2611. doi: 10.1007/s13202-020-00889-2.

Patel, D., Shah, D., & Shah, M. (2020b). The intertwine of brain and body: A quantitative analysis on how big data influences the system of sports. *Annals of Data Science, 7*(1), 1–16. doi: 10.1007/s40745-019-00239-y.

Pathan, M., Patel, N., Yagnik, H., & Shah, M. (2020). Artificial cognition for applications in smart agriculture: A comprehensive review. *Artificial Intelligence in Agriculture, 4*, 81–95. doi: 10.1016/j.aiia.2020.06.001.

Rigollier, C., Lefèvre, M., Blanc, P., & Wald, L. (2002). The operational calibration of images taken in the visible channel of the Meteosat series of satellites. *Journal of Atmospheric and Oceanic Technology, 19*(9), 1285–1293. doi: 10.1175/1520-0426(2002)019<1285:tocoit>2.0.co;2.

Rigollier, C., Lefèvre, M., & Wald, L. (2004). The method heliosat-2 for deriving shortwave solar radiation from satellite images. *Solar Energy, 77*(2), 159–169. doi: 10.1016/j.solener.2004.04.017.

Schulz, J., Albert, P., Behr, H., Caprion, D., Deneke, H., Dewitte, S., Dürr, B., Fuchs, P., Gratzki, A., Hechler, P., Hollmann, R., Johnston, S., Karlsson, K., Manninen, T., Müller, R., Reuter, M., Riihelä, A., Roebeling, R., Selbach, N., … Zelenka, A. (2009). Operational climate monitoring from space: The EUMETSAT satellite application facility on climate monitoring (CM-SAF). *Atmospheric Chemistry and Physics, 9*(5), 1687–1709. doi: 10.5194/acp-9-1687-2009.

Shah, D., Dixit, R., Shah, A., Shah, P., & Shah, M. (2020). A comprehensive analysis regarding several breakthroughs based on computer intelligence targeting various syndromes. *Augmented Human Research, 5*(1), 14. doi: 10.1007/s41133-020-00033-z.

Sukhadia, A., Upadhyay, K., Gundeti, M., Shah, S., & Shah, M. (2020). Optimization of smart traffic governance system using artificial intelligence. *Augmented Human Research, 5*(1), 13. doi: 10.1007/s41133-020-00035-x.

Talaviya, T., Shah, D., Patel, N., Yagnik, H., & Shah, M. (2020). Implementation of artificial intelligence in agriculture for optimisation of irrigation and application of pesticides and herbicides. *Artificial Intelligence in Agriculture, 4*, 58–73. doi: 10.1016/j.aiia.2020.04.002.

Zelenka, A. (2001). Estimating insolation over snow covered mountains with Meteosat VIS-channel: a time series approach. *Proceedings of the Eumetsat Meteorological Satellite Data Users Conference*, Eumetsat Publishing, EUM P, Vol. 33., pp. 346–352.

20 Trends in Processing, Preservation of Tomatoes and Its Allied Products
A Review

Abiola Folakemi Olaniran and Omorefosa Osarenkhoe Osemwegie
Landmark University

Oluwaseun Peter Bamidele
University of Pretoria

Yetunde Mary Iranloye and Clinton Emeka Okonkwo
Landmark University

Charles Oluwaseun Adetunji
Edo State University Uzairue

Oluwakemi Christianah Erinle
Landmark University

CONTENTS

20.1   INTRODUCTION

Tomato (*Solanum lycopersicum* L.), which comes in different varieties, is a universal essential vegetable that is known for its positive health, culinary, economic, and industrial benefits (Alenazi et al., 2020). Tomato is popularly consumed in different forms (raw, cooked, and processed). Tomato is a high source of antioxidants, vitamins (A, B, C, E, and K), lycopene, carotenoids, phenolics, minerals (potassium, folate, and sodium), cholesterol, and saturated fat levels are low (Olaniran et al., 2015; Migliori et al., 2017). Approximately 160 million tons of fresh tomatoes are produced annually on an international scale, with China (56.8 million tons) and India (18.7 million tons)

DOI: 10.1201/9781003268468-20

ranking as the highest producers. While no African nation ranked among the 12 highest producers of tomato in 2019, China's export of tomato sauce accounts for one-third of the total global tomato export (Mendonça et al., 2020). In contrast, three times as many potatoes and six times as many rice are cultivated globally than tomatoes (FAOSTAT, 2016). However, about a fourth of those 160 million tons are cultivated for industrial production, making tomatoes the world's most processed crop. Every year, nearly 40 million tons of tomatoes are produced in some of the world's largest food industries around 3.7 million metric tons of processed tomatoes were produced (FAOSTAT, 2020; WPTC, 2019). The production of processed tomatoes was around 3.7 million metric tons (FAOSTAT, 2020; WPTC, 2019) at the international level in the year 2019. The main global tomato production was linked to the temperate zones of the northern hemisphere (40th parallels North) between July and December, an average of 91% of the world's crop is processed. However, between January and June, the remaining 9% is stored in the Southern Hemisphere. Brazil (4.3 million tons annually) which ranked ninth among the top producing countries in 2019 is an exception by being in the Southern hemisphere. Although tomato processing industries exist in many countries, production is extremely limited with 83% of annual global production from the ten largest producing countries. However, outside of these ten nations, the amount of food processed has constantly increased in recent years and the shared global contributions from the top ten producers tend to decrease (Boccia et al., 2019). There are two types of cultivated tomatoes, those for fresh consumption and those for industrial transformation (processing), both of which are usually grown in the field. Tomatoes come in a number of ways for consumers, including paste, sauce, ketchup, dried, powdered, frozen, and juice. They are available for consumption outside of the usual harvest season due to their production in the food industry. Consumption of tomato products rich in lycopene has been reported to reduce the risk of breast cancer, ovarian cancer, cardiovascular diseases, etc. in humans (Forbes-Hernandez et al., 2016; Kulczyński et al., 2017).

Tomato processing is huge a business in many parts of the world with ready markets both internationally and nationally. The production of tomatoes in many developing nations is threatened by disease, agronomic, socioeconomic, and preservation challenges (Alenazi et al., 2020). The review of processing, preservation of tomatoes and allied products is therefore important to purposely provide the knowledge-based inspirations required to potentiate a proportionate improvement in the tomato value-chain trend for sustainable alleviation of poverty and hunger in the developing nations.

Tomatoes are eaten in large quantities, either fresh or as concentrates in the form of juices and pastes. Although thermally processed juices are microbiologically stable, they are easily susceptible to microbiological degradation and lycopene bioavailability (Fattore et al., 2016). Processes including heat treatment may impact cell reinforcements profile and various micronutrients of tomatoes. Manufacturing techniques could affect the color estimates of tomato pastes (Kelebek et al., 2017). Due to unregulated conditions during manufacturing, Brix estimates of homemade tomato pastes were far higher than those of commercial equivalents. To decrease the level of microbial contamination and changes in physical properties, non-thermal technologies such as high-pressure homogenization (HPH), pulsed electric field (PEF), ultrasonic (US), and irradiation have been widely used in tomato juice processing (Gao et al., 2019). The fruit of the tomato (Solanum lycopersicum) is very important in human nutrition because it contains many bioactive compounds, and the value of these compounds has made it valuable for use in industrial processes (Mohammed et al., 2017). Because they are a staple food and a rich source of high-quality proteins, amino acids, fatty acids, and fiber, processed tomato products have a significant impact on the basic efficiency of patterns of eating of the population (Shao et al., 2015). Emerging technologies such as high pressure–low temperature, modified atmospheres, the addition of antioxidants, pulsed electric fields, etc. are applied in processing, preservation and packaging of tomato products (Viuda-Martos et al., 2014). High-pressure homogenization (HPH) is a valuable technology for achieving changes in physical properties that are desired of tomato products such as pulp sedimentation behavior, particle size distribution, turbidity, color, and microstructure (Wang et al., 2018). There are different processing methods during the production of tomato and its allied products that may be broadly divided into thermal and non-thermal methods (Jayathunge et al., 2017;

Mehta et al., 2019). Thermal processing of tomatoes may be carried out by pasteurization, blanching, cooking, frying, canning, dehydration, or drying (Martínez-Hernández et al., 2016). Thermal processing though reported as beneficial lycopene is liberated and solubilized in free, ester structure due to the interference of food mediums such as cell walls and membranes during production. This improved the carotenoid's bioavailability and can retain its total phenolic, flavonoids, and total antioxidant activity in spite of the loss of vitamin C at 88°C (Martínez-Hernández et al., 2016). The highest rates of lycopene abasement have been reported for non-thermal processes such as striping and freeze-drying. The use of PEF (pulsed electric field) technology and HHP (high-hydrostatic-pressure) processing to boost the health-promoting properties of tomato products due to lower lycopene degradation or increased isomerization of cis-lycopene isomers, which have improved properties than trans-isomers, was used (Mehta et al., 2019). In addition, ultrasound has been used to remove lycopene from tomato products and waste (Rahimpour and Dinani, 2018; Rahimi and Mikani, 2019). Other operations or steps such as chopping, grinding, milling, and homogenization are also crucial in sustaining the viscosity of the tomato product (Xu et al., 2018).

20.2   PRODUCTS FROM TOMATO PROCESSING

Tomatoes are cultivated all year round in greenhouses mainly for distribution fresh to processing industries, exports, and local marketing outlets. Several tomato products, such as paste, dried vegetables, canned foods, tomato juice, and other related items, cater to a wide range of customer tastes and are available for consumption outside of the regular production season. Tomato fruits used for commercial processing are usually ripe, firm, and unblemished. Technological processing of tomatoes, such as cooking and baking, has little effect on the lycopene content of finished products, and any lycopene losses are typically small (Górecka et al., 2020).

20.2.1   TOMATO PASTE

Tomato paste contains nourishing ingredients and it is a common source of natural antioxidants that are most consumed virtually in many countries including Nigeria due to its unique taste, flavor, and aroma (Olaniran et al., 2013). Tomato paste industries constitute one of the largest market and product output in the world (Debastiani et al., 2021). It's also a significant food-related industrial product that's been produced on a small scale in many tomato-producing countries homes. It is valued along with preservation and green-house technologies in breaking the seasonality of tomato fruits and guarantees their sustainable supply all the year-round. The redness of tomatoes to be processed into a paste is a major criterion that can affect consumer preference hence it must be completely red to elicit appeal from consumers. The level of maturity, water-soluble dry matter content, low Brix value, high dry matter content, elemental composition, sugar contents, disease resistance, and mold growth resistance are some other fruit qualities. Tomato paste made at home is usually processed under more permissive conditions, such as a long cooking period at atmospheric pressure or exposure to sunlight (Kelebek et al., 2017). Tomato paste on the shelf is typically made through a series of steps that include thermodynamic treatment for pulp preservation, filtration, pasteurization, and evaporation. The color and flavor characteristics of the paste are improved and preserved during industrial processing by a vacuum-assisted evaporation process that produces low pressure and low temperatures. Tomato pastes on the shelf have been shown to contribute more to total micronutrient intakes, such as total phenolic, antioxidants, and total flavonoids, than homemade tomato pastes.

20.2.2   TOMATO JUICE

Unlike tomato paste, the juice of tomato is a common beverage that can be consumed for a variety of reasons. It's typically made in a continuous process, beginning with a hot break into tomato paste

and then diluting with water to the desired concentration (2–5 °Brix) (Håkansson, 2019). The juice of tomato is an un-concentrated pasteurized product consisting of liquid expressed from ripe tomato together with a substantial proportion of the pulp. Different technologies have been reported and recommended for tomato juice production. This involves the use of high-pressure homogenizers in the manufacture of tomato juice to split cells of tomato and viscosity increase (Innings et al., 2020). Furthermore, the low-pressure homogenization using hydrodynamic cavitation (HC) technology to improve the microbial stability of tomato juice (Hilares et al., 2019), Concentration of sugar aqueous solution in tomato juice using a clathrate hydrate-based procedure (Ghiasi et al., 2020) have equally been used in tomato juice production. However, the short shelf life, rapid alteration of its organoleptic and physical properties of tomato juice by fermentation, and general environmental impact coupled with growing toxicological concerns of chemical preservatives have affected the development of its sustainable commercialization (Parajuli et al., 2021). Thermal processing, homogenization, plant-based phenolic additives and fermentation techniques have been reported in improving, their sensory qualities, appearance, texture, taste, or nutritional attributes (Rojas et al., 2019; Lu et al., 2020).

20.2.3 Tomato Ketchup

This is unique among leafy food products after tomato paste that is being manufactured and sold on the global market. Ketchup is prepared to a fairly thick consistency by heating up the mash of ripe tomatoes with flavours, sugar, vinegar, and salt before being packed in glass bottles. Although the technique of preparation is straightforward on a basic level just as in the procedure, there are some problems associated with its storage and delivery. Some of these include the growth of a dark circle called "Black neck" close to the surface of the ketchup inside the tapered-bottled neck, thin liquid from the ketchup, and developments of molds. These deformities can effectively be overawed by adopting standard methods and using tomatoes of good quality.

In tomato ketchup production, healthy and fully ripe tomato fruits that are well-developed, red coloured are sorted and washed systematically in freshwater. The marred and green parts should be removed with a knife made of stainless steel, the sound portion cut into smaller bits. The fruit prepared should be placed in a stainless-steel or aluminum open basin and press with a scoop made of wood. The crushed mass should be cooked until it reaches a temperature of 80 to 85 °C or till the skins are loosened from the flesh to form the pomace. By rubbing softly with the lower end of an enameled cup, strain through a fine mosquito net or a 1 mm mesh stainless steel sieve. Ketchup is produced using stressed tomato juice or mash and flavours, salt, sugar, and vinegar, with or without onion and garlic, and contains at the very least 12% tomato solids and 25% complete solids. To prevent vitamin loss and browning during subsequent stockpiling, the ketchup should be filled hot (around 88°C). It doesn't spoil for a long time if it's made with good tomatoes and the right quantities of sugar, salt, vinegar, and flavours, and if it's kept in a cold, dry place after opening the secure bottled if the latter is kept in a cold, dry place. Adding 0.025% sodium benzoate to the item prior to packaging is also a good idea and afterward purify the bottles as a precautionary measure against decay during the three to about a month that the ketchup stays in the opened container before it is utilized (Belović et al., 2018).

20.2.4 Tomato Sauces

Any combination of fruits or vegetables can be used in making sauces. In reality, however most market in several countries sells chili sauce, tomato sauce and small amounts of fruit sauces such as 'Worcester' sauce, which consist of dates and apples in addition to tomatoes. Chutneys are made of diverse vegetables, sugar, fruits and at times vinegar, the mixtures are thick and look like jam. To balance the sugary taste of chutney, any edible sour fruit can be used as a base. Depending on the natural acidity, the addition of vinegar may not be necessary due to the high sugar content of

chutneys which has a preservative influence, elemental composition, and development of the fruits that are used. Most chutneys yield a caramelized syrup when boiled in most international cuisines, which changes the colour, thickness, and taste of the product. Boiling is also a means of preserving the product through pasteurization.

The processing method for the preparation of tomato sauce is related to tomato ketchup except that all of the seeds and unfiltered mash are utilized and that the seeds are not excluded. The products are filled hot into cans and bottles and treated in the water at 85°C–90°C for 30 minutes. In several ways, thicker ketchup is closer to tomato sauce in structure.

20.3 TOMATO PRESERVATION

New food processing and distribution technologies are being implemented by both developed and developing countries in response to the increase in demands for functional products, changing consumer preferences, and patterns of action. This is in addition to the growing interest in minimizing economic scale wastage streams of perishable fruits and vegetables while simultaneously maximizing the fruits and vegetables supply chain values towards attaining the global set standard for their consumption (Amit et al., 2017; Belović et al., 2018). The tomato industry is one of the most important agri-food sectors of the world with food industries involved with its production generating huge amounts of wastes and more than 40 million tons of tomatoes at the global level. This represents about 2.5% of incoming raw material (Boccia et al., 2019). Tomatoes are generally susceptible to mildew rot during storage, particularly when stored under high temperature and humidity conditions. Therefore, there is a need for the application of preservative(s) to minimize the rate of deterioration and extend the storage period of tomatoes (Olaniran et al., 2013). General fresh-keeping methods for tomato are cold storage, mechanical refrigeration, cooling devices, natural preservative agents (essential oils, extracts, paste, etc), the addition of chemical preservative, mud-pot storage (Wu et al., 2015). Microencapsulation technique was found to be more effective in maintaining a constant supply of the oil with bacteriostatic effect, the stability of flavour and anticorrosion in the whole tomato, its concentrates, and derivatives like tomato ketchup (Wu et al., 2015; Souza et al., 2018; Corrêa-Filho et al., 2019). The use of plant extracts as a bioactive film to preserve the tomatoes (Cejudo Bastante, 2019; Banu et al., 2020). The use of Pulsed Electric Field and Modified Atomspheric technique to preserve the tomato (Leng et al., 2020; Paulsen et al., 2019). The use of Ozone; a non-thermal technology is becoming appealing in the tomato value chain as an alternative preservative method due to its quick disintegration capacity with little or no residual toxicological effect when such is exploited for tomato juice production (Pandiselvam et al., 2019). Natural materials such as salt, sugar, vinegar, ginger, cinnamon, garlic and benzoates, nitrate, sulphites which are chemical preservatives are used commonly on a commercial scale for tomato paste preservation (Olaniran et al., 2015; George-Okafor et al., 2020). Even though chemical preservatives are essential for reducing spoilage organisms and ensuring the availability of healthy, affordable foods, their negative effect on food safety has become a widespread concern among consumers. Some of these preservatives have been linked to serious health problems including allergies, asthma, neurological damage, and tissue and organ cancer. (George-Okafor et al., 2020). Application of hurdle technology which is based on a reasonable combination of microbial limiting variables or maneuverability of their interaction in ensuring multiple target disruption of potential microorganisms cum biochemical changes that undermine food safety and nutrient quality is common in developed countries (Giannakourou et al., 2020; Tapia et al., 2020).

REFERENCES

Alenazi, M. M., Shafiq, M., Alsadon, A. A., Alhelal, I. M., Alhamdan, A. M., Solieman, T. H., Ibrahim, A.A., Shady, M.R. & Al-Selwey, W. A. (2020). Improved functional and nutritional properties of tomato fruit during cold storage. *Saudi Journal of Biological Sciences*, 27(6), 1467–1474.doi: 10.1016/j.sjbs.2020.03.026.

Amit, S. K., Uddin, M. M., Rahman, R., Islam, S. M., & Khan, M. S. (2017). A review on mechanisms and commercial aspects of food preservation and processing. *Agriculture & Food Security*, 6(1). doi: 10.1186/s40066-017-0130-8.

Banu, A. T., Ramani, P. S., & Murugan, A. (2020). Effect of seaweed coating on quality characteristics and shelf life of tomato (Lycopersicon esculentum mill). *Food Science and Human Wellness*, 9(2), 176–183. doi: 10.1016/j.fshw.2020.03.002.

Belović, M., Torbica, A., Pajić Lijaković, I., Tomić, J., Lončarević, I., & Petrović, J. (2018). Tomato pomace powder as a raw material for ketchup production. *Food Bioscience*, 26, 193–199. doi: 10.1016/j.fbio.2018.10.013.

Boccia, F., Di Donato, P., Covino, D., & Poli, A. (2019). Food waste and bio-economy: A scenario for the Italian tomato market. *Journal of Cleaner Production*, 227, 424–433. doi: 10.1016/j.jclepro.2019.04.180.

Cejudo Bastante, C., Casas Cardoso, L., Fernández-Ponce, M., Mantell Serrano, C., & Martínez de la Ossa, E. (2019). Supercritical impregnation of olive leaf extract to obtain bioactive films effective in cherry tomato preservation. *Food Packaging and Shelf Life*, 21, 100338. doi: 10.1016/j.fpsl.2019.100338.

Corrêa-Filho, L., Lourenço, S., Duarte, D., Moldão-Martins, M., & Alves, V. (2019). Microencapsulation of tomato (Solanum lycopersicum L.) pomace Ethanolic extract by spray drying: Optimization of process conditions. *Applied Sciences*, 9(3), 612. doi: 10.3390/app9030612.

Debastiani, R., Iochims dos Santos, C. E., Maciel Ramos, M., Sobrosa Souza, V., Yoneama, M. L., Amaral, L., & Ferraz Dias, J. (2021). Elemental concentration of tomato paste and respective packages through particle-induced X-ray emission. *Journal of Food Composition and Analysis*, 97, 103770. doi: 10.1016/j.jfca.2020.103770.

FAO-STAT (Food and Agriculture Organization of the United Nations). (2020). Statistics Division. Rome, dataset available at http://www.fao.org/faostat/en/#home (verified, February 2020).

FAOSTAT (Food and Agriculture Organization of the United Nations). (2016). http://www.fao.org/faostat/en/#data/QC [last accessed in January 2018].

Fattore, M., Montesano, D., Pagano, E., Teta, R., Borrelli, F., Mangoni, A., Seccia, S., & Albrizio, S. (2016). Carotenoid and flavonoid profile and antioxidant activity in "Pomodorino Vesuviano" tomatoes. *Journal of Food Composition and Analysis*, 53, 61–68. doi: 10.1016/j.jfca.2016.08.008.

Forbes-Hernandez, T. Y., Gasparrini, M., Afrin, S., Bompadre, S., Mezzetti, B., Quiles, J. L., Giampieri, F., & Battino, M. (2016). The healthy effects of strawberry polyphenols: Which strategy behind antioxidant capacity? *Critical Reviews in Food Science and Nutrition*, 56(1), S46–S59. doi: 10.1080/10408398.2015.1051919.

Gao, R., Ye, F., Wang, Y., Lu, Z., Yuan, M., & Zhao, G. (2019). The spatial-temporal working pattern of cold ultrasound treatment in improving the sensory, nutritional and safe quality of unpasteurized raw tomato juice. *Ultrasonics Sonochemistry*, 56, 240–253. doi: 10.1016/j.ultsonch.2019.04.013.

George-Okafor, U., Ozoani, U., Tasie, F., & Mba-Omeje, K. (2020). The efficacy of cell-free supernatants from lactobacillus plantarum Cs and lactobacillus acidophilus ATCC 314 for the preservation of home-processed tomato-paste. *Scientific African*, 8, e00395. doi: 10.1016/j.sciaf.2020.e00395.

Ghiasi, M. M., Mohammadi, A. H., & Zendehboudi, S. (2020). Clathrate hydrate based approach for concentration of sugar aqueous solution, orange juice, and tomato juice: Phase equilibrium modeling using a thermodynamic framework. *Fluid Phase Equilibria*, 512, 112460. doi:10.1016/j.fluid.2020.112460.

Giannakourou, M. C., Stavropoulou, N., Tsironi, T., Lougovois, V., Kyrana, V., Konteles, S. J., & Sinanoglou, V. J. (2020). Application of hurdle technology for the shelf life extension of European eel (Anguilla Anguilla) fillets. *Aquaculture and Fisheries*. doi: 10.1016/j.aaf.2020.10.003.

Górecka, D., Wawrzyniak, A., Jędrusek-Golińska, A., Dziedzic, K., Hamułka, J., Kowalczewski, P. Ł., & Walkowiak, J. (2020). Lycopene in tomatoes and tomato products. *Open chemistry*, 18(1), 752–756. doi.org/10.1515/chem-2020-0050.

Håkansson, A. (2019). Can high-pressure homogenization cause thermal degradation to nutrients? *Journal of Food Engineering*, 240, 133–144. doi: 10.1016/j.jfoodeng.2018.07.024.

Hilares, R. T., Dos Santos, J. G., Shiguematsu, N. B., Ahmed, M. A., da Silva, S. S., & Santos, J. C. (2019). Low-pressure homogenization of tomato juice using hydrodynamic cavitation technology: effects on physical properties and stability of bioactive compounds. *Ultrasonics Sonochemistry*, 54, 192–197. doi:10.1016/j.ultsonch.2019.01.039.

Innings, F., Alameri, M., Koppmaier, U. H., & Håkansson, A. (2020). A mechanistic investigation of cell breakup in tomato juice homogenization. *Journal of Food Engineering*, 272, 109858. doi:10.1016/j.jfoodeng.2019.109858.

Jayathunge, K., Stratakos, A. C., Cregenzán-Albertia, O., Grant, I. R., Lyng, J., & Koidis, A. (2017). Enhancing the lycopene in vitro bioaccessibility of tomato juice synergistically applying thermal and non-thermal processing technologies. *Food Chemistry*, 221, 698–705. doi: 10.1016/j.foodchem.2016.11.117.

Kelebek, H., Selli, S., Kadiroğlu, P., Kola, O., Kesen, S., Uçar, B., & Çetiner, B. (2017). Bioactive compounds and antioxidant potential in tomato pastes as affected by hot and cold break process. *Food Chemistry, 220*, 31–41. doi:10.1016/j.foodchem.2016.09.190.

Kulczyński, B., Gramza-Michałowska, A., Kobus-Cisowska, J., & Kmiecik, D. (2017). The role of carotenoids in the prevention and treatment of cardiovascular disease–Current state of knowledge. *Journal of Functional Foods, 38*, 45–65. doi:10.1016/j.jff.2017.09.001.

Leng, J., Mukhopadhyay, S., Sokorai, K., Ukuku, D. O., Fan, X., Olanya, M., & Juneja, V. (2020). Inactivation of Salmonella in cherry tomato stem scars and quality preservation by pulsed light treatment and antimicrobial wash. *Food Control, 110*, 107005. doi: 10.1016/j.foodcont.2019.107005.

Lu, C., Ding, J., Park, H. K., & Feng, H. (2020). High intensity ultrasound as a physical elicitor affects secondary metabolites and antioxidant capacity of tomato fruits. *Food Control, 113*, 107176. doi:10.1016/j.foodcont.2020.107176.

Martínez-Hernández, G. B., Boluda-Aguilar, M., Taboada-Rodríguez, A., Soto-Jover, S., Marín-Iniesta, F., & López-Gómez, A. (2016). Processing, packaging, and storage of tomato products: Influence on the Lycopene content. *Food Engineering Reviews, 8*(1), 52–75. doi: 10.1007/s12393-015-9113-3.

Mehta, D., Sharma, N., Bansal, V., Sangwan, R. S., & Yadav, S. K. (2019). Impact of ultrasonication, ultraviolet and atmospheric cold plasma processing on quality parameters of tomato-based beverage in comparison with thermal processing. *Innovative Food Science & Emerging Technologies, 52*, 343–349. doi: 10.1016/j.ifset.2019.01.015.

Mendonça, T. G., Silva, M. B. D., Pires, R. C. D. M., & Souza, C. F. (2020). Deficit irrigation of subsurface drip-irrigated grape tomato. *Engenharia Agrícola, 40*, 453–461. doi: 10.1590/1809-4430-Eng. Agric.v40n4p453-461/2020.

Migliori, C. A., Salvati, L., Di Cesare, L. F., Lo Scalzo, R., & Parisi, M. (2017). Effects of preharvest applications of natural antimicrobial products on tomato fruit decay and quality during long-term storage. *Scientia Horticulturae, 222*, 193–202. doi: 10.1016/j.scienta.2017.04.030.

Mohammed, S.M., Abdurrahman A.A, and Attahiru, M. (2017). Proximate analysis and total lycopene content of some tomato cultivers obtained from Kano State, Nigeria. *Chemistry Research Journal, 8*(1), 64–69, doi:10.4314/csj.v8i1.9.

Olaniran, A. F., Abiose, S. H., & Adeniran, A. H. (2015). Biopreservative Effect of Ginger (Zingiber officinale) and Garlic Powder (Allium sativum) on Tomato Paste. *Journal of Food Safety, 35*(4), 440–452. doi:10.1111/jfs.12193.

Olaniran, A. F., Abiose, S. H., & Gbadamosi, S. O. (2013). Effect of ginger and garlic as biopreservatives on proximate composition and antioxidant activity of tomato paste. *Ife Journal of Technology, 22*(1), 15–20.

Pandiselvam, R., Subhashini, S., Banuu Priya, E., Kothakota, A., Ramesh, S., & Shahir, S. (2019). Ozone based food preservation: A promising green technology for enhanced food safety. *Ozone: Science & Engineering, 41*(1), 17–34. doi: 10.1080/01919512.2018.1490636.

Parajuli, R., Matlock, M. D., & Thoma, G. (2021). Cradle to grave environmental impact evaluation of the consumption of potato and tomato products. *Science of the Total Environment, 758*, 143662. doi: 10.1016/j.scitotenv.2020.143662.

Paulsen, E., Barrios, S., & Lema, P. (2019). Ready-to-eat cherry tomatoes: Passive modified atmosphere packaging conditions for shelf life extension. *Food Packaging and Shelf Life, 22*, 100407. doi: 10.1016/j.fpsl.2019.100407.

Rahimi, S., & Mikani, M. (2019). Lycopene green ultrasound-assisted extraction using edible oil accompany with response surface methodology (RSM) optimization performance: Application in tomato processing wastes. *Microchemical Journal, 146*, 1033–1042. doi: 10.1016/j.microc.2019.02.039.

Rahimpour, S., & Taghian Dinani, S. (2018). Lycopene extraction from tomato processing waste using ultrasound and cell-wall degrading enzymes. *Journal of Food Measurement and Characterization, 12*(4), 2394–2403. doi: 10.1007/s11694-018-9856-7.

Rojas-Gracia, P., Roque, E., Medina, M., López-Martín, M. J., Cañas, L. A., Beltrán, J. P., & Gómez-Mena, C. (2019). The DOF transcription factor SlDOF10 regulates vascular tissue formation during ovary development in tomato. *Frontiers in Plant Science, 10*, 216. doi:10.3389/fpls.2019.00216.

Shao, D., Venkitasamy, C., Li, X., Pan, Z., Shi, J., Wang, B., Teh, H.E., & McHugh, T. H. (2015). Thermal and storage characteristics of tomato seed oil. *LWT-Food Science and Technology, 63*(1), 191–197. doi:10.1016/j.lwt.2015.03.010.

Souza, A. L., Hidalgo-Chávez, D. W., Pontes, S. M., Gomes, F. S., Cabral, L. M., & Tonon, R. V. (2018). Microencapsulation by spray drying of a lycopene-rich tomato concentrate: Characterization and stability. *LWT, 91*, 286–292. doi: 10.1016/j.lwt.2018.01.053.

Tapia, M. S., Alzamora, S. M., & Chirife, J. (2020). Effects of water activity (a W) on microbial stability as a hurdle in food preservation. *Water Activity in Foods*, 323–355. doi: 10.1002/9781118765982.ch14.

Viuda-Martos, M., Sanchez-Zapata, E., Sayas-Barberá, E., Sendra, E., Pérez-Álvarez, J. A., & Fernández-López, J. (2014). Tomato and tomato byproducts. Human health benefits of Lycopene and its application to meat products: A review. *Critical Reviews in Food Science and Nutrition*, *54*(8), 1032–1049. doi: 10.1080/10408398.2011.623799.

Wang, Y., Sun, P., Li, H., Adhikari, B. P., & Li, D. (2018). Rheological behavior of tomato fiber suspensions produced by high shear and high pressure homogenization and their application in tomato products. *International Journal of Analytical Chemistry*, *2018*, 1–12. doi: 10.1155/2018/5081938.

World Processing Tomato Council (WPTC) (2019). WPTC Crop Update and World Production Estimate as of 23 October 2019. Available online at https://www.wptc.to/release.Wptc.php (verified Februaty 2020).

Wu, H., Xue, N., Hou, C., Feng, J., & Zhang, X. (2015). Microcapsule preparation of allyl isothiocyanate and its application on mature green tomato preservation. *Food Chemistry*, *175*, 344–349. doi: 10.1016/j.foodchem.2014.11.149.

Xu, Q., Adyatni, I., & Reuhs, B. (2018). Effect of processing methods on the quality of tomato products. *Food and Nutrition Sciences*, *9*(02), 86–98. doi: 10.4236/fns.2018.92007.

21 Relevance of Natural Bioresources and Their Application in Pharmaceutical, Food and Environment

Olugbemi T. Olaniyan
Rhema University

Charles Oluwaseun Adetunji
Edo State University Uzairue

CONTENTS

21.1 INTRODUCTION

The global population has been forecasted to increase drastically to 9.6 billion in the year 2050. Therefore, there is a need to guarantee the adequate provision of food to ever increasing population. There are several increased pressure on the agriculture sector to boost the production of agriculture and ameliorates the challenges boarding around food insecurity and malnutrition. The goal of feeding the ever increasing population could only be possible by making more land to be more available as well as increasing the level of soil fertility, remediation of heavily polluted soil. This could only be possible through the application of eco-friendly tools through the application of natural recourses that could enhance food security and increase in food production. Majority of the numerous agricultural practices depend on several agricultural practices through the usage of synthetic pesticides such as insecticides, herbicides and fungicides as well as over irrigation, intensive tillage (Singh, 2011).

This technique has assisted many poor farmers in the developing nation to need the demand of their population but there are numerous adverse effects that are associated with such techniques. Some of these challenges associated with such problems include higher cost of these agricultural pesticides, overuse of water and land resources, declining of soil fertility. Therefore, there is a need to search for a sustainable agricultural techniques that help in meeting the current and imminent food requests of the population without any within the accessible limited resources, deprived of weakening the environmental excellence (Singh and Strong, 2016).

DOI: 10.1201/9781003268468-21

The application of beneficial microorganism have been established to portend the capacity to increase the generation of sustainable green energy production as well as improves the level of soil fertility (Koller et al., 2012). Typically some have validated the fact that cyanobacterial biomass could be applied for food production, biofertilizer as well as food supplement production (Benson et al., 2014). This could go along way in prevention of GHG emissions and improves the assurance of food security (Singh et al., 2011).

The constant application of synthetic chemical, food preservation and agrochemicals have led to an increase in agricultural produce but entails several; challenges involved. These concern entails health and environmental challenges associated with these agricultural products. Therefore, the application of natural products is more safer for food preservation and cheaper agrochemical (Dayan, et al., 2009, US EPA, 1996, Njoku et al., 2017). Most of these natural products could help prevention pest invasion resulting to uninterrupted food supply (Casida, 2010).

Moreover, it has been established that pesticides could react with the ecosystem and plant genome. But most of these pesticides could get metabolized most especially organisms' tissues leaving disastrous end-products. Also, most of these metabolites could cause adverse effect most especially to ecosystem, target organism and mankind (Rohan et al., 2012; Chiu et al., 2015)

Therefore, it has become a factual fact food preservation and crop protection are very crucial toward the achievement of global food sustainability. There are several factors that caused food spoilage such microbial attacks, climatic conditions and insect infestation. Hence, the utilization of natural preservatives has been applied in agrochemical industries, pharmaceutical, food (Mogoşanu et al., 2017, Santos-Sánchez et al., 2017). Typical example of Phytochemicals that could be applied in the production of virucides, pesticides, insecticides, herbicides, entails polyketides, alkaloids, amino acids, phenylpropanoids, terpenoids, lipids (Thorat et al., 2017, Vora et al., 2018, Yan et al., 2018)

Therefore, this chapter intends to provide a comprehensive review the application of natural product bioresources and their application in pharmaceutical, food and environment

21.2 NATURAL BIORESOURCES AND THEIR APPLICATION IN PHARMACEUTICAL, FOOD AND ENVIRONMENT

Several minerals from biomass sever as a source of raw materials for food, medicine and bioagents (Godfray et al., 2010). In their study, it was discovered that the global environmental and economic crisis has prompted enormous research output in the area of bioresources for the generation of bio-based products for the industrial revolution. The authors noted that microbial polysaccharides mainly cellulose serve as emerging and potential raw materials in bio-based industries for the production of paper, food, pharmaceutical products, biomedical products, tissue engineering, dentistry and cosmetics. It is noted that cellulose is very abundant, homogenous, non-degradable, excellent physiochemical properties and renewable natural products. It is worthy to note that many microbes are capable of synthesizing cellulose including plants. Studies have reported that microbial cellulose has several advantages over plants' cellulose in terms of lack of pectin, lignin and hemicellulose products capable of affecting the structure.

Arevalo et al. (2016) demonstrated that several techniques have been deployed in the manufacturing of bio-based products such as surface casting, pultrusion, ultrasonic-assisted casting, extrusion, press molding, injection molding, hand lay-up, sheet molding compounding, enzymatic grafting and filament winding. Suffice to mention that many synthetic products are known to cause severe health risks like respiratory dysfunctions and skin irritation but bio-based products have low health risk, non-toxic, eco-friendly and easy handling. Arevalo et al. (2016) also revealed that cellulose-based materials are currently utilized for diverse products such as paper, food, tissue engineering, packaging, electronics, pharmaceuticals, biocomposites, medicine and dentistry. Many bacteria strains are utilized for the production of biomedical materials and value added products. In food and pharmaceutical industries, the health benefits of microbial cellulose derived from some species like *Acetobacter xylinum* have been established using pineapple water-based and coconut water-based culture mediums. There is a great need for further exploitation, sustainable

conservation and utilization of so many bioresources available for the survival and development of mankind through the production of agricultural products, environmental protection, therapeutics, biofuel, food, drugs, cosmetics, bioenergy and other value-added products .

21.3 PLANT BIORESOURCES

Due to the serious population explosion creating severe food insecurity, environmental pollution, climate change and nutritional deficiency. Thus, there is serious need to increase the capacity for the growing demand for food, nutritional security, clothing, health and shelter globally. Recently, there have been a lot of efforts to explore traditional medicine as bioactive molecules derived from plants for health needs. Plant diversity has been reported to play a critical role in addressing nutritional security, food and medicinal compounds.

Dulloo et al. (2014) revealed that plant species serve as sources of essential and fundamental bioresources for food, clothes, medicines, building materials, pharmaceuticals, ornaments, biochemical products, bioactive metabolites, nutraceutical and biomedical devices (Bharucha and Pretty, 2010). Many plants derived anticancer agents have been reported for plant-based therapies. Many of the phytochemicals such as vinblastine, alkaloids, vinorelbine and vincristine have been reported for the antidiabetic, anticancer and fertility properties. Also, marine-derived anticancer agents from Africa like invertebrates have been reported to possess powerful anticancer properties such as hemiasterlins, proanthocyanidins and dolastatins. Plant derived anti-plasmodial compounds such as anthraquinones, pseudoguaianolides, isoquinolines, morindone, pentacyclic triterpenes and diterpenes. Briskin (2000) reported that plants are currently being utilized to tackle non-communicable and communicable diseases particularly in Asia and Africa which provides a cheaper and assessable means of disease management. The authors revealed that these medicinal plants are rich in bioactive fractions with capacity to manage chronic and infectious diseases physiologically through nutrigenomic mechanisms.

Ayurveda science includes all natural bioresources that can be utilized for medicine and environmental protection. In the study, the author grouped bioresources into different areas such as primary, secondary, tertiary and quaternary bioresources. Furthermore, primary bioresources were described to entail a specific purpose of generating bioresources for products, food and energy. Secondary bioresources are produced during the primary process as a form of byproducts or secondary residue. Tertiary bioresources were described as raw materials during the production cycle. Quaternary bioresources are generated after the utilization of a product. Dnyaneshwar (2017) concluded that there should be an establishment of traditional bioresources value to cater for the ongoing global development.

Bioresources are very important renewable resources such as bio-based raw materials in agriculture, forestry, fisheries and aquaculture, food, biofuels and bioenergy available for the production of essential value chain products for chemical and plastic industry, building and construction industry, pharmaceutical industry, technological sector, textile, wastewater treatment and many more. Many authors have revealed that bioresources are under threat due to unsustainable harvesting, thus the need for protection and other multi-dimensional interventions. Currently, the world's bio-based polymer and chemical production has been projected to grow exponentially to about 113 million tonnes by the year 2050 which is serving as a substitute for petrochemical products.

Jamshidi-Kia et al. (2018) revealed that environmental resources serve as valuable materials for human as sources of food and medicine. Information about so many medicinal plants has been passed from one generation to the other. Medicinal plants are sources of active plant-derived compounds for the treatment of many notable human diseases, maintenance of good health and development. These phytochemical compounds are source of drug discovery agents and pharmaceutical products.

Africa is very rich in diverse arrays of medicinal plants which are currently reemerging for health aid and bioprospecting of new plant-derived pharmaceutical drugs. Many rely on the use of traditional medicine in developing countries, thus extract from these plants serves as chemotherapeutics

for the treatment of ailments and maintenance of personal health. Global the market for plant-derived bioresources, fragrance, pharmaceuticals, ingredients, flavors and chemicals exceeds several billion of dollars annually (UNESCO, 1998).

21.4 MARINE BIORESOURCES

Collins et al. (2020) reported that the growing rate of blue bioeconomy has generated a lot of contributions and economic gains to the sustainable development goals of the United Nations. Even though there are numerous challenges facing the sector, pharmaceutical, medical, agricultural, biotechnology and nutraceutical sector seems to experience the worst hit relating to the issue of technology, market, supply, sustainable, mass generation of bioactive molecules/other biomaterials. The marine ecosystem provides numerous opportunities for economic and environmental services to humanity. The oceanic economy like seas, oceans, and coasts offers gas, energy, fisheries, transport, food, shipping, offshore oil, tourism, extracellular enzymes like pectinase from marine fungi, wind, minerals, and other bioresources. Collins et al. (2020) reported that marine bioresources can be utilized for biotechnological applications and the estimated global market has been projected to be around 6.5 billion USD by the year 2024. Marine bioresources and bioeconomy can produce economic growth, support to public health system, jobs, environmental protection and tourism.

Collins et al. (2020) reported that many marine organisms are potential sources of bioresources like fungi, bacteria, microorganisms, sponges, macro/microalgae, cyanobacteria, fishes, plants, invertebrates, mollusks, and animals generating different compounds for pharmaceuticals, cosmetics, nutraceuticals, biotechnology, agricultural and many others. Scientists are now utilizing many of these marine bioresources to generate drugs, medicines, enzymes, foods, paints, supplements, biomaterials, antiaging products, co-adjuvants, enhancers, bioplastics, and many others. It should be noted that the scientific knowledge for marine bioresources is increasing for commercialization and generation of value-added chains of products. Though many parts of our blue bioeconomic sector remained unexploited when compared with the terrestrial environment, thus many of the emerging blue bioeconomy for biotechnology to be very small and underdeveloped (Jaspars et al., 2016).

21.5 EDIBLE INSECTS BIORESOURCES

Studies have shown the role of edible insects like crickets, male moths, prawns, grasshoppers, bees, locusts, cockroaches, mealworms, melon and sorghum bugs, weaver ants, as a potential target for the generation of many valuable industrial biotechnological value-added chains of products. Across the globe, insects are utilized for food, animal feed as a result of high protein content form part of the diets of not less than 2 billion individuals (Ayieko et al., 2012).

Many valuable products are derived from edible insects such as honey, silk, oil, beeswax, propolis, therapeutic enzymes, toxins and venoms, dyes, and ornaments. The environmental benefits are largely based on pollination processes, reduction in environmental contamination, pest control and conservation of ecosystem. Edible insects are highly nutritious with a lot of vitamins, fat, fiber, chitin, proteins and minerals. Insect-based animal feed products have recently emerged as a major aquacultural, livestock and poultry production with established niche markets (Acuña et al., 2011).

21.6 AGRICULTURAL-BASED BIORESOURCES

Sadh et al. (2018) reported that many agricultural residues that are released into the environment as waste products which are underutilized, untreated forms constitute a major source of environmental pollution. These agricultural residues can be converted into an alternative source of value-added products like biofuel, enzymes, animal feeds, biogas, antibiotics, antioxidants, vitamins, biochemical and raw materials. The authors noted that juice industries generate a huge amount of waste like peels, stubbles, stalks, shells, straws, seeds, bagasse, pulp, wood shavings and husks contributing

to about 147.2 million metric tons of fiber sources all over the world. Also, many of these agricultural wastes like banana, pomegranate lemon and walnut peels are natural sources of antimicrobial agents, nutritional factors, fermented products, oil, grease, substrate, metabolites and other biomolecules. Recently, renewed interest of many scientists in the utilization of agricultural bioresources as sources of raw materials for the production of valued added products has been witnessed. Some of the products include enzymes, biosurfactants, organic acids, bioethanol, biofertilizers, aroma compounds, biopesticides, pigments, antibiotics, feeds, and vitamins. Agro-industrial residues or waste have been shown to be very rich sources of nutrition, bioactive molecules, with a lot of compositions like sugars, minerals and proteins (Sukan et al., 2014).

21.7 CONCLUSION AND FUTURE RECOMMENDATION TO KNOWLEDGE

This chapter has provided detailed information on the application of natural bioresources and their application in pharmaceuticals, food and environment. Moreover, several relevant information on the relevance of several natural bioresources were also highlighted such as plant bioresources, marine bioresources, edible insects bioresources, agricultural-based bioresources. There is a need for government and policymakers to pay attention to several available natural resources that could boost the increase of food production and maintenance of health environment.

REFERENCES

Acuña, A. M., Caso, L., Aliphat, M. M., & Vergara, C. H. (2011). Edible insects as part of the traditional food system of the Popoloca town of los Reyes Metzontla, Mexico. *Journal of Ethnobiology*, *31*(1), 150–169. https://doi.org/10.2993/0278-0771-31.1.150.

Ayieko, M. A., Kinyuru, J. N., Ndong'a, M. F., & Kenji, G. M. (2012). Nutritional value and consumption of black ants (*Carebara vidua Smith*) from the Lake Victoria region in Kenya. *Advance Journal of Food Science and Technology*, *4*(1), 39–45.

Benson, D., Kerry, K., & Malin, G. (2014). Algal biofuels: Impact significance and implications for EU multi-level governance. *Journal of Cleaner Production*, *72*, 4–13. https://doi.org/10.1016/j.jclepro.2014.02.060.

Bharucha, Z., & Pretty, J. (2010). The roles and values of wild foods in agricultural systems. *Philosophical Transactions of the Royal Society B: Biological Sciences*, *365*(1554), 2913–2926. https://doi.org/10.1098/rstb.2010.0123.

Briskin, D. P. (2000). Medicinal plants and phytomedicines. Linking plant biochemistry and physiology to human health. *Plant Physiology*, *124*(2), 507–514. https://doi.org/10.1104/pp.124.2.507.

Casida, J. E. (2010). Pest toxicology: The primary mechanisms of pesticide action. In: Krieger, R. (ed.) *Hayes' Handbook of Pesticide Toxicology*. Elsevier B.V., Amsterdam, pp. 103–117.

Chiu, Y., Afeiche, M., Gaskins, A., Williams, P., Petrozza, J., Tanrikut, C., Hauser, R., & Chavarro, J. (2015). Fruit and vegetable intake and their pesticide residues in relation to semen quality among men from a fertility clinic. *Human Reproduction*, *30*(6), 1342–1351. https://doi.org/10.1093/humrep/dev064.

Collins, J. E., Vanagt, T., Huys, I., & Vieira, H. (2020). Marine bioresource development – Stakeholder's challenges, implementable actions, and business models. *Frontiers in Marine Science*, *7*, 62. https://doi.org/10.3389/fmars.2020.00062.

Dayan, F. E., Cantrell, C. L., & Duke, S. O. (2009). Natural products in crop protection. *Bioorganic & Medicinal Chemistry*, *17*(12), 4022–4034. https://doi.org/10.1016/j.bmc.2009.01.046.

Godfray, H. C., Beddington, J. R., Crute, I. R., Haddad, L., Lawrence, D., Muir, J. F., Pretty, J., Robinson, S., Thomas, S. M., & Toulmin, C. (2010). Food security: The challenge of feeding 9 billion people. *Science*, *327*(5967), 812–818. https://doi.org/10.1126/science.1185383.

Jamshidi-Kia, F., Lorigooini, Z., & Amini-Khoei, H. (2018). Medicinal plants: Past history and future perspective. *Journal of Herbmed Pharmacology*, *7*(1), 1–7. https://doi.org/10.15171/jhp.2018.01.

Jaspars, M., De Pascale, D., Andersen, J. H., Reyes, F., Crawford, A. D., & Ianora, A. (2016). The marine biodiscovery pipeline and ocean medicines of tomorrow. *Journal of the Marine Biological Association of the United Kingdom*, *96*(1), 151–158. https://doi.org/10.1017/s0025315415002106.

Koller, M., Salerno, A., Tuffner, P., Koinigg, M., Böchzelt, H., Schober, S., Pieber, S., Schnitzer, H., Mittelbach, M., & Braunegg, G. (2012). Characteristics and potential of micro algal cultivation strategies: A review. *Journal of Cleaner Production*, *37*, 377–388. https://doi.org/10.1016/j.jclepro.2012.07.044.

Mogoşanu, G. D., Grumezescu, A. M., Bejenaru, C., & Bejenaru, L. E. (2017). Natural products used for food preservation. *Food Preservation*, 365–411. https://doi.org/10.1016/b978-0-12-804303-5.00011-0.

Njoku, K., Ezeh, C., Obidi, F., & Akinola, M. (2017). Assessment of pesticide residue levels in vegetables sold in some markets in Lagos state, Nigeria. *Nigerian Journal of Biotechnology*, 32(1), 53. https://doi.org/10.4314/njb.v32i1.8.

Rohan, D., et al. (2012). Pesticide residue analysis of fruits and vegetables. *Journal of Environmental Chemistry and Ecotoxicology*, 4(2), 19–28. https://doi.org/10.5897/jece11.072.

Sadh, P. K., Duhan, S., & Duhan, J. S. (2018). Agro-industrial wastes and their utilization using solid state fermentation: A review. *Bioresources and Bioprocessing*, 5(1). https://doi.org/10.1186/s40643-017-0187-z.

Santos-Sánchez, N. F., Salas-Coronado, R., Valadez-Blanco, R., Hernández-Carlos, B., & Guadarrama-Mendoza, P. C. (2017). Natural antioxidant extracts as food preservatives. *Acta Scientiarum Polonorum Technologia Alimentaria*, 16(4), 361–370. https://doi.org/10.17306/j.afs.0530.

Singh, J. S. (2011). Methanotrophs: The potential biological sink to mitigate the global methane load. *Current Science*, 100, 29–30.

Singh, J. S., & Strong, P. (2016). Biologically derived fertilizer: A multifaceted bio-tool in methane mitigation. *Ecotoxicology and Environmental Safety*, 124, 267–276. https://doi.org/10.1016/j.ecoenv.2015.10.018.

Singh, J. S., Pandey, V. C., & Singh, D. (2011). Efficient soil microorganisms: A new dimension for sustainable agriculture and environmental development. *Agriculture, Ecosystems & Environment*, 140(3–4), 339–353. https://doi.org/10.1016/j.agee.2011.01.017.

Sukan, A., Roy, I., & Keshavarz, T. (2014). Agro-industrial waste materials as substrates for the production of poly(3-hydroxybutyric acid). *Journal of Biomaterials and Nanobiotechnology*, 05(04), 229–240. https://doi.org/10.4236/jbnb.2014.54027.

Thorat, P.P., Kshirsagar, R.B., Sawate, A.A., & Patil, B.M. (2017). Effect of lemongrass powder on proximate and phytochemical content of herbal cookies. *Journal of Pharmacognosy and Phytochemistry*, 155(66), 155–159.

UNESCO (1998). FIT/504-RAF-48 Terminal Report: Promotion of Ethnobotany and the Sustainable Use of Plant Resources in Africa, pgs. 60, Paris.

US EPA (1996). *Food Quality Protection Act: Draft Implementation Plan*. US EPA, Washington, DC.

Vora, J., Srivastava, A., & Modi, H. (2018). Antibacterial and antioxidant strategies for acne treatment through plant extracts. *Informatics in Medicine Unlocked*, 13, 128–132. https://doi.org/10.1016/j.imu.2017.10.005.

Yan, Y., Liu, Q., Jacobsen, S. E., & Tang, Y. (2018). The impact and prospect of natural product discovery in agriculture. *EMBO Reports*, 19(11). https://doi.org/10.15252/embr.201846824.

22 Genetically Modified and Wild Potatoes

Depository of Biologically Active Compounds and Essential Nutrients

Charles Oluwaseun Adetunji
Edo State University Uzairue

Muhammad Akram
Government College University Faisalabad

Olugbenga Samuel Michael
Bowen University

Benjamin Ewa Ubi
Ebonyi State University

Oluwaseyi Paul Olaniyan
Osun State University

Rabia Zahid
Government College University Faisalabad

Ruth Ebunoluwa Bodunrinde
Federal University of Technology Akure

Rumaisa Ansari
Government College University Faisalabad

Juliana Bunmi Adetunji
Osun State University

Areeba Imtiaz
Government College University Faisalabad

Osahon Itohan Roli
Edo State University Uzairue

DOI: 10.1201/9781003268468-22

Wadzani Palnam Dauda
Federal University Gashua

Neera Bhalla Sarin
Jawaharlal Nehru University

CONTENTS

22.1 INTRODUCTION

Potato designated by binomial nomenclature as *Solanum tuberosum* belongs to the family Solanaceae, comprising about 90 genera and 2,800 species. It is known as Aalu in Bengali and Urdu, Batata in Gujarati, Alu in Hindi, Urulaikkilangnku in Tamil, Potato in English, and Alugedde in Kannada. Potato is an annual non-woody (herbaceous) plant, mainly reproduced vegetatively via tubers and typically by botanical seeds, i.e., True Potato seeds. Besides leaves, stem and floral parts tubers are the major part of the plant. The potato tuber is an enlarged part of an underground stem from which new shoots are produced. The tuber is morphologically a fleshy stem, carrying buds and eyes in the axil of small-scale-like leaves. There is the presence of eyes on the apical end of the tuber, with a small number near the stolon or basal end. The presence of eye number and their distribution are used in the characteristics of the various variety. Being the most favorite and main component of food items it lies fourth among the crops after wheat, maize, and rice. It is frequently used as fries, chips, canned and mashed potatoes, and ready meals (Tierno et al., 2015).

Potato has been identified as one of the most widely consumed vegetables endowed with several benefits. It is a highly nutritious crop with numerous components such as carbohydrates (22%), proteins (2%), fats (0.1%), water (74%) along with minerals and traces elements viz. potassium, sodium, iodine and magnesium, folic acid, pyridoxine, vitamin C, ascorbic acid and iron. The presence of other biological activities such as antibacterial, antifungal, antiviral, and anti-inflammatory properties has been reported. Moreover. The availability of mineral and salt content has also been proven to enhance its anti-inflammatory capability most especially when applied burn-induced wounds using its peel and pulp that has been crushed and converted to paste for easy application. Also,

some other research has affirmed its antioxidant property by reducing free oxygen radicles which makes potato the second most effective antioxidant food after broccoli (Chandrasekara and Kumar, 2016). Furthermore, some other biological activities include anti-irritating, soothing, anti-cancer, anti-ulcer, anti-LDL peroxidation, de-congesting, and antiaging properties. Due to its low calories it is a good alternative for cereals and grains in the obese and overweight populations that's why it is considered an anti-obesity agent. Some other work has shown its significance in the treatment of gastrointestinal tract ailments which has been used in the treatment of constipation and hemorrhoids.

Solanum tuberosum has been identified as one of the important crops worldwide with high nutritional values (Millam, 2006). Potatoes are mainly used as staple foods but they are also very significant in the medicinal field (Kuete, 2014). According to the consumption of food crops around the world, potatoes lie on the third number in importance, following wheat and rice (Camire et al., 2009). Potato and its components are known to be important in the control of several cardio metabolic measures which include lowering cholesterol levels, improving lipid profile, preventing cardiovascular diseases, and lowering inflammation markers (McGill et al., 2013).

Throughout the past few years, molecular tools have helped a lot to isolate much of the genes taking part in the synthesis of methionine and cysteine, both of which are sulfur-containing amino acids. The development of transgenic plants that are modified in the activity of individual genes is made easy with the help of plant transformation technology. Transgenic or genetically modified form of *Solanum tuberosum* or potato contains methionine which plays an active role in the reduction of serum lipid levels and the activity of genetically modified or transgenic potato as an anti-hyperlipidemic agent (Nikiforova et al., 2002)

Genetically modified organisms are designated as those having their genetic makeup changed and organized followed by the researcher desire. Genetic modification is a product of genetic engineering which can be designed either by deleting the gene of non-interest or by adding the gene of interest in the DNA strands or RNA in the case of microbes. The purpose of generating genetically modified organisms (GMO) is to introduce the desired characteristics in the phenotype of an organism via incorporating genes of interest in genotype or to diminish the less desired characteristics from phenotype by removing genes of non-interest. Another third technique is the replacement of an existing gene with the gene of interest in DNA (Lee and Gelvin, 2008; Park, 2007; Cabot et al., 2001; Kita et al., 2013). The targeted organism for gene modification includes a variety of animals, plants, and microorganisms and several applications are derived from such organisms for therapeutic, industrial, nutritional purposes. The introduction of genetic modification has gained keen interest among the medicine as it generates superb consequences regarding the treatment and cure of various diseases especially that of genetic origin. Plants are a major focus as they are a rich and vast source of food as well. Among plants, potato remained a point of interest due to the presence of characteristics specified by Gregor Mendel to have in compulsion in an organism for genetic intervention. The presence of tuber plays a key role as this part contained all the active phytochemicals in it in rich concentration compared to other parts. Research presented the Benzoic acids, Cinnamic acid, flavonoids, Anthocyanins, alkaloids, proteins, amino acids, carbohydrates especially starch, vitamins, 2-Carboxyarabinitol-1- phosphate, carotenoids, and phytic acid-rich tubers of potato (Umadevi et al., 2013).

Genetically modified organisms are being used in research for being a major source of healthy food as well as to produce things other than food. The term genetically modified organism is very similar to the technical legal term defined in the Cartagena Protocol on Biosafety as, 'living modified organism'. It reflects the regulation of international trade in living genetically modified organisms more precisely, "any living organism, carrying an innovative combination of genetic material achieved through the use of modern biotechnology" (Sahar et al., 2017; Kita et al., 2013). Genetically modified organisms have been studied for their various applications. They have been used in research of both medical and biological types as well as in agriculture and the production of experimental medicine and pharmacological drugs (Falck-Zepeda et al., 2000; Shipitalo et al.,

2008; Christou et al., 2006; Hutchison et al., 2010). The use of genetic engineering in the production of food has amplified not only the quantity but also the quality of food (Izquierdo, 2000).

Plants are specially used for genetic modifications in higher organisms. Among the plants potatoes specifically represent a better model for genetic modifications for several reasons. One of the most important reasons is the presence of tubers in potato plants. Tubers are the storage organs for the plant in which the plant can store nutrients that they use during winter and dry seasons for growth. Also because potato plants reproduce asexually by the process of vegetative propagation if modified genetically the feature of this modification will be conserved for quite a long period of time. On the contrary, a tetraploid genome, high variability, and low numbers of produced mutants contribute to the drawbacks of using genetic modification in potato plants. In Europe, Cultivar Desirée has been used to carry out studies targeted at expressing the foreign genes in plants (Davies, 1996). Potato grows through vegetative propagation by planting small pieces of dormant buds or eyes containing potato tubers or they are grown from botanical seeds. The dormant buds of potato tubers develop into new shoots when provided with suitable environmental conditions. These tubers are a good source of carbon and nitrogen. They contain storage proteins and starch.

According to the reports published by Food and Agricultural Organization Statistics (FAO statistics), after rice, wheat and corn, the fourth majorly yielded crop is potato and in terms of area under cultivation, it is number eight. Because of the presence of tubers and high productivity per unit area, the potato plant represents one of the best options to lessen the shortages of food. Potato is a perennial herbaceous plant belonging to the *Solanaceae* family and is scientifically known as *Solanum tuberosum* L. Potato has yellow stamens with white to purple flowers. Some cultivars of potato contain green seeds, small in size and containing several seeds of up to 300 in number.

In view of all the aforementioned, this chapter intends to provide a holistic review of the nutritional and other health benefits of genetically modified potatoes. Moreover, recent advances in the application of biotechnology for the enhancement of biological components available in the mass production of genetically modified potatoes were provided in detail.

22.2 BIOLOGICALLY ACTIVE COMPOUNDS OF GENETICALLY MODIFIED POTATOES

Potato tubers contain total carbohydrates in the ranges of 1.0–7.0 g/kg. The reducing sugars such as glucose and fructose although in high amounts initially, are reduced greatly near the season of cultivation. Therefore, young tubers have large amounts of reduced tubers as compared to matured ones. The primary carbohydrate component of the dry matter of potatoes is starch and consists of amylase and amylopectin. The functional properties of various foods have been enhanced by adding starch to them. The characteristics of the starch structure and ratio of amylase to amylopectins differ in various cultivars of potatoes. Characteristics of starch and its content in potatoes greatly affect the quality of processing and nutrition of potato products. Several methods have been adopted to improve the processing operation of the starch present in potatoes. These modifications are of chemical, enzymatic and physical types.

In agriculture, extensive use of chemical pesticides has led to increased resistance in pests against these pesticides. Because of this research have been carried out on the production of genetically modified crops which can successfully resist these pests without the use of chemical pesticides and can be a lot more beneficial (Slater et al., 2003). Potato beetle (*Leptinotarsa decemlineata*) is one of the potato plant pests which can cause serious consequences and becomes resistant to chemical insecticides. Cry3A gene which has been derived from the bacteria Bacillus thuringiensis has been introduced into the potato plants to control this beetle. This gene produces a toxic protein in the leaves of the potato plant and when these leaves are ingested by the beetle, the protein passes on to the bowels of the beetle and causes its death due to toxicity. It is interesting to note that this protein is toxic to the potato beetles of all developmental stages however their natural enemies are not affected by it (Perlak et al., 1993).

Cry3A gene is selective only to the Colorado potato beetles and does not affect other plant pests. In regards to this other modifications of potatoes have been developed such as wheat α-amylase inhibitors (WAI), snowdrop lectins (GNA), and bean chitinases (BCH). Peach-potato aphids also called *Myzus persicae* have been used to study the insecticidal properties of these modified plants. The insecticidal property of snowdrop lectins (GNA) was recorded as best. These exerted inhibitory effects on the development of insects by impairing their fertility and thus leading to decreased propagation of the population of insects (Gatehouse et al., 1996). This has been further tested by another study in which the modified plants containing the GNA were tested against the larvae moth *Lacanobia olearacea* which is also a vector of various viral infections while also being a pest of potato plants. The plants expressing GNA were found to have increased resistance against the moth and further supported the evidence that lectins GNA when genetically introduced in potato plants protect them from several insect pests of the plants (Gatehouse et al., 1997). A major drawback of introducing the lectin GNA in plants was observed to be a decrease in the production of foliar glycoalkaloids of the plants which are indigestible to various types of insects and mammals. As a consequence of lectin gene introduction in potato plants, while it may protect the plant from some pests it may make it more attractive for some others which initially did not target the plant for their food (Birch et al., 2002). Another disadvantage of this procedure was shown in a study performed by feeding the GNA-modified potatoes to the rats with results in the proliferation of their intestinal cells and changes in the gastric mucosa. The changes in the gastric mucosa were considered to be due to the presence of lectins GNA while the intestinal proliferation was due to genetic transformation (Ewen and Pusztai, 1999). It raised questions about the safety of these GNA-modified potatoes for ingestion as food in humans. However, no other published data about this has been found.

Temporin A gene has been introduced into the plants to control potato blight. Studies have confirmed that the plants which express Temporin A are resistant to most fungal pathogens such as *P. erytroseptica* and *P. infestans* (Osusky et al., 2004). Potato blight has also been controlled by the plants producing hydrogen peroxide. It has been achieved by the introduction of glucose oxidase gene from *Aspergillus niger* in potato plants. Glucose oxidase catalyses the oxidation of β-D-glucose with the result in the release of hydrogen peroxide along with gluconic acid (Wu et al., 1995). Similarly, plants modified with the introduction of the coat protein of PLRV (potato leafroll luteovirus) showed increased resistance to luteovirus. It was found that the transgenic potato plants contained a decreased amount of the viral antigen as compared to the control (van der Wilk et al., 1991).

Potato plants have also been modified to increase their nutritional value. Potato plants modified by the introduction of a gene for a non-allergenic protein AmA1 from *A. hypochondriacus* have been produced and shows an increased amount of all amino acids in the tubers of potato plants as compared to nontransgenic potatoes. The important amino acids produced in increased amounts include lysine, cysteine, methionine, and tyrosine which are in very small amounts in nontransgenic potato plants. This is highly beneficial as it increases the nutritional value of the potato plants and has been used to decrease malnutrition in Indian poor children (Chakraborty et al., 2000).

22.3 LIST OF PHYTOCHEMICALS AND BIOLOGICALLY ACTIVE COMPONENTS PRESENT IN POTATOES

The presence of antioxidants and phytochemical properties extracted from plants has been of great use in the reduction of diseases (Lee et al., 2017). Potato (*Solanum tuberosum*) tuber cultivated in over 125 countries, constitutes one of the highest consumption followed by rice, maize, and wheat in the world (Jansky et al., 2019). The phytonutrient component of Potato is rich in carbohydrate, starch, vitamin C and B- complex, niacin, potassium, sodium, iodine and magnesium, folic acid, pyridoxine, and Iron, however, with fairly low protein content and fat but has an excellent biological value of 90–100 (Kanter and Elkin, 2019; Navarre, 2019; Beals, 2019).

Potato contains several secondary metabolites as a biologically defensive mechanism that is desirable in human nutrients, among them include phenolics, flavonoids, polyamines, polyphenols,

anthocyanins, and carotenoids. The biologically- active substances of potatoes cannot be overemphasis because of their health-promoting attributes (Kosieradzka et al., 2008). Moreover, Secondary products found in the plant are regarded as plant metabolism which implicated human health as antioxidants.

An in vivo experiment conducted by Chaparro et al. (2018) revealed another approach of metabolomics and Economics to exploring the bioactive compound present in Potato tuber. Another finding on comparative phytochemicals composition of different colored potatoes (Solanum tubers L.) tubers by Lee et al. (2017) confirmed the metabolic association-related biochemical pathway between metabolite characteristics and color difference in tubers. Rodriguez-Perez et al. (2018), in vivo study of comprehensive metabolite profiling of potato (Solanum tuberosum L.), leaves by HPLC-ESI-QTOF-MS the presence of 109 compounds present in potato leaves, including organic acids, amino acids, and derivatives, phenolic acids, flavonoids, iridoids, oxylipins and other polar and semi-polar compounds. The research further ascertain that 45% of quinic acid and its derivatives, were found in the quantified leave extract (Rodriguez-Perez et al., 2018)

Countess finding has led to proving the potentials of several phytochemicals present in Potato (Solanum tubers L.) whole plant and present improvement trend is genetically modified to annex the availability of the plant all season as consumption and extraction of biologically active compounds, as shown in Table 22.1.

22.3.1 Anti-Inflammatory In Vitro and In Vivo

Several scientific findings have revealed that secondary metabolites synthesize from plants have the potential effect on the treatment of diseases as complementary medicine (Bagad et al., 2013).

TABLE 22.1
Phytochemicals Present in Potato (*Solanum tubers* L.)

S/N	Secondary Metabolites	Name of the Compound	Source	Biological Activities	Reference
1	Carbohydrates	Tuberonic alpha Glucosidase	Leaves, Tuber	GI lowering Activities.	Das et al. (2017)
2	Vitamins	α-Tocopherol(vitamin E), Folic acid (Vitamin B9)	Tubers	Antioxidant activity	Anjum Sahair et al. (2018)
3	Protein	Patin	Tuber	The antioxidant or antiradical activity	Anjum Sahair et al. (2018)
4	Folates	Solanine and Chaconine	leaves	Anti –inflammatory	FAO (2008)
5	Anthocyanin's	Petunidins, Malvidin, Pelargonidin glycosides, Peonidin glycosides.	Tuber	Antitumor activity, antibacterial activity, strong antioxidative activity, anti-influenza virus activity, and anti-stomach cancer activity	Mohammad et al. (2016); Amanpour et al. (2015)
6	carotenoid	β-Carotene, Cryptoxanthin, lutein, Zeaxanthin, Violaxanthin, Antheraxanthin, Neoxanthin	Tuber	Antioxidant	Lachman et al. (2016)
7	Phenolic Acid	Benzoic acids and Cinnamic acid	Peel	Anti-inflammation	Anjum Sahair et al. (2018)
8	Alkaloids	α-solanine and α-chaconine	Tuber	Anticancer and anti-inflammatory	Das et al. (2017)
9	Flavonoid	Catechin, cyaniding, delphinidin, malvidin, malvidin-3-(p-coumaroyl rutinoside), pelargonidin, peonidin, petunidin, rutin	Tuber	Anti-inflammatory	Akyol et al. (2016)

Inflammation, a pathologic condition that intersects a wide range of diseases such as diabetes, cardiovascular diseases, rheumatic, immune-mediated condition, etc., is one of the targets using plants' active extraction to eradicate. Many drugs have been synthesized from a plant such as steroids, steroids, anti-inflammatory drugs, and immunosuppressants in the treatment of an inflammatory crisis. Inflammatory response is a defense mechanism to hazardous stimuli such as allergens and/ or injury to the tissue (Ghasemian et al., 2016). The secondary metabolites in Potato such as glycoalkaloid contain resistant starch, fiber, and anthocyanin which possess anti-inflammatory components (Bibi et al., 2019; Reddivari et al., 2019). Anthocyanin characterized the pigment of potatoes and has antioxidant effects.

An *in vivo* experiment by Choi and Koo (2005) reveals that the ethanolic extract from potato tuber at doses of 100 and 200 mg/kg produced a significant anti-inflammatory property. Kenny et al. (2013) finding suggested that the sub-cytotoxic concentration of potato glycoalkaloids and potato peel extract possess anti-inflammatory effects in vitro. Kenny et al. (2013) findings further suggested that sub-cytotoxic concentrations of potato glycoalkaloids and potato peel extracts possess anti-inflammatory effects *in-vitro*. The content of biologically-active substances, secondary metabolites, changed upon transgenesis may contribute not only to the differentiated degree of nutrients utilization but also yield health-promoting or detrimental as well as immune-stimulatory and anti-inflammatory effects or may induce disturbances in redox balance- a change in the oxidative status of body cells, degradation of DNA.

22.3.2 CARDIOVASCULAR EFFECT

Cardiovascular disease is associated with so many risk factors which impair the health condition of individuals and has led to so many experiments to ascertain mechanics through which the disease condition can be adverted. Having this in mind, Tang et al. (2017) investigated on several vegetables and their significant effect in ameliorating cardiovascular disease. The researchers observed that high consumption of potatoes could be effective in reducing the risk of CVDs however, their review showed that aqueous extract of African potatoes (APE) had deleterious inotropic effects when experimented on electrical driven left atrial muscle of guinea pig and a corresponding deleterious chronotropic effects on rhythmic beating right atrial. While validating the effect of APE on bands of atrial muscles, Tang et al. (2017) a dependent decrease in positive inotropic and chronotropic reactions of these muscles in guinea pig induced with noradreline. Their further investigation revealed that APE was found to have influenced the actions of the portal veins in experimented rat models through interrupting the periodic, impulsive, and myogenic contractions of the portal veins and its concentration. Their experiment was able to validate that using a dose-dependent APE, a remarkable decrease of systemic arterial blood pressure and heart rates of hypertensive rats was observed. Their investigation therefore proof that APE has the ability to act as a natural therapeutic measure in ameliorating cardiac dysfunction and hypertension. To buttress their findings, the authors revealed that levels of thiobarbituric acid reactive substances (TBARS) found in the heart was reduced with a corresponding increase in vitamin plasma level which they attributed to antioxidants present in potatoes indicating that potatoes could ameliorate CVDs.

22.3.3 NEUROPROTECTIVE EFFECT

As an age-related disease, Alzheimer's disease has been linked with several known neuronal pathology associated with forgetful commemorations, twisting of the neurofibrillary, with an extensive degeneration of neurons and synapse loss in the brain. Ye et al. (2009) investigation revealed that forget commemorations observed after administration of amyloid-beta peptide (Aβ 1-42) was able to induce Alzheimer's disease. The authors study validated that presence of natural phenolic antioxidant phytochemical present in PSPA was effective in reducing neurotoxicity caused by Aβ of which 3(4,5-dimethylthiazol-2yl)2,5-diphenyl-2Htetrazolium bromide (MTT assays) was used to

attest for the neurotoxicity of Aβ after which, neuroprotective effect of PSPA was tested. To this fact, the authors observed that high dose PSPA was found to have a neuroprotective effect based on the possession of a strong antioxidant property. However, they equated the neuroprotective ability of PSPA to that of a strong antioxidant found in α-tocopherol. The authors made use of oxidative stress to further buttress their point showing that PSPA could ameliorate Aβ-induced neurotoxicity were they observed superficial elevation of intracellular oxidative stress when measured with oxidative stress-sensitive dyes Dichlorofluorescin diacetate (DCF-DA) was detected buttressing a point that Aβ could ameliorate neurotoxicity. Aβ found in PSPA has a characteristic feature of lessening cellular hydroxyl radicals and superoxide anion. However, administration of PSPA at a high concentration produced a resultant effect on the treated experimental models which was almost equivalent to the result gotten from positive control. The authors observed PSPA is a renowned natural phenolic phytochemical antioxidant that has a neuroprotective effect through the inhibition of neurotoxicity caused by Aβ. Investigating with the use of MTT assays, the authors observed that the use of high-dose PSPA has the same effect as administration of a strong antioxidant α-tocopherol which was confirmed by the protective effect of PSPA against Aβ-induced neurotoxicity. To this, the authors observed an obvious elevation of intracellular oxidative stress which was measured with the aid of oxidative stress-sensitive dyes DCF-DA. According to the authors, the presence of phenolic phytochemicals in PSPA, scavenging properties that led to the decrease in ROS and LPO in both in vivo and in vitro studies shows the neuroprotective ability of PSPA, with a mechanism that involves antioxidant properties of PSPA having attributes of eight different mechanisms with high active DPPHI scavenging, superoxide anion radical scavenging, hydrogen peroxide scavenging, total reducing power and metal chelating on ferrous ions activities. Their investigation revealed the neuroprotective ability of PSPA which is due to antioxidant properties present in PSPA possessed with the ability to block apoptosis activated by ROS, with a specific apoptosis induced by Aβ neurotoxicity showing DNA fragmentation. Hence, for PSPA to improve the brain against neurotoxicity, it has to inhibit DNA fragmentation. Furthermore, their investigation revealed that stimulation of apoptogenic signaling pathway has the significant effect of blocking the antioxidant ability severing the proper functioning of the brain with an alternative increase in calcium levels of the brain due to free radical-mediated by mitochondria. However, this was attenuated after administration of PSPA showing the pretreatment of PC cells with PSPA has a significant effect on improving mitochondrial function. Deduced from the authors' study, neurotoxicity triggered by Aβ led to the stimulation of capsade-3 affecting the normal function of the brain, administration of PSPA was found to have ameliorated capsade-3 effect and improved the functioning of the brain. Hence, the study indicate the neuroprotective effect of genetically modified potatoes has the ability to alter neurotoxicity while improving the brain functioning.

Ji et al. (2012) investigation of potato clones revealed the presence of proton NMR spectra containing a high percentage of phenolic compounds in potato tuber extracted from water while potato tuber extracted from methanol has variable phenolic components. The authors observed that the colors of the potatoes correlate with the percentage phenolic content where purple and red potato clones were found to be present with high phenolic content than the yellow or un-pigmented clones. The authors further investigated the chlorogenic acid content found in cloned potatoes and observed a value of 0.33–8.67 mg/g and 0.04–4.26 mg/g as levels of chlorogenic acid content found in dry and tuber potatoes respectively while a value of 0.01–3.14 mg/g was found in granule potatoes, which buttress their point that chlorogenic acid is a main phenolic compound found in cloned potatoes that could range from 50% to70%. To ascertain the anthocyanins content that could be found in cloned potatoes, the authors employed two different techniques which resulted in different results based on the colors and forms in which it was experimented on. They further observed that anthocyanins levels were higher in peeled cloned potatoes than the tuber cloned potatoes. Furthermore, the authors validated the glycoalkaloids contents of the various forms of cloned potatoes with the result that peeled potatoes contained a higher form of glycoalkaloids than tubers while granulated potatoes contains lesser glycoalkaloids than tubers linking the loss of glycoalkaloids in granulated

potatoes to have resulted from the granulation process, however, these potatoes still retain their phenolic content. An antioxidant activity test was carried out using DPPH radical scavenging assay with a result showing a higher percentage of antioxidant activity in the peeled potato clone than in any other potato clone. After these examinations of potato clones to attest to their protective properties, their neuroprotective properties were conducted in primary cortical neurons culture primed from E16 CD1 mice to reveal the polyphenolic-enhanced potato extracts found in DIV8. Their observation revealed that neuronal apoptosis was measured with the use of a cytotoxicity detection kit plus. Deduced from their observation, the authors observed that cloned granulated potato extract had a neuroprotective effect by showing a remarkable protective effect on the cortical neurons that have undergone apoptosis induced by oxygen-glucose deprivation (OGD). The authors' experiment revealed that from all cloned potatoes experimented upon, granulated potato extracts were found to possess neuroprotective ability which was accrued to a modification of potato component after granulation hence, neurons thrive better with granulated cloned potato. Inferred from their study it could be observed that genetically modified potato has a neuroprotective ability.

Youm et al. (2005) observed that neuritic plaques and neurofibrillary tangles are symptoms that accompany Alzheimer's disease (AD) with memory loss and cognitive deficiencies that affected mainly the aged. The authors were able to determine the causative mechanism behind neuritic plaques and neurofibrillary tangles, and they observed that genetic and pathological indicator all points to amyloid cascade as a suggestive means of indicating AD. They further buttress their point stating that the build-up and collection of beta-amyloid (Aβ) has a significant effect by damaging the structure and morphology of neurons eventually causing disintegration of synapses leading to neuronal cell apoptosis. However, the authors noticed an increase in research on different drugs and food substances to ameliorate AD which has little or no substantial effect in ameliorating the disease condition. To this fact, Youm et al. (2005) investigated transgenic potatoes known to be genetically modified food and their significant effect in ameliorating AD. Deduced from their study, transgenic plants could serve as a substitute for showcasing recombinant antibodies, antigens, and therapeutics. The authors' study indicated that antigens found in plants were extracted successfully and administered to animals with a resultant effect of increased immunogenicity. However, their research was able to validate the component of these antigens which comprises of disease-related virus capsid and surface proteins, as well as pathogenic bacterial antigens. However, their study showed that inculcating these antigens into food substances could be of maximum benefits aside from increasing production of the plant available to the consumer. Their study made use of plant-derived human Aβ, where they breed transgenic potatoes that possesses 5 tandem recurrences of Aβ1-42 (5Aβ42). Their study was able to produce potato plants with no abnormalities indicating that the β-amyloid protein that can be found in potato does not possess an antagonistic effect on shoots and roots of the plants in an in vitro study. However, the authors observed the presence of a double 35S promoter which was more in number than any other promoter which shows the presence of a detailed antigen that could be affected by the antigen nature. Deduced from the authors' study, administration of orally 5Ab potato extract with an adjuvant to mice stirred a primary resistant response which elapses at seven weeks, which Youm et al. (2005) linked their findings to others study that the phenomenon could be traced to memory immune cells that was observed after oral immunization producing an instantaneous and resilient secondary antibody reaction. However, the authors observed that immunization carried out on Tg2576 transgenic mouse, resulted in reduced adaptive immune reaction that was linked with Aβ1-40. Introducing CTB as an adjuvant which is known as a non-toxic fragment of the toxin cholera, the authors observed that CTB has the activity of a non-T helper (Th) 1-inducing adjuvant. Deduced from the authors' study, a vaccine can be produced from transgenic potatoes which are edible in nature and can be used for a therapeutic purpose to ameliorate Alzheimer's disease which is linked to the production of Aβ in transgenic potatoes, revealing that genetically modified plants are when properly harness, are present with neuroprotective properties.

Having understood the significant effect of oxidative stress and its connection with neurodegeneration Shan et al. (2009) investigation was to understand how genetically modified potatoes could

ameliorate the decline in memory and cognitive function in the aged mouse. To this effect, Shan et al. (2009) made use of a genetically modified potato called purple sweet potato and extracted anthocyanins from the potato which was administered to their experimental model testing the activities of anthocyanins. However, their study has been able to indicate that anthocyanins have some positive effect that includes anti-inflammation, antioxidation, and antimutagen properties. Understanding the morphological and functional changes associated with the aging brain, it was deduced from their study that these changes result in degeneration of motor and cognitive abilities in the brain of aged people which according to the authors lined the changes in the brain as neurodegeneration having similar characteristic with other forms such as amyotrophic lateral sclerosis (ALS), Alzheimer's disease (AD), and Parkinson's disease (PD). The authors observed that D-galactose is an example of reducing sugar with a metabolism rate that could be easily achieved at normal concentration level and give a resultant result of aldose and hydroperoxide stimulated by the catalyst galactose oxidase to produce superoxide anion and oxygen-derived free radicals in high concentration levels. The author's investigation proved that the reaction of D-galactose with free amino acids found in proteins as well as peptides in an in vivo experiment produced Advanced glycation endproducts (AGEs) which furthers activate inflammation of neurons by stimulating NF-kB pathway prompting several neurological impairments that includes behavioral impairment of learning and memory and decreased antioxidant enzyme activities in the brain, establishing a hypothesis that AGEs has a neurotoxic effect on the brain. Furthermore, the authors observed that to extract these toxins from the brain, flavonoids which are potential antioxidants could mitigate the effect of neurotoxins injected into the brain by AGEs due to the anti-aging and anti-inflammatory properties of flavonoids. Understanding the importance of flavonoids, the authors were able to extract and observe the presence of essential flavonoids on purple sweet potato colour (PSPC) using HPLC analysis, flavonoids such as cyanidin acyl glucosides and peonidin acyl glucosides (peonidin 3-O-(6-O-(E)-caffeoyl2-O-β-D-glucopyranosyl-β-D-glucopyranoside)-5-O-β-Dglucoside amongst others were detected. The authors were able to detect from all flavonoids and discovered that cyanidin acyl glucosides and peonidin acyl glucosides were both present with a higher concentration of antioxidant as well as anti-inflammatory properties. However, Shan et al. (2009) were able to observe that PSPC was present with two major flavonoids namely; acylated cyanidin and peonidin which after administration to their experimented mouse with induced neurotoxicity caused by D-galactose was reduced through the process of regulating the levels of ROS in mouse brain. Subjecting the experimental animals to behavioral tests, it was inferred from their study that D-galactose severe the learning and memory function in mice whereas the authors used open-field and passive avoidance tasks, it was deduced from their study that PSPC has the ability to enhance brain functions in aging models while showing a remarkable repressed effect in during behavioral deterioration induced by D-galactose. Experimenting with the aid of western blotting to ascertain the molecular mechanism through which PSPC treatment could aid the aging brain from the neurotoxicologically effect of D-galactose, the authors observed that while D-galactose was known to reduce significantly Cu/Zn-SOD and CAT antioxidant enzyme activity as well as the buildup of ROS. Administration of PSPC was able to reinstate Cu/Zn-SOD and CAT activity while also improving the antioxidant activity of the body. Having known that increased levels of MDA activities are an index of brain oxidative damage, administration of PSPC was found to have decreased the MDA levels that were increased from D-galactose administration and as well decreasing the secretion of free radicals revealing the ability of PSPC to arrest brain aging. Deduced from their study, it could be observed that genetically modified potatoes have the ability to improve an aging brain thus, reducing the susceptibility of encountering neurodegenerative diseases.

22.4   ANTIDIABETIC ACTIVITY OF GENETICALLY MODIFIED POTATOES

Potatoes have been genetically modified to act as a mean to reduce hyperglycemia of diabetes mellitus. The reason why potatoes are preferable delivery systems for the anti-diabetic agents is in their

cost effectiveness, large cultivation and ability to pass on the genetic modifications through vegetative propagation. Type 1 diabetes mellitus also called as insulin dependent diabetes mellitus results from the progressive destruction of the pancreatic B cells which produce insulin in our body. This destruction is autoimmune and is mediated by the T lymphocytes which become auto-reactive and results over long periods of time characterizing the chronicity of the disease. A number of auto antigens which target the islet cells have been identified to be insulin, glutamic acid decarboxylase (GAD), and tyrosine phosphatase-like IA-2. These play an important role in the initiating and maintaining the events leading to type 1 diabetes mellitus (Atkinson and Maclaren, 1994; Bosi and Botazzo, 1995). In a diabetic mouse model which is non obese and is very similar to human type 1 diabetes mellitus, there is a period preceding the over diabetes in which there is an infiltration of islets by lymphocytes. If intervention is done in this period it can alter the disease process. The results observed the non-obese diabetic mouse model have suggested a potential for prevention of diabetes by immunological intervention in clinical practice. Even though immunosuppressive drugs such as cyclosporine A have been used to alter T-lymphocyte immune responses to inhibit the development of diabetes in both humans and non-obese diabetic mice, interventions targeted at suppressing specifically the immune responses associated with the disease and not the overall immune responses are much safer and much preferable. The general immunosuppressive drugs adversely lead to increased risk of infections and cancer. It has been shown that reduction of peripheral immune responses to proteins administered orally can help in keeping the immune system intact to fight off pathogens. This immune regulatory system which is endogenous and the mechanisms involved in it are collectively termed oral immune tolerance. This method has been used to approach the autoantigens such as insulin and glutamic acid decarboxylase (GAD) targeting the islet cells to prevent the development of diabetes in non-obese diabetic mice. It has been demonstrated that transgenic plants can be effectively used as a mode of production and delivery of autoantigen, mouse GAD67 associated with diabetes. Genetically modified plants are not only cost-effective but they can successfully express GAD in a form that is immunologically active. Feeding these GAD-containing genetically modified plants can result in the prevention of diabetes in non-obese diabetic mice (Ma et al., 1997). Genetically modified plants are a cost-effective means for the large-scale production of recombinant proteins. A study was carried out in which GAD-induced genetically modified potato tubers were introduced to some non-obese diabetic mice to find out whether it would affect the proliferation of T-lymphocytes. The dose of 1 mg/day of GAD was administered in one mouse which is very high as compared to what would be administered to humans. After four weeks of supplementation these mice were administered with highly purified recombinant GAD67 derived from E. coli in hind footpads. The results were obtained after sacrificing the mice ten days later. Lymph node and spleen T cells were isolated to study their proliferation in response to GAD67 in vitro. The transgenic plant-fed mice showed decreased proliferation of T cells as compared to control plant-fed mice. A two-fold increase in the anti-GAD antibodies was also observed in the mice fed by GAD67 genetically modified plant as compared to the control. Similarly, potato tubers modified to contain small amounts of CTB-insulin infusion proteins when fed to non-obese diabetic mice showed a decrease in insulitis. The study demonstrates the use of transgenic plants especially transgenic potatoes as an effective delivery system to induce oral immune tolerance against type 1 diabetes mellitus in animals (Ma and Jevnikar, 2002). Adiponectin is a cytokine produced by the adipocytes and is a 30kDa protein that is involved in breaking down fatty acids as well as regulating glucose levels in humans. Adiponectin is present in normal amounts of 5–30mg/mL in blood plasma and attributes to 0.01% of all plasma proteins (Arita et al., 1999). Low levels of adiponectin in plasma have been associated with hypertension, obesity, and type 2 diabetes mellitus. Sweet potatoes have been genetically modified by the introduction of mouse adiponectin cDNA. DNA gel blot analysis and PCR has been used to detect the protein in transgenic sweet potatoes. Expression of the mouse adiponectin in genetically modified plants was confirmed in the study. It was suggested that full length adiponectin produced by the mammals can be used to treat and control increased levels of glucose as observed in type 2 diabetes but the adiponectin from E. coli failed to show any such activity (Berberich et al., 2005).

22.5 ANTIFUNGAL ACTIVITY

The use of expressed protein of potato can be used against fungal infection. This could be done by the use of transgenic plants carrying gene ac2 from amaranth – *Amaranthus caudatus* (Liapkova et al., 2001). Protein resulting from the expression of this gene is highly homologous with cysteine/-glycine-rich domains in the chitin-binding proteins (Broekaert et al., 1992). These proteins mostly bind to chitin present in internal fungal cell walls causing alteration of their polarity and inhibiting of growth of the fungi (Selitrennikoff, 2001).

Fungal disease of potato plants is potato blight (*Phytophthora infestans*). Temporin A protein-producing potato plants display resistance to this disease. Temporin A is a small naturally occurring antimicrobial peptide, which enhances plant resistance to potato blight, and is also of bacterial origin. This is a disease caused by the fungus *P. erytroseptica* and bacterium *Erwinia carotovora*. The results obtained confirm that transgenic potato plants that express temporin A can serve as a good tool for control of most significant fungal pathogens such *P. infestans* and *P. erytroseptica* (Osusky et al., 2004).

22.6 ANTIVIRAL ACTIVITY

Potato plants transformed with *Phytolacca insularis* antiviral protein (PIP) were resistant to infection of Potato Virus Y (PVY), Potato virus X (PVX), and *Potato leafroll virus* (PLRV) (Moon et al., 1997). Nonetheless, because of their toxicity, all these RIPs generated a severe phenotype in transgenic plants. This led to several attempts where RIPs with lower toxicity toward host plants were used, including PAP II, which is a low-toxic isoform of PAP isolated from leaves (Wang et al., 1998).

Luteovirus (*Poleovirus* sp.) is known to be a potato viral pathogen, spread by aphids; this pathogen causes disease with the signs of potato leafroll luteovirus (PLRV) infection. van der Wilk et al. (1991) investigated the resistance of potato plants against this virus after inserting the coat protein gene of PLRV into the genome of potato plants. A detectable level of coat protein was not gathered in any of the tested virus-infected transgenic plants that contained markedly lower levels of viral antigen than control plants; this resulted from a reduced rate of virus multiplication in transgenic plants (van der Wilk et al., 1991).

22.6.1 IN VIVO ANTIVIRAL ACTIVITY

The causal agent of epidemic gastroenteritis in a human was treated with transgenic potatoes carrying a gene for the capsid protein of Norwalk virus – NVCP. Capsid protein was expressed in potato tubers, in the amount of approximately 0.37% of total protein. Immunogenicity of transgenic potato plants was tested in mice; IgG antibodies against recombinant Norwalk virus were detected in them (Mason et al., 1996). Capability of this "edible vaccine" to activate the immune system was tested also in human volunteers; the immune response was activated in the majority (95%) of the people (Tacket et al., 2000).

Zhou et al. (2003) used the e extract from potato tubers containing S1 protein on mice after administration of three doses of the extract. Similar results were recorded in chickens and they were completely protected against IBV virus after the third application of the genetically modified plant transgenic potatoes (Zhou et al., 2003). This research team was also the first (Zhou et al., 2004) to prepare a transgenic plant expressing the full-length S protein of IBV. Transgenic potato plants were used for oral and intramuscular immunization of chickens. The results demonstrated a high titer of anti-IBV antibodies, which protected the experimental animals from the infection of the virulent IBV in the trial. The results show the possibility of transgenic potato plants expressing the S protein of IBV to be used in the control of infectious bronchitis

The modified potato plant extract was injected into mice. This was used to test the immunogenicity of recombinant protein; antibodies against VP6 protein (Matsumura et al., 2002). Mouse

rotavirus was used as a source of VP6 protein in another study. Protein effectivity was again tested in mice. These were fed with transgenic potato tubers as a source of protein. Oral immunization of mice stimulated the production of IgG and IgA antibodies against capsid protein and thus represented a progress in the development of the rotavirus vaccine by means of agricultural plants (Yu and Langridge, 2003). It was also confirmed in mice, intraperitoneally and orally that extracted from transgenic potatoes (Gomez et al., 2000).

22.7 ANTIBACTERIAL ACTIVITY

Modified potato plant was used against enterotoxigenic strains of E. coli (ETEC) A thermolabile enterotoxin B (LT-B) from bacteria E. coli was expressed in potato plants. The toxin is produced by enterotoxigenic strains of E. coli (ETEC) that colonize small intestine and cause acute watery diarrhea. Faecal IgA and serum anti-LT-B IgG antibodies were noticed in mice that had been given three doses of transgenic potatoes. Despite the infection in mice was induced, full immunity against the bacterial disease was not developed in this case . This modification mentioned was also investigated by Lauterslager et al., in one of their studies, which was focused on testing the capability of the recombinant LT-B protein to produce mucosal and total systemic antibody response in mice. Mice used in the experiment were either preimmunized or non-immunized (so that there would not be any form of false result). Whilst no anti-LT antibody formation was detected in the non-immunized mice, anti-LT IgA antibodies were produced by the immunized mice. Higher antibody response was found after LTB-protein administration through intestinal intubation than after oral administration (Lauterslager et al., 2001).

Lactoferrin is one of the human proteins inserted into the genome of potato plants. Lactoferrin has a protective function because it can bind iron, which makes it inaccessible to bacteria. The potential for lactoferrin to act both as an antimicrobial and an immune regulatory agent in addition to its nutritional and pharmaceutical value has led to development of transgenic potato plants carrying a cDNA fragment encoding human lactoferrin (hLF). The biological qualities of lactoferrin, formed by transgenic potatoes were confirmed in four different human pathogenic bacterial strains. GM potato plants were shown to be able to produce human lactoferrin that maintains biological, bacteriostatic, and bactericidal qualities against various pathogenic bacteria (Chong and Langridge, 2000).

Further genetic modification was the insertion of the gene for human interferon-α-2b (HuIFN-α-2b) and α-8 (HuIFN-α-8) into the genome of potato plants. Interferons are classified as antiviral cytokines responsible for various cytotoxic effects including anti-tumor activity. The HuIFN-α genes introduced into the potato plant were correctly translated and transcribed in plant cells and their biological activity was verified by inhibition of vesicular stomatitis virus (VSV) replication on a human amniotic cell line. It is supposed that potato plants carrying genes for human interferons will be used as food additives or as additional substances for the treatment of infectious diseases and decreased immunity (Ohya et al., 2001). 4.2.3. Production of human tumor necrotizing factor (HuTNF-α) Ohya et al. (2002) produced transgenic potato plants carrying a gene for human tumor necro- Review Article Veterinarni Medicina, 51, 2006 (5): 212–223 220 tizing factor-α (HuTNF-α). This cytokinin is produced by stimulated cells of the immune system and can improve the inflammatory immune response of the organism or cause in vitro lysis of tumor cells. Transgenic potato plants were shown to produce HuTNF-α, and the extract from transformed plants causes a cytotoxic effect. This capability together with a relatively high protein yield will predestine transgenic plants producing TNF-α to be used in human and veterinary medicine (Ohya et al., 2002). Oral vaccines are potentially usable for the production of any vaccine, which comprises or contains subunit components. Oral administration of subunit vaccines is particularly suitable for stimulation of immunity against the pathogens that enter the body via the intestines.

Oral vaccines are also important for the control of pathogenic infections of other mucosal surfaces (hepatitis B, HIV) due to the shared origin of the mucosal immune system. As plants can

produce large amounts of subunit vaccines, they may be particularly used for the control of diseases affecting large populations of people. Production at low costs, stability of plant vaccines during storage at ambient temperature, reduction of material and nursing staff (administration of injections) predestines edible vaccines to be particularly used in developing countries.

Other types of genetic modifications of potato plants Genetic engineering extends the potential use of natural plant materials, e.g. in the production of modified starch or synthesis of novel polymers. Plant biomaterials, characterized by their renewability and biodegradability, might replace synthetic plastic materials and elastomers made of petroleum in the future.

22.8   ANTIULCER ACTIVITY OF GENETICALLY MODIFIED POTATO

Potatoes have been genetically modified to act as a means to treat ulcers. The reason for the selection of potatoes includes delivery systems for the anti-ulcerogenic agents along with their cost effectiveness, large cultivation, and ability to pass on the genetic modifications through vegetative propagation, short and fast life cycle.

Peptic ulcer is an aggregate of most normal heterogeneous issue, present as a pit in the coating of the gastrointestinal tract (GIT) mucosa in light of acid, pepsin, bile acid, pancreatic compound, and microscopic organisms. It is because of awkwardness between destructive(corrosive and pepsin) and protective (bicarbonates, mucin, and so forth.) factors. Peptic ulcer ailment likewise happens because of organization of NSAIDs, stress, H. pylori, or obsessive condition, for example, Zollinger-Ellison disorder (Wegener et al., 2015). NSAIDs cause disintegrations, petechiae, ulceration as well as gastritis and blend with impedance of ulcer mending. Further, they likewise initiate harm to the mucosa with awkwardness among forceful and protective elements. Even though an excellent number of enemies of ulcers medications, for example, antisecretory drugs, H2 receptor opponents, proton inhibitors, antimuscarinic, cytoprotectants, and prostaglandins analogs are accessible, the adverse effects related to these medications limit their utilization. Numerous home-grown medications from natural plants and arrangement of medication are upheld for the administration of peptic ulcer. Natural prescriptions utilized as entire plant powders/extricates from various parts are currently a day's considered as sheltered medicine for the treatment of various illnesses as it is a general idea that plant based medications are more secure with no reactions. Potato has assessed tentatively by in vitro and in vivo conventions for hostile to ulcer movement.

An examination was led to survey the capability of S. tuberosum for the treatment of ulcers. Ranitidine is utilized as a standard reference to assess hostile to ulcer development in rat models, for example, pylorus ligation model and stress-instigated ulcers by chilly water inundation. At the point when alcoholic extract of tubers of S. tuberosum (AETST) and aqueous extract of tubers of S. tuberosum (AQETST) were oppressed for LD50 learn at the concentration of 2,000 mg/kg body weight (Atoui et al., 2005). The concentration was chosen as low (100 mg/kg), medium (200 mg/kg) and high (400 mg/kg), and the dosages of both the extracts essentially decreased the ulcer ($P < 0.05$, 0.01 and 0.001). The investigation uncovered that both the AETST and AQETST had antiulcer potential. Phytochemical constituents, for example, tannins, flavonoids, and triterpenes have just been accounted for as their enemy of ulcer induction. These phytochemical constituents were available in both the extracts and, thus, liable for the watched suppression (Emad, 2006).

Following the progression of hereditary designing instruments, a few potato cultivars with wanted yield, dry issue, protein, and cancer prevention agent quality, coking surface, (for example, waxy, floury), tissue shading, and abiotic stress tolerant plant have likewise been created. The interest in starches with extraordinary properties valuable for mechanical nourishment handling has prompted the presentation of altered starches utilizing hereditary designing procedures. The hereditarily altered potato is likewise utilized for remedial reasons for ulcer treatment. Hereditarily adjusted potatoes involved dietary things contained lactin Galanthus nivalis agglutinin (GNA), a mannose explicit lectin, articulation which is administered by GNA quality inclusion in potato (Chauhan, 2014)

GNA has observable effects on gastro intestinal tract supported by *in vivo* experimental studies. The GNA transgene causes the mucosal proliferation of the stomach which could be the key reason for ulcer treatment. It also has effects on the small as well as the large intestine. Genetically modified potatoes (GMP) have been evaluated for effects on mucosa of mammals. By genetic modification of potato GNA a mannose-specific lectin, has been expressed under the action of CaMV35s promotors that boosts the resistance of insects and worms in the abdomen. Due to the ability of GNAto bind mannose of epithelium on the villi of the small intestine. This mechanism takes time to give response and is proven by *in vivo* rat model where response was produced after 10 days of ingestion of GMP and compared among control and experimental groups vis parent potato, GNA-GM potato and parent line potatoes with GNA, where GM potatoes are further divided in to raw and boiled GNA-GM potatoes. The difference lies in the expression of GNA in raw and boiled GM potatoes justified by ELISA test. The research revealed that stimulation of stomach epithelium refers to the expression of GNA whereas the proliferation enhancement of jejunal epithelium with raw potato and antagonistic proliferation suppression of cecal epithelium with boiled potato ingestion is not much dependent on GNA expression. The mechanism lies in that binding of GNA to intestinal mucosal cells is followed by penetration in to the mucosal cells along with some other components of GM-GNA potato. The crypt hyperplasia proved the epithelial proliferation of jejunum by GNA expression. Also the enhanced T-lymphocytes infiltration is suspected to eliminate destructive enterocytes (Stanley and Pusztai, 1999).

22.9   ANTI-HYPERLIPIDEMIC ACTIVITY

Methionine is hypothesized to have effect on cholesterol mechanism in order to suppress serum lipid levels. It has two metabolic pathways; *transamination pathway* and *transsulfuration pathway*. The metabolism of methionine via transsulfuration pathway yields cysteine and taurine, both of which are liable for plasma cholesterol lowering activity of methionine containing transgenic potato (SugiYama et al., 1985)

HDL (High Density Lipoprotein) or good cholesterol carries almost 50% of serum cholesterol. According to recent studies, the amount of reduction in serum cholesterol level when high methionine transgenic potato is taken in diet indicates that methionine affects the regulation of Apo A-I secretion from liver into blood stream, which is a major protein component of HDL (Morita et al., 1997).

Preferably, for lowering serum lipid levels, potatoes containing low fat and high sodium content should be prepared, comparatively higher potassium levels are necessary to neutralize the effect of sodium and shield against accumulation of fat or cholesterol in vessels (Nicolle et al., 2004).

22.10   ANTICANCER ACTIVITY

Research studies revealed that non communicable diseases such as cancers are seen rare in the people using fruits and vegetables regularly. Recent studies have explored the medical advantages of potatoes and related plants, for example, anti-infection, anticancer, and cancer prevention agent properties (Roleira et al., 2015). Following the progression of hereditary designing instruments, a few potato cultivars with wanted yield, dry issue, protein and cancer prevention agent quality, cooking surface, (for example, waxy, floury), tissue shading, and abiotic stress tolerant plant have likewise been created. The interest for starches with extraordinary properties valuable for mechanical nourishment handling has prompted the presentation of altered starches utilizing the hereditary designing procedures. The hereditarily altered potato likewise utilized in remedial reason for cancer treatment. The activity of anthocyanins contained in the Solanum tuberosum L. in both breast and hematological tumors were researched. The biomedical exercises of anthocyanin remove got from the Vitelotte cultivar were resolved. Sub-atomic genotyping was performed to appropriately distinguish this exceptional genotype in contrast with other potato types and to advance the usage

of this hereditary asset by plant raisers. Besides, cell and sub-atomic portrayal of the activity of anthocyanin separate in malignant growth cells uncovered that adjustment of cell cycle controllers happens upon treatment. Just as inciting apoptotic players, for example, TRAIL in malignancy frameworks, anthocyanin separate restrained Akt-mTOR flagging in this manner actuating development of intense myeloid leukemia cells. These outcomes are of enthusiasm for perspective on the effect on nourishment utilization and as practical nourishment segments on potential malignant growth treatment and avoidance by *Solanum tuberosum* L.

K-State's Soyoung Lim, doctoral understudy in human sustenance, Manhattan, worked with George Wang, partner teacher of human nourishment at K-State, to comprehend the color impacts of a Kansas-reared purple sweet potato on disease prevention. The potatoes were isolated by different characteristics dependent on substance pigmentation and fiber substance. The investigation established that the Kansas-reproduced potato had altogether higher anthocyanin substance contrasted with different potatoes. In the study, it was found that two subordinates of anthocyanin were prevailing: cyanidin and peonidin, which demonstrated huge, cell development hindrance for the malignancy cells, yet there were no noteworthy changes in the cell cycle (Lim et al., 2013). Ohya et al. (2002) delivered transgenic potato plants conveying a quality for human tumor necrotizing factor-α (HuTNF-α) (Ohya et al., 2001). This cytokinin is delivered by invigorated cells of the insusceptible framework and can improve the incendiary resistant reaction of the living being or cause in vitro lysis of tumor cells. Transgenic potato plants were appeared to create HuTNF-α, and the concentrate from changed plants causes a cytotoxic impact. These capacities together with a generally high protein yield will fate transgenic plants delivering TNF-α to be utilized in human and veterinary medication (Ohya et al., 2001). Another study has demonstrated that dietary anthocyanins gastrointestinal tract carcinomas, and when topically applied, these molecules repress skin malignant growths. Ongoing research exhibits that the induction of differentiation obtained using anthocyanin extract in both leukemic and breast carcinoma systems is complemented by c-Myc (regulator genes that code for transcription factors) down regulation, and balance of p90 and ERK (extracellular signal-regulating kinase) phosphorylation, proposing that these signal transduction pathways may partially add to the foundation of the differentiated phenotype. Strikingly, a comparative 'situation' has been recently reported as mediated by the inhibition of PKB (protein kinase B)/mTOR (mammalian objective of rapamycin) pathways incited by histone deacetylase (HDAC) inhibitors in leukemia (Nishioka et al., 2008). Some latest studies have validated that the adequacy of various dosages of the anthocyanin extract was surveyed in solid and hematological cancer cell lines. The crude anthocyanin extract showed anti- proliferative activity in some solid malignant growth cell lines like HeLa (cervical carcinoma), MCF7, 3, MDA-MB231 (breast carcinoma) cells and LnCaP (prostate carcinoma). Besides, this anti-proliferative activity, could likewise be appeared in myeloid leukemia cells such as U937 and NB4 the model of intense promyelocytic leukemia (APL), an acute myeloid leukemia (AML) which carries the chromosomal translocation prompting the outflow of the combination protein PML-RAR alpha (Ieri et al., 2011)

In another in vitro study, purple potato extracts were tested on four cancer cell lines consuming crystal violet and tetrazolium MTT assays. The viability of treated cells was controlled by the crystal violet assays, where the concentration of the dye relates with the quantity of attached viable cells. Results acquired showed that the concentrate was proficient in lessening Caco-2 (colorectal adenocarcinoma) cell viability (Hansen et al., 1989).

A current in vitro study exhibited that in vitro purple potato extract essentially smothered multiplication in CSCs (cancer stem cells). The extract additionally upregulated proteins engaged with mitochondria-intervened apoptotic pathway and downregulated proteins associated with the Wnt/β-catenin signaling pathway. Purple potato eradicated colon cancer stem cells with nuclear β-catenin in vivo through induction of apoptosis and stifled tumor incidence in mice with azoxymethane (AOM)-induced colon cancer giving boost to the anti-cancer properties of purple potato(Lala et al., 2006).

Moreover, anthocyanins prepared from purple potatoes incite apoptosis in cultured human stomach cancer KATO III cells. Morphological changes showing apoptotic bodies and DNA

fragmentations were seen in cells treated with potato anthocyanins. As the anthocyanins have no particular side effects, we can use them persistently to prevent the occurrence of carcinomas. Besides, as potato anthocyanins are normally colored, those potato pigments give a tasteful special visualization on foodstuff and drink (Hayashi et al., 2006).

22.11 ANTI-ALLERGEN

Rahnama et al. (2017) carried out an immunologic study on feeding on transgenic *Solanum tuberosun* on Wistar rats. The transgenic *Solanum tuberosun* possess cry1Ab and nptII and some marker genes which serves as target. It was observed in an in-silico study that cry1Ab and nptII gene does not share same protein sequence to toxins or allergens already known. During the experimental approach, Wistar rats were maintained on a diet composed of 20% transgenic *Solanum tuberosun* or its parental control, non-transgenic *Solanum tuberosun*, for 90 days. The results revealed that ingestion of transgenic *Solanum tuberosun* does not have any effect on parameters like food intake, growth rate, food efficiency, and general health outlook of the Wistar rats (Rahnama et al., 2017). Consequently, Rahnama and colleagues also reported that there was no observed significance in the level of immunoglobulin -A, -G, -M (IgA, IgG and IgM), IFN-c and interleukin 6 (IL-6) among the transgenic and non-transgenic fed rats. The authors therefore summaries that consumption of transgenic *Solanum tuberosun* had no antagonistic effect on Wistar rat immune system

22.12 ANTIOXIDANT PROPERTIES

Lee et al (2019) reported on the beneficial effect of polyphenol and anthocyanin from sweet potato cultivars on mopping up oxidative stress markers and its cytoprotection. It was observed that the polyphenolic contents of the purple-fleshed cultivar were high in the different extract of distilled water, fermented ethanol, and ethanol extracts, respectively (39, 68, and 71 µg gallic acid equivalent/g). The purple-fleshed potato cultivar revealed that it had 29 mg/100 g of anthocyanin a key player in increasing the concentrations of polyphenols. It was also reported that the purple cultivar had an increased antioxidant activity than the other cultivars. Meanwhile, the purple cultivar extract was able to improve cellular oxidative stress marker levels in HepG2 cells stimulated by tert-butyl hydroperoxide to a normal level. The authors conclude that anthocyanin-enriched purple cultivar potato had robust antioxidant activities which can be used to suppress oxidative damage for health improvement.

22.13 CONCLUSION AND FUTURE RECOMMENDATION

This chapter has provided a detailed report about the benefits of genetically modified potatoes which includes their health and nutritional usefulness. Furthermore, more emphasis was a lad on the application of genetically modified potatoes as a typical examples of nutraceutical foods endowed with several pharmacological components and diverse biologically active compounds. This chapter also affirms that the consumption of transgenic potatoes has the potential to be effectively used to suppress and prevent diabetes mellitus in animal models such as non-obese diabetic mice. However, the lack of human trials in testing the efficiency of the use of transgenic plants as modes to deliver anti-diabetic agents in humans limits the applications in humans. Therefore, the utilization of genetically modified potatoes in in-vivo animal studies suggested their application as a source of medicine in humans to treat various diseases especially diabetes, ulcer, cancer, and other foodborne diseases (Adetunji et al., 2020a,b; Adetunji and Anani 2020; Adetunji and Varma, 2020; Olaniyan and Adetunji, 2021; Inobeme et al., 2021a, b; Anani et al., 2021; Adetunji et al., 2021a,b,c,d,e,f,g,h; Jeevanandam et al., 2021; Adetunji et al., 2022a,b). Therefore, there is an urgent need to establish the level of toxicity of these genetically modified potatoes so as to establish their softy level.

REFERENCES

Adetunji C.O., and Anani O. (2020). Bio-fertilizer from Trichoderma: Boom for Agriculture Production and Management of Soil- and Root-Borne Plant Pathogens. In: *Innovations in Food Technology: Current Perspectives and Future Goals*, Mishra P, Mishra RR, Adetunji CO (Eds.). pp. 245–256. DOI:10.1007/978-981-15-6121-4

Adetunji C.O., and Varma A. (2020). Biotechnological application of *Trichoderma*: A powerful fungal isolate with diverse potentials for the attainment of food safety, management of pest and diseases, healthy planet, and sustainable agriculture. In: *Trichoderma: Agricultural Applications and Beyond. Soil Biology*, vol 61. Manoharachary C, Singh HB, Varma A. (Eds.). Cham: Springer. DOI:10.1007/978-3-030-54758-5_12

Adetunji C.O., Roli O.I., and Adetunji J.B. (2020a). Exopolysaccharides derived from beneficial microorganisms: Antimicrobial, food, and health benefits. In: *Innovations in Food Technology*, Mishra P, Mishra RR, Adetunji CO. (Eds.). Singapore: Springer. DOI:10.1007/978-981-15-6121-4_10

Adetunji C.O., Akram M., Imtiaz A., Bertha E.C., Sohail A., Olaniyan O.P., Zahid R., Adetunji J.B., Enoyoze G.E., and Sarin N.B. (2020b). Genetically modified cassava; the last hope that could help to feed the world: recent advances. In: *Genetically Modified Crops - Current Status, Prospects and Challenges*, Kishor PBK, Rajam MV, Pullaiah T. (Eds.). https://www.springer.com/gp/book/9789811559310.

Adetunji C.O., Inobeme A., Anani O.A., Jeevanandam J., Yerima M.B., Thangadurai D., Islam S., Oyawoye O.M., Oloke J.K., Olaniyan O.T. (2021a). Isolation, Screening, and Characterization of Biosurfactant-Producing Microorganism That Can Biodegrade Heavily Polluted Soil Using Molecular Techniques. In *Green Sustainable Process for Chemical and Environmental Engineering and Science*. Biosurfactants for the Bioremediation of Polluted Environments, pp. 53–68. https://doi.org/10.1016/B978-0-12-822696-4.00016-4

Adetunji C.O., Jeevanandam J., Anani O.A., Inobeme A., Thangadurai D, Islam S, and Olaniyan OT. (2021b). Strain improvement methodology and genetic engineering that could lead to an increase in the production of biosurfactants. In: *Green Sustainable Process for Chemical and Environmental Engineering and Science*, Inamuddin, Adetunji CO, Asiri AM (Eds,), pp. 299–315. London: Elsevier.

Adetunji C.O., Jeevanandam J., Inobeme A., Olaniyan O.T., Anani O.A., Thangadurai D., and Islam S. (2021c). Application of biosurfactant for the production of adjuvant and their synergetic effects when combined with different agro-pesticides. In: *Green Sustainable Process for Chemical and Environmental Engineering and Science*, Inamuddin, Adetunji CO, Asiri AM (Eds.), pp. 255–277. London: Elsevier.

Adetunji C.O., Olaniyan O.T., Anani O.A., Inobeme A., Ukhurebor K.E, Bodunrinde R.E., Adetunji J.B., Singh K.R.B., Nayak V., Palnam W.D., and Singh R.P. (2021d). Bionanomaterials for green bionanotechnology. In: *Bionanomaterials: Fundamentals and Biomedical Applications*. IOP Publishing. DOI:10.1088/978-0-7503-3767-0ch10

Adetunji C.O., Inobeme A., Olaniyan O.T., Olisaka F.N., Bodunrinde R.E., Ahamed M.I. (2021e). Microbial desalination. In: *Sustainable Materials and Systems for Water Desalination. Advances in Science, Technology & Innovation (IEREK Interdisciplinary Series for Sustainable Development)*, Inamuddin, Khan A. (Eds.). Cham: Springer. DOI:10.1007/978-3-030-72873-1_13

Adetunji J.B., Adetunji C.O., and Olaniyan O.T. (2021f). African walnuts: A natural depository of nutritional and bioactive compounds essential for food and nutritional security in Africa. In: *Food Security and Safety*, Babalola OO (Eds.). Cham: Springer. DOI:10.1007/978-3-030-50672-8_19

Adetunji C.O., Olaniyan O.T., Anani O.A., Olisaka F.N., Inobeme A., Bodunrinde R.E., Adetunji J.B., Singh K.R.B., Palnam W.D., and Singh R.P. (2021g). Current scenario of nanomaterials in the environmental, agricultural, and biomedical fields. nanomaterials in bionanotechnology. In: *Nanomaterials in Bionanotechnology: Fundamentals and Applications*. 1st Edn., Chapter 6. CRC Press. DOI:10.1201/9781003139744-6

Adetunji C.O., Kremer R.J., Makanjuola R., and Sarin N.B. (2021h). Application of molecular biotechnology to manage biotic stress affecting crop enhancement and sustainable agriculture. *Advances in Agronomy*, 168, 39–81. https://www.sciencedirect.com/science/article/pii/S0065211321000304?dgcid=author

Adetunji C.O., Olaniyan O.T., Adetunji J.B., Osemwegie O.O., and Ubi B.E. (2022a). African mushrooms as functional foods and nutraceuticals. In: *Fermentation and Algal Biotechnologies for the Food, Beverage and Other Bioproduct Industries Edition*, 1st Edn, p. 19. First Published 2022. CRC Press. eBook ISBN 9781003178378. DOI:10.1201/9781003178378-12

Adetunji C.O., Ukhurebor K.E., Olaniyan O.T., Ubi B.E., Oloke J.K., Dauda W.P., and Hefft D.I. (2022b). Recent advances in molecular techniques for the enhancement of crop production. In: *Agricultural Biotechnology, Biodiversity and Bioresources Conservation and Utilization*, 1st Edn., p. 20. First Published 2022. CRC Press. eBook ISBN 9781003178880. DOI:10.1201/9781003178880-12.

Amanpour R., Abbasi-Maleki S., Neyriz-Naghadehi M., and Asadi-Samani M. (2015). Antibacterial effects of Solanum tuberosum peel ethanol extract in vitro. *Journal of HerbMed Pharmacology*, 4(2), 45–48.

Anani OA, Jeevanandam J, Adetunji CO, Inobeme A, Oloke JK, Yerima MB, Thangadurai D, Islam S, Oyawoye OM, Olaniyan OT. (2021). Application of biosurfactant as a noninvasive stimulant to enhance the degradation activities of indigenous hydrocarbon degraders in the soil. In: *Green Sustainable Process for Chemical and Environmental Engineering and Science. Biosurfactants for the Bioremediation of Polluted Environments*, 2021, pp. 69–87. DOI:10.1016/B978-0-12-822696-4.00019-X

Anjum Sahair R., Sneha S., Raghu N., Gopenath T.S., Karthikeyan M., Gnanasekaran A., Chandrashekrappa G.K., and Basalingappa K.M (2018). Solanum tuberosum L: Botanical, phytochemical, pharmacological and nutritional significance. *International Journal of Phytomedicine*, 10(3), 115–124. DOI:10.5138/09750185.225.

Arita Y., Kihara S., Ouchi N., Takahashi M., Maeda K., Miyagawa J., Hotta K., Shimomura I., Nakamura T., Miyaoka K., Kuriyama H., Nishida M., Yamashita S., Okubo K., Matsubara K., Muraguchi M., Ohmoto Y., Funahashi T., and Matsuzawa Y. (1999). Paradoxical decrease of an adipocyte specific protein, adiponectin, in obesity. *Biochemical and Biophysical Research Communications*, 257, 79–83.

Atkinson M.A. and Maclaren N.K. (1994). The pathogenesis ofinsulin-dependent diabetes mellitus. *New England Journal of Medicine*, 331, 1428–1436.

Atoui A.K., Mansouri A., Boskou G., and Kelfalas P. (2005). Tea and herbal infusions: Their antioxidant activity and phenolic profile. *Food Chemistry*, 89, 27–36.

Bagad A.S., Joseph J.A., Bhaskaran N., and Agawal A. (2013). Comparative evaluation of anti-inflammatory activity of curcuminoids, turmerones and aqueous extract of *Curcuma onga*. *Advances in Pharmacological Sciences*, Article ID 805756, 7 pages.

Beals K. (2019). Potatoes, nutrition, and health. *American Journal of Potato Research*, 96(2), 102–110.

Berberich T., Takagi, T., Miyazaki, A., Otani, M., Shimada, T., and Kusano, T. (2005). Production of mouse adiponectin, an anti-diabetic protein, in transgenic sweet potato plants. *Journal of Plant Physiology*, 162(10), 1169–1176.

Bibi S., Navarre D., Sun X., Du, M., Rasco B., and Zhu M. (2019). Effect of potato consumption on gut microbiota and intestinal epithelial health. *American Journal of Potato Research*, 96(2), 170–176.

Birch A., Geoghegan I.E., Griffiths D.W., and McNicol J.W. (2002). The effect of genetic transformations for pest resistance on foliar solanidine-based glycoalkaloids of potato (Solanum tuberosum). *Annals Applied Biology*, 140, 143–149.

Bosi E. and Botazzo G.F. (1995). Autoimmunity in insulin-dependent diabetes mellitus. *Clinical Immunotheraphy*, 3, 125–135.

Broekaert W.F., Marien W., Terras F.R., De Bolle M.F., Proost P., Van Damme J., Dillen, L., Claeys M., Rees S.B., and Vanderleyden J. (1992). Antimicrobial peptides from Amaranthus caudatus seeds with sequence homology to the cysteine/glycine-rich domain of chitinbinding proteins. *Biochemistry*, 31, 4308–4314.

Cabot R.A., Kühholzer B., Chan A., et al. (2001). Transgenic pigs produced using in vitro matured oocytes infected with a retroviral vector. *Animal Biotechnology*, 12(2), 205–214.

Camire M.E., Kubow S., and Donnelly D.J., (2009). Potatoes and human health. *Critical Reviews in Food Science and Nutrition*, 49(10), 823–840.

Chakraborty S., Chakraborty N., and Datta A. (2000). Increased nutritive value of transgenic potato by expressing a nonallergenic seed albumin gene from Amaranthus hypochondriacus. *Proceedings of the National Academy of Sciences of the United States of America*, 97, 3724–3729.

Chandrasekara A. and Kumar T.J. (2016). Roots and tuber crops as functional foods: A review on phytochemical constituents and their potential health benefits. *International Journal of Food Science*, 2016, 2016, 3631647. DOI:10.1155/2016/3631647.

Chaparro J.M., Holm D.G., Broeckling C.D., Prenni J.E., and Heuberger A.L. (2018). Metabolomics and ionomics of potato tuber reveals an influence of cultivar and market class on human nutrient a bioactive compound. *Frontiers in Nutrition*, 5, 36. DOI:10.3389/fnut.2018.00036.

Chauhan K. (2014). Study of Antioxidant potential of Solanum tuberosum peel extracts. *Journal of Integrated Science and Technology*, 2(1), 27–31.

Choi E. and Koo S. (2005). Anti-nociceptive and anti- inflammatory effects of the ethanolic extract of potato (Solanum tubeerlosum). *Food and Agricultural Immunology*, 16(1), 29–39. DOI:10.1080/09540100500064320.

Chong D.K. and Langridge W.H. (2000). Expression of full length bioactive antimicrobial human lactoferrin in potato plants. *Transgenic Research*, 9, 71–78.

Christou P., Capell T., Kohli A., et al. (2006). Recent developments and future prospects in insect pest control in transgenic crops. *Trends in Plant Science*, 11(6), 302–308.

Das K, Krishna P, Sarkar A, Ilangovan SS, and Sen S. (2017). A review on pharmacological properties of Solanum tuberosum. *Research Journal of Pharmacy and Technology*, 10(5), 1517–1522.

Davies H.V. (1996). Recent developments in our knowledge of potato transgenic biology. *Potato Research*, 39, 411–427.

Emad S. (2006). Antioxidant effect of extracts from red grape seed and peel on lipid oxidation in oil of Sunflower. *LWT - Food Science and Technology*, 39, 883–892.

Ewen S.W.B. and Pusztai A. (1999). Effect of diets containing genetically modified potatoes expressing Galanthus nivalis lectin on rat small intestine. *Lancet*, 354, 1353–1354.

Falck-Zepeda J.B., Traxler G., and Nelson R.G. (2000). Surplus distribution from the introduction of a biotechnology innovation. *American Journal of Agricultural Economics*, 82(2), 360–369.

Gatehouse A.M.R., Davison G.M., Newell C.A., Merryweather A., Hamilton W.D.O., Burgess E.P.J., Gilbert R.J.C., and Gatehouse J.A. (1997). Transgenic potato plants with enhanced resistance to the tomato moth, Lacanobia oleracea: Growth room trials. *Molecular Breeding*, 3, 49–63.

Gatehouse A.M.R., Down R.E., Powell K.S., Sauvion N., Rahbe Y., Newell C.A., Merryweather A., Hamilton W.D.O., and Gatehouse J.A. (1996). Transgenic potato plants with enhanced resistance to the peach-potato aphid Myzus persicae. *Entomologia Experimetalis et Applicata*, 79, 295–307.

Ghasemian M., Owlia S., and Owlia M.B. (2016). Review of anti-inflammatory herbal medicines. *Advances in Pharmacological Sciences*. Article ID 9130979, 11 pages.

Gomez N., Wigdorovitz A., Castanon S., Gil F., Ordas R., Borca M.V., and Escribano J.M. (2000). Oral immunogenicity of the plant derived spike protein from swine transmissible gastroenteritis coronavirus. *Archives of Virology*, 145, 1725–1732.

Hansen M.B., Nielsen S.E., and Berg K. (1989). Re-examination and further development of a precise and rapid dye method for measuring cell growth/cell kill. *Journal of Immunological Methods*, 119(2), 203–210.

Hayashi K., Hibasami H., Murakami T., et al. (2006). Induction of apoptosis in cultured human stomach cancer cells by potato anthocyanins and its inhibitory effects on growth of stomach cancer in mice. *Food Science and Technology Research*, 12(1), 22–26.

Hutchison W.D., Burkness E.C., Mitchell P.D., et al. (2010). Areawide suppression of European corn borer with Bt maize reaps savings to non-Bt maize growers. *Science*, 330(6001), 222–225.

Ieri F., Innocenti M., Andrenelli L., et al. (2011). Rapid HPLC/DAD/MS method to determine phenolic acids, glycoalkaloids and anthocyanins in pigmented potatoes (Solanum tuberosum L.) and correlations with variety and geographical origin. *Food Chemistry*, 125(2), 750–759.

Inobeme A., Jeevanandam J., Adetunji C.O., Anani O.A., Thangadurai D., Islam S., Oyawoye O.M., Oloke J.K., Yerima M.B., and Olaniyan O.T. (2021a). Ecorestoration of soil treated with biosurfactant during greenhouse and field trials. In: *Green Sustainable Process for Chemical and Environmental Engineering and Science. Biosurfactants for the Bioremediation of Polluted Environments*, 2021, pp. 89–105. DOI:10.1016/B978-0-12-822696-4.00010-3

Inobeme A., Anani O.A., Jeevanandam J., Yerima M.B., Thangadurai D., Islam S., Oyawoye O.M., Oloke J.K., Olaniyan O.T. (2021b). Isolation, screening, and characterization of biosurfactant-producing microorganism that can biodegrade heavily polluted soil using molecular techniques. In: *Green Sustainable Process for Chemical and Environmental Engineering and Science. Biosurfactants for the Bioremediation of Polluted Environments*, 2021, pp. 53–68. https://doi.org/10.1016/B978-0-12-822696-4.00016-4

Izquierdo J. (2000). Plant biotechnology and food security in Latin America and the Caribbean. *Plant Biotechnology*, 3(1), 2225–2532.

Jansky S., Navarre R., and Bamberg J. (2019). Introduction to the special issue on the nutritional value of potato. *American Journal of Potato Research*, 96, 95–97. DOI:10.1007/s12230-018-09708-1.

Jeevanandam J., Adetunji C.O., Selvam J.D., Anani O.A., Inobeme A., Islam S., Thangadurai D., Olaniyan O.T. (2021). High industrial beneficial microorganisms for effective production of a high quantity of biosurfactant. In: *Green Sustainable Process for Chemical and Environmental Engineering and Science,* 2021, pp. 279–297. London: Elsevier.

Ji X., Rivers L., Zielinski Z., Xu M., MacDougall E., Stephen, J., and Zhang J. (2012). Quantitative analysis of phenolic components and glycoalkaloids from 20 potato clones and in vitro evaluation of antioxidant, cholesterol uptake, and neuroprotective activities. *Food Chemistry*, 133(4), 1177–1187. DOI:10.1016/j.foodchem.2011.08.065.

Kanter M., and Elkin C. (2019). Potato as a Source of Nutrition for Physical Performance. *American Journal of Potato Research*, 96, 201–205. DOI:10.1007/s12230-018-09701-8.

Kenny O.M., McCarthy C.M., Brunton N.P., Hossain M.B., Rai D.K., Colins S.G., Jones P.W., Maguire A.R., and O'Brien N.M. (2013). Anti-inflammatory properties of potato glycoalkaloids in stimulated Jurkat and Raw 264.7 mouse macrophages. *Life Sciences*, 92(13), 775–782. DOI:10.1016/j.lfs.2013.02.006.

Kita A., Bakowska-Barczak A., Hamouz K., Kulakowska K., and Lisinska G. (2013). The effect of frying on anthocyanin stability and antioxidant activity of crisps from red- and purplefleshed potatoes (Solanum tuberosum L.). *Journal of Food Composition and Analysis*, 32(2), 169–175.

Kosieradzka I., Sawosz E., Szopa J., and Bielecki W. (2008). Potato genetically modified by 14-3-3 protien repression in growing rat diest. Part II: Health status of experiment animal. *Polish Journal of Food and Nutrition Sciences*, 58(3).

Kuete V. (2014). Physical, hematological, and histopathological signs of toxicity induced by African medicinal plants. In *Toxicological Survey of African Medicinal Plants*. Victor K. (Ed), Elsevier, pp. 635–657. University of Dschang, Cameroon.

Lachman J., Hamouz K., Orsak M., and Kotikova Z. (2016). Carotenoids in potatoes – a short overview. *Plant Soil and Environment*, 62(10), 474–481. DOI:10.17221/459/2016-PSE.

Lala G., Malik M., Zaho C., et al. (2006). Anthocyanin-rich extracts inhibit multiple biomarkers of colon cancer in rats. *Nutrition and Cancer*, 54(1), 84–93.

Lauterslager T.G., Florack D.E., van der Wal T.J., Molthoff J.W., Langeveld J.P., Bosch D., Boersma W.J., and Hilgers L.A. (2001). Oral immunisation of naive and primed animals with transgenic potato tubers expressing LT-B. *Vaccine*, 19, 2749–2755.

Lee J.H., Woo K.S., Lee H.U., Nam S.S., Lee B.W., Lee Y.Y., Lee B., and Kim H.J. (2019) Intracellular reactive oxygen species (ROS) removal and cytoprotection effects of sweet potatoes of various flesh colors and their polyphenols, including anthocyanin. *Preventive Nutrition in Food Science*, 24(3), 293–298.

Lee L.Y and Gelvin S.B. (2008). T-DNA binary vectors and systems. *Plant Physiology*, 146(2), 325–332.

Lee W., Yeo Y, Oh S., Cho K.S., Park Y.E., Park S.K., Lee S.M., Cho H.S., and Park S.Y. (2017). Compositional analyses of diverse phytochemicals and polar metabolites from different-colored potato (Solanum tubersum L.) tubers. *Food Science and Biotechnology*, 26(5), 1379–1389. DOI:10.1007/s10068-017-0167-2.

Liapkova N.S., Loskutova N.A., Maisurian A.N., Mazin V.V., Korableva N.P., Platonova T.A., Ladyzhenskaia E.P., and Evsiunina A.S. (2001). Isolation of genetically modified potato plant containing the gene of defensive peptide from Amaranthus (in Russia). *Applied Biochemistry and Microbiology*, 37, 349–354.

Lim S., Xu J., Kim J., et al. (2013). Role of anthocyanin-enriched purple-fleshed sweet potato p40 in colorectal cancer prevention. *Molecular Nutrition & Food Research*, 57(11), 1908–1917.

Ma S. and Jevnikar, A.M. (2002). Suppression of autoimmune diabetes by the use of transgenic plants expressing autoantigens to induce oral tolerance. In: *Molecular Farming of Plants and Animals for Human and Veterinary Medicine*, Erickson L., Yu W.J., Brandle J., Rymerson R. (Eds), pp. 179–196. Dordrecht: Springer. DOI:10.1007/978-94-017-2317-6_8.

Ma S.W., Zhao D.L., Yin Z.Q, and Mukherjee M. (1997). Transgenic plants expressing autoantigens fed to mice to induce oral immune tolerance. *Nature Medicine*, 3, 793–796.

Mason H.S., Ball J.M., Shi J.J., Jiang X., Estes M.K., and Arntzen C.J. (1996). Expression of Norwalk virus capsid protein in transgenic tobacco and potato and its oral immunogenicity in mice. *Proceedings of the National Academy of Sciences of the United States of America*, 93, 5335–5340.

Matsumura T., Itchoda N., and Tsunemitsu H. (2002). Production of immunogenic VP6 protein of bovine group A rotavirus in transgenic potato plants. *Archives of Virology*, 147, 1263–1270.

McGill C.R., Kurilich A.C., and Davignon J. (2013). The role of potatoes and potato components in cardio-metabolic health: A review. *Annals of Medicine*, 45(7), 467–473.

Millam S. (2006). Potato (Solanum tuberosum L.). *Methods in Molecular Biology*, 344, 25–36. DOI:10.1385/-1-59745-131-2:25. PMID: 17033048.

Mohammad B., Hossain Nigel P., and Brunton Rai D.K. (2016). Effect of drying methods on the steroidal alkaloid content of potato peels, shoots and berries. *Molecules*, 21(4), 403.

Morita T., Oh-hashi, A., Takei, K., Ikai, M., Kasaoka, S., and Kiriyama, S. (1997). Cholesterol-lowering effects of soybean, potato and rice proteins depend on their low methionine contents in rats fed a cholesterol-free purified diet. *The Journal of Nutrition*, 127(3), 470–477.

Navarre R. (2019). Potato vitamins, minerals, and phytonutrients from a plant biology perspective. *American Journal of Potato Research*, 96(2), 111–126.

Nicolle C., Simon G., Rock E., Amouroux P., and Rémésy C. (2004). Genetic variability influences carotenoid, vitamin, phenolic, and mineral content in white, yellow, purple, orange, and dark-orange carrot cultivars. *Journal of the American Society for Horticultural Science*, 129(4), 523–529.

Nikiforova V., Kempa S., Zeh M., Maimann S., Kreft O., Casazza A.P., Riedel K., Tauberger E., Hoefgen R., and Hesse H. (2002). Engineering of cysteine and methionine biosynthesis in potato. *Amino Acids*, 22(3), 259–278.

Nishioka C., Ikezoe T., Yang J., et al. (2008). Blockade of mTOR signaling potentiates the ability of histone deacetylase inhibitor to induce growth arrest and differentiation of acute myelogenous leukemia cells. *Leukemia*, 22(12), 2159.

Ohya K., Itchoda N., Ohashi K., Onuma M., Sugimoto C., and Matsumura T. (2002). Expression of biologically active human tumor necrosis factor-alpha in transgenic potato plant. *Journal of Interferon and Cytokine Research*, 22, 371–378.

Ohya K., Matsumura T., Ohashi K., Onuma M., and Sugimoto C. (2001). Expression of two subtypes of human IFN-alpha in transgenic potato plants. *Journal of Interferon and Cytokine Research*, 21(8), 595–602.

Olaniyan O.T., and Adetunji, C.O. (2021). Biochemical Role of Beneficial Microorganisms: An Overview on Recent Development in Environmental and Agro-Science. In: *Microbial Rejuvenation of Polluted Environment. Microorganisms for Sustainability,* Adetunji, C.O., Panpatte, D.G., Jhala, Y.K. (Eds), vol 27. Singapore: Springer. DOI:10.1007/978-981-15-7459-7_2.

Osusky M., Osuska L., Hancock R.E., Kay W.W., and Misra S. (2004). Transgenic potatoes expressing a novel cationic peptide are resistant to late blight and pink rot. *Transgenic Research*, 13, 181–190.

Park F. (2007). Lentiviral vectors: Are they the future of animal transgenesis? *Physiol Genomics*, 31(2), 159–173.

Perlak F.J., Stone T.B., Muskopf Y.M., Petersen L.J., Parker G.B., McPherson S.A., Wyman J., Love S., Reed G., Biever D., and Fischhoff D.A. (1993). Genetically improved potatoes: Protection from damage by Colorado potato beetles. *Plant Molecular Biology*, 22, 313–321.

Reddivari L., Wang T., Wu B., and Li S. (2019). Potato: An anti-inflammatory food. *American Journal of Potato Research*, 96(2), 164–169.

Rodriguez-Perez C., Gomez-Caravaca A.M., Guerra-Hernandez E., Cerretani L., Garcia-Villanova B., and Verardo V. (2018). Comprehensive metabolites profiling of Solanum tuberosum L. (potato) leaves by HPLC-ESI- QTOF-MS. *Food Research International*, 112, 390–399.

Roleira F.M., Tavares-da-Silva E.J., Varela C.L., et al. (2015). Plant derived and dietary phenolic antioxidants: Anticancer properties. *Food Chemistry*, 183, 235–258.

Sahar A.A., Al-Saadi M., Alutbi S.D., and Madhi Z.J. (2017). The effects of *in vitro* culture on the Leaf Anatomy of Potato (Solanum tuberosum L. CV. Arizaona). *International Journal of Current Research*, 9(7), 54337–54342.

Selitrennikoff C.P. (2001). Antifungal proteins. Review. *Applied and Environmental Microbiology*, 67, 2883–2894.

Shan Q., Lu J., Zheng Y., Li J., Zhou Z., Hu B., and Ma D. (2009). Purple sweet potato color ameliorates cognition deficits and attenuates oxidative damage and inflammation in aging mouse brain induced by D-Galactose. *Journal of Biomedicine and Biotechnology*, 1–9. DOI:10.1155/2009/564737.

Shipitalo M.J., Malone R.W., and Owens L.B. (2008). Impact of glyphosatetolerant soybean and glufosinate-tolerant corn production on herbicide losses in surface runoff. *Journal of Environmental Quality*, 37(2), 401–408.

Slater A., Scott N.W., Fowler M.R. (eds.) (2003). *Plant Biotechnology, the Genetic Manipulation of Plants*, 1st ed. Oxford University Press Inc., New York, 346 pp.

Stanley W.B. and Pusztai E.A. (1999). Effect of diets containing genetically modified potatoes expressing Galanthus nivalis lectin on rat small intestine. *The Lancet*, 354(9187), 1353–1354.

SugiYama K., Kushima, Y., and Muramatsu, K., (1985). Effects of sulfur-containing amino acids and glycine on plasma cholesterol level in rats fed on a high cholesterol diet. *Agricultural and Biological Chemistry*, 49(12), 3455–3461.

Tacket C.O., Mason H.S., Losonsky G., Estes M.K., Levine M.M., and Arntzen C.J. (2000). Human immune responses to a novel Norwalk virus vaccine delivered in transgenic potatoes. *Journal of Infectious Diseases*, 182, 302–305.

Tang G.-Y., Meng X., Li Y., Zhao C.-N., Liu Q., and Li H.-B. (2017). Effects of vegetables on cardiovascular diseases and related mechanisms. *Nutrients*, 9(8), 857. DOI:10.3390/nu9080857.

Tierno R, Lopez A, Riga P, Arazuri S, Jaren C, Benedicto L., and Ruiz de Galarreta J. (2015). Phytochemicals determination and classification in purple and red fleshed potato tubers by analytical methods and near infrared spectroscopy. *Journal of the Science of Food and Agriculture*, 96, 1888–1899.

Umadevi M., Sampath Kumar P.K., Bhowmik D., and Duraivel S. (2103). Health benefits and cons of Solanum tuberosum. *Journal of Medicinal Plants Studies*, 1(1), 16–25.

Van der Wilk F., Posthumus-Lutke Willink D., Huisman M.J., Huttinga H., and Goldbach R. (1991). Expression of the potato leafroll luteovirus coat protein gene in transgenic potato plants inhibits viral infection. *Plant Molecular Biology*, 17, 431–439.

Wegener C.B., Jansen G., and Jurgens H. (2015). Bioactive compounds in potatoes: Accumulation under drought stress conditions. *Functional Foods in Health and Disease*, 5(3), 108–116.

Wu G., Shortt B.J., Lawrence E.B., Levine E.B., Fitzsimmons K.C., and Shah D.M. (1995). Disease resistance conferred by expression of a gene encoding H_2O_2-generating glucose oxidase in transgenic potato plants. *Plant Cell*, 7, 1357–1368.

Ye J., Meng X., Yan C., and Wang C. (2009). Effect of purple sweet potato anthocyanins on β-amyloid-mediated PC-12 cells death by inhibition of oxidative stress. *Neurochemical Research*, 35(3), 357–365. DOI:10.1007/s11064-009-0063-0.

Youm J.W., Kim H., Han J.H.L., Jang C.H., Ha H.J., Mook-Jung I., and Joung H. (2005). Transgenic potato expressing Aβ reduce Aβ burden in Alzheimer's disease mouse model. *FEBS Letters*, 579(30), 6737–6744. DOI:10.1016/j.febslet.2005.11.003.

Yu J. and Langridge W. (2003). Expression of rotavirus capsid protein VP6 in transgenic potato and its oral immunogenicity in mice. *Transgenic Research*, 12, 163–169.

Zhou J.Y., Cheng L.Q., Zheng X.J., Wu J.X., Shang S.B., Wang J.Y., and Chen J.G. (2004). Generation of the transgenic potato expressing full-length spike protein of infectious bronchitis virus. *Journal of Biotechnology*, 111, 121–130.

Zhou J.Y., Wu J.X., Cheng L.Q., Zheng X.J., Gong H., Shang S.B., and Zhou E.M. (2003). Expression of immunogenic S1 glycoprotein of infectious bronchitis virus in transgenic potatoes. *Journal of Virology*, 77, 9090–9093.

23 Recent Advances in the Application of Metagenomic in Promoting Food Security, Human Health, and Environmental Sustainability

Olugbemi T. Olaniyan
Rhema University

Charles Oluwaseun Adetunji
Edo State University Uzairue

CONTENTS

23.1 INTRODUCTION

Metagenomics serves as a molecular technique that could be applied in the investigation of microorganism through the evaluation of their DNA obtained from environmental samples without a need to obtain a more purified form of the organism. This technology permits the evaluation of the DNA of the total population of the microorganisms available in that environment. It gives a golden opportunity to identify the type of a particular microorganism present and provide a better understanding of the type of metabolites and functional activities of the microorganisms in a particular population (Langille et al., 2013). It has been discovered that the usage of high-throughput sequencing technologies has increased the capability to evaluate and quantify the amount of genomic material available in particular material (Aarestrup et al., 2012). The usage of techniques requires a lot of improvement in the nearest future.

Moreover, the application of metagenomic evaluation using untargeted sequencing has been given considerable attention that might be linked to the high throughput of current sequencing technologies which permit the assessment of complex samples (Smits et al., 2015). The application of metagenomics together with multiplex real-time polymerase chain reaction (RT-PCR) protocols could help in the detection of several pathogenic diseases and detection of several pathogens that could serve as a threat to human and environment (Binnicker, 2015).

Recent studies have reported that food security is a big challenge, particularly among developing and developed countries. Chronic hunger and malnutrition have been shown to be a major health

and social challenge in these countries resulting in persistent and severe undernourishment. Several projections have been made predicting that Africa's population in increase geometrically to about 2.4 billion by the year 2050. The resultant effect of this population explosion on food security have been predicted to be negative thereby making the global food security challenge to be worrisome (Abubakar et al., 2019; Adetunji et al., 2018a, b, c; Adetunji et al., 2017; Adetunji et al., 2011c; Adetunji et al., 2019; Adetunji et al., 2014, Adetunji et al., 2013a, b; Adetunji et al., 2011a, b; Adetunji and Olaleye, 2011).

23.2   RECENT ADVANCES IN THE APPLICATION OF METAGENOMIC IN PROMOTING FOOD SECURITY

Wu et al. (2015) revealed that homology-dependent and independent algorithms and next-generation sequencing are the recent advances in technology that can be utilized in the identification of viroids and viruses. These technologies enable fast and easy management of diseases in tissues through viral metagenomics. Plant metagenomics and plant virus metagenomics are sophisticated technologies that utilize next-generational sequencing for advances in plant biodiversity and ecological studies generating a very large data set for deeper evolutionary and ecological analyses.

Humblot et al. (2014) reported that metagenomics can be applied to the production of fermented food, thus facilitating increased nutrition, shelf life and overall transformation of food substances. Through metagenomics, commercialization and optimization can be achieved by increased characterization of the microbial community, thus providing a faster and more efficient approach. The authors described the metagenomics approach as the massive application of bioinformatics tools and sequencing techniques targeting nucleic acids like (RNA/DNA). Also, other techniques that target the genes have also been established such as fingerprinting, enzyme restriction, PCR amplification and electrophoretic techniques for microbial community profiling. These techniques have been utilized to target the genetic content of many microbes. Humblot et al. (2014) reported metagenomics approaches to provide the ability to study complex fermentative ecosystems in many fermented food products by targeting the genetic contents, thus providing adequate information on the biological activity, physiochemical properties, gene expression study, toxinogens, survival of pathogenic organisms, safety, concentration and diversity. The authors suggested that many several technical challenges should be considered to avoid bias during data generation.

Coughlan et al. (2015) reported that microorganisms are very useful as a biotechnological tool for the functional metagenomic analysis and functional-based screening of pharmaceutical and food products. The authors revealed that genomic data provides valuable information about the functionality of these microbes. The functional-based screening like gene expression and cloning of novel proteins, DNA metagenomic for industrial applications. Coughlan et al. (2015) reported many novel enzymes and bioactives derived from beneficial microbes that are currently being utilized in pharmaceutical and food processing. Some of them serve as biocatalysts for large-scale food production, brewing, fermentation and degradation, baking, and flavoring.

Prayogo et al. (2020) revealed that novel enzymes for diverse industrial applications can be produced from natural habitat utilizing metagenomic applications. The authors reported that many microbes are known to generate several metabolites, enzymes like cellulases, lipases, AHL-lactonase. BRPD, oxoflavin-degrading enzyme, chitinases, transaminases and proteases sand other essential compounds but unfortunately many of these microbes are yet to be discovered. They reported that the application of functional metagenomic techniques can help to uncover these microbes, metabolic pathways, and their genes.

23.3   RECENT ADVANCES IN THE APPLICATION OF METAGENOMIC IN PROMOTING HUMAN HEALTH

Gu et al. (2019) reported that virtually all infectious pathogenic agents have RNA and DNA genomes that can be utilized for genomic sequencing through the utilization of high-throughput or

metagenomic next-generation sequencing for diverse technological and algorithm platform. These platforms allow fast detection, characterization, analysis of microorganisms from patient samples. Recently conventional diagnostics have been reported to have several limitations, particularly in efficient tracking of several infectious diseases. Sustainable improvement and enhancement in health care system involve the production and utilization of metagenomics to improve breeding and production techniques in plants and animals (Arthikala et al., 2014).

Metagenomics techniques have been shown to facilitate diagnostics, biomedical engineering, gene therapy, and pharmacogenomics (McCullum et al., 2003). Donovan et al. (2012) have utilized metagenomic approaches to study the functional relation between food microbiota and human gut system. Their results demonstrated that many of metabolic diseases like obesity are linked to a particular microbiome.

Lobiuc et al. (2019) reported that gut microbiome in type II diabetes can be analyzed utilizing the metagenomic techniques. In their study, it was revealed that physiological conditions in disease and health of the gut can be modulated by gut microbiota which may be characterized with high throughput metagenomic sequencing techniques. From their results it was discovered that probiotics modulated change in the composition and microbiota physiology and metabolism. Several reports have suggested that whole-genome shotgun sequencing (WGS) provides higher yield and sensitivity in variant detection, disease type and diagnostic role in microbial diversity study. Also it was recommended that routine utilization of multiomics platforms, genomics techniques, single-cell technologies, and integration of systems biology in medical research will contribute significantly to the design of several precision interventions with specific consideration of the species, disease-specific factors, and host (Kiousi et al., 2021).

Ninian et al. (2021) reported that metagenomics techniques are useful in novel enzyme bioprospecting such as endoglucanases in second-generation biofuel production. The authors suggested that the identification of novel cellulolytic gene mining and lignocellulosic biomass-degrading enzymes can be done through the utilization of metagenomics.

23.4 RECENT ADVANCES IN THE APPLICATION OF METAGENOMIC IN PROMOTING ENVIRONMENTAL SUSTAINABILITY

Ghosh et al. (2018) reported that metagenomics techniques are useful in the detection and analysis diverse microbial species in the environment. The authors disclosed that several uncharacterized microbes that can be studied using metagenomics from different samples. The metabolic, dynamic, functional activities, diversity and complex characteristics of the microbiota can be understudy using functional metagenomics.

Omics approaches particularly metagenomics have been utilized to detect and analyzed microbes like *L. monocytogenes*, *Staphylococcus aureus*, *E. coli* O157:H7, *Campylobacter jejuni*, *Shigella* species and *Salmonella* species that are responsible for various food spoilage, contaminations, and foodborne pathogenesis. The authors reported that foodomics techniques like genomics, proteomics, metagenomics, and transcriptomics are currently being utilized for food safety, analysis, and quality.

Barooah et al. (2021) highlighted the role of freshwater ecosystem in regulating the level of global carbon cycle, thereby maintaining the global climate. The authors revealed that through high-throughput state-of-the-art technologies like functional metagenomics, Shotgun metagenomic sequencing, metaproteomics, next-generation sequencing and metatranscriptomics microbial diversities can be studied in the environment without isolation and culturing. Also, many of these high throughput technologies give adequate insight into the poorly understood biological system and taxonomic profiling. The field of metagenomics is a novel area in which investigation are directed to understand the genes of environmental microorganisms within the microbial community. Some of the functional metagenomic investigations carried out recently in the field of

environmental physiology include niche-specialized low complexity communities, labeling technologies and reactor enrichments. The newest areas in the field of metagenomics are metaproteomics and metatranscriptomics which arc being utilized for medical biotechnology (Brulc et al., 2009). Many authors have reported diverse applications of metagenomics such as food safety, industrial waste management, bioremediation processes in a polluted environment, forensic science and diagnostic physiology. Effendi et al. (2019) utilized high throughput sequencing of 16S rRNA gene to analyze soil bacteria, richness, composition and conditions utilizing Illumina platform. From their results, it was discovered that *Cyanobacteria and Proteobacteria* were found lower abundance in the soil.

23.5 CONCLUSION AND FUTURE RECOMMENDATION TO KNOWLEDGE

This chapter has provided detailed information on the metagenomic in promoting food security, promoting human health and environmental sustainability. Therefore, there is a need to create more awareness on the application of other biotechnological techniques, synthetic biology, nanotechnology, proteomics and genomics and metabolomics most especially to diverse people from different fields. This will go a long way in mitigating against several challenges facing mankind in diverse fields of their endeavor.

REFERENCES

Aarestrup, F.M., Brown, E.W., Detter, C., Gerner-Smidt, P., Gilmour, M.W., Harmsen, D., et al. (2012). Integrating genome-based informatics to modernize global disease monitoring, information sharing, and response. *Emerging Infectious Disease* 18, e1. https://doi.org/10.3201/eid/1811.120453.

Abubakar, Y., Tijjani, H., Egbuna, C., Adetunji, C.O., Kala, S., Kryeziu, T.L., Ifemeje, J.C. and Patrick-Iwuanyanwu, K.C.. (2019). Pesticides, history and classification. In *Natural Remedy for Pest, Diseases and Weed Control*, Elsevier, USA, pp. 1–18.

Adetunji, C.O. and Olaleye, O.O. (2011). Phytochemical screening and antimicrobial activity of the plant extracts of *Vitellaria paradoxa* against selected microbes. *Journal of Research in Biosciences* 7(1), 64–69.

Adetunji, C.O., Kolawole, O.M., Afolayan, S.S., Olaleye, O.O., Umanah, J.T., and Anjorin, E. (2011a). Preliminary phytochemical and antibacterial properties of *Pseudocedrela kotschyi*: A potential medicinal plant. *Journal of Research in Bioscience. African Journal of Bioscience* 4(1), 47–50. ISSN: 2141-0100.

Adetunji, C.O., Arowora, K.A., Afolayan, S.S., Olaleye, O.O., and Olatilewa, M.O. (2011b). Evaluation of antibacterial activity of leaf extract of *Chromolaena odorata*. *Science Focus* 16(1), 1–6. ISSN: 1598-7026. www.sciencefocusngr.org. Published by Faculty of Pure and Applied Sciences, LAUTECH.

Adetunji, C.O., Olaleye, O.O., Adetunji, J.B., Oyebanji, A.O., Olaleye, O.O., and Olatilewa, M.O. (2011c). Studies on the antimicrobial properties and phytochemical screening of methanolic extracts of *Bambusa vulgaris* leaf. *International Journal of Biochemistry* 3(1).

Adetunji, C.O. Fawole, O.B., Arowora, K.A., Nwaubani, S.I., Oloke, J.K., Adepoju, A.O., Adetunji, J.B., and Ajani, A.O. (2013a). Performance of edible coatings from carboxymethylcellulose(CMC) and corn starch(CS) incorporated with *Moringa oleifera* extract on *Citrus sinensis* stored at ambient temperature. *Agrosearch* 13(1), 77–85.

Adetunji, C.O., Fawole, O.B., Arowora, K.A., Nwaubani, S.I., Ajayi, E.S., Oloke, J.K., Aina, J.A., Adetunji, J.B., and Ajani, A.O. (2013b). Postharvest quality and safety maintenance of the physical properties of Daucus *carota L.* fruits by Neem oil and Moringa oil treatment: A new edible coatings. *Agrosearch.* 13(1), 131–141.

Adetunji, C.O., Ogundare, M.O., Ogunkunle, A.T.J., Kolawole, O.M., and Adetunji, J.B. (2014). Effects of edible coatings from xanthum gum produced from *Xanthomonas campestris* pammel on the shelf life of *Carica papaya* linn fruits. *Asian Journal of Agriculture and Biology* 2(1), 8–13.

Adetunji, C.O., Oloke, J.K., Prasad, G., and Akpor, O.B. (2017). Environmental influence of cultural medium on bioherbicidal activities of *Pseudomonas aeruginosa* C1501 on mono and dico weeds. *Polish Journal of Natural Sciences* 32(4), 659–670.

Adetunji, C.O., Adejumo, I.O., Oloke, J.K., and Akpor, O.B.. (2018a). Production of phytotoxic metabolites with bioherbicidal activities from *Lasiodiplodia pseudotheobromae* produced on different agricultural

wastes using solid-state fermentation. *Iranian Journal of Science and Technology, Transactions A: Science* 42(3), 1163–1175. https://doi.org/10.1007/s40995-017-0369-8.

Adetunji, C.O., Adejumo, I.O., Afolabi, I.S., Adetunji, J.B., and Ajisejiri, E.S. (2018b). Prolonging the shelf-life of 'Agege Sweet' Orange with chitosan-rhamnolipid coating. *Horticulture, Environment, and Biotechnology*, 59(5), 687–697. https://doi.org/10.1007/s13580-018-0083-2.

Adetunji, C.O., Paomipem, P., and Neera, B.S. (2018c). Production of ecofriendly biofertilizers produced from crude and immobilized enzymes from *Bacillus subtilis* CH008 and their effect on the growth of *Solanum lycopersicum*. *Plant Achieve* 18(2), 1455–1462.

Adetunji, C.O., Afolabi, I.S., and Adetunji, J.B. (2019). Effect of Rhamnolipid-*Aloe vera* gel edible coating on post-harvest control of rot and quality parameters of 'Agege Sweet' Orange. *Agriculture and Natural Resources* 53, 364–372.

Arthikala, M., Nanjareddy, K., Lara, M., and Sreevathsa, R. (2014). Utility of a tissue culture-independent agrobacterium-mediated in planta transformation strategy in bell pepper to develop fungal disease resistant plants. *Scientia Horticulturae* 170(0), 61–69.

Barooah, M., Goswami, G., Hazarika, D.J., and Kangabam, R. (2021). Chapter 24: High-throughput analysis to decipher bacterial diversity and their functional properties in freshwater bodies. In *Microbial Metatranscriptomics Belowground*, pp. 511–541. https://doi.org/10.1007/978-981-15-9758-9_24.

Binnicker, M.J. (2015). Multiplex molecular panels for diagnosis of gastrointestinal infection: Performance, result interpretation, and cost-effectiveness. *Journal of Clinical Microbiology* 53, 3723–3728. https://doi.org/10.1128/JCM.02103-15.

Brulc, J.M., Antonopoulos, D.A., Miller, M.E., Wilson, M.K., Yannarell, A.C., Dinsdale, E.A., Edwards, R.E., Frank, E.D., Emerson, J.B., Wacklin, P., Coutinho, P.M., Henrissat, B., Nelson, K.E., and White, B.A. (2009). Gene-centric metagenomics of the fiber-adherent bovine rumen microbiome reveals forage specific glycoside hydrolases. *Proceedings of National Academy of Sciences USA* 106, 1948–1953.

Coughlan, L.M., Cotter, P.D., Hill, C., and Alvarez-Ordóñez, A. (2015). Biotechnological applications of functional metagenomics in the food and pharmaceutical industries. *Frontiers in Microbiology* 6, 672. https://doi.org/10.3389/fmicb.2015.00672.

Donovan, S.M., Wang, M., Li, M., Friedberg, I., Schwartz, S.L., and Chapkin, R.S. (2012). Host-microbe interactions in the neonatal intestine: Role of human milk ligosaccharides. *Advances in Nutrition* 3, 450S–455S.

Effendi, Y., Pambudi, A., Sasaerila, Y., and WijiHastuti, R.S. (2019). Metagenomic analysis of diversity and composition of soil bacteria under intercropping system *Hevea brasiliensis* and *Canna indica*. *Annual Conference on Environmental Science, Society and its Application IOP Conf. Series: Earth and Environmental Science* 391, 012023. IOP Publishing. https://doi.org/10.1088/1755-1315/391/1/012023.

Ghosh, A., Mehta, A., and Khan, A.M. (2018). Metagenomic analysis and its applications. *Reference Module in Life Sciences*. https://doi.org/10.1016/b978-0-12-809633-8.20.

Gu, W., Miller, S., and Chiu, C.Y. (2019). Clinical metagenomic next-generation sequencing for pathogen detection. *Annual Review of Pathology: Mechanisms of Disease* 14, 319–338. https://doi.org/10.1146/annurev-pathmechdis-012418-012751.

Humblot, C., Turpin, W., Chevalier, F., Picq, C., Rochette, I., and Guyot, J.P. (2014). Determination of expression and activity of genes involved in starch metabolism in *Lactobacillus plantarum* A6 during fermentation of a cereal-based gruel. *International Journal of Food Microbiology* 185, 103–111.

Kiousi, D.E., Rathosi, M., Tsifintaris, M., Chondrou, P., and Galanis, A. (2021). Pro-biomics: Omics technologies to unravel the role of probiotics in health and disease. *Advances in Nutrition* 12(5), 1–19. https://doi.org/10.1093/advances/nmab014.

Langille, M. G. I., Zaneveld, J., Caporaso, J. G., McDonald, D., Knights, D., Reyes, J. A., et al. (2013). Predictive functional profiling of microbial communities using 16S rRNA marker gene sequences. *Nature Biotechnology* 31, 814–821. https://doi.org/10.1038/nbt.2676.

Lobiuc, A., Pavel, I., Toderean, R., Avatamanitei, S., and Covasa, M. (2019). Metagenomic insights on the role of gut microbiota in type-2 diabetes. In *The 7th IEEE International Conference on E-Health and Bioengineering - EHB 2019*, Iasi, Romania.

McCullum, C., Benbrook, C., Knowles, L., Roberts, S., and Schryver, T. (2003). Application of modern biotechnology to food and agriculture: Food systems perspective. *Journal of Nutrition Education and Behavior* 35(6), 319–332.

Prayogo, F.A., Budiharjo, A., Kusumaningrum, H.P., Wijanarka, W., Suprihadi, A., and Nurhayati, N. (2020). Metagenomic applications in exploration and development of novel enzymes from nature: A review. *Journal of Genetic Engineering and Biotechnology* 18, 39. https://doi.org/10.1186/s43141-020-00043-9.

Agricultural Biotechnology: Food Security Hot Spots

Smits, S.L., Bodewes, R., Ruiz-González, A., Baumgärtner, W., Koopmans, M.P., Osterhaus, A.D.M.E., et al. (2015). Recovering full-length viral genomes from metagenomes. *Frontiers in Microbiology* 6, 1069. https://doi.org/10.3389/fmicb.2015.01069.
Wu, Q., Ding, S.W., Zhang, Y., and Zhu, S. (2015). Identification of viruses and viroids by next-generation sequencing and homology-dependent and homology-independent algorithms. *Annual Review of Phytopathology* 53, 425–444. https://doi.org/10.1146/annurev-phyto-080614-120030.

Index

For Product Safety Concerns and Information please contact our EU
representative GPSR@taylorandfrancis.com
Taylor & Francis Verlag GmbH, Kaufingerstraße 24, 80331 München, Germany

www.ingramcontent.com/pod-product-compliance
Lightning Source LLC
Chambersburg PA
CBHW080139220326
41598CB00032B/5111